Health Informatics

This series is directed to healthcare professionals leading the transformation of healthcare by using information and knowledge. For over 20 years, Health Informatics has offered a broad range of titles: some address specific professions such as nursing, medicine, and health administration; others cover special areas of practice such as trauma and radiology; still other books in the series focus on interdisciplinary issues, such as the computer based patient record, electronic health records, and networked healthcare systems. Editors and authors, eminent experts in their fields, offer their accounts of innovations in health informatics. Increasingly, these accounts go beyond hardware and software to address the role of information in influencing the transformation of healthcare delivery systems around the world. The series also increasingly focuses on the users of the information and systems: the organizational, behavioral, and societal changes that accompany the diffusion of information technology in health services environments.

Developments in healthcare delivery are constant; in recent years, bioinformatics has emerged as a new field in health informatics to support emerging and ongoing developments in molecular biology. At the same time, further evolution of the field of health informatics is reflected in the introduction of concepts at the macro or health systems delivery level with major national initiatives related to electronic health records (EHR), data standards, and public health informatics.

These changes will continue to shape health services in the twenty-first century. By making full and creative use of the technology to tame data and to transform information, Health Informatics will foster the development and use of new knowledge in healthcare

More information about this series at http://www.springer.com/series/1114

Connie W. Delaney • Charlotte A. Weaver
Judith J. Warren • Thomas R. Clancy
Roy L. Simpson
Editors

Big Data-Enabled Nursing

Education, Research and Practice

 Springer

Editors
Connie W. Delaney
School of Nursing
University of Minnesota School of Nursing
Minneapolis
Minnesota
USA

Charlotte A. Weaver
Issaquah
Washington
USA

Judith J. Warren
School of Nursing
University of Kansas School of Nursing
Plattsmouth
Nebraska
USA

Thomas R. Clancy
School of Nursing
University of Minnesota
Minneapolis
Minnesota
USA

Roy L. Simpson
Nell Hodgson Woodruff School of Nursing
Emory University
Atlanta
Georgia
USA

ISSN 1431-1917 ISSN 2197-3741 (electronic)
Health Informatics
ISBN 978-3-319-53299-8 ISBN 978-3-319-53300-1 (eBook)
DOI 10.1007/978-3-319-53300-1

Library of Congress Control Number: 2017945758

Printed on acid-free paper

This Springer imprint is published by Springer Nature
The registered company is Springer International Publishing AG
The registered company address is: Gewerbestrasse 11, 6330 Cham, Switzerland

To my family; and nursing's honor of the bold authentic voices of the patients, families and communities.

Connie W. Delaney

In memory of Betty R. Weaver, a mother who supported her daughter to follow her dream.

Charlotte A. Weaver

To Edward, my loving husband, who supported, listened, critiqued, and edited.

Judith J. Warren

To my father, John Clancy, who convinced me to become a nurse 40 years ago and it has been a wonderful journey ever since.

Thomas R. Clancy

To the patients and caregivers whose frustration with technology creates opportunities for big data visualization to enhance the effectiveness and efficiency of healthcare quality, including its processes, structures and outcomes.

Roy L. Simpson

Foreword 1

The idea of big data brings up the old saying "water, water, everywhere and not a drop to drink". It may be wise to add that "drowning in data" is also of concern. The sheer volume and velocity of data has exploded with electronic information systems, the global internet, software and hardware technology, the cloud and use by almost everyone. The idea of big data has grown rapidly. Even with the advances in methods of collection, analysis, and reporting, the usefulness of data has not grown as rapidly. The vision of how data will add to knowledge and wisdom has initiated high interest and dreams of answering major questions to guide multiple current issues, challenges and decisions. Most professional and general public publications include discussions of big data addressed from perspectives of clinical care, science, business, education, travel, finance, marketing, and quality determinants.

In this important book pioneering authors address several important considerations and hopes for the creation, generation and use of big data particularly from the perspectives of the science and delivery of health care. The book is written as basic information for those who are becoming interested and for those who already have an interest in how big data can add to information and knowledge of importance to them. More specifically the authors describe how each individual, group, or nation might conquer, contribute to, and use big data to inform their own questions about science, practice, education, policy, organization, resources, and quality.

Researchers have and will continue to generate large amounts of data in many formats, at all levels and in every corner of the world. There is an increasing and urgent need to better understand how to capture, store, manage, analyze and share data to further knowledge and foster new discovery. Likewise, those who deliver and pay for care continue to generate large amounts of clinical, administrative, policy, and cost data to inform the decisions made by each individual or group. And the recipients of the decisions made by these groups have keen interest as buyers, patients, funders, and policy makers. Thus both those seeking the big picture and those seeking more specific targets can easily be blurred by the thousands of data elements.

Big data is of importance to many types of stakeholders. Biomedical, clinical, health services, economics, and population are examples of researchers. Serious expectations are proposed for precision medicine with the aim to cure major diseases like cancer, Alzheimer's, AIDS and genetic disorders. Physicians, nurses, physical therapists, dentists, psychologists and other health professionals have need for specific data of importance to their practice. This need for specific data only increases when specialty areas are considered. Further stakeholders are hospitals, home care, long-term care, mental health, public health and hospice organizations. Patients and consumers are major stakeholders with needs for data tailored to specific conditions and situations. Ownership, governance, payers, and policy makers increase the complexity of essential data. Recently, community social determinants data have been identified.

A few years ago I was invited to present at the First International Interdisciplinary Conference on Big Data held in Singapore. I proposed that big data was not big enough especially from the perspective of nursing practice and nursing science and from the perspective of patients, families and communities. Our experience with data for decades has been focused on classifications of diseases and medical procedures. Data about the assessment, problems, interventions, and outcomes of nursing care are invisible. It was and still is a challenge to find essential patient data from the perspective of nurses in electronic information systems, data warehouses, repositories and the cloud. There was surprise and agreement from engineers and computer scientists at the conference. Unfortunately, this state exists even with extensive work on a Nursing Minimum Data Set and a Nursing Minimum Management Data Set as well as work on international and national development of nursing terminologies and vocabularies. This is a serious limitation and challenge to meeting the goal of advancing nursing science and nursing practice. It is also a serious limitation for all interdisciplinary efforts to generate and use data.

The potential value of big data has created intense interest from the creators and users to be sure that what is of importance to each is included in big data. A current challenge is the desire to have a longitudinal plan of care for each patient. Each stakeholder wants to include a very specific type of data important to a specific practice. Data of importance to another is often considered clutter and makes the information/decision support system time consuming and of little use. Also a challenge and of keen interest is that each piece of data is clearly defined, and is valid and reliable. Highly reliable evidence-based care to ensure quality, safety, and value in health care clinical decisions needs to be supported by accurate, timely and up to date clinical information. Missing data on care not delivered or care data not delivered is of great importance as part of the search for best care. All stakeholders need to accelerate the integration of best knowledge into a care decision.

SNOMED CT and LOINC represent international and nationwide work on reference and clinical terminology that are helping achieve the goal of having standardized interoperable data. This should help the goal that each stakeholder uses the same data element with the same concept. Researchers and clinicians often use different terminology. This quest is no small challenge because of the many areas of science, technology, clinical care, administration, policy development, and the broad variety of health delivery systems and the populations who are the participants in health care.

Each part and chapter of this book has comprehensive descriptions of the evolution of data and knowledge discovery methods that span qualitative as well as quantitative data mining and other methods. Multiple examples are included. There are opportunities and challenges as the data grows in scale, complexity, volume, variety and velocity. Opportunities also expand with the rapid growth of new approaches to data management, analysis, and sharing and with further development of technology (hardware and software).

When all is said and done, the value of big data to patients will depend on how well their care can be delivered and received. A most important question may be precision medical diagnosis and treatment using a wide range of data. Or it may be that cost or community safety is a top question. On the other hand, important concerns may include questions regarding their immediate and long-term care. Big data will not be of much interest to patients having such very personal problems as unmanaged pain, pressure ulcers, lack of information to inform self- management, nausea, vomiting, falls with injury, hospital acquired infections, and uncoordinated care if not included in big data sets.

There is still much to learn about what really contributes to outcomes of value to stakeholders. Use of big data holds promise for advancing health care, research methods, education and policy. This book is a major contribution to that learning.

Norma M. Lang, RN, PhD, FAAN, FRCN
School of Nursing
University of Pennsylvania
Philadelphia, PA, USA

University of Wisconsin–Milwaukee
Milwaukee, WI, USA

Foreword 2

Big data science has to be the concern of all nurses. The 21st-century question for every health professional is: How do you promote transformational change in which the emphasis is not on transitory, isolated performance improvements by individuals, but on sustained, integrated, comprehensive advancement of the whole? The turn-of-the-century focus on bridging the quality divide, with its emphasis on re-engineered care processes and effective use of information technology, has been replaced by the wish to create the continuously learning healthcare systems (LHS) described by the Institute of Medicine. The LHS can analyze all aspects of the care experience for real-time decision-making by patients and clinicians alike. In the space of a dozen years, the stress has moved from simply establishing and using the electronic health record to analyzing all aspects of the care experience for new insights which might involve mining a host of previously unconnected data bases (e.g., quality-safety benchmarks, cost accounting, environmental hazards, admission-discharge records, and so forth).

Big data is not only of importance to nurse informaticians and health services researchers, but to all who aspire to leadership positions in practice, education, research, and policy. Nursing leaders have highlighted the need for nurses to know enough about big data that they can appreciate its relevance to care coordination. Nurses need to be cognizant of technologic developments if they are educators or deans so they can be futuristic about program planning and faculty recruitment. Nurses need to understand enough about how big data can give them insights into health risk differences. Consider, for example, the challenge of designing population-based care for urban and rural women in a particular country. Nurses need to recognize the promise and potential perils of big data if their research is concerned with advancing symptom management.

As we know, leadership isn't just what you can do yourself, but encompasses what you can get done. Do you know enough about trends and the changing nature of health care to hire people with the right skills sets and ask pithy questions of experts who are telling you that your organization should be moving in a particular direction? Do you know how technology-mediated interventions might increase patient engagement and adherence in your clinical setting, so that you can start to move your setting in that direction? Nurse leaders in all settings and at all levels need to be familiar with big data science, and big data needs to be shaped so it asks the questions of concern to care giving. This major volume on big data science can, therefore, be of use to those nurses already concerned with these matters, and I believe it can also be valuable as an immersion in futures thinking for those who don't know enough currently about how to address the informatics revolution underway. If you are trying to get your head around how to handle the tidal wave of data, the increasing concern about figuring out the social determinants of health, or the transformation taking place in how we think and work, then this book is must reading.

Angela Barron McBride, PhD, RN
School of Nursing, Indiana University
Bloomington, IN, USA

Preface

This book's purpose is to engage all of nursing in the potential that big data analytics holds for advancing our profession and the discipline of nursing spanning practice, operations, research, academics, industry, and policy. The book includes big data state-of-the-art-and-science reviews, as well as applied chapters and case study exemplars in nursing using big data analytic methods and technology. In this book, we celebrate the early adopters and the transformative initiatives in play at healthcare organizations, vendors, payers and academia. We also aim to present the opportunities for nursing's impact in this new, emerging knowledge-driven world.

Nursing research historically adopted qualitative methodologies with purposive sampling and quantitative methodologies with small sample sizes because access to patients or large study populations was constrained. Clinical trials, bench research, epidemiology studies and large data methods were in the medical domain and used traditional biostatical analyses. However, the digitization of medical records and payers' claims data has redefined population studies and made large databases available to all disciplines. In the United States, large payer data has been amassed and organizations have been created to welcome scientists to explore these data to advance knowledge discovery. Health systems' electronic health records (EHRs) have now matured to generate massive databases with longitudinal trending. The learning health system infrastructure is maturing, and being advanced by health information exchanges (HIEs) with multiple organizations blending their data, or enabling distributed computing. The evolution of knowledge discovery methods that use quantitative data mining and new analytic methods, including the development of complex data visualization, are enabling sophisticated discovery not previously possible. These developments present new opportunities for nursing, and call for skills in research methodologies that can best be further enabled by forging partnerships with data science expertise spanning all sectors. Recognizing that these new opportunities also call for reassessment of all levels of academic preparation of nursing professionals from pre-licensure through post-doctoral training, parts of this book are dedicated to nursing education and competencies needed at all levels.

This book represents the first big data/data science book in nursing to be published worldwide. It succinctly captures the state of big data and societal context, provides exemplars to establish a foundation for nursing's response to the big data science frontier and provides multiple pathways for driving nursing's future. Accordingly, we organized the book into five parts with the goal of introducing the core concepts of big data and data science in Part I with examples that relate to nursing as well as other industries. Part II brings in the new and emerging technologies that make big data analytics possible, and illustrates through case studies and references to initiatives currently happening. These two foundational parts also provide state-of-the-art/science reviews that are written by fellow nurses with an eye to demystifying and removing any intimidation that might surround this field.

Introduced throughout all five parts is the important principle of using partnerships and building teams that include big data analytics experts and data scientists in order to have the clinical and technical skill mix needed. The days of the single researcher, analyst, or single domain team are being called into question for their relevancy and efficiency. Recognizing that all missions—academic, research, practice, policy—are transformed by big data, Part III focuses on research. Specifically, this part dives into the complexity of disease, advancement of networks to increase access to large data capacity, and actual application of data analytics to drive transformation of the healthcare system. Taken together, Part III's chapters show the potential of nursing's engagement in big data science to transform the science by the new knowledge generated and its application in practice, education and policy.

The last two parts attend to applied current state exemplars for nurse executives to have reference roadmaps, competencies needed at all levels, and a look at the near future impact for healthcare delivery, education and research. Throughout Part IV and V, "readiness" is directed at those who own change across the sectors: those who teach our next generation of nurses; the health policymakers who support change through regulations, guidance and funding; and nurse executives who define care strategies within their healthcare organizations. Front and center to all these sectors within the near future big data world is the critical state of the nursing workforce. Part V includes a description of quantity, emergent roles, education and appropriate certification and credentialing that "readiness" for the changes afoot will require.

A theme throughout the book is the goal of having "sharable and comparable" nursing data, and the need for standards to make this possible. While nursing is making progress on having adequately matured, codified terminologies to represent nursing concepts, actions and outcomes across all care domains, we are not there yet. The tactics used to compensate for this current state are reflected in the chapters and case studies presented in Part IV and V. Interoperability and data standards are the key challenge of our times and will continue to have intense focus. Standards that work for all are not just U.S. challenges, but rather extend worldwide; and thus, the significance of a global world permeates these invitations for engagement, transformation and empowered nursing.

In summary, this book is applicable to all nurses and interprofessional colleagues in all roles. We deliberately constructed the content and selected the applied examples and case studies so that the book can serve as a technology reference, or a "101 Intro" to big data for all nurses, and most importantly, a "how to" guide for planning your own big data initiatives. We hope that you will use the book broadly for continuing education purposes as well as for educational curricula; but above all, we hope that you read and enjoy the book!

Connie W. Delaney
Charlotte A. Weaver
Judith J. Warren
Thomas R. Clancy
Roy L. Simpson

Acknowledgments

The idea for this book grew out of many nursing and interprofessional dialogues about big data and data science's growing presence throughout the healthcare industry including research, education and policy. The need for a "big data and nursing" book emerged from multiple conversations that happened in the context of the University of Minnesota School of Nursing's annual Big Data and Nursing conferences that started in June 2013. From these rich interchanges, we recognized a pressing urgency to bring the potential of big data analytics into all the domains of nursing—practice, operations, research, academics, industry, and policy. More sobering was the discovery that there were no books in the marketplace in 2016 that specifically focused on nursing and big data. So our biggest "thank you" goes to all the authors and co-authors of the many chapters and case studies that make up this book. As early contributors in this first nursing publication on the topic worldwide, these authors bravely stepped forward to share their work and to lay out candidly how they are bringing big data applications into their respective domains.

In addition, we owe thanks to many others for helping us bring this body of work to you. To start at the beginning, we want to acknowledge the debt we have to the annual Big Data and Nursing Knowledge Conference Steering Committee Members who have worked diligently to ensure that we had this annual gathering without which the ideas, content and contributors for this book would not have happened. Bonnie Westra and Lisiane Pruinelli served as the organizing principals, aided by Susan Matney, Joyce Sensmeier, Daniel Pesut, Nancy Ulvestad and four of the book editors. A number of organizations also contributed by sponsoring representatives to the annual conferences and those include: American Nurses Association, American Association of Colleges of Nursing, American Medical Informatics Association, National Institute of Nursing Research, Trinity Health, Hospital Corporation of America, and University of Minnesota Medical Center. Cerner Corporation and OptumLabs'™ Scott Regenstein, Greta Bagshaw and William Crown were instrumental partners as fellow researchers, book contributors and data science experts who willingly shared knowledge and guidance.

We give our greatest thanks to Dixie Berg, consulting editor, for helping us track all the details, keeping all the editors organized and on task, and for doing the final book manuscript preparation. Behind the scenes, we have Toni Bennett from Cerner Corporation and Dana Hurley from the University of Minnesota School of Nursing who gave immeasurable assistance with planning and organization for conference calls across time zones and difficult schedules. Many of our colleagues behind the scenes encouraged us and contributed their ideas, time and creativity. We are indebted to Melanie Dreher, Nancy Dunton, Catherine Ivory, Tess Settergren, Russ McDonough, and Jeannine Rivet. A special note of appreciation goes to Grant Weston, who as our Springer editor was a strong supporter of this book and a constant source of guidance throughout this exciting journey.

Connie W. Delaney
Charlotte A. Weaver
Judith J. Warren
Thomas R. Clancy
Roy L. Simpson

Contents

List of Contributors

C.F. Aliferis, MD, PhD, FACMI Institute for Health Informatics, University of Minnesota, Minneapolis, MN, USA

Amberly Barry, RN, PHN Analytics & Transformation, IBM Global Healthcare, Armonk, NY, USA

Murielle S. Beene, DNP, MBA, MPH, MS, RN-BC, PMP Department of Veterans Affairs, Veterans Health Administration, Washington, DC, USA

Bobbie Berkowitz, PhD, RN, NEA-BC, FAAN Columbia University School of Nursing and Columbia University Medical Center, New York, NY, USA

Marlene A. Bober, RN, MS Acute Enterprise Care Management, Advocate Health Care, Chicago, IL, USA

Philip E. Bourne, PhD Data Science Institute, University of Virginia, Charlottesville, VA, USA

Barbara Brandt, PhD National Center for Interprofessional Practice and Education, University of Minnesota, Minneapolis, MN, USA

Frank Cerra, MD National Center for Interprofessional Practice and Education, University of Minnesota, Minneapolis, MN, USA

Chih-Lin Chi, PhD, MBA OptumLabs™, Cambridge, MA, USA

School of Nursing, University of Minnesota, Minneapolis, MN, USA

Institute for Health Informatics, University of Minnesota, Minneapolis, MN, USA

Marilyn Chow, RN, PhD Kaiser Permanente, Oakland, CA, USA

Beverly Christie, DNP, RN Fairview Health Services, Minneapolis, MN, USA

Thomas R. Clancy, PhD, MBA, RN, FAAN School of Nursing, University of Minnesota, Minneapolis, MN, USA

Rosaly Correa-de-Araujo, MD, MSc, PhD U.S. Department of Health and Human Services, Division of Geriatrics and Clinical Gerontology, National Institute on Aging, National Institutes of Health, Bethesda, MD, USA

William Crown, MD OptumLabs, Minneapolis, MN, USA

Sandra Daack-Hirsch, PhD, RN The University of Iowa, Iowa City, IA, USA

Connie W. Delaney, PhD, RN, FAAN, FACMI School of Nursing, University of Minnesota, Minneapolis, MN, USA

Melanie Dreher, PhD, RN, FAAN Trinity Health System, Livonia, MI, USA

Patricia Eckardt, PhD, RN Heilbrunn Family Center for Research Nursing, Rockefeller University, New York, NY, USA

Jane Englebright, PhD, RN, CENP, FAAN Hospital Corporation of America, Nashville, TN, USA

Cynthia Gadd, PhD, MBA, MS, FACMI Department of Biomedical Informatics, Vanderbilt University, Nashville, TN, USA

Grace Gao, DNP, RN-BC School of Nursing, University of Minnesota, Minneapolis, MN, USA

April Giard, MSN, NP-BC, NEA-BC Eastern Maine Health System, Brewer, ME, USA

Bryan Gibson, DPT, PhD Department of Biomedical Informatics, School of Medicine, University of Utah, Salt Lake City, UT, USA

Miriam Halimi Trinity Health System, Livonia, MI, USA

Lynda Hardy, PhD, RN, FAAN College of Nursing, The Ohio State University, Columbus, OH, USA

Ellen Harper, DNP, RN-BC, MBA, FAAN School of Nursing, University of Minnesota, Minneapolis, MN, USA

Susan Henly, PhD, RN, FAAN School of Nursing, University of Minnesota, Minneapolis, MN, USA

Edmund Jackson, PhD Hospital Corporation of America, Nashville, TN, USA

Amy Jarabek, MSA, MEd National Center for Interprofessional Practice and Education, University of Minnesota, Minneapolis, MN, USA

Peggy Jenkins, PhD, RN College of Nursing, University of Colorado, Denver, CO, USA

Steven G. Johnson, PhD-C, PhD Institute for Health Informatics, University of Minnesota, Minneapolis, MN, USA

Katherine Kim, PhD, MPH, MBA Betty Irene Moore School of Nursing, University of California–Davis, Davis, CA, USA

Anne LaFlamme, DNP, RN University of Minnesota Medical Center–Fairview Health Services, Minneapolis, MN, USA

Joanne LaFluer, PhD VA IDEAS Center of Innovation, Salt Lake City, UT, USA
Department of Pharmacotherapy, School of Pharmacy, University of Utah, Salt Lake City, UT, USA

Stephanie Lambrecht, MS, RN-BC, MHA Elsevier Clinical Solutions, USA

Gay Landstrom, PhD, RN, NEA-BC Ascension St. John Providence, Warren, MI, USA

Norma M. Lang, RN, PhD, FAAN, FRCN School of Nursing, University of Pennsylvania, Philadelphia, PA, USA
University of Wisconsin–Milwaukee, Milwaukee, WI, USA

May Nawal Lutfiyya, PhD, FACE National Center for Interprofessional Practice and Education, University of Minnesota, Minneapolis, MN, USA

Satish M. Mahajan, PhD, MSat, MEng, RN Betty Irene Moore School of Nursing, University of California–Davis, Davis, CA, USA

E. LaVerne Manos, DNP RN-BC School of Nursing, University of Kansas, Kansas City, KS, USA

Donna Mayo, MSN, RN Elsevier Clinical Solutions, USA

Angela Barron McBride, PhD, RN School of Nursing, Indiana University, Indianapolis, IN, USA

Linda A. McCauley, RN, PhD, FAAN, FAAOHN Nell Hodgson Woodruff School of Nursing, Emory University, Atlanta, GA, USA

Douglas McNair, MD, PhD Cerner Math, Kansas City, MO, USA

Julie A. Miller, PhD Patient-Centered Outcomes Research Institute, Washington, DC, USA

Judy Murphy, RN, FACMI, FAAN, FHIMSS IBM Global Healthcare, Armonk, NY, USA

Sarah N. Musy, PhD Institute of Nursing Science, University of Basel & Interselspita Bern University Hospital, Basel, Switzerland

Robert Nieves, BSN, RN, MBA, MPA, JD Elsevier Clinical Solutions, USA

Ann O'Brien, RN, MSN National Patient Care Services, Kaiser Permanente, Pleasanton, CA, USA

Jung In Park, PhD-C, RN School of Nursing, University of Minnesota, Minneapolis, MN, USA

Judith Pechacek, DNP School of Nursing, University of Minnesota, Minneapolis, MN, USA

Paula M. Procter, RN, PGCE, MSc, SFHEA, FBCS, CITP Department of Nursing and Midwifery, Sheffield Hallam University, Sheffield, UK

Lisiane Pruinelli, PhD, RN School of Nursing, University of Minnesota, Minneapolis, MN, USA

Piper A. Ranallo, PhD Six Aims for Behavioral Health, Minneapolis, MN, USA

Teresa Schicker, MPA National Center for Interprofessional Practice and Education, University of Minnesota, Minneapolis, MN, USA

Joe V. Selby, MD, MPH Patient-Centered Outcomes Research Institute (PCORI), Washington, DC, USA

Walter Sermeus Department of Public Health & Primary Care, KU Leuven, Leuven, Belgium

Lisa Shah, MSN, RN The University of Iowa, Iowa City, IA, USA

Suzan G. Sherman, PhD, RN Nursing and Allied Health, Fairview Health Services, Minneapolis, MN, USA

Michael Simon, PhD University of Basel & Interselspita Bern University Hospital, Basel, Switzerland

Roy L. Simpson, DNP, RN, FAAN, FACMI, DPNAP Technology Management, Nell Hodgson Woodruff School of Nursing, Emory University, Atlanta, Georgia, USA

Kevin Smith, BSN, RN, CNML, CVRN-BC Naples Healthcare Systems, Naples, FL, USA

Vicki Snavely, RN Cerner Corporation, Kansas City, MO, USA

Rayne Soriano, PhD, RN Clinical Informatics and Clinical Transformation Program, Kaiser Permanente, Sacramento, CA, USA

Stuart Speedie, PhD, FACMI Institute for Health Informatics, University of Minnesota, Minneapolis, MN, USA

Darinda Sutton, MSN, RN-BC Cerner Corporation, Kansas City, MO, USA

Nora Triola, PhD, RN, NEA-BC Trinity Health System, Livonia, MI, USA

Michelle Troseth, MSN, RN, DPNAP, FAAN Elsevier Clinical Solutions, USA

Jin Wang, MS, MHI Institute for Health Informatics, University of Minnesota, Minneapolis, MN, USA

OptumLabs, Minneapolis, MN, USA

Judith J. Warren, PhD, RN, FAAN, FACMI School of Nursing, University of Kansas, Kansas City, KS, USA

Charlotte A. Weaver, RN, PhD, FAAN, FHIMSS Retired Healthcare Executive, Board Director, Issaquah, Washington, USA

Charlene Weir, PhD Department of Biomedical Informatics, University of Utah, Salt Lake City, Utah, USA

IDEAS Center of Innovation, Veterans Affairs Health Services Research and Development Service, Salt Lake City, Utah, USA

John M. Welton, PhD, RN, FAAN Health Systems Research, University of Colorado College of Nursing, Denver, CO, USA

Bonnie L. Westra, PhD, RN, FAAN, FACMI Center for Nursing Informatics, School of Nursing, University of Minnesota, Minneapolis, MN, USA

Marisa L. Wilson, DNSc, MHSc, RN-BC, CPHIMS, FAAN Department of Family, Community and Health Systems, School of Nursing, The University of Alabama at Birmingham, Birmingham, AL, USA

Qing Treitler Zeng, PhD, FACMI VA IDEAS Center of Innovation, Salt Lake City, UT, USA

Biomedical Informatics, School of Medicine, University of Utah, Salt Lake City, UT, USA

Part I
The New and Exciting World of "Big Data"

Judith J. Warren

The purpose of this section is to introduce the concepts and approaches to the use of big data. While we have always analyzed large data sets, the internet and other new technologies have created new sources of data, new types of data at a speed unheard of before. It is projected that information will double in 2020 every 12 hours and therefore increase the need to make sense of it. Yet we also have new programming languages, statistical languages, visualization graphics, and database management systems that will help us keep up with this new data challenge. Chapter 1 will discuss the promise of big data in nursing and invite nurses to meet this new challenge and opportunity. Chapter 2 will provide an overview of the current healthcare landscape and its use of big data. Chapter 3 will explore this new world of big data, describe its sources, its analytics, and its applications. Big data will be differentiated from the traditional very large data sets. Resources will be shared on where to go to learn more about big data and the technologies and methods evolving to manage big data.

Chapter 1
Why Big Data?: Why Nursing?

Connie W. Delaney and Roy L. Simpson

Abstract Attention on "big data" spans nursing and the health sciences, and extends as well to engineering/computer sciences through to the liberal arts in professional literature. A current Google search (3 Nov 2016) of "big data" yields 288 million entries. A focused search of "big data and nursing" yields more than 3.9 million entries. Thus we ask, "Why big data? Why nursing?" The focus of this chapter is to provide an overview of why big data has emerged now and to make the case for how big data has the capacity to change health, healthcare systems, and nursing. This chapter lays a foundation for the chapters and case studies to follow that explore what data, knowledge, and transformation processes are needed to put information and knowledge into the hands of nursing wherever nurses are working. In this chapter we examine the big data sources within and beyond nursing and healthcare that can be collected and analyzed to improve nursing and patient, family and community health. This chapter entices the reader to examine "Why big data now?" and "Why big data in the future?" This chapter is meant to stir curiosity for "Why should I be knowledgeable?" Whether the reader's role is in clinical practice, education, research, industry, or policy, the applied uses of big data analytics are empowering change at an exponential speed across all domains. Big data has the capacity to illuminate nursing's discovery of new knowledge and best practices that are safe, effective and lead to improved outcomes including well-being of providers; it also can expand nursing's vision and future possibilities through increasing awareness of what nursing doesn't know. The importance of nursing's lens on the new discoveries obtained through big data and data science is critical to the transformation of health and healthcare systems. This transformation completes the challenge of placing the *person* at the center of all care initiatives and actions.

C.W. Delaney, PhD, RN, FAAN, FACMI (✉)
School of Nursing, University of Minnesota, Minneapolis, MN, USA
e-mail: Delaney@umn.edu

R.L. Simpson, DNP, RN, FAAN, FACMI, DPNAP
Technology Management, Nell Hodgson Woodruff School of Nursing,
Emory University, Atlanta, GA, USA

© Springer International Publishing AG 2017
C.W. Delaney et al. (eds.), *Big Data-Enabled Nursing*, Health Informatics,
DOI 10.1007/978-3-319-53300-1_1

Keywords Nursing • Big data • Nursing informatics • Analytics • Knowledge discovery • Triple Aim

1.1   Why Big Data?

Big data has been described as a revolution. Mayer-Schönberger and Cukier (2013) define big data as the new raw material for discovery and innovation. As such, big data is one of the hottest trends in technology and is predicted to have a dramatic impact on the economy, science, and society at large. The authors cite the powerful example of Google's prediction of the spread of the winter flu (H1N1) in the United States. Analytics demonstrated that a combination of 45 search terms (from the 50 million search terms), processed by a mathematical model, resulted in a strong correlation between the Google prediction and the U.S. Centers for Disease Control and Prevention's (CDC) figures nationwide. The difference, that is value added, is that the use of big data through Google could tell such spread of disease in near real time, as opposed to the CDC's lag time of months.

The Google flu example illustrates a palpable paradigm shift that incorporates the fairly new technology-enabled developments of relatively unlimited information availability and robust computing power. These new tools and technology developments profoundly extend the human capacity for discovering and translating knowledge into any applied decision, and for healthcare specifically to improve care quality and safety, lower costs, and address provider satisfaction. The knowledge society has evolved from historic areas of agricultural, industrial, and information societies. Exponential growth in computer processing speed, the digitization of everything, the Internet and Internet of Things (IoT), recombination of existing technologies, as well as the electronic health record, health insurance claims, the quantified self movement, geo-spatial data, social media, eMobile Health, and whole gene sequencing has launched us into the knowledge paradigm. These exponential changes are complemented by a bold shift in empowerment of people/patients and a complementary commitment from health systems to person-centric care (Topol 2015).

A look at key trends magnifies the important and fast evolution of big data in defining our world, how we think and do, and our ways of supporting discovery. Chris Stelland (2015) recently identified the biggest trends in this space. Stelland notes that not only the expansion of the Internet of Things (IoT) but the coming together of the IoT phenomenon with cloud technologies and big data analytic tools will be a driver of the data explosion in petabyte volumes. Companies such as Google, Amazon Web Services, and Microsoft are advancing services that support data moving seamlessly to cloud-based analytics engines (Wallace 2015). The increasing adoption of NoSQL (Not only SQL) technologies (see Chaps. 3 and 4), commonly associated with unstructured data, is changing enterprise IT landscapes where the benefits of schema-less database concepts become more pronounced (Heudecker et al. 2016).

The ability to increase the speed of data processing is a key innovation. One example of this trend is the ascendance of the open source support of Apache Spark (Feng 2014). The growth of data ecosystems to support big data platforms in enterprises is a must and is clearly being addressed by the installation of Hadoop server architecture [an open source, Java-based programming framework (part of Apache project) that supports the processing and storage of extremely large data sets in a distributed computing environment]. In a recent survey of 2200 Hadoop customers (Wheatley 2015), only 3% of respondents anticipated they would be doing less with Hadoop in the next 12 months. Seventy-six percent of those who already use Hadoop planned on doing more within the next three months and, finally, almost half of the companies planned deployment within the next 12 months. The same survey also found Tableau to be the leading data visualization tool for companies using or planning to use Hadoop, as well as those furthest along in Hadoop maturity. Security, another key need and trend, is receiving critical attention. For example, Hadoop is adding security to its enterprise standards package. The Apache Sentry project provides a system for enforcing fine-grained, role-based authorization to data and metadata stored on a Hadoop cluster (Henschen 2013).

Beyond the enterprise, end users have a growing demand for the ability to analyze and explore this explosion of data. This activity will further blur the lines behind the "traditional" business intelligence concepts and the world of "big data" and people empowerment (Feng). With the demands from end-users for increased access to data, as well as more analytic capacity, there is an ever growing need for preparing end-users to discover and analyze all forms of data. Self-service data preparation tools (Alteryx, Trifacta, Paxata, Lavastorm, Informatica) are exploding in popularity because they reduce time and complexity to analyze data for applied research teams and businesses (Kaelber et al. 2012). And lastly, another robust trend is the growth of the cloud warehouse function as an on-demand cloud data warehouse used by Redshift, BigQuery, and others (Brust 2015).

The ability to measure a phenomenon and transform it into a quantified format (the ability to record information, convert analog information to binary code) for tabulation and analysis is the new evolution in the era of knowledge discovery. The capacity that enables diversity of data, prediction and planning, and enhances the value of moving data from an ancillary resource for business or narrow areas of science to all data is very valuable. Moreover, the importance and relevance of historical data are also being reinforced.

As healthcare and nursing engage in the era of big data, sensitivity and commitment to transcending challenges are paramount. Big data is messy; the likelihood of errors, inconsistency in formatting, and challenges in selecting appropriate mathematical models increase as the volume of data increases. Imperfections of the tools used to measure, record, and analyze information will be reflected in big data as they were in smaller data sets. While decision makers generally welcome data being available in real time, increased thoughtfulness to the possible risks of arriving at fast decisions is paramount. Issues of privacy, accountability of data users, and social justice to guarantee human freedom to act mandate that institutional and professional attention be focused on advocating for all people, especially those who might be harmed by and/or not understand big data.

For the professionals, the challenge is the adoption of a new mental model for using expanded data analytics. Professionals are challenged to work and discover in real time, versus the double blind clinical trial research designs that provide traditional discovery pathways of clinical trials. Operational leaders, researchers and educators must bridge practice and research; big data is one such bridge. The interdisciplinary teams will increasingly depend upon communications and teamwork to generate questions and inform analytics as part of the way they work. Big data offers such a focus. Moreover, effectively amassing and using big data requires strong partnerships to extend beyond the usual healthcare team in practice, research and education. Such partnerships are in play today with participants from vendors, data analytic experts, industry, and policy developers (see Chaps. 4, 6, 7, 8, 18, and 21). In summary the era of big data offers healthcare, nursing professionals, and interprofessional teams immeasurable opportunities to do things better, innovate, and most meaningfully to co-design our fundamental shifts in basic understanding of humans and their world.

1.2 Why Big Data in Nursing?

Improving care quality and safety, lowering costs, addressing provider satisfaction, and advancing our knowledge in nursing depend on engaging big data, enhancing data capacity and integration and using cutting-edge analytic methods to support discovery. Brennan and Bakken (2015) note that contemporary big data initiatives in health care will benefit from greater integration with nursing science and nursing practice. They acknowledge that big data and data science in nursing have the potential to provide greater richness in understanding patient phenomena and in tailoring interventional strategies that are personalized to the patient. Along with others (Chatterjee 2012; Kurzweil 2006; Hey 2009; NIH), they agree that the knowledge era we are addressing encompasses data that exceeds human comprehension and management by standard computer systems, and brings a level of imprecision not found in traditional inquiry.

A literature review conducted by Westra et al. (in press) identified, analyzed and synthesized exemplars of big data nursing research applied to practice and disseminated in key informatics and nursing research journals between 2009 and 2015. Seventeen studies were identified. This literature review confirmed that nursing scientists have engaged in big data and data science over the past decade. While this finding is promising, multiple challenges were also clear. The majority of studies focused on clinical practice with little evidence of specificity in both description of the practice environment or addressing nursing clinical practice. There is urgent capacity to expand big data science to include nursing data in health promotion, acute and chronic care, symptom management, health systems, and care across the continuum from home health, to intensive care, to acute and chronic care for children, adults, families and communities. All of these domains hold clear opportunities to reveal and enhance the availability of nursing sensitive data. For example, the variables endorsed by the American Nurses Association recognized terminologies and LOINC's standard representation of the Nursing Management Minimum Data Set, which captures personnel, context of care, and other workforce significant factors to understand contextual impact on care management, outcomes, costs, and patient satisfaction, are absent from traditional data discovery. Big data's ability to retrieve data from multiple database sources may provide access to this data. In general, there was little evidence based on Westra's review that nursing was accessing the plethora of data resources available and applicable for nursing research. Such sources would include the following: clinic, urgent care and school settings, administrative claims data from a health plan/insurer perspective; laboratory or imaging data; social media; data originating from wearable technologies; molecular biology; and data in combination with other data available from EHRs. While there are multiple networks and consortia supporting access to data from a multitude of settings, sites, and collaborating partners (e.g., Patient-Centered Outcomes Research Institute (PCORI) and Clinical and Translational Science Awards (CTSAs)), Westra's literature review found no evidence of the use of these consortia and data resources. These sources can be used in big data analytics through the use of the Hadoop server architecture and new analytic models from data science.

The special issue of the Western Journal of Nursing Research "Big Data: Data Science in Nursing" (Delaney and Westra 2017) also provides evidence of nursing engagement in big data. This special issue focuses on nursing big data science and uses the NIH definition of "biomedical big data". It includes 11 exemplars of using big data science to drive cutting-edge research and identify emerging methods in nursing big data science. These studies focused on the architecture and standardization of data to enable big data science (the bench science of nursing and health informatics) (Matney et al.); use of EHR data (Wilkie et al.; Monsen et al.), administrative/claims data (Chi et al.), and sensor data (Phillips et al.; Hertzberg et al.). A variety of data analytics was evident, including visualization (Docherty et al.; Monsen et al.), natural language processing (NLP) (Topaz et al.); and standardized vocabularies (Kim et al.; Monsen et al.), and word adjacency (Miller et al.).

The National Institutes of Health/National Institute of Nursing Research (NIH/NINR) has fostered nursing research in the above cutting-edge initiatives. NINR Director Patricia Grady consistently emphasizes the importance of nurse scientists in leading change through evidence-based practice and fostering cross-disciplinary teams and collaboration. The NINR promotes research to translate discoveries into practice and stimulates new directions in nursing research, such as creating a blueprint for genomic nursing science, data science opportunities in nursing, the role of nursing in the Precision Medicine Initiative Cohort Program, and NINR's five-year strategic plan, "NINR Strategic Plan: Advancing Science, Improving Lives". These initiatives share dependency on EHRs, genomics, sophisticated data science, and mobile health technologies.

1.3 Summary

Nursing's engagement in big data and data science exists, is empowering nursing's contribution to advancements in human health, and illustrates capacity for growth. Many challenges have been highlighted in this chapter and more are shared in this book and its case studies. NIH notes that "big data is more than just very large data or a large number of data sources" (NIH 2015). Data science is an interdisciplinary field, which focuses on creating models that capture the underlying patterns of complex systems and codifying models into working applications (Vasant 2013). In contrast but complementary, big data focuses on collection and management of large amounts of varied data. Indeed this field of study extends beyond a new method; it is the emergence of the fourth paradigm of science—eScience (Hey et al. 2009). Such emergence invites us to in-depth discernment of Kurzweil's concept of singularity (Kurzweil 2006), a point at which the combination of human and machine intelligence transcends the limits of a human mind. The key challenge is determining how nursing and nurses partner with the data science field to transcend human intellectual limitations to advance the health of people, families, and populations.

References

Brennan PF, Bakken S. Nursing needs big data and big data needs nursing. J Nurs Scholarsh. 2015;47(5):477–84. doi:10.1111/jnu.12159.

Brust A. Cloud data warehouse race heats up. ZDNet. 2015, Jun 26. http://www.zdnet.com/article/cloud-data-warehouse-race-heats-up/. Accessed 10 Nov 2016.

Chatterjee AB. Intrinsic limitations of the human mind. Int J Basic Appl Sci. 2012;1(4):578–83. doi:10.14419/ijbas.v1i4.418.

Chi C, Wang J, Clancy T, Robinson J, Tonellato P, Adam T. Big data cohort extraction to facilitate machine learning to improve statin treatment. In Delaney C and Westra B. Big data: data science in nursing. West J Nurs Res (Special Issue). 2017.

Delaney C, Westra B. Big data: data science in nursing. West J Nurs Res (Special Issue). 2017;39(1).

Docherty S, Vorderstrasse A, Brandon D, Johnson C. Visualization of multidimensional data in nursing science. In Delaney C and Westra B. Big data: data science in nursing. West J Nurs Res (Special Issue). 2017.

Feng J. 5 best practices for Tableau and Hadoop. http://www.tableau.com/learn/whitepapers/5-best-practices-tableau-hadoop. Accessed 10 Nov 2016.

Feng J. Databricks application spotlight: Tableau software. 2014. https://databricks.com/blog/2014/10/15/application-spotlight-tableau-software.html. Accessed 10 Nov 2016.

Henschen D. Cloudera brings role-based security to Hadoop. Information Week. 2013, Jul 24. http://www.informationweek.com/big-data/software-platforms/cloudera-brings-role-based-security-to-hadoop/d/d-id/1110903. Accessed 10 Nov 2016.

Hertzberg V, Mac V, Elon L, Mutic A, Peterman K, Mutic N, Tovar-Aguilar JA, Economos J, Flocks J, McCauley L. Novel analytic methods needed for real-time continuous core body temperature data. In Delaney C and Westra B. Big data: data science in nursing. West J Nurs Res (Special Issue). 2017.

Heudecker N, Feinberg D, Adrian M, Palanca T, Greenwald R. Gartner magic quadrant for operational database management systems. 2016. https://info.microsoft.com/CO-SQL-CNTNT-FY16-09Sep-14-MQOperational-Register.html. Accessed 10 Nov 2016.

Hey T, Tansley S, Toll K, editors. The fourth paradigm: data-intensive scientific discovery. Seattle (WA): Microsoft Corporation; 2009.

Kaelber D, Foster W, Gilder J, Love T, Jain A. Patient characteristics associated with venous thromboembolic events: a cohort study using pooled electronic health record data. J Am Med Inform Assoc. 2012;19(6):965–72. doi:10.1136/amiajnl-2011-000782.

Kim H, Jang I, Quach J, Richardson A, Kim J, Choi J. Explorative analyses of nursing research data. In Delaney C and Westra B. Big data: data science in nursing. West J Nurs Res (Special Issue). 2017.

Kurzweil K. The singularity is near: when humans transcend biology. Westminster, London: Penguin Books; 2006.

Matney S, Settergren T, Carrington J, Richesson R, Sheide A, Westra B. In Delaney C and Westra B. Big data: data science in nursing. West J Nurs Res (Special Issue). 2017

Mayer-Schönberger V, Cukier K. Big data: a revolution that will change how we live, work and think. London: John Murray; 2013.

Miller W, Groves D, Knopf A, Otte J, Silverman R. Word adjacency graph modeling: separating signal from noise in big data. In Delaney C and Westra B. Big data: data science in nursing. West J Nurs Res (Special Issue). 2017.

Monsen K, Peterson J, Mathiason M, Kim E, Votava B, Pieczkiewicz D. Discovering public health nurse-specific family home visiting intervention patterns using visualization technique. In Delaney C and Westra B. Big data: data science in nursing. West J Nurs Res (Special Issue). 2017.

National Institutes of Health/National Institute of Nursing Research (NIH/NINR). https://www.ninr.nih.gov/researchandfunding/datascience. Accessed 10 Nov 2016.

NIH. https://datascience.nih.gov/; https://datascience.nih.gov/bd2k. Accessed 10 Nov 2016.

NINR. NINR strategic plan: advancing science, improving lives. https://nihrecord.nih.gov/newsletters/2016/07_15_2016/story7.htm. Accessed 10 Nov 2016.

Phillips L, Deroche C, Rantz M, Alexander G, Skubic M, Despins L, Casanova-Abbott C, Harris B, Galambos C, Koopman R. Using embedded sensors in independent living to predict gait changes and falls. In Delaney C and Westra B. Big data: data science in nursing. West J Nurs Res (Special Issue). 2017.

Stelland C. Top 8 big data trends for 2016. 2015. http://www.tableau.com/asset/top-8-trends-big-data-2016?utm_campaign=Prospecting-BGDATA-ALL-ALL&utm_medium=Paid+Search&utm_source=Google+Search&utm_language=EN&utm_country=USCA&kw=%2Btop%20%2B8%20%2Bbig%20%2Bdata%20%2Btrends&adgroup=CTX-Big+Data-Big+Data+All-B&adused=106945857375&matchtype=b&placement=&kcid=512788ef-9c83-4023-9baf-541715034e38&gclid=CJGpiaj31dACFUYbaQodYYgHqA. Accessed 10 Nov 2016.

Topaz M, Radhakrishnan K, Blackley S, Lei V, Lai K, Zhou L. Studying associations between heart failure self-management and rehospitalizations using natural language processing. In Delaney C and Westra B. Big data: data science in nursing. West J Nurs Res (Special Issue). 2017.

Topol E. The patient will see you now. New York: Basic Books; 2015.

Vasant D. Data science & prediction. Commun ACM. 2013;56(12):64–73.

Wallace M. All the things: data visualization in a world of connected devices. 2015. http://www.tableau.com/about/blog/2015/1/all-things-data-visualization-world-connected-devices-36393. Accessed 10 Nov 2016.

Westra, BL, Sylvia M, Weinfurter EF, Pruinelli L, Park JI, Dodd D, Keenan GM, Senk P, Richesson RL, Baukner V, Cruz C, Gao G, Whittenburg L, Delaney CW (2017). Big Data Science: A Literature Review of Nursing Research Exemplars. Nursing Outlook. 2016 Dec 8. poi: S0029-6554(16)30396-7. doi: 10.1016/j.outlook.2016.11.021. [Epub ahead of print]

Wheatley M. AtScale's Hadoop maturity survey highlights big data's relentless growth. 2015. http://siliconangle.com/blog/2015/09/17/atscales-hadoop-maturity-survey-highlights-big-datas-relentless-growth/. Accessed 10 Nov 2016.

Wilkie D, Khokhar A, Lodhi M, Yao Y, Ansari R, Keenan G. Framework for mining and analysis of standardized nursing care plan data. In Delaney C and Westra B. Big data: data science in nursing. West J Nurs Res (Special Issue). 2017.

Chapter 2
Big Data in Healthcare: A Wide Look at a Broad Subject

Marisa L. Wilson, Charlotte A. Weaver, Paula M. Procter, and Murielle S. Beene

Abstract Healthcare, as an industry, is being held accountable for outcomes that were never before considered. No longer is it a fee for service industry but one in which providers are accountable for quality, performance, and satisfaction. How can healthcare utilize the sea of data collected to optimize care for quality, performance, and satisfaction? One answer is to harness the power of that data and to manage the ability to perform meaningful analyses. Clinicians and ancillary providers collect data as they interact with patients each and every day. Data is brought into patient records from laboratories, radiology readings, pathology reports, transfers, and consults where it is integrated and reviewed in order to make informed decisions related to care. Patient and consumer generated data is being gathered and stored on millions of mobile applications or in personal health records. Some patients are passing this data to their care providers for inclusion in care decision-making. Data is coming from external care providers in disparate systems and is passed from one provider to another as patients traverse a trajectory of care that may or may not be seamless. At each juncture, the data is reviewed, updated, and then transferred onto the next provider or caregiver. At some point in the not too distant future, data describing our homes, schools, workplaces, and communities will be considered as these social determinants of health contribute greatly to the overall health of persons and communities. This will encompass data and information describing poverty, employment, food security, housing, education, incarceration and institutionalization, access to care, environment, and literacy. All of this data

M.L. Wilson, D.N.Sc., M.H.Sc., R.N.-B.C., C.P.H.I.M.S., F.A.A.N. (✉)
Department of Family, Community, and Health Systems, The University of Alabama at Birmingham School of Nursing, Birmingham, AL, USA
e-mail: mwilsoa@uab.edu

C.A. Weaver, RN, PhD, FAAN, FHIMSS
Retired Healthcare Executive, Board Director, Issaquah, WA, USA

P.M. Procter, R.N., P.G.C.E., M.Sc., S.F.H.E.A., F.B.C.S., C.I.T.P., F.I.M.I.A.N.I.
Department of Nursing and Midwifery, Sheffield Hallam University, Sheffield, UK

M.S. Beene, D.N.P., M.B.A., M.P.H., M.S., R.N.-B.C., P.M.P., F.A.A.N.
Department of Veterans Affairs, Veterans Health Administration, Washington, DC, USA

© Springer International Publishing AG 2017
C.W. Delaney et al. (eds.), *Big Data-Enabled Nursing*, Health Informatics,
DOI 10.1007/978-3-319-53300-1_2

11

will need to be extracted from dissimilar storage points, cleaned and transformed for use, and loaded into a storage repository so that clinicians will all be able to gain a fuller picture of how to care better and to delivery that quality.

Keywords Big data • Analytics • Veterans administration • National health service • Home health • Population health

2.1   Reaching the Tipping Point: Big Data and Healthcare

Today, the terms big data and analytics often are heard around healthcare venues. Speakers, presenters, and conferences are dedicated to this topic. But for the direct care provider, what is big data and what are the implications for practice? First, it means that the data a direct care provider collects and enters into a system will not solely be attached to that one patient/provider interaction or encounter but that much of that data will eventually migrate to and be included in a large pool of data coming from many patient and provider interactions. Big data refers to the phenomenon of exponential growth in the availability of both structured and unstructured data coming from multiple sources such as databases, transactions, records, log files, digital documents, and image and audio files. The "big" in big data is characterized by the sheer volume, the ever increasing velocity or speed, and the amazing variety of data that is available to manage, synthesize, and analyze. The term "big data" is usually tied to the term "data analytics" because it is through this data science that it is possible to examine the immense collection of raw data being stored and to manipulate the data with the express purpose of creating information and knowledge from what the human mind cannot do unaided. Analytics is the method for making sense of big data so that we can use what we have from multiple sources to describe, predict, and transform care to address quality, satisfaction, and outcomes.

Big data and analytic processes are not new. Big data, along with sophisticated analytic tools and techniques, has been used by many industry leaders, outside of healthcare, to make informed business decisions. Big data and analytics have also been used in the sciences to verify or disprove existing models or theories. Consider the current use of big data within the retail industry. It is well known that brick and mortal retailers are struggling as they adapt to the impact of internet competition. These retailers are working hard to compete by utilizing big data and analytics to better understand their customers, their needs, and their purchasing behaviors. Brick and mortar retailers have a distinct advantage over their online counterparts in that they can physically interact with and monitor their customers in ways that internet-based retailers cannot. Not only can they gather data at the point of a transaction regarding a purchase and the purchaser but they can monitor where a shopper goes in a store, what they look at, how they react to merchandise (Marr 2015). The bar coded card carried and swiped at the time of purchase in order to receive a sales price tells the retailer exactly what is purchased, how often a specific item is purchased, the category of items customers may purchase in the future, and specifically where customers shop. Through the retailer's use of big data and analytics, customers then receive targeted advertisements to lure them back into the store for future purchases that satisfy their needs. The retailers are doing this in order to gain an appreciation of what their customer wants to experience when entering a store by using multiple sources of big data to predict trends, to forecast demand, and to optimize pricing. Could healthcare use some of those same techniques to determine what potential patients are looking for in a satisfactory healthcare experience?

It is not just retailers using big data and analytic tools to understand the buying habits of potential customers. Some industries are using data and analytics to help us out of seemingly dire predicaments. Take the case of a hotel chain using big data and analytics to help frustrated travelers avoid having to sleep on a row of chairs in an airport during a storm. Bad weather reduces travel and, generally, impacts overnight lodging demands as people cancel plans. However, one hotel chain created a way around this. The hotel chain recognized that cancelled flights leave many travelers in a bind and in need of a place to sleep. The company takes advantage of big data opportunities and receives freely available weather and flight cancellation information that is organized by combinations of hotel and airport locations. The company then uses algorithms that factor in weather severity, travel conditions, time of the day, and cancellation rates by airport and airline among other variables (Schaeffer 2015). With these big data insights and a recognition that most travelers use mobile devices, the hotel chain then uses targeted mobile campaigns to deliver ads to weary and stranded travelers to make it easier for them to book a nearby hotel. The end result for the hotel chain is compelling. Flight cancellations average 1–3% daily, which translates into 150–500 cancelled flights or around 25,000–90,000 stranded passengers each day (Schaeffer 2015). With its big data and geo-based mobile marketing campaigns, one hotel chain achieved a 10% business increase from 2013 to 2014 (Schaeffer 2015). Could healthcare enterprises use some of the same data types and techniques to better target populations for specific interventions to either keep them healthy or to meet their needs during times of stress? With the data and technology available and with the demand to improve healthcare experiences and outcomes, the answer to these questions are "can" and "should".

However, healthcare has been very resistant to these changes and has lagged behind other industries. There is often resistance to change from making treatment decisions based on individual clinical judgment to decisions based on protocols driven by big data and analytics (Groves et al. 2013). Recent federal mandates have spurred the implementation of electronic health records (EHR) and are moving the industry away from paper based charting. Moreover, consumer use of mobile devices with health related applications is steadily increasing. Krebs and Duncan (2015) examined heath related applications (app) use among mobile phone owners in the United States and found that over 58% had downloaded a health related app with fitness and nutrition being the most common categories. With the increased use of EHRs over the last few years spurred on by the Meaningful Use mandates impacting all types of facilities, with individuals and populations increasingly utilizing patient facing technologies to quantify and digitize their data, and with the need to incorporate socio-behavioral data from non-medical sources in order to truly understand the health of patients, the industry is at a tipping point (Office of the National Coordinator [ONC] 2015). Experts in the analytics of healthcare data need to use

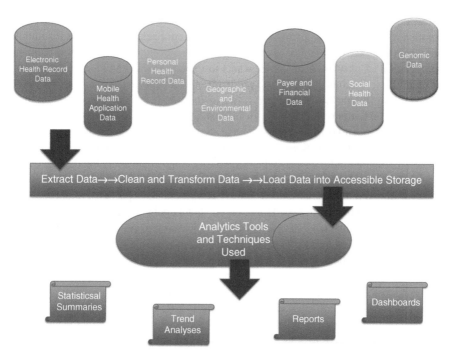

Fig. 2.1 Data from multiple sources will need to be extracted from the original database, examined for use, transformed, loaded into another database, and then extracted to create reports, dashboards, and visual graphics for analysis and pattern recognition

this sea of potential information, coming from multiple sources, to address the mandates of value based care versus just reimbursing for a service.

Collecting the data from a variety of sources, managing and storing the data, and determining the appropriate analytics will not be easy tasks (Fig. 2.1). The process is complex requiring a thorough understanding of the context and content of the care provided during which the data is being collected and used. It will be up to clinical care providers to take the leadership of the process so that they can efficiently use all of the big data available to determine how to rethink and redesign the care given now. It will be up to savvy informaticians and information technology professionals to build the transformational processes, the storage mechanisms, and analytic tools necessary to make the process a reality for those at the front lines.

2.2 Big Data and Analytics Enabling Innovation in Population Health

Care providers are being asked to consider the needs of not just one patient but of the populations served beyond their facility walls. Healthcare enterprises must be innovative if they are to be held accountable for the health of their covered lives or populations served beyond a single interaction. But, what exactly is population health and how does it differ from public health, a concept generally more familiar in the age of Ebola and Zika epidemics? What does population health have to do with big data and analytics? First, population health is part of but different from public health, which is more familiar to most clinicians. Population health has been extended beyond the realm of public health and brought into use by healthcare systems so that there is often confusion on what is meant by the term and who owns responsibility for it (Institute of Medicine [IOM] 2014, 2015). While public health can be defined at a national, state or community level, it includes all the inhabitants of a defined geographic space, their socio-economic descriptors, and data on the level and distribution of disease, functional status and well being. In its simplest form, public health focuses on the collection and use of data for monitoring and surveillance of reportable conditions, and carries the responsibility of using data to mount rapid response and guidance to protect the public's health within defined geographic areas. In comparison, for healthcare systems and payers, population health refers to managing the populations of patients served or individuals insured though managed care who are diagnosed with chronic diseases to achieve certain costs and quality metrics (IOM 2014, 2015).

The focus on population health has caused healthcare systems to shift focus to selected subpopulations of patients with the introduction of payments linked to population outcomes rather than the standard "fee for service" for an individual patient. Fee-for-service provides an incentive for providers to do more to earn more. While payment for population outcomes requires healthcare facilities and providers to know their target groups in order to design appropriate services, the only way to do that is with big data.

Throughout the first two decades of the twenty-first century, the Centers for Medicare and Medicaid Services (CMS) rolled out stages for changing reimbursement to hospitals and physician practices based on best practice metrics and quality outcomes for patients within given diagnostic groupings. By 2010, CMS and insurance payers had aligned their reimbursement levels to incentivize and reward quality outcomes and to penalize poor quality by decreasing payments.

The quality metric levels needed to qualify for full payment and to avoid penalties increased annually throughout the second decade and hospital systems strained to adapt to these regulatory mandates. These reimbursement reforms called for information technology that would allow healthcare systems to identify, track and manage at-risk populations not just patients seen during an interaction with a clinician. Thus, CMS' reimbursement reform also impacted the electronic health record (EHR) systems industry and ushered in a new era of looking beyond centralized EHR data collected from individual patients during discreet hospital stays and doctor office visits. EHR providers are now called upon to support broader data exchange and sharing as well as new functionality and data analytics capabilities needed by health systems to identify patients by risk categories and disease groupings, to support care interventions and track metrics. With the shift away from "fee for service" to "value-based purchasing" healthcare systems and payers had to adopt population health and data analytics capabilities to adapt and stay in business in this new world. Consequently by

2010, most healthcare providers were in the population health and the big data business whether they realized it or not.

This move towards the need to manage population outcomes ushered in a need for a new systems architecture. As mentioned above, EHR systems in the market today and used by acute, ambulatory and post-acute care organizations were not designed to support population health functions nor built on architectural platforms that allow for data sharing. While the EHR vendors responded to the market changes by providing registry functions by disease types, broader querying capabilities, and building external data repositories to support data aggregation and analytics, nothing changed in their platforms. The EHR approach uses a centralized repository model in which all data is imported, cleaned and standardized for analyses. And as the early efforts in Accountable Care Organizations (ACO) demonstration projects illustrated, to effectively manage interventions for sub-populations with targeted chronic diseases and at-risk populations, health care providers also needed to be able to query and assimilate data from multiple different and often disparate sources (Conn 2015). These included external laboratories, diagnostic radiology, ambulatory, community pharmacies, post-acute care services as well as other acute care organizations outside their network. Predominantly, the technological approach adopted has been to bring all the disparate data into a centralized data repository, do all the standardization work to make the data usable, and then to perform analytics on this detailed data. This approach has come under criticism for being costly, difficult and not enabling the sharing and comparing of data, evaluative and cost effectiveness analyses as well as creating silos of health data (IOM 2015; Diamond et al. 2009; Friedman and Parrish 2010). While the centralized repository represents the dominant technological model used even today, it has been widely criticized in favor of a more flexible, nimble and inexpensive approach of networked, distributed data systems.

Most influential is the Markel Foundation's *Connecting for Health* series of reports and recommendations (2007). Known for their proposed nationwide framework for a secure "network of networks" to achieve an environment that supports data access and sharing needs at the individual level to the general population for better health (2004, 2006), Markel Foundation further extended their framework to population health in 2007. In keeping with their "network of networks" approach, they proposed a system architecture model that would bypass current state obstacles and would serve the multiple users and stakeholders of population health data (Markel 2007, 2016). In their "First Principles" is the bold recommendation that a population health network should be composed of a large network of distributed data sources that can be queried without needing a central structure (2007). The current state of isolated data silos and duplicative efforts by each organization to build repositories and analytic systems undermine our health systems, consumers and population health goals for a flexible, nimble and inexpensive health information environment available to all for the greater public benefit.

As an innovation, the Markel Foundation's vision includes access to this big data for users of a population health distributed network and extends this access to consumers, families, health professionals, policy makers, public health officials, researchers, health administrators and others. It is an inclusive definition of population health, its uses, and need for access, stating: "The technical and policy framework for the network should anticipate the diverse requirements of this array of users—much like the Internet itself" (2007, p. 3). Clearly, whether serving the needs of public health entities, health systems, payers, policy makers or individuals, these data applications equate to big data. In addition, new technologies (cloud computing, web-based data query, APIs—application programming interfaces) and data science methods are quickly enabling these new solutions to be built to support nimbleness and quick response, enabling public health and healthcare systems to move into the forefront of healthcare in response to external threats. For public health, this has played out in the form of highly infectious diseases that have the ability to rapidly escape national boundaries given our tightly connected, global world. Most recently, the occurrence of Ebola and Zika have made national and international public health systems' disease tracking and surveillance capabilities front page news (Markel Foundation 2007) and demanded flexible and robust data analytics capabilities that assimilate world-wide data, track trends and allow for responsiveness to the threats. Just as pressing for healthcare systems is the need to get specific answers to questions on effective clinical strategies by levels of patients at risk so that appropriate and effective care is delivered when needed.

2.2.1 Blending in the Social Determinants

In this journey towards population health, there has to be a consideration of those factors apart from just interactions with the medical system that contribute to the health of the groups targeted. These factors are the social determinants of health and are those complex pathways that influence health outside of sporadic interactions with medical care. There is a large body of evidence to support the powerful role of social determinants in shaping health across of variety of indicators and across settings (Braveman and Gottlieb 2014). The limited impact of medical care alone on overall health has been appreciated as far back as the mid-nineteenth century when Scottish physician Thomas McKeown pointed out that mortality from multiple causes had fallen precipitously and steadily decades before the availability of modern medical care (Braveman and Gottlieb 2014). So as healthcare struggles with managing value, quality, and satisfaction within populations, there will be a need to blend data coming from outside of the medical interaction along with that generated from within EHRs. This data will describe concepts beyond basic demographics of the populations and will stretch into areas such as occupation, working conditions, geography, environment, crime, and conflict.

How analytics tools will help to layer social determinants data is still to be determined. However, there are several existing sources of data and systems that need to be considered. As an example, geographic information systems (GIS) generated data is available for those managing big data to merge and layer with health data in order to determine how environment impacts health of individuals and populations. GIS is a system designed to capture, store, manipulate, analyze, manage, and present all types of spatial or geographical data. GIS allows analysts to map person specific or population events to geographic location. With GIS you can determine if a cluster of individuals with poor health resides in a location with known risk factors, such as crime, high lead levels, air pollution, or food deserts. GIS enables providers to see risk and opportunity in clearly delineated geographic breakdown so that healthcare planned can go directly to those in need to provide services.

2.3 Big Data in Action

The previous sections covered big data and analytics from a philosophical and mostly theoretical perspective. This section will address several real world examples of current activities related to big data and analytics as an introduction to how care providers will be affected by this tipping point. A brief glimpse into some of the activities that are currently underway in regard to big data and analytics within the United Stated Department of Veterans Affairs, home health, and in the National Health Service in Great Britain will provide a glimpse into the future of the tipping point.

2.3.1 The Department of Veterans Affairs

The Department of Veterans Affairs' Veterans Health Administration (VHA) has three decades of experience collecting data about the veterans it serves nationwide through locally developed information systems that use a common electronic health record (EHR). As one of the largest health systems in the United States, the VHA offers veterans a full spectrum of inpatient, outpatient, mental health, rehabilitation, and long-term care services, linked by an EHR platform (Fihn et al. 2014). The VHA provides direct health services to more than six million veterans throughout the United States and Puerto Rico.

The VHA's information infrastructure, the Veterans Information Systems Technology Architecture (VistA), became operational in 1985. VistA is comprised of multiple applications accessed using a graphical user interface, the Computerized Patient Record System (CPRS). Ground-breaking when first introduced, CPRS/VistA includes features similar to those now found in commercially available EHR systems, such as electronic navigation tabs; dialog boxes; decision support; and customizable, drop-down menus. In 2004, CPRS/VistA became the system for clinical care delivery--documenting all routine clinical activities; retrieving results (tests, diagnostic procedures, and imaging); and entering orders for medications, procedures, and consultations (Fihn, et al. 2014).

2.3.1.1 Advanced Analytics in the VHA

The VHA has three decades of experience collecting data about the veterans it serves. In the past, the focus was primarily on retrieving data about individual patients to support direct care delivery and secondarily on generating basic summary reports on quality of health care and operational metrics for managers' use (IOM 2013). VHA has undertaken the task of building an advanced analytics infrastructure and applications to allow sophisticated, real-time analysis of the data it has collected. The VHA recognized the need to standardize reporting on a national scale and improve efficiency, so the agency developed its human resources and IT systems to permit the generation of routine, authoritative, national reports summarizing performance at the national, regional, local facility, and provider levels (Lazer et al. 2014).

Before the VHA could tackle the problems with its existing measures and approach to decision support, it needed to reexamine the fundamentals of how it records, stores, and reports data. The original architecture for CPRS/VistA was decentralized (that is, development occurred at medical and IT centers throughout the country)—and data applications implemented across different sites have invariably undergone local adaptations that do not necessarily conform to national standards (Hanson et al. 2014). Furthermore, a veteran seen at more than one VHA facility may have data residing on multiple CPRS/VistA systems. For example, a retired veteran with diabetes may obtain his flu shot in Florida and the rest of his medical care in New York. While remote CPRS/VistA systems may be queried for data on an individual patient, this approach is untenable when aggregating the clinical data of many patients for administrative, quality improvement, and research needs.

VHA's evolution toward "big data," has been defined as the rapid evolution of applying advanced tools and approaches to large, complex, and rapidly changing data sets. Despite the potential for rich clinical data to support continuous learning and improving population health, few large, integrated health care delivery systems have successfully employed their electronic health records (EHRs) for this purpose (Powell et al. 2012).

To provide a reliable source of standardized national data, the VHA began construction of the Corporate Data Warehouse (CDW) in 2006. The CDW is a repository for patient-level data aggregated from across the VHA's national health delivery system, within a business-driven logic structure that enforces higher data quality and interoperability. The CDW does not include all local CPRS/VistA data, but consolidates over sixty domains of key clinical and operational data (such as demographics, laboratory results, medications dispensed from outpatient pharmacies, immunizations, and vital signs). Selection of those domains was based upon priorities established through a governing council that represents clinical and operational leaders and subject-matter experts.

The CDW features 4000 central processing units (CPUs), 1.5 petabytes of storage, twenty million unique patient records, 1000 separate data tables, 20,000 columns, eighty billion rows, and a range of data elements (Fihn et al. 2014). The CDW is refreshed nightly with new data from the CPRS/VistA systems moving to a refresh frequency of every four hours, permitting "near real-time" analysis and reporting. Massive data storage alone does not define big data (Ward and Barker 2013). That designation also requires the ability to access and act upon a vast amount of data using a variety of advanced tools. Consistent with those expectations, the CDW was built to include advanced data management, statistical, graphical, and business analysis software. More than 20,000 analysts, program managers, researchers, and others can access the CDW for a variety of initiatives. The common goal of those initiatives is to assist clinical and operational decision makers by providing information and insights that are not feasible using solely local data.

Though the VHA has long used EHRs and tracked performance of its health care delivery system, its transition to big data is a rather recent development, made possible by the creation of the CDW's high-performance, accessible computing environment. In the early years of that transition, the VHA encountered a variety of issues that remain organizational priorities and that other large health systems transitioning to big data are likely to confront. The primary reason for consolidating decentralized data resources are: (1) improving data governance, particularly related to data quality and data access; (2) continued growth of data sources; integrating analytics into routine clinical workflow; and (3) building capacity for advanced analytics such as clinical prediction (Asch et al. 2004).

2.3.1.2 Consolidating Data Sources

Key to developing high-level analytics is access to a wide range of data sources, many of which were never designed to be compatible with national or organization wide standards. The VHA hosts vast amounts of legacy-system data. Each legacy system has its own idiosyncratic data rules, definitions, and structures. Prior attempts to standardize across all of these systems proved unrealistic, given the slow process of applying common standards (for example, HL7, which is a standard for exchanging data between medical applications) and the exponentially increasing volume of data. In contrast, the launch of the CDW allowed the VHA to stream data selectively from CPRS/VistA and to organize data fields and tables in order to minimize redundancy and make errors more readily apparent (Asch et al. 2004). This process allows the CDW to be rapidly populated and available for immediate use as opposed to weeks or months later, as was the case with the legacy systems.

2.3.1.3 Data Governance, Access and Quality

The VHA's expedition into big data has also created new unforeseen organizational challenges. Managers and clinical end users of data now must help determine which reports and analyses are most critical to patient care and which data elements need to be prioritized for standardization and validation at the national level. This joint approach to prioritization ensures that the most critical data in the CDW (such as patient identifiers and medications dispensed) are carefully managed centrally. In contrast, other CDW data (such as vital signs and laboratory codes) are cleaned, documented, and validated over time by data users across the VHA who share their insights in a wiki-type environment.

Aggregating data within the CDW has greatly reduced impediments to issuing reports based on complete data and that meet the VHA's corporate standards. Any VHA employee now has access to basic reporting functions and, with appropriate approvals and training, can potentially access the CDW itself. Additionally, the consolidation of reporting activities allows the VHA to generate authoritative analyses with more consistent results. This has been enabled by the development of an internal cadre of analysts who understand where data are housed and how they are coded, as well as a national program to train data users throughout the organization in both basic skills (for example, use of spreadsheets and automated reports) and advanced techniques (such as the use of structured query language [SQL] programming for ad hoc analyses and reports) (Box et al. 2010).

Even so, only a small proportion of the VHA's workforce will ever complete advanced analytics training. Future investments will, therefore, need to include improved business intelligence software tools that can automate sophisticated analyses, thereby eliminating the need for some basic training (Zeng et al. 2012).

2.3.1.4 Advanced Analytics

Advanced analytics, such as development and deployment of accurate, multivariable risk models, may lead to more context-sensitive decision support at the point of care. Such context-sensitive decision aids are expected to promote interventions that are more likely to improve health outcomes and minimize adverse events for patients than current broad population recommendations for prevention or screening (Zulman et al. 2008). However, the computational resources required to run such models on large numbers of patients in real time are considerable and can encroach upon more routine but essential analytic tasks, such as calculating performance measures. Furthermore, how to deliver probabilistic information to clinicians and patients in a manner that improves decision-making and outcomes requires extensive research (Toll et al. 2008). Most important, if these models are to be used in routine patient care, they must be computationally efficient, and their accuracy and reliability must be assured and monitored. An overarching problem in examining massive data sets for predictive analytics, comparative effectiveness, and program evaluation is bias. Bias by indication is a particular problem and occurs because treatments and tests are not administered randomly (that is, they might systematically be given to sicker or to healthier patients), and resulting relationships with outcomes may be mistakenly presumed to be causal. Equally noxious is that many observed associations in large data sets are actually random noise that achieves statistical significance because of large sample sizes, poorly specified measurements, "over fitting" (that is, when a statistical model describes random error rather than a true underlying relationship), or quirks in the detection algorithm (Lazer et al. 2014).

2.3.1.5   Expanding Data Sources

Growth in health related data continues at rapid pace, fueled by new sources that include patient-generated data, clinical information systems for intensive care units and surgery, and radio-frequency identification (RFID) systems for tracking the locations and movement of patients and medical devices. Much of the medical equipment issued by the VHA (such as continuous positive airway pressure, or CPAP, machines; scales; and blood pressure monitors) or implanted in veterans (such as pacemakers and defibrillators) can transmit data. In addition, programs to collect data from patients via Internet portals and mobile devices are rapidly proliferating.

Another emerging and potentially large data source of data is genomics. The VHA has undertaken an ambitious initiative to enroll a million veterans in a longitudinal cohort study and establish a database with information on genomics, lifestyle, military exposure, and health (Box et al. 2010). Creation of useful analytic data sets from the voluminous data streams requires substantial investment of time by clinical content experts, analysts, and programmers.

Moreover, all of this must be accomplished within a chaotic financial and political environment, often with intrusive oversight (Tsai et al. 2013). Health systems, such as the VHA, had to evolve and develop skills to manage individual and organizational change and to identify, assess, and mitigate risk at the whole enterprise level harnessing big data.

2.3.1.6   User Impact

The challenges of information overload and distraction on the user's part have already been mentioned. In a recent survey of VHA primary care staff, data overload was cited as a major source of dissatisfaction (Byrd et al. 2013). One solution has been to develop transactional systems that simultaneously support clinical documentation, collection of information for quality tracking, and context-sensitive decision support. The VHA has successfully demonstrated such an approach in its Clinical Assessment, Reporting, and Tracking (CART) system, which operates in all seventy-nine VHA cardiac catheterization labs. CART is used for clinical documentation but simultaneously tracks the quality and safety of invasive cardiac procedures, permitting near-real-time investigation of serious adverse events and monitoring of device failures, radiation exposure, conscious sedation, and device inventory (Box et al. 2010; Tsai et al. 2013).

2.3.1.7   The VHA Pushes to Use Big Data

The VHA has made substantial strides in creating an infrastructure to employ its immense data resources in advanced analytics and to integrate those products into direct patient care and program evaluation. The organization's advancement into big data has produced notable successes, while exposing new challenges in such areas as management of data access and quality, deployment of new software applications, and prevention of information overload among busy clinicians and managers. The VHA's experience indicates that most of these problems can be addressed through better governance, active engagement of clinical teams, improved reporting platforms within the EHR, and other strategies that have been described. As recent events have illustrated, reliance on "big data" without effectively implementing these other strategies can have disastrous effects.

2.3.2   A View from Home Health

In the home healthcare sector, electronic health record systems are built to: accommodate the minimum data set collection required by CMS, generate mandated quality metric reports, and to be able to bill for services. However, home health providers must be able to ensure that they are serving those who need home based care and that they are providing appropriate care in dispersed areas to patients who are coming out of different health systems. As a result, there are a number of questions that need answers and big data manipulated by an appropriate analytics approach can help to craft home care services appropriate to those who can benefit most. The ability for a home health executive to get questions answered by pulling information from financial and clinical data, resource use, and outcome data is, however, quite difficult.

However, in two examples presented here, the clinical executive of a large, national home health service provider partnered with nurse researchers from different universities to be able to pose the questions, conduct the data analyses, and obtain answers that informed specific clinical strategies with evidence that supported the business case for the expenditures. The partnerships involved data analysts, nurse informaticians and IT resources from the home healthcare organization to identify data elements needed from patient records, to de-identify the data, address missing values or errors, export and send the data to the university's research faculty partner. The universities' team consisted of domain experts, nurse researchers, data scientists, and an epidemiologist. The important message here is that these big data initiatives required partnerships and multidiscipline teams with a data scientist as a member of each team.

2.3.2.1   Telemonitoring to Prevent Hospital Readmission from the Home

One of the main objectives of home health is to keep the patient in the home, of course. This is not always easy. In home healthcare, the use of telemonitoring for patients at risk for hospital readmission was considered a best practice and came to be widely adopted across the industry by 2008. However, evidence of the effectiveness, how best to use it, who to use it on, and what difference it made was basically not available as patients can be readmitted from the home to multiple hospitals. Since the use of telemonitoring is not reimbursed by payers, the clinical executive of a home health agency needed clear evidence that supported adding the cost of telemonitoring. In looking at how to get answers to the questions of who and who not to use monitoring on and what differences could be demonstrated for outcomes, especially rehospitalization rates, the opportunity to partner with a university research team to analyze big data presented a solution. An analysis of 552 Medicare patients identified with the diagnosis of congestive heart failure who had been admitted into home healthcare directly from a hospital discharge was conducted by Kang, McHugh, Chittams, and Bowles (2016). The analysis helped to identify the specific CHF patients who were most at risk for rehospitalization. The clinical picture was of a person who had skin issues (open wound or lesion), experienced pain levels that interfered with activity or movement, poor overall health status, and a high number of therapy visits needed. The highest predictive variable was presence of a skin lesion at home healthcare admission. Severe pain that occurred with frequency to interfere with movement regardless of skin lesion, health status or presence of a caretaker in the home also had a high risk for rehospitalization. These risk profiles helped to define a targeted group and a specific clinical strategy for the appropriate use of telemonitoring with the highest risk CHF patients in home healthcare and for this organization to develop programs and data tracking and reporting that helped to achieve significant lowering of their rehospitalization rates, as well as reporting to the Accountable Care Organization partners.

2.3.2.2 Improving Outcomes of Home Healthcare Patients with Wounds

An opportunity was presented to the chief clinical officer (CCO) of a home health and hospice organization to participate in a large national study that was focused on answering the question of differences in outcomes for wound care patients in home healthcare (HH) between nurses who held certification by the Wound, Ostomy and Continence Nursing (WOCN) Certification Board or from a WOCN-accredited nursing education program and those nurses without this wound certification (Westra et al. 2013). The relevance of this question to the CCO was that nurses with certifications tend to have bachelor degrees or higher and command higher salaries. Thus, in determining if a hiring strategy for every home healthcare program should include one or more WOCN wound certified nurses, the business case and clinical appropriateness for this strategy needed evidence. This 2008–2009 big data study looked at almost 450,000 episodes of care from over 780 home healthcare organizations and found higher incidence of wounds, incontinence, and urinary tract infections (UTI) for agencies with no WOC nurse. Additionally, those home health agencies with WOC nurses had significantly better improvement outcomes for pressure ulcers, lower extremity ulcers, surgical wounds, urinary and bowel incontinence, and UTIs (Westra et al. 2013). With this evidence, the CCO could justify putting in place a new clinical strategy for both hiring WOC nurses as well as funding for staff nurses interested in obtaining their WOCN certification.

2.3.2.3 Home Health Powered by Big Data

Both of these home health case examples demonstrate the power of big data and appropriate analytics to inform clinical strategies for best outcomes as well as directing expenditures that are clinically appropriate and will lead to a more effective healthcare system in both quality outcomes and costs. In terms of methods, both are examples of bringing data from multiple sources into a central repository, doing all the difficult work to standardize the data, and create a stand-alone silo of health data. However, for most home healthcare organizations, the resources and information technology infrastructures are fairly modest. The opportunity to partner with organizations experienced in the analytics of big data, such as university research teams, provides opportunity to create evidence, guide clinical strategies, and inform practice. Not ideal, but best that could be done with resources available while we wait for our state of the art to move to the Markel Foundation's vision of a "network of networks" that enables data sharing, access, and evaluative comparisons available to all.

2.3.3 The Spine: A United Kingdom Big Data Endeavor

United Kingdom (UK) has engaged with health data collection in an enthusiastic way due mainly to the way in which the National Health Service (NHS) is funded and the need to deploy resources to where they will be most effective. The NHS was established in 1948 with the principle that 'access to NHS services is based on clinical need, not an individual's ability to pay' (Department of Health 2015, p. 3). This does not mean that the NHS is free, but for UK citizen, it is offered at 100% discount at the point of contact. Following the devolution policy at the turn of the century (Parliament.uk 2015) the NHS is now divided into four independent, although interrelated, regions: NHS England, NHS Wales, NHS Scotland and NHS Northern Ireland. NHS England is the largest of the four regions covering a population of around 60 million people with a budget of £116 billion ($170 billion) in 2015 (NHS Confederation 2015) with a per capita cost of around £3300 ($4900) per year (Organisation for Economic Cooperation and Development 2015). Working in conjunction with NHS England is Public Health England (PHE) which has a mandate to promote health and wellbeing and to reduce inequalities. Alongside the NHS is a private care system funded through personal or organization insurance, accounting for around 1% of population expenditure on healthcare (Organisation for Economic Cooperation and Development 2015).

2.3.3.1 Healthcare in England

Healthcare in England through the NHS provides primary, secondary and tertiary services via community care, emergency care, hospital care, long term conditions care and end-of-life care. People generally register with their local general practitioner center and use this route to access specific hospital services through an electronic referral system and the general practitioner is considered the 'gate-keeper' and is also the commissioner for care in their geographical area. Where there is an emergency, individuals can access care through the Accident and Emergency (A&E/ER) route without the need to start the journey via the general practitioner. To remove inequalities, the National Institute for Health and Care Excellence (NICE) issues guidelines for the processes and targets for hospital service referrals usually by condition type. Most post-hospital recovery and long term condition care is provided in the community, with occasional appointments with either a medical or nurse consultant to monitor progress. There is growth in the use of telehealth for monitoring of those with long term conditions to reduce the need for hospital appointments.

A core principle of the NHS is that the patient is at the center of all services (Department of Health 2015, p. 3) built around actual need. However as projected demographics change, there is a growing necessity to ensure that the population can continue to avail themselves of required healthcare within financial limits. Thus big data and data analytics are needed to insure continued deployment of resources.

2.3.3.2   The NHS and Health Information Technology

The NHS has been a pioneer in the use of technology to support clinical practice, with the first computerized systems in use during the 1960s. The developments followed that initial work had the goal of understanding the actual processes of care delivery through the collection of data. In preparation for a significant move forward in the gathering and use of NHS data, the unique NHS Number was introduced in 1996 and is now standard across all four UK countries. Having a unique identifier has enabled many further developments all of which are increasing the knowledge around health-care data; one development in particular is the NHS Spine which was introduced in 2002. According to a recent information release (Health and Social Care Information Centre 2015a) "the Spine supports the NHS in the exchange of information across national and local information systems", the release goes on to state:

- The Spine connects over 28,000 healthcare IT systems in 21,000 organizations
- Handles six billion messages every year
- Has 900,000 registered users (around 70% of the total NHS workforce)
- Typically has 250,000 users accessing the service at any one time
- Holds over 500 million records and documents
- In peak periods—can handle 1500 messages per second.

With the Spine operational, the infrastructure for realistic data collection and use at local and national levels is in place. The creation of a data/information industry has occurred within both NHS England and Public Health England resulting in the expansion of the Health and Social Care Information Centre (HSCIC) which was established in 2013. The HSCIC 2015–2020 Strategy states: "Better use of health and care data will help those involved to manage the system more effectively, commission better services, understand public health trends in more detail, develop new treatments and monitor the safety and effectiveness of care providers. We also need to make health and care data really useful to citizens in their daily lives" (HSCIC 2015b, p. 16).

The HSCIC 2015–2020 strategy aligns with two further influential documents that maintain the NHS principles in an information intensive era. These are the NHS Five Year Forward View (NHS England 2014) and Personalized Health and Care 2020: Using Data and Technology to Transform Outcomes for Patients and Citizens—A Framework for Action (HM Government 2014). These policies revealed a strong understanding that fiscal change and financial savings can be made if the roles and functions of the NHS and PHE are better known through the collections of data.

2.3.3.3   The NHS and Big Data

Across NHS England there are many different computerized systems in use for the recording of health care, and most are now shared across professional boundaries and between primary, secondary and tertiary health care, allowing a coordinated approach to managing a patient's journey. In 2018, all NHS England citizens will have online access to their full health record using protocols maintain the governance of the information.

The availability of data and its analysis have created a sustainable route to precision (or personalized) medicine along with greater citizen individual responsibility. There are two main outlets publishing collected, collated and validated information. These are: the NHS Health and Social Care Information Centre (HSCIC), National Institute of Health and Care Excellence (NICE 2016) and NHS Choices (NHS Choices 2016) with the latter being directed towards publically available information.

With the ever increasing interest in individual use of the Internet of Things, especially where this links sensors to machines, there is going to be an expansion in available health data. The sensors are to be individual based and will monitor human vital signs, exercise regimes, and information communicating significant processes. The NHS is aware that it must be ready to support both individuals and the services offered in a balanced and sustainable way, and that this requires partnerships and understanding. The partnerships are in personal, population and public health and the understanding requires all those concerned with health and social care to have the knowledge and skills to use data from reliable sources competently and confidently. Such competence and confidence requires education so that errors in data input can be reduced. Once directly sensed information is pushed through machines using cloud based technologies then the physical and person-based input errors will be reduced, leaving only highly technically oriented issues degrading the resultant data masses.

NHS England has a mission to promote health as well as respond to illness. It is only in the last few years that health promotion has really been possible especially targeted to ensure the best potential results for an ever decreasing budget. Under the auspices of Public Health England (PHE 2016), there are matrixed sets of data and visualized statistics with open access which can be used by individuals and healthcare professionals alike to assess such issues as health inequalities in their community at Clinical Commissioning Group level right down to individual general practice centers. The way in which the data is presented is clear and generally graphical to add further comprehension and comparison.

At the other end of the scale are collections of open access national data available through HSCIC, such as the Clinical Commissioning Group Outcomes Indicator Set (HSCIC 2015c) or NHS Safety Thermometer: Patient Harms and Harm Free Care: England December 2014–December 2015, official statistics (HSCIC 2016). Clearly this type of data collection and analysis may not be considered by some as Big Data in the widest sense, but this type of data is helping in monitoring the delivery of care, helping in targeting scarce resources to improve care and reduce health inequalities and providing transparency for citizens in England and the wider UK through the other three countries.

In alignment with the NHS Constitution individuals have the opportunity to comment upon the health services they receive and the results of their comments are published through an open web portal called NHS Choices. Such information is offered as part of meeting the NHS Constitution's pledge to offer patients and citizens informed choice in their care needs. An example of information available through the portal for citizens is where they can find local general practice facilities including all contact details and user ratings, recommendations and services offered. It is not only the health professionals who are moving forward with better information; it is also those who require the services.

In 1989, the UK Department of Health published 'A Strategy for Nursing' in which the following paragraph should be included in our thoughts, "Accountable practitioners must be more than passive recipients of information. They will need to acquire the analytical skills to ask the right questions, to know where to seek answers to them, and to reach informed decisions on the basis of the fullest knowledge available" (Poole 1989). With this knowledge, gained through the information collected, analyzed, and synthesized, informed decisions may be made delivering the right healthcare to the right individuals at the right time through a collaborative/partnership that is affordable and reasonable.

Through the developments in information and communications technology the NHS has enabled the next level of collective data to commence which is expected to result in transformation of services offered through better use of scarce resources and targeted provision; improved co-ordination and logistics of care through understanding demand from information gathering and reporting and the development of robust public health programs that meet the increasing diversity of our population and help individuals take on personal responsibility for care. To manage big data needs big thinking: England has focused healthcare, capable of standard reporting using informatics and terminology standards and quality management.

2.4   Summary

Healthcare has quickly moved into a time when the practitioners have to think outside of their individual interactions with patients. Healthcare organizations of all types and in many locations are being held to account for the health of their populations not just for the provision of medical services to an individual patient. Juxtaposed on this mandate is the ever-growing digitalization of many aspects of the lives of people served. To meet the needs and expectations of payers and those served, the data gathered has to be managed and analyzed. Practitioners will need to be at the forefront of development of the tools needed to be able efficiently and effectively responds to the mandate. For health care to become truly "healthcare" and not just the provision of episodic medical services, practitioners will need to be as proficient at utilizing big data sources and analytics tools as they are in managing an individual record.

References

Asch SM, McGlynn EA, Hogan MM, Hayward RA, Shekelle P, Rubenstein L, et al. Comparison of quality of care for patients in the Veterans Health Administration and patients in a national sample. Ann Intern Med. 2004;141(12):938–45.

Box TL, McDonell M, Helfrich CD, Jesse RL, Fihn SD, Rumsfeld JS. Strategies from a nationwide health information technology implementation: the VA CART story. J Gen Intern Med. 2010;25(Suppl 1):72–6.

Braveman P, Gottlieb L. The social determinants of health: it's time to consider the causes of the causes. Public Health Rep. 2014;129(Suppl 2):19–31.

Byrd JB, Vigen R, Plomondon ME, Rumsfeld JS, Box TL, Fihn SD, et al. Data quality of an electronic health record tool to support VA cardiac catheterization laboratory quality improvement: the VA Clinical Assessment, Reporting, and Tracking System for Cath Labs (CART) program. Am Heart J. 2013;165(3):434–40.

Conn J. ACOs make progress in using big data to improve care. Modern Healthcare. 2015. http://www.modernhealthcare.com/article/20150117/MAGAZINE/301179981. Accessed 27 Apr 2016.

Department of Health. NHS Constitution for England. 2015. https://www.gov.uk/government/publications/the-nhs-constitution-for-england. Accessed 28 Apr 2016.

Diamond CC, Mostashari F, Shirky C. Collecting and sharing data for population health: a new paradigm. Health Aff. 2009;28(2):454–66.

Fihn SD, Francis J, Clancy C, Nielson C, Nelson K, Rumsfeld J, Cullen T, Bates J, Graham G. Insights from advanced analytics at the Veterans Health Administration. Health Aff. 2014;33(7):1203–11.

Friedman DJ, Parrish RG. The population health record: concepts, definition, design and implementation. JAMIA. 2010;17:359–66.

Groves, P., Kayyali B, Knott D, & Van Kuiken, S. The 'big data' revolution in healthcare: accelerating value and innovation. Center for US Health System Reform. Business Technology Office. 2013.

Hanson MM, Miron-Shatz T, Lau YS, Paton C. Big data in science and healthcare: a review of recent literature and perspectives. Contribution of the IMIA Social Media Working Group. Yearb Med Inform. 2014 Aug 15;9:21–6. doi:10.15265/IY-2014-0004.

Health and Social Care Information Centre. The Spine. 2015a. http://systems.hscic.gov.uk/ddc/spine/spine.pdf. Accessed 15 Mar 2016.

Health and Social Care Information Centre. Information and technology for better care. 2015b. http://www.hscic.gov.uk/hscicstrategy. Accessed 28 April 2016.

Health and Social Care Information Centre. Clinical commissioning group outcomes indicator set: England. 2015c. http://www.hscic.gov.uk/catalogue/PUB19278/ccg-ind-dec-15-comm.pdf. Accessed Dec 2015.

Health and Social Care Information Centre. NHS safety thermometer: patient harms and harm free care: England December 2014–December 2015. http://www.hscic.gov.uk/catalogue/PUB19566/nhs-safe-rep-dec-2014-dec-2015.pdf. 2016, Jan. Accessed 15 Mar 2016.

Health and Social Care Information Centre Strategy 2015–2020, HSCIC, March 2015.

HM Government. Personalised health and care 2020: using data and technology to transform outcomes for patients and citizens—a framework for action, NHS National Information Board. 2014. https://www.gov.uk/government/uploads/system/uploads/attachment_data/file/384650/NIB_Report.pdf. Accessed 15 Mar 2016.

IOM. Best care at lower cost: the path to continuously learning health care in America. Washington, DC: National Academies Press; 2013.

IOM. Working definition of population health. 2014. http://iom.nationalacademies.org/~/media/Files/Activity%20Files/PublicHealth/PopulationHealthImprovementRT/Pop%20Health%20RT%20Population%20Health%20Working%20Definition.pdf. Accessed 12 Feb 2016.

IOM. Spread, scale, and sustainability in population health—workshop in Brief. 2015. http://www.nap.edu/catalog/21711/spread-scale-and-sustainability-in-population-health-workshop-in-brief. Accessed 16 Feb 2016.

Kang Y, McHugh MD, Chittams J, Bowles KH. Utilizing home healthcare electronic health records for telehomecare patients with heart failure. Comput Inform Nurs. 2016;34(4):175–82.

Kaplan B, Harris-Salamone KD. Health IT success and failure: recommendations from literature and an AMIA workshop. J Am Med Inform Assoc. 2009;16(3):291–9.

Krebs P, Duncan DT. Health app use among US mobile phone owners: a national survey. JMIR mHealth uHealth. 2015;3(4):e101. doi:10.2196/mhealth.4924.

Lazer D, Kennedy R, King G, Vespignani A. Big data. The parable of Google Flu: traps in big data analysis. Science. 2014;343(6176):1203–5.

Leeds Care Record. 2015. http://www.leedscarerecord.org/. Accessed 15 Mar 2016.

Markel. Connecting for health. Decision-making for population health—"first principles". 2007. http://www.connectingforhealth.org/resources/first_principles.pdf. Accessed 12 Feb 2016

Markel. Markle common framework: connecting consumers. 2016. http://www.markle.org/health/markle-common-framework. Accessed 27 Apr 2016.

Marr, C. Big data: a game changer in the retail sector. Forbes.Tech. 2015. http://www.forbes.com/sites/bernardmarr/2015/11/10/big-data-a-game-changer-in-the-retail-sector/#7350dbf7678a. Accessed 20 Feb 2016.

National Institute for Health and Care Excellence. Improving health and social care through evidence-based guidelines. https://www.nice.org.uk. Accessed 15 Mar 2016.

NHS. Choices. http://www.nhs.uk/pages/home.aspx. Accessed 15 Mar 2016.

NHS Confederation. Key statistics on the NHS. 2015. http://www.nhsconfed.org/resources/key-statistics-on-the-nhs. Accessed 15 Mar 2016.

NHS England. Five year forward view. 2014. https://www.england.nhs.uk/wp-content/uploads/2014/10/5yfv-web.pdf. Accessed 15 Mar 2016.

Office of the National Coordinator. Connecting health and care for the nation: a shared nationwide interoperability roadmap. 2015. https://www.healthit.gov/sites/default/files/hie-interoperability/nationwide-interoperability-roadmap-final-version-1.0.pdf. Accessed 2 Feb 2016.

Organisation for Economic Cooperation and Development. Focus on health spending. 2015, Jul. http://www.oecd.org/health/health-systems/Focus-Health-Spending-2015.pdf. Accessed 15 Mar 2016.

Parliament.uk. Devolved parliaments and assemblies. 2015. http://www.parliament.uk/about/how/role/devolved. Accessed 15 Mar 2016.

Poole A. A strategy for nursing: a report of the steering committee. London: Department of Health; 1989.

Powell AA, White KM, Partin MR, Halek K, Christianson JB, Neil B, et al. Unintended consequences of implementing a national performance measurement system into local practice. J Gen Intern Med. 2012;27(4):405–12.

Public Health England. Guidance: PHE data and analysis tools. https://www.gov.uk/guidance/phe-data-and-analysis-tools. Accessed 15 Mar 2016.

Schaeffer C. Big data in retail examples. CRMsearch. Retrieved at Five retail big data examples with big payback. 2015. http://www.crmsearch.com/retail-big-data.php. Accessed Feb 2016.

Toll DB, Janssen KJ, Vergouwe Y, Moons KG. Validation, updating, and impact of clinical prediction rules: a review. J Clin Epidemiol. 2008;61(11):1085–94.

Tsai TT, Box TL, Gethoffer H, Noonan G, Varosy VD, Maddox TM, et al. Feasibility of proactive medical device surveillance: the VA Clinical Assessment Reporting and Tracking (CART) in catheterization laboratories pilot program. Med Care. 2013;51(3 Suppl 1):S57–61.

University of Kansas. A community toolbox. Chapter 3. What is GIS. http://ctb.ku.edu/en/table-of-contents/assessment/assessing-community-needs-and-resources/geographic-information-systems/main. Accessed 2 Feb 2016.

Ward JS, Barker A. Undefined by data: a survey of big data definitions. Ithaca, NY: Cornell University Library; 2013, Sep 20. http://arxiv.org/pdf/1309.5821v1.pdf. Accessed 5 Mar 2016.

Westra BL, Bliss DZ, Savik K, Hou Y, Borchert A. Effectiveness of wound, ostomy, and continence nurses on agency-level wound and incontinence outcomes in home care. J Wound Ostomy Continence Nurs. 2013;40(1):25–33.

Zeng QT, Redd D, Rindflesch T, Nebeker J. Synonym, topic model, and predicate-based query expansion for retrieving clinical documents. AMIA Annu Symp Proc. 2012;2012:1050–9.

Zulman DM, Vijan S, Omenn GS, Hayward RA. The relative merits of population-based and targeted prevention strategies. Milbank Q. 2008;86(4):557–80.

Chapter 3
A Big Data Primer

Judith J. Warren

Abstract The aim of this chapter is to describe the history of big data and its characteristics—variety, velocity, and volume—and to serve as a big data primer. Many organizations are using big data to improve their operations and/or create new products and services. Methods for generating data, how data is sensed, and then stored, in other words data collection, will be described. Mobile and internet technologies have transformed data collection for these companies and new sources are emerging at an unheard of speed. Due to the explosion of data, the teams needed to manage the data have evolved to include data scientists, domain experts, computer scientists, visualization experts, and more. The ideas of intellectual property are also changing. Who owns the data, the products generated from the data, and applications of the data? Challenges and tools for data analytics and data visualization of big data will be described, thus, setting the foundation for the rest of the book.

Keywords Big data • Data science • Data scientist • Data visualization • Digitization • Datafication • Privacy risks • Hadoop • NoSQL • Internet of Things

3.1 What Is Big Data?

"Big Data refers to things one can do at a large scale that cannot be done at a smaller one, to extract new insights or create new forms of value, in ways that change markets, or organizations, the relationship between citizens and governments, and more" (Mayer-Schonberger and Cukier 2013). Big data occurs when the size of the data becomes the major concern for the data analyst. New methods of data

J.J. Warren, Ph.D., R.N., F.A.A.N., F.A.C.M.I. (✉)
University of Kansas School of Nursing, Kansas City, KS, USA

Warren Associates, LLC, Plattsmouth, NE, USA
e-mail: jjwarren@live.com

© Springer International Publishing AG 2017
C.W. Delaney et al. (eds.), *Big Data-Enabled Nursing*, Health Informatics,
DOI 10.1007/978-3-319-53300-1_3

collection have evolved where the data input becomes passive. The data can come from social network posts, web server logs, traffic flow sensors, satellite imagery, audio streams, online searches, online purchases, banking transactions, music downloads, uploaded photographs and videos, web page content, scans of documents, GPS location information, telemetry from machines and equipment, financial market data, medical telemetry, online gaming, athletic shoes, and many more. The volume of data is so large, so fast, and so distributed that it cannot be moved. With big data, the processing capacity of a traditional database is exceeded (Dumbill 2012a, b). Fortunately, new methods of storage, access, processing, and analysis have been developed.

Big data has transformed how we analyze information and how we make meaning in our world. Three major shifts in our thinking occur while dealing with big data (Mayer-Schonberger and Cukier 2013). The first shift is about sampling data from all the data or the population for analysis to understand our world. With big data, we no longer need to sample from the population. We can collect, store and analyze the population. Due to innovations in computer memory storage, server design, and new software approaches, we can analyze all the data collected about a topic rather than be forced to only look at a sample. For all of analytic history, from the cave paintings in Lascaux to record the movements of animal herds (Encyclopedia of Stone Age Art 2016) to the cuneiform tablets used to record harvest and grain sales (Mark 2011) to statistical formulas from the 1700s and 1800s used to describe behavior (Stigler 1990), we were only able to collect, record, and analyze a sample of the population. Collecting data was a manual process and thus very labor intensive and expensive. [See Box 3.1].

Determining a sample and collecting a limited data set was the answer to this labor-intensive data collection process. Statistical sampling helped to select a representative set of data and to control for error in measurement. Now the innovations in computer science, data science, and data visualization create an opportunity to analyze all the data. As Mayer-Schonberger and Cukier summarize, n = all (2013). Analyzing all the data facilitates exploring subcategories/submarkets—to see the variations within the

---

Box 3.1 The Domesday Book of 1086

William the Conqueror mandated a tally of English people, land, and property to know what he ruled and how to assess taxes. Scribes were sent across England to interview and collect information about his subjects. It took years to collect and analyze the data. It was the first major census of its kind and served to document and datafy people's rights to property and land, and the ability to give military service. The book was used to award titles and land to worthy individuals. It has been used over a thousand years to settle disputes. It was last used in a British Court in 1966. The United Kingdom's National Archives have datafied the Domesday Book (National Archives 2016) and have digitized it so that it may be searched (Domesday Book Online 2013).

population. Google used their large database based on millions of Google searches and the Centers for Disease Control's (CDC) flu outbreak database to develop an algorithm that could predict flu outbreaks in near real-time (Google Flu Trends 2014). Google collected 50 million common search terms and compared with CDC data of spread of seasonal flu between 2003 and 2008. They processed 450 million different math models using machine learning to create an algorithm. Then Google compared the algorithm's prediction against actual flu outbreaks in 2007 and 2008. The model, comprised of 45 search terms, was used to create real time flu reporting in 2009. Google searches are powered by the big data effort to find connections between web pages and the search engine. As the searches are conducted and new connections are made, data is created. Amazon uses big data to recommend books and products to their customers. The data is collected every time a customer searches and purchases new books and products. These companies use big data to understand behaviors and make predictions about future behavior. This increased sophistication in the analysis and use of that data created the foundation of data science (Chartier 2014). Data science is based on computer science, statistics, and machine learning-based algorithms. With the advent of the Internet and the Internet of Things, data is collected as a byproduct of people seeking online services that is recorded as digitized behavior to be analyzed. Digitization is so pervasive that in 2010, 98% of the United States economy was impacted by digitization (Manyika et al. 2015).

The second shift in thought created by big data is the ability to embrace the messiness of data, to eliminate the need to be perfect without error (Mayer-Schonberger and Cukier 2013). Having the population of data means that the need for exactness in a sample lessens. With less error from sampling, we can accept more measurement error. Big data varies in quality since collection is not supervised nor controlled. Data is generated by online clicks, computerized sensors, likes and rankings by people, smart phone use, or perhaps credit card use. The messiness is managed through the sheer volume of data—the population (n = all). Data is also distributed among numerous data warehouses and servers. Bringing the distributed data together for analysis has its own challenges with exactness. Combining different types of data from different sources causes inconsistency due to different formatting structures. Cleaning this messiness in the data has led to evolution of a new role in big data—the data wrangler.

The third shift in thought created by big data is to move to thinking only about correlation, not causality (Mayer-Schonberger and Cukier 2013). Yet, mankind has a need to understand the world and jumps to thinking in causal terms to satisfy this need. In big data, the gold is in the patterns and correlations that can lead to novel and valuable insights. The use of big data and data science doesn't reveal WHY something is happening, but reveals THAT something is happening. Our creative need to combine data sets and to use all the data to create new algorithms for understanding the data leads us away from thinking of a dataset developed for a single purpose to thinking about what does the value of this dataset have by itself and in combination with other datasets (Chartier 2014). Data becomes a reusable resource, not a static collection point in time. A note of caution about correlation: very large data sets can lead to ridiculous correlations. Interpretation of results needs to be investigated by a domain expert to insure an analysis that truly leads to knowledge and insight. The focus on correlation creates data-driven decisions instead of hypothesis-drive decisions.

3.1.1 Datafication and Digitization

To understand the innovation of big data, data itself needs to be explored. What makes data? The making of data occurred when man first measured and recorded a phenomenon. Early man in Mesopotamia counted grain production, recorded its sale, and analyzed it to calculate taxes owed to the king. So to datafy a phenomenon is to measure it and put it into a quantified format so that it can be tabulated and analyzed. Datafication made it possible to record human activity so that the activity can be replicated, predicted, and planned. Modern examples of datafication are email and social media where relationships, experiences, and moods are recorded. The purpose of the Internet of Things is to datafy everyday things.

With the advent of computers, we can also digitize our data. Digitization turns analog information into a format that computers can read, store, and process. To accomplish this, data is converted into the zeros and ones of binary code. For example, a scanned document is datafied but once it is processed by optical character recognition (OCR), it becomes digitized. The Gartner reports 4.9 billion connected things are currently in use in 2015 and by 2020, 25 billion connected things will be in use (Gartner 2014). These connected objects will have a "digital voice" and the ability to create and deliver a stream of data reflecting their status and their environment. This disruptive innovation radically changes value proposition, creates new services and usage scenarios, and drives new business models. The analysis of this big data will change the way we see our world.

3.1.2 Resources for Evaluating Big Data Technology

With the disruptive changes of big data, new products and services are needed for the storage, retrieval, and analysis of big data. Fortunately, companies are creating reports that list the services available and their penetration in the marketplace. Consumers new to the field should study these reports before investing in new servers, software, and consultants. Both Gartner and Forrester have rated the products of companies engaged in big data hardware, software, and consulting services. Both of these consulting firms provide a service to consumers by providing information on the status of big data as a new trend that is making an impact in industry.

Gartner has rated big data on their Magic Quadrant (2016a, b). The magic Quadrant is a two-by-two matrix with axes rating the ability to execute and the completeness of vision. The quadrants where the products are rated depict the challengers, leaders, niche players, and visionaries. The quadrant gives a view of market competitors and how well they are functioning. Critical Capabilities is a deeper dive into the Magic Quadrant (Gartner 2016a). The next tool is the famous Gartner Hype Cycle (2016c). The axes are visibility vs. maturity. The graph formed depicts what is readily available and what is still a dream, thus informing of where the hype and adoption are for the trend. The five sections of the graph or the lifecycle of the trend are technology trigger, peak of inflated expectations, trough of disillusionment, slope of enlightenment, and plateau of productivity. Big data has a Hype Cycle of its own that breaks out the components and technologies of big data (GilPress 2012).

Forrester rates products on the Forrester Wave (Forrester Research 2016; Gualtieri and Curran 2015; Gualtieri et al. 2016; Yuhanna 2015). The Wave is a graph with two axes, current offering vs. strategy. Market presence is then plotted in the graph using concentric circles (waves) to show vendor penetration in the market. This information helps the customer to select the product best for their purpose.

3.2 The V's: Volume, Variety, Velocity

Big data is characterized by the "Three V's." The three V's can be used to understand the different aspects of the data that comprise big data and the software platforms needed to explore and analyze big data. Some experts will add a fourth "V", value. Big data is focused on building data products of value to solve real world problems.

3.2.1 Volume

Data volume is quantified by a unit of storage that holds a single character, or one byte. One byte is composed of eight bits. One bit is a single binary digit (1 or 0). Table 3.1 depicts the names and amounts of memory storage. In 2012, the digital

Table 3.1 Names and amounts of memory storage

Name	Symbol	Binary measurement	Decimal measurement	Number of bytes	Equal to
Kilobyte	KB	2^{10}	10^3	1024	1024 bytes
Megabyte	MB	2^{20}	10^6	1,048,576	1024 KB
Gigabyte	GB	2^{30}	10^9	1,073,741,824	1024 MB
Terabyte	TB	2^{40}	10^{12}	1,099,511,627,776	1024 GB
Petabyte	PB	2^{50}	10^{15}	1,125,899,906,842,624	1024 TB
Exabyte	EB	2^{60}	10^{18}	1,152,921,504,606,846,976	1024 PB
Zettabyte	ZB	2^{70}	10^{21}	1,180,591,620,717,411,303,424	1024 EB
Yottabyte	YB	2^{80}	10^{24}	1,208,925,819,614,629,174,706,176	1024 ZB

universe consisted of one trillion gigabytes (1 zettabyte). This amount will double every two years and, by 2020, will consist of 40 trillion gigabytes (40 zettabytes or 5200 gigabytes per person) (Mearian 2012).

As data storage has become cheaper, as predicted by Moore's Law, the ability to keep everything has become a principle for information technology (Moore 2016). In fact, it is sometimes easier and cheaper to keep everything than it is to identify and keep the data of current interest. Big data is demonstrating that the reuse and analysis of all data and the combinations of data can lead to new insights and new data products that were previously not imagined. Some examples will demonstrate the volume of data that exists in big data initiatives:

- Google processes 24 petabytes of data per day. A volume that is thousands of times the quantity of all printed material in the US Library of Congress (Gunelius 2014)
- Facebook users upload 300 million new photos every hour; the like button or comment is used three billion times a day (Chan 2012)
- YouTube has over one billion users who watch hundreds of millions of hours of video per day (YouTube 2016).
- Twitter has over 100 million users log in per day; with over 500 million tweets per day Twitter Usage Statistics 2016).
- IBM estimates that 2.5 quintillion bytes of data (2.3 trillion gigabytes) is created daily; 90% has been created in the last two years (IBM 2015, 2016).

First, different approaches to storing these very large data sets have made big data possible. The foremost tool is Hadoop that efficiently stores and processes large quantities of data. Hadoop's unique capabilities support new ways of thinking about how we use data and analytics to explore the data. Hadoop is an open-source distributed data storage and analysis platform that can be used on large clusters of servers. Hadoop uses Google's MapReduce algorithm to divide a large query into multiple smaller queries. MapReduce then sends those queries (the Map) to different processing nodes and then combines (the Reduce) those results back into one query. Hadoop also uses YARN (Yet Another Resource Negotiator) and HDFS (Hadoop Distributed File System) to complete its processing foundation (Miner 2016). YARN is a management system that keeps track of CPU, RAM, and disk space and insures that processing runs smoothly. HDFS is a file system that stores data on multiple computers or servers. The design of HDFS facilitates a high throughput and scalable processing of data. Hadoop also refers to a set of tools that enhance the storage and analytic components: Hive, Pig, Spark, and HBase are the common ones (Apache Software Foundation 2016). Hive is a SQL-like query language for use in Hadoop. Pig is also a query language optimized for use with MapReduce. While Spark is a framework for general purpose cluster computing, HBase is a data store that runs on top of the Hadoop distributed file storage system and is known as a NoSQL database. NoSQL databases are used when the volume of data exceeds the capacity of a relational database. To be able to engage in big data work, it is essential that these tools are understood by the entire big data team (Grus 2015).

3.2.2 Variety

The data in big data is characterized by its variety (Dumbill 2012a, b). The data is not ordered, due to its source or collection strategy, and it is not ready for processing (characteristics of structured data in a relational database). Even the data sources are highly diverse: text data from social networks, images, or raw data from a sensor. Big data is known as messy data with error and inconsistency abounding. The processing of big data uses this unstructured data and extracts ordered meaning. Over 80% of data is unstructured or structured in different formats. Initially, data input was very structured, mostly using spreadsheets and data bases, and collected in a way for analytics software to process. Now, data input has changed dramatically due to technological innovation and the interconnectedness of the Internet. Data can be text from emails, texting, tweets, postings, and documents. Data can come from sensors in cars, athletic shoes, bridge stress, mobile phones, pressure readings, number of stairs climbed, or blood glucose levels. Data from financial transactions such as stock purchases, credit cards, and grocery purchases with bar codes. Location data is recorded via the global position satellites (GPS) residing on our smart phones know where they are and communicate this to the owners of the software. Videos and photographs are digitized and uploaded to a variety of locations. Digitized music and speech are shared across many platforms. Mouse clicks are recorded for every Internet and program use (think of the number of times you are asked if the program can use your location). Hadoop and its family of software products have been created to explore these different unstructured data types without the rigidity required by traditional spreadsheet and database processing.

3.2.3 Velocity

The Internet and mobile devices have increased the flow of data to users. Data flows into systems and is processed in batch, periodic, near real time, or real time (Soubra 2012). Before big data, companies usually analyzed their data using batch processing. This strategy worked when data was coming in at a slow rate. With new data sources, such as social media and mobile devices, the data input speed picks up and batch processing no longer satisfies the customer. So as the need for near real time or real time data processing increases, new ways of handling the data velocity come into play. However, it is not just the velocity of incoming data, but the importance of how quickly the data can be processed, analyzed and returned to the consumer who is making a data-driven decision. This feedback loop is critical in big data. The company that can shorten this loop has a big competitive advantage. Key-value stores and columnar databases (also known as NoSQL databases) that are optimized for the fast retrieval of precomputed information have been developed to satisfy this need. This family of NoSQL databases was created for when relational databases are unable to handle the volume and velocity of the data.

3.3   Data Science

3.3.1   What Is Data Science?

The phrase data science is linked with big data and is the analysis portion of the innovation. While there is no widely accepted definition of data science, several experts have made an effort. Loukides (2012) says that using data isn't, by itself, data science. Data science is using data to **create a data application** that acquires the value from the data itself and creates more data or a **data product**. Data science combines math, programming, and scientific instinct. Dumbill says that big data and data science create "the challenges of massive data flows, and the erosion of hierarchy and boundaries, will lead us to the statistical approaches, systems thinking and machine learning we need to cope with the future we're inventing" (2012b, p. 17). Conway defines data science using a Venn diagram consisting of three overlapping circles. The circles are math (linear algebra) and statistical knowledge, hacking skills (computer science), and substantive expertise (domain expertise). The intersection between hacking skills and math knowledge is machine learning. The intersection between math knowledge and expertise is traditional research. The intersection between expertise and hacking skills is a danger zone (i.e., knowing enough to be dangerous and to misinterpret the results). Data science resides at the center of all the intersections (Conway 2010). O'Neil and Schutt add the following skills to their description of data science: computer science, math, statistics, machine learning, domain expertise, communication and presentation skills, and data visualization (2014). Yet, the American Society of Statistics weigh in on data science by saying it is the technical extension of statistics and not a separate discipline (O'Neil and Schutt 2014). The key points in thinking about data science, especially in arguing for a separateness from statistics, are mathematics and statistical knowledge, computer science knowledge, and domain knowledge. A further distinction about data science is that the product of engaging in data science is creating a **data product** that feeds data back into the system for another iteration of analysis, a practical endeavor not traditional research. A more formal definition of data science proposed by O'Neil and Schutt is, "a set of best practices used in tech companies, working within a broad space of problems that could be solved with data" (2014, p. 351).

3.3.2   The Data Science Process

The data science process closely parallels the scientific process while including a feedback loop. Each step of the process feeds to the next one but also has feedback loops. First, the real world exists and creates data. Second the data is collected. Third, the data is processed. In the fourth step data cleaning occurs and feeds into machine learning/algorithms, statistical models, and communication/visualization/ reports. The fifth step is exploratory analysis but also feeds back into data collection. The sixth step is creating models with machine learning, algorithms, and statistics but also feeds into building a data product. The seventh step is to communicate the results, develop data visualizations, write reports and feeds back into decision making about the data. The eighth step is to build a data product. This data product is then released into the real world, thus closing the overall feedback loop.

A data scientist collects data from a multitude of big data sources as described in the previous section on Variety. However, the data scientist needs to have thought about the problem of interest and determine what kind of data is needed to find solutions for the problem or to gain insight into the problem. This is the step that uses Hadoop and its associated toolbox-HDFS, MapReduce, YARN, and others. This data is unprocessed and cleaning it for analysis consumes about 80% of the data scientist's time (Trifacta 2015). Programming tools, such as Python, R, SQL, are used to get the data ready for analysis. This cleaning and formatting process is called data munging, wrangling, joining, or scraping the data from the distributed databases (Provost and Fawcett 2013; Rattenbury et al. 2015). Common tools for this process, other than programming language, are Beautiful Soup, XML parsers, and machine learning techniques. Quality of the data must also be assessed, especially handling missing data and incongruence of data. Natural language processing tools may be used for this activity. Once the data is in a desired format, then analysis, interpretation, and decision-making using the data can occur.

The data scientist can then begin to explore the data using data visualization and sense-making of the data (the human expertise). The beginning step, keeping in mind the problem of interest, in working with the data is to conduct an exploratory data analysis (EDA; Tukey 1977). Graphing the data helps to visualize what the data is representing. The analyst creates scatterplots and histograms from different perspectives to get "a feel" for the data (Jones 2014). The graphs will help to know how and what probability distributions (curves plotted on an x and y axes) to calculate as the data is explored (remember to look at correlation and not causality). EDA may reveal a need for more data, so this becomes an iterative process. Experience determines when to stop and proceed to the next step. A firm grasp of linear algebra is essential in this step.

Next, use the data to "fit a model" using the parameters or variables that have been discovered (this uses statistical knowledge). Caution, do not overfit the model (a danger zone event described by Conway 2010). The model is then optimized using one of the two preferred programming languages in data science: Python and R. Python is usually preferred by those whose strength is in computer science, while R is preferred by statisticians. MapReduce may also be used at this step. Algorithms, from statistics, used to design the model may be linear regression, Naïve Bayes, k-nearest neighbor, clustering, and so forth. The algorithm selection is determined by the problem being solved: classification, cluster, prediction, or description. Machine learning may also be used at this point in analysis and uses approaches from computer science. Machine learning leads to data products that contain image recognition, speech recognition, ranking, recommendations, and personalization of content.

The next step is to interpret and visualize the results (data visualization will be discussed later in the chapter). Communication is the key activity in this step.

Informal and formal reports are written and given. Presentations are made to customers and stakeholders about the implications and interpretations of the data. Presentation skills are critical. Remember in presenting complex data, a picture is worth a thousand words. Numerous tables of numbers and scatterplots confuse and obscure meaning for the customer. A visual designer can be a valuable member to design new data visualization approaches or infographics (Knaflic 2015).

The final step is to create a data product from the analysis of the data and return it to the world of raw data. Well known products are spam filters, search ranking algorithms, or recommendation systems. A data product may focus on health by collecting data and returning health recommendation to the individual. Research productivity may be communicated through publications, citations of work, and names of researchers following your work as ResearchGate endeavors to do (ResearchGate is an online community of nine million researchers; 2016). As these products are used in the big data world, they contribute to ongoing data resources. The data science process creates a feedback loop. It is this process that makes data science unique and distinct from statistics.

3.4   Visualizing the Data

Data visualization has always been important for its ability to show at a glance very complicated relationships and insights. Data represents the real world but it is only a snapshot covering a point in time or a single time series. Visualization is an abstraction of the data and represents its variability, uncertainty, and context in a way that the human brain can apprehend (Yau 2013). Data visualization occurs prominently in three steps of the data science process: step four data cleaning, step five exploratory data analysis, and step seven communication (O'Neil and Schutt 2014). Graphing and plotting the data in step four depicts outliers and anomalies that a data wrangler may want to explore to see if there are issues with the data. The issues could be with format, missing data, or inconsistencies. During exploratory data analysis, the graphing of data may demonstrate insights, inconsistencies, or the need for more data. In step seven the results of the data science project are communicated requiring more complex graphs that depict multiple variables.

In designing a visualization of data, there are four components to consider (Yau 2013). The first component is the use of visual cues to encode the data in the visualization. The major cues are shape, color, size, and placement in the visualization. The second component is selecting the appropriate coordinate system. There are three main systems from which to choose: a Cartesian system (x and y axes), a polar system (points are on a radius at an angle, Nightingale used this in her graphic of British solder deaths in the Crimean War), or a geographic system (maps, longitude, latitude). The third component is the use of scale defined by mathematical functions. The most common scales are numeric, categorical and time. The last component is the context that helps to understand the who, what, where, when, and why of the data. Data must be interpreted in context and the visualization must demonstrate this context to the viewer. Doing data visualization well is understanding that the task is to map the data to a geometry and color thus creating a representation of the data. The viewer of the data visualization must be able to go back and forth between what the visual is and what it represents—to see the pattern in the data.

A good visual designer and data scientist follow a process to develop the visualization (Yau 2013). As with the analysis process the team must have some questions to guide the visualization process. First the data collected and cleaned must be graphed to enable the team to know what kind of data they have. Tools, such as Excel, R (though R is a programming language, it can generate graphics as well), Tableau, or SAS, can be used to describe the data with scatterplots, bar charts, line graphs, pie charts, polar graphs, treemaps, or other basic ways to display data. This step must be continued until the team "knows" the data they have. The second step is to determine what you want to know about the data. What story do you want to tell with the data? The third step is to determine the appropriate visualization method. The nature of the data and the models used in analysis will guide this step. The data must be visualized with these assumptions and the previous four components of visualization design in mind. The last step is to look at the visualization and determine if it makes sense. This step may take many iterations until the visualizations convey the meaning of the data in an intuitive way to the viewer or customer of the analysis. Using a sound, reproducible process for creating the visualization insures that the complexity and art of creating representations of data are accurate and understood.

Three of the most well-known data visualizations of all times are Nightingale's Mortality in the Crimean War (ims5 2008; Yau 2013), Minard's Napoleon March on Moscow (Sandberg 2013; Tufte 1983); and Rosling's Gapminder (2008; Tableau 2016). These visualizations depict multiple variables and their interrelationships. They demonstrate well thought-out strategies for depicting data using more than simple graphs. Nightingale invented the coxcomb graph to depict the causes of death in the Crimean war. The graph displays time, preventable deaths, deaths from wounds, and death from other causes. Her graph is said to be the second best graph ever drawn (Tableau 2016). Minard depicted Napoleon's march on Russia by displaying geography, time, temperature, the course and direction of the army, and the number of troops remaining. He reduced numerous tables and charts into one graphic that Tufte (1983) called the best statistical graph ever drawn. Rosling's bubble graph depicts the interaction between time, income per person, country, and life expectancy. The great data visualizations go beyond the basic graphing approaches to depict complex relationships within the data (Tufte 1990, 1997).

3.5 Big Data Is a Team Sport

Doing data science requires a team as no one person can have the all the skills needed to collect, clean, analyze, model, visualize, and communicate the data. Teams need to have technical expertise in a discipline, curiosity with a need to understand a problem, the ability to tell a story with data and to communicate effectively, and the ability to view a problem from different perspectives (Patil 2011). When pulling together a team consider the following people: programmers with skill in Python, R, and other query languages; database managers who can deploy and manage Hadoop and other NoSQL databases; information technology (IT) professionals who know how to manage servers, build data pipelines, data security, and other IT hardware; software engineers who know how to implement machine learning and develop applications; data wranglers who know how to clean and transform the data; visual designers who know how to depict data that tell a story and to use visualization software; scientists who are well versed in crafting questions and searching for answers; statisticians who are well versed in developing models, designing experiments, and creating algorithms; informaticians who understand data engineering; and experts in the domain being explored (O'Neil and Schutt 2014).

The team must determine who has which skills and how to collaborate and enhance these skills to create the best data product for the organization. The organizational culture must be one that supports and embraces data science to its fullest for the greatest success (Anderson 2015; Patil and Mason 2015). As organizations begin to use data science in their product development initiatives, a certain level of data science maturity is required. Guerra and Borne have identified ten signs of a mature data science capability (2016). A mature data science organization makes all data available to their teams; access is critical and silos are not allowed. An agile approach drives the methodology for data product development (The Agile Movement 2008). Crowd sourcing and collaboration are leveraged and promoted. A rigorous scientific methodology is followed to insure sound problem solving and decision making. Diverse team members are recruited and given the freedom to explore; they are not micromanaged. The teams and the organization ask the right questions and search for the next question of interest. They celebrate a fast-fail collaborative culture that encourages the iterative nature of data science. The teams show insights through illustrations and storytelling than encourage asking "what if" questions that require more that simple scatterplots and bar charts. Teams build proof of values, not proof of concepts. Developing proof of value focuses on value leading to solving the unknowns, not just that it is a good thing to do. Finally, the organization promulgates data science as a way of doing things, not a thing to do. Data science drives all functions in the organization and shifts how organizations operate.

As with any team that develops products, intellectual property is a key concern. When data is used to develop data products, the ethics of data ownership and privacy become critical issues. Traditional data governance approaches and privacy laws and regulations don't completely guide practice when big data is ubiquitous and practically free. With big data, no one organization owns all the data they need. New models of collaboration and data sharing are emerging. As these models evolve, new questions emerge about data ownership, especially if it is collected as a byproduct of conducting business—banking, buying groceries, searching the Internet, or engaging in relationships on social media. Ownership of this type of data may not always be clear. Nor is it clear who can use and reuse the data. Numerous questions arise from exploring this grey area. If the data is generated from a transaction, who controls and owns that? Who owns the clicks generated from cruising the Internet? To add to this confusion, consumers are now wanting to control or prevent collection of the data they generate—the privacy issue (Pentland 2012). There have been situations where individuals have been re-identified from anonymized data. The White House, in response to this concern, has drafted a Consumer Privacy Bill of Rights Act (2015). The draft acknowledges a "rapid growth in the volume and variety of personal data being generated, collected, stored, and analyzed." Though the use of big data has the potential to create knowledge, increase technological innovation, and improve economic growth, big data has the potential to harm individual privacy and freedom. The bill urges that laws must keep current with technology and business innovation. As the practice of using big data and data science becomes more mainstream, the ethical issues and their solutions will appear. The Data Science Association has a Code of Conduct for their members (2016). This Code speaks to conflict of interest, data and evidence quality, and confidentially of the data. For a good discussion of the ethics of big data, read Martin's (2015) article on the ethical issues of the big data industry.

3.6 Conclusion

In January 2009, Hal Varian, the chief economist at Google, said in an interview that, "The ability to take data—to be able to understand it, to process it, to extract value from it, to visualize it, to communicate it—is going to be a hugely important skill in the next decades..." (Yau 2009). Varian says this skill is also important for elementary school, high school and college kids because data is ubiquitous and free. The ability to understand that data and extract value from it is now a scarce commodity. Varian believes that being a data scientist and working with big data will be a 'sexiest' job around. The Internet of Things has created disruption in the way we think about data as it is coming at us from everywhere and interconnected. In a period of combinatorial innovation, we must use the components of software, protocols, languages, capabilities to create totally new inventions. Remember, however, that big data is not the solution. Patterns and clues can be found in the data, yet have no meaning nor usefulness. The key to success is to decide what problem you want to solve, then use big data and data science to help solve the problem and meet your goals (Dumbill 2012). Finally, big data is just one more step in the continuation of mankind's ancient quest to measure, record, and analyze the world (Mayer-Schonberger and Cukier 2013).

Case Study 3.1: Big Data Resources—A Learning Module

Judith J. Warren and E. LaVerne Manos

Abstract This case study is a compilation of resources for a learner to explore to gain beginning knowledge and skill in big data, data science, and data visualization. The resources focus on acquiring knowledge through books, white papers, videos, conferences, and online learning opportunities. There are also resources for learning about the hardware and software needed to engage in big data.

Keywords Big data • Data science • Data visualization • Data wrangling • Hadoop/mapreduce • Data analytics • Data scientist • Data science teams • Volume/variety/velocity of data sets • Data products

3.1.1   Introduction

The volume, variety, and velocity of big data exceed the volume of datasets common in health care research and operations. New technologies created to manage and analyze big data are being developed and tested at a rapid rate. This life cycle process is happening so fast that it is difficult to learn the technology and approaches much less keep up on the latest innovations. The phases of this life cycle are development, testing, discarding, testing, adopting, combining, using and discarding/reworking. These phases transpire in swift iterative cycles, and data scientists who utilize the tools work with a toolbox composed of well-developed software to niche software designed for specific uses, many of which are open source.

Today we are overwhelmed with an unprecedented amount of information and data. Big data comes from all kinds of sources: global positioning devices (GPS), loyalty shopping cards, online searches and selections, genomic information, traffic and weather information, health data from all sorts of personal devices (person generated health data), as well as data created from healthcare during inpatient and outpatient visits. Data is collected every second of every day. These types of data, including unstructured raw data, have been used in other industries to understand their business and create new products. Healthcare has been slower to adopt the use of big data in this way. The 2013 report by McKinsey Global Institute proposes that the effective use of big data in healthcare could create large value for the healthcare industry, over $300 billion every year (Kayyali B, Knott D, Van Kuiken, S. The big-data revolution in US health care: Accelerating value and innovation. Mc Kinsey & Company. 2013. Accessed at http://www.mckinsey.com/industries/healthcare-systems-and-services/our-insights/the-big-data-revolution-in-us-health-care).

J.J. Warren, Ph.D., R.N., F.A.A.N., F.A.C.M.I. (✉)
University of Kansas School of Nursing, Kansas City, KS, USA

Warren Associates, LLC, Plattsmouth, NE, USA

E. LaVerne Manos, D.N.P., R.N.-B.C.
University of Kansas School of Nursing, Kansas City, KS, USA

The effective use of big data requires a data science approach to find and analyze subsets of data that administrators, clinicians, and researchers will find usable. Unlike a relational database where writing a query is fairly straight forward, gathering data from multiple data stores/warehouses of big data is much more complex. The ability to manage an incoming data stream of extraordinary volume, velocity and variety of data requires the expertise of a team. This case study provides a beginning resource for learning about big data, data science, and data visualization.

3.1.2   Resources for Big Data

As big data has caught the imagination of corporations and health care, the resources have exploded and most are readily available on the Internet. The following resources have been selected for learners who are just beginning their exploration of big data and a few that will stretch their knowledge towards competence. As you do your own searches, you will find many more. This listing will get you into the field and Internet space to find more resources that fit your learning style.

3.1.2.1   Big Data Conferences

Conferences are good places to explore a new field or gain more understanding of a field with which you have expertise. Networking is key at these events and can link you to others for future project work. These are just the tip of the iceberg of conferences, so enjoy looking for new ones near you.

1. In 2013, the University of Minnesota School of Nursing convened the first conference called Nursing Knowledge: Big Data Science, http://www.nursing. umn.edu/icnp/center-projects/big-data/index.htm. The first conference was invitational and explored the potential of big data for the improvement of patient outcomes as the result of nursing care. The conference was so successful, it has been held annually and been open to all registrants. Nursing Knowledge: Big Data Science is a working conference with many workgroups creating projects that are making an impact in Nursing research, education, and practice.
2. "Big Data 2 Knowledge" hosted by the National Institutes of Health (NIH) also has conferences, training sessions, and webinars. These events are geared towards creating a research cohort that is expert in big data and data analytics.
3. The Strata + Hadoop World Big Data conference is a meeting where business decision makers, strategists, architects, developers, and analysts gather to discuss big data and data science. At the conference you explore big data and hear what is emerging in the industry (http://conferences.oreilly.com/strata/hadoop-big-data-ca). O'Reilly Media and others put on this conference and afterwards post all the presentations to their web site. So even if you can't attend, you can hear about cutting-edge big data.

3.1.2.2   Big Data Books and Articles

A tried and traditional way to learn about any knowledge is through books, journal articles, and white papers. The following are basic references to get you started in the big data initiative.

1. Anderson C. Creating a Data-Driven Organization. Sebastopol, CA: O'Reilly Media; 2015. http://shop.oreilly.com/product/0636920035848.do
2. Betts R, Hugg, J. Fast Data: Smart and at Scale. Sebastopol, CA: O'Reilly Media; 2015. https://voltdb.com/blog/introducing-fast-data-smart-and-scale-voltdbs-new-recipes-ebook
3. Brennan PF, Bakken S. Nursing needs big data and big data needs nursing. Journal of Nursing Scholarship, 2015;47: 477–484.
4. Chartier, T. Big Data: How Data Analytics Is Transforming the World. Chantilly, VA: The Great Courses. 2014. (includes video lectures). (http://www.thegreatcourses.com/courses/big-data-how-data-analytics-is-transforming-the-world.html)
5. Davenport T, Dyche J. Big data in big companies. 2013. http://www.sas.com/reg/gen/corp/2266746. Accessed 15 Dec 2015.
6. Mayer-Schonberger V, Cukier K. Big Data: A revolution that will transform how we live, work, and think. New York: Hought Mifflin Harcourt Publishing. 2013.
7. O'Reilly Radar Team. Planning for big data: A CIO's handbook to the changing data landscape. Sebastopol, CA: O'Reilly Media. 2012. http://www.oreilly.com/data/free/planning-for-big-data.csp
8. O'Reilly Team. Big data now. Sebastopol, CA: O'Reilly Media. 2012.
9. Patil DJ, Mason H. Data driven: Creating a data culture. Sebastopol, CA: O'Reilly Media. 2015. http://datasciencereport.com/2015/07/31/free-ebook-data-driven-creating-a-data-culture-by-chief-data-scientists-dj-patil-hilary-mason/#.Vp6x33n2bL8

3.1.2.3   Big Data Videos

For those who need to see and hear, videos are great. Below are some from YouTube and other websites. Don't forget to look at Tim Chartier's work listed in the Books section. Great Courses combine expert faculty, a book, and video lectures. Explore TED Talks for more information about Big Data.

1. Big Data Tutorials and TED Talks, http://www.analyticsvidhya.com/blog/2015/07/big-data-analytics-youtube-ted-resources/
2. Kenneth Cukier: Big data is better data, https://www.youtube.com/watch?v=8pHzROP1D-w
3. The Secret Life of Big Data | Intel, https://www.youtube.com/watch?v=CNoi-XqwJnA (a good overview of the history of Big Data, a must watch)
4. What is Big Data? https://www.youtube.com/watch?v=c4BwefH5Ve8

5. What is BIG DATA? BIG DATA Tutorial for Beginners, https://www.youtube.
   com/watch?v=2NLyIqU-xwg
6. What Is Apache Hadoop? http://hadoop.apache.org/
7. What is Big Data and Hadoop? https://www.youtube.com/watch?v=FHVuRxJpiwI
8. What Does The Internet of Things Mean? https://www.youtube.com/
   watch?v=Q3ur8wzzhBU
9. MapReduce, https://hadoop.apache.org/docs/r1.2.1/mapred_tutorial.html

3.1.2.4 Big Data Web Sites

Many companies, with a web site, provide information, free books, white papers, tutorials, and free trial software. These sites are a rich resource. Gartner and Forrester are companies that evaluate and rate emerging companies and products in the big data industry.

1. IBM Big Data and Analytics platform, now known as IBM Watson Foundations, http://www.ibmbigdataanalytics.com
2. Forrester Wave, https://www.forrester.com/The+Forrester+Wave+Big+Data+H adoop+Distributions+Q1+2016/fulltext/-/E-res121574#AST1022630
3. Gartner,   http://www.gartner.com/technology/research/methodologies/ research_mq.jsp

   (a) Magic Quadrant
   (b) HypeCycle
   (c) Critical Capabilities

4. Intel Processors, http://www.intel.com/content/www/us/en/homepage.html

   (a) The Butterfly Dress, https://www.youtube.com/watch?v=6ELuq3CzJys (a bit of fun with data and technology)
   (b) 50th Anniversary of Moore's Law, http://newsroom.intel.com/docs/DOC-6429 (if you are in informatics, you must know about Moore's Law)
   (c) How Intel Gave Stephen Hawking his Voice, http://www.wired. com/2015/01/intel-gave-stephen-hawking-voice;   https://www.youtube. com/watch?v=JA0AZUj2lOs

5. Kaggle, www.kaggle.com
6. O'Reilly Media, https://www.oreilly.com/topics/data
7. SAS, http://www.sas.com/en_us/insights/big-data.html
8. VoltDB, https://voltdb.com
9. Yuhanna, N. (August 3, 2015). The Forrester Wave: In-Memory Database Platforms, Q3 2015. http://go.sap.com/docs/download/2015/08/4481ad9e-3a7c-0010-82c7-eda71af511fa.pdf
10. Zaloni. http://www.zaloni.com/health-and-life-sciences

3.1.3   Resources for Data Science

Data science is composed of data wrangling and data analysis. Data wrangling is the process of cleaning and mapping data from one "raw" form into another format. Then algorithms can be applied to make sense of big data. The following resources have been selected for learners who are just beginning their exploration of data science and a few that will stretch their knowledge towards competence. As you do your own searches, you will find many more. This listing will get you into the field and Internet space to find more resources that fit your learning style.

3.1.3.1   Data Science Conferences

Conferences are good places to explore a new field or gain more understanding of a field with which you have expertise. Networking is key at these events and can link you to others for future project work.

1. "Big Data 2 Knowledge" hosted by the National Institutes of Health (NIH) also has conference, training sessions, and webinars. These events are geared towards creating a research cohort that is expert in Big Data and Data Analytics.
2. The Data Science Conference, http://www.thedatascienceconference.com.

3.1.3.2   Data Science Books and Articles

A tried and traditional way to learn about any knowledge is through books, journal articles, and white papers. The following are basic references to get you started in Data Science.

1. Ghavami, PK. Clinical Intelligence: The big data analytics revolution in healthcare: A framework for clinical and business intelligence. CreateSpace Independent Publishing Platform. 2014.
2. Grus, J. Data science from scratch: First principles with python. Sebastopol, CA: O'Reilly Media. 2015.
3. Gualtieri M, Curran R. The Forrester Wave: Big data predictive analytics solutions, Q2, 2015. April 1, 2015. https://www.sas.com/content/dam/SAS/en_us/doc/analystreport/forrester-wave-predictive-analytics-106811.pdf
4. Janert, PK. Data analysis with open source tools: A hands-on guide for programmers and data scientists. Sebastopol, CA: O'Reilly Media. 2010.
5. Loukides, M. What is data science? Sebastopol, CA: O'Reilly Media. 2012.
6. Marconi K, Lehmann H. Big data and health analytics. Boca Raton, FL: CRC Press. 2015.
7. O'Neil C, Schutt R. Doing data science: Straight talk from the frontline. Sebastopol, CA: O'Reilly Media. 2015.
8. Optum. Getting from big data to good data: Creating a foundation for actionable analytics. 2015. https://www.optum.com/content/dam/optum/CMOSpark%20

Hub%20Resources/White%20Papers/OPT_WhitePaper_ClinicalAnalytics_ONLINE_031414.pdf

9. Patil DJ. Building data science teams: The skills, tools, and perspectives behind great data science groups. Sebastopol, CA: O'Reilly Media. 2011.

10. Provost F, Fawcett T. Data science for business: What you need to know about data mining and data-analytic thinking. Sebastopol, CA: O'Reilly Media. 2013.

11. Rattenbury T, Hellerstein JM, Heer J, Kandel S. Data wrangling: Techniques and concepts for agile analysts. Sebastopol, CA: O'Reilly Media. 2015.

12. Tailor K. The patient revolution: How big data and analytics are transforming the health care experience. Hoboken, NJ: John Wiley and Sons. 2016.

13. Trifacta. Six Core Data Wrangling Activities. 2015. https://www.trifacta.com/wp-content/uploads/2015/11/six-core-data-wrangling-activities-ebook.pdf. Accessed 15 Jan 2016.

3.1.3.3 Data Science Videos

For those who need to see and hear, videos are great. Below are some from YouTube and other websites. Explore TED Talks for more information about Big Data.

1. Analytics 2013—Keynote—Jim Goodnight, SAS, https://www.youtube.com/watch?v=AEI0fBQYJ1c
2. Big Data Analytics: The Revolution Has Just Begun, https://www.youtube.com/watch?v=ceeiUAmbfZk
3. Building Data Science Teams, https://www.youtube.com/watch?v=98NrsLE6ot4
4. Deep Learning: Intelligence from Big Data, https://www.youtube.com/watch?v=czLI3oLDe8M
5. The Future of Data Science—Data Science @ Stanford, https://www.youtube.com/watch?v=hxXIJnjC_HI
6. The Patient Revolution: How Big Data and Analytics Are Transforming the Health Care Experience, https://www.youtube.com/watch?v=oDztVSDUbxo

3.1.3.4 Data Science Web Sites

Many companies, with a web site, provide information, free books, white papers, tutorials, and free trial software. These sites are a rich resource.

1. Alteryx, http://www.alteryx.com.
2. Data Science at NIH, https://datascience.nih.gov/bd2k
3. IBM, http://www.ibmbigdataanalytics.com.
4. Kaggle-the Home of Data Science, https://www.kaggle.com
5. Python Programming Language, https://www.python.org/
6. R Programming language, https://www.r-project.org/about.html
7. SAS, https://www.sas.com/en_us/home.html.
8. Trifacta, https://www.trifacta.com/support.

3.1.4   Resources for Data Visualization

Data visualization is the third part of big data. Humans can absorb more data when it is depicted in images or graphs. The following resources have been selected for learners who are just beginning their exploration of data visualization and a few that will stretch their knowledge towards competence. As you do your own searches, you will find many more. This listing will get you into the field and Internet space to find more resources that fit your learning style.

3.1.4.1   Data Visualization Conferences

Conferences are good places to explore a new field or gain more understanding of a field with which you have expertise. Networking is key at these events and can link you to others for future project work. Most conferences on big data and data science include presentations on data visualization.

3.1.4.2   Data Visualization Books and Articles

A tried and traditional way to learn about any knowledge is through books, journal articles, and white papers. The following are basic references to get you started in the data visualization.

1. Beegel J. Infographics for dummies. Hoboken, NJ: John Wiley & Sons. 2014.
2. Few S. Now you see it: Simple visualization techniques for quantitative analysis. Oakland, CA: Analytics Press. 2009.
3. Harris RL. Information graphics: A comprehensive reference. Atlanta, GA: Management Graphics. 1996.
4. Jones B. Communicating data with tableau: Designing, developing, and delivering data visualization. Sebastopol, CA: O'Reilly Media. 2014. http://cdn.oreillystatic.com/oreilly/booksamplers/9781449372026_sampler.pdf
5. Knaflic CN. Storytelling with data: A data visualization guide for business professionals. Hoboken, NJ: John Wiley & Sons. 2015.
6. Tufte ER. Envisioning information. Cheshire, CN: Graphics Press. 1990.
7. Tufte ER. The visual display of quantitative information. Cheshire, CN: Graphics Press. 1983. (This is the classic text in visualization.)
8. Tufte ER. Visual explanations: Images and quantities, evidence and narrative. Cheshire, CN: Graphics Press. 1997.
9. Yau N. Data points: Visualization that means something. Indianapolis, IN: John Wiley & Sons; 2013.
10. Yau N. Visualize this: The FlowingData guide to design, visualization, and statistics. Indianapolis, IN: John Wiley & Sons; 2011.

3.1.4.3   Data Visualization Videos

For those who need to see and hear, videos are great. Below are some from YouTube and other websites. Explore TED Talks for more information about Big Data.

1. The beauty of data visualization, https://www.youtube.com/watch?v=5Zg-C8AAIGg
2. The best stats you've ever seen, https://www.youtube.com/watch?v=usdJgEwMinM
3. Designing Data Visualizations, https://www.youtube.com/watch?v=lTAeMU2XI4U
4. The Future of Data Visualization, https://www.youtube.com/watch?v=vc1bq0qIKoA
5. Introduction to Data Visualization, https://www.youtube.com/watch?v=XIgjTuDGXYY

3.1.4.4   Data Visualization Web Sites

Many companies, with a web site, provide information, free books, white papers, tutorials, and free software.

1. FlowingData, https://flowingdata.com.
2. SAS, http://www.sas.com/en_us/home.html

   (a) Data visualization and why it is important, http://www.sas.com/en_us/insights/big-data/data-visualization.html

3. Tableau, http://www.tableau.com/

   (a) Tableau. (2015). The 5 Most Influential Data Visualizations of All Time. http://www.tableau.com/top-5-most-influential-data-visualizations   (note Florence Nightingale is the number two graph)
   (b) Visual Analysis Best Practices: Simple Techniques for Making Every Data Visualization Useful and Beautiful, http://get.tableau.com/asset/10-tips-to-create-useful-beautiful-visualizations.html

4. Trifacta, https://www.trifacta.com

3.1.5   Organizations of Interest

As the field of big data, data science and data visualization evolve, professional organizations will be formed. Listservs and blogs will be created. Academia will offer courses and degree programs. Certification and accreditation organizations will help to establish quality programs and individual performance. The following are just a sampling of what exists.

3.1.5.1 Professional Associations

Professionals will form professional organizations as they define their discipline. The organizations provide a forum for discussing practice, competencies, education, and the future.

1. American Statistics Association, http://www.amstat.org/
2. American Association of Big Data Professionals, https://aabdp.org/

    (a) Offers certification in various Big Data roles, https://aabdp.org/certifications.html

3. Data Science Association, http://www.datascienceassn.org/
4. Digital Analytics Association, http://www.digitalanalyticsassociation.org/

3.1.5.2 Listservs: A Sampling

Most web sites, organizations, industry, and publishers have listservs. This is a very efficient way to keep up with what is happening in these areas. The listserv is pushed to your email and enables you to see the latest thoughts, conferences, books, and software an industry that is evolving rapidly.

1. 10 Data Science Newsletters To Subscribe To, https://datascience.berkeley.edu/10-data-science-newsletters-subscribe
2. Information Management, http://www.information-management.com/news/big-data-analytics/Big-Data-Scientist-Careers-10026908-1.html
3. O'Reilly Data Newsletter, http://www.oreilly.com/data/newsletter.html. Sign up to get the latest information about Big Data, Data Analytics, Data Visualization, and Conferences.

3.1.5.3 Certificates and Training: A Sampling

As jobs in these fields become more widely available, the demand for these skills will grow. Online education and formal degrees will become important for employers to consider. Certification may make a difference for employment.

1. Data Science at Coursera, https://www.coursera.org/specializations/jhu-data-science
2. Data at Coursera, https://www.coursera.org/specializations/big-dataQ
3. SAS Certification program, http://support.sas.com/certify/index.html
4. MIT Professional Education, https://mitprofessionalx.mit.edu/about
5. R Programming, https://www.coursera.org/learn/r-programming

3.1.5.4   Degree Programs: A Sampling

Degree programs are proliferating as the demand for big data professionals and data scientists increases. It will be important to select well before investing time and money into the programs. Always look for programs that are accredited. The University/College must be accredited by the US Department of Education. Even the department/school they reside in must be accredited by the appropriate accreditor. Accreditation assures the quality of the education.

1. 23 Great Schools with Master's Programs in Data Science, http://www.mastersin-datascience.org/schools/23-great-schools-with-masters-programs-in-data-science
2. Carnegie Mellon University, http://www.cmu.edu/graduate/data-science/
3. Harvard, http://online-learning.harvard.edu/course/big-data-analytics
4. List of Graduate Programs in Big Data & Data Science, http://www.amstat.org/education/bigdata.cfm
5. Map of University Programs in Big Data Analytics, http://data-informed.com/bigdata_university_map/
6. Northwestern Kellogg School of Management, http://www.kellogg.northwestern.edu/execed/programs/bigdata.aspx?gclid=CLTa_Jf5u8oCFYVFaQodCpwHag

3.1.6   Assessment of Competencies

Teachers and students have used Bloom's Taxonomy to create objectives that specify what is to be learned. The levels of Bloom can also be used to guide evaluation of the attainment of these objectives by the student. In 2002, Bloom's was revised to reflect cognitive processes as well as knowledge attainment (http://www.unco.edu/cetl/sir/stating_outcome/documents/Krathwohl.pdf). The new taxonomic hierarchy is as follows (Krathwohl, 2002, p215):

1. "Remember—retrieving relevant knowledge from long-term memory
2. Understand—determining the meaning of information
3. Apply—using a procedure in a given situation
4. Analyze—breaking material into its constituent parts and detecting the relationships between the parts and the whole
5. Evaluate—making judgements based on criteria
6. Create—putting elements together to form a coherent whole or make a product."

For big data and data science assignments the graduate student should be able to master the levels of "remember, understand, and apply" by engaging with the above resources. Objective assessments, in the form of tests, can then be used to determine mastery. Performance assessments are used to evaluate the achievement of the higher levels of Bloom-- analyze, evaluate and create. Performance assessments are conducted by experts and faculty through the use of case studies, simulations, projects, presentations, or portfolios.

3.1.7 Learning Activities

The following are several learning activities designed to help you apply the knowledge and skills learned from the above resources. The Bloom level for each activity is listed.

1. Conduct a web search on HADOOP and data warehouses. What did you learn about big data? What are the issues in storing and accessing data that has volume, velocity, and variety? Define Oozie, PIG, Zookeeper, Hive, MapReduce, and Spark. How are they used in big data initiatives? (Bloom level—Understand)
2. A good source of data to practice wrangling, analysis and visualization is DATA. gov, http://www.data.gov. Download a file and then one of the free trial software packages and try different things. Trifacta lets you work on data wrangling. Excel can help with analysis. Tableau can help with visualization. Other sources of data are

   (a) https://r-dir.com/reference/datasets,
   (b) https://www.kaggle.com/datasets and
   (c) http://www.pewresearch.org/data/download-datasets.(Bloom level—Apply)

3. Take a data set and graph the data five different ways, e.g. scatter plot, histogram, radar chart, or other types of graphs. What insight did you get looking at the graphs? What analytic questions do you have that you would like to pursue based on the graphs? Were the graphs consistent? Was there one that represented the data best and why? (Bloom level—Analyze)
4. Keep a log of data that you personally generate through online use, mobile devices, smart phones, email, music, videos, pictures, financial transactions, and fitness/health apps. What format is this data in? Conduct an exploratory data analysis. Visualize the results several ways. Evaluate the visualizations using Yau's (2013) four components: visual cues, coordinate system, scale, and context. (Bloom level—Evaluate)
5. Create a list of keywords and a glossary for a document using Python. Download Python 3.4.4.msi (https://www.python.org/downmoads) and numpy-1.11.0.zip (http://www.numpy.org). Select a document and save it as a' .txt' file (if the name of the file contains a /U, then replace that with //U so the name will parse; Python uses /U as a code). Develop a Python script to determine word frequency in the document (http://programminghistorian.org/lessons/counting-frequencies). Wrangle the data so that only words are left ad remove stop words. From the remaining list select keywords and glossary words. (Bloom level—Create)

3.1.8   Guidance for Learners and Faculty Using the Module

This case study has provided learning resources for faculty and students to learn about big data, data science, and data visualization. The best strategy is to select some of the resources that best match your learning style—visual, audio, and tactile—and interact with them first. You may also want to use various search engines to search for other information about big data, data science, and data visualization. All online resources were accessed in January or February 2016. Download some programs and data and explore the process of wrangling, analysis and visualization.

References

The Agile Movement. 2008, Oct 23. http://agilemethodology.org. Accessed 25 Jan 2016.

Anderson C. Creating a data-driven organization. Sebastopol, CA: O'Reilly Media; 2015. http://shop.oreilly.com/product/0636920035848.do

Apache Software Foundation. Welcome to Apache Hadoop. 2016. http://hadoop.apache.org. Accessed 15 Jan 2016.

Chan C. What Facebook deals with every day: 2.7 billion likes, 300 million photos uploaded and 5—terabytes of data. 2012, Aug 22. http://gizmodo.com/5937143/what-facebook-deals-with-everyday-27-billion-likes-300-million-photos-uploaded-and-500-terabytes-of-data. Accessed 18 Jan 2016.

Chartier T. Big data: how data analytics is transforming the world. Chantilly, VA: The Great Courses (includes video lectures); 2014. http://www.thegreatcourses.com/courses/big-data-how-data-analytics-is-transforming-the-world.html

Conway D. The data science Venn diagram. 2010, Sept 30. http://drewconway.com/zia/2013/3/26/the-data-science-venn-diagram. Accessed 30 Jan 2016.

Data Science Association. Code of conduct. 2016. http://www.datascienceassn.org/code-of-conduct.html. Accessed 15 Apr 2016.

The Domesday Book Online. 2013. http://www.domesdaybook.co.uk. Accessed 10 Jan 2016.

Dumbill E. What is big data? In: O'Reilly Team, editor. Big data now. Sebastopol, CA: O'Reilly Media; 2012a. p. 3–10.

Dumbill E. Why big data is big: the digital nervous system. In: O'Reilly Team, editor. Big data now. Sebastopol, CA: O'Reilly Media; 2012b. p. 15–7.

Encyclopedia of Stone Age Art: Lascaux Cave Paintings. 2016. http://www.visual-arts-cork.com/prehistoric/lascaux-cave-paintings.htm. Accessed 10 Jan 2016.

Forrester Research. 2016. https://www.forrester.com/home. Accessed 6 Jan 2016.

Gartner. Gartner says 4.9 billion connected "things" will be in use in 2015. 2014, Nov 11. http://www.gartner.com/newsroom/id/2905717. Accessed 10 Jan 2016.

Gartner. Gartner magic quadrant. 2016a. http://www.gartner.com/technology/research/methodologies/research_mq.jsp. Accessed 5 Jan 2016.

Gartner. Gartner critical capabilities. 2016b. http://www.gartner.com/technology/research/methodologies/research_critcap.jsp. Accessed 5 Jan 2016.

Gartner. Gartner hype cycle. 2016c. http://www.gartner.com/technology/research/methodologies/hype-cycle.jsp. Accessed 5 Jan 2016.

GilPress. Gartner's hype cycle for big data. 2012, Oct. https://whatsthebigdata.com/2012/08/16/gartners-hype-cycle-for-big-data. Accessed 5 Jan 2016.

Google Flu Trends. 2014. https://www.google.org/flutrends/about. Accessed 15 Jan 2016.

Grus J. Data science from scratch: first principles with python. Sebastopol, CA: O'Reilly Media; 2015.

Gualtieri M, Curran R. The Forrester Wave: big data predictive analytics solutions, Q2, 2015. 2015, Apr 1. https://www.sas.com/content/dam/SAS/en_us/doc/analystreport/forrester-wave-predictive-analytics-106811.pdf. Accessed 18 Jan 2016.

Gualtieri M, Yuhanna N, Kisker H, Curran, R, Purcell B, Christakis S, Warrier S, Izzi M. The Forrester Wave™: big data Hadoop distributions, Q1 2016. 2016, Jan 19. https://www.forrester.com/report/The+Forrester+Wave+Big+Data+Hadoop+Distributions+Q1+2016/-/E-RES121574#AST1022630, Accessed 25 Jan 2016.

Guerra P, Borne K. Ten signs of data science maturity. Sebastpol, CA: O'Reilly Media; 2016.

Gunelius S. The data explosion in 2014 minute by minute—infographic. 2014, Jul 12. http://aci.info/2014/07/12/the-data-explosion-in-2014-minute-by-minute-infographic. Accessed 18 Jan 2016.

IBM. The four V's of big data. 2015. http://www.ibmbigdatahub.com/sites/default/files/infographic_file/4-Vs-of-big-data.jpg?cm_mc_uid=24189083104014574569048&cm_mc_sid_50200000=1457456904

IBM. What is big data? 2016. http://www-01.ibm.com/software/data/bigdata/what-is-big-data.html. Accessed 18 Jan 2016.

Ims5. Nightingale's coxcombs. 2008, May 11. http://understandinguncertainty.org/coxcombs. Accessed 5 Feb 2016.

Jones B. Communicating data with tableau: designing, developing, and delivering data visualization. Sebastopol, CA: O'Reilly Media; 2014. http://cdn.oreillystatic.com/oreilly/booksamplers/9781449372026_sampler.pdf

Knaflic CN. Storytelling with data: a data visualization guide for business professionals. Hoboken, NJ: Wiley; 2015.

Loukides M. What is data science? Sebastopol, CA: O'Reilly Media; 2012.

Manyika J, Ramaswamy S, Khanna S, Sazzazin, H, Pinkus, G, Sethupathy G, Yaffe A. Digital America: a tale of the haves and have-mores. 2015, Dec. http://www.mckinsey.com/industries/high-tech/our-insights/digital-america-a-tale-of-the-haves-and-have-mores. Accessed 15 Apr 2016.

Mark JJ. Cuneiform. In: Ancient history encyclopedia. 2011. http://www.ancient.eu/cuneiform. Accessed 10 Jan 2016.

Martin KE. Ethical issues in the big data industry. MIS Q Exec. 2015;14(2):67–85.

Mayer-Schonberger V, Cukier K. Big data: a revolution that will transform how we live, work, and think. New York: Hought Mifflin Harcourt Publishing; 2013.

Mearian L. By 2020, There will be 5,200 GB of data for every person on Earth. Computer World. 2012, Dec 11. http://www.computerworld.com/article/2493701/data-center/by-2020--there-will-be-5-200-gb-of-data-for-every-person-on-earth.html. Accessed 20 Feb 2016.

Miner D. Hadoop: what you need to know. Sebastopol, CA: O'Reilly Media; 2016.

Moore GE. Moore's Law. 2016. http://www.mooreslaw.org. Accessed 18 Apr 2016.

National Archives: Domesday book. 2016. http://www.nationalarchives.gov.uk/museum/item.asp?item_id=1. Accessed 10 Jan 2016.

O'Neil C, Schutt R. Doing data science: straight talk from the frontline. Sebastopol, CA: O'Reilly Media; 2014.

Patil DJ, Mason H. Data driven: creating a data culture. Sebastopol, CA: O'Reilly Media; 2015. http://datasciencereport.com/2015/07/31/free-ebook-data-driven-creating-a-data-culture-by-chief-data-scientists-dj-patil-hilary-mason/#.Vp6x33n2bL8

Patil DJ. Building data science teams: the skills, tools, and perspectives behind great data science groups. Sebastopol, CA: O'Reilly Media; 2011.

Pentland AS. Big data's biggest obstacles. Harvard Business Review Insight Center Report. The promise and challenge of big data supplement. 2012, Oct 2. p. 17–8.

Provost F, Fawcett T. Data science for business: what you need to know about data mining and data-analytic thinking. Sebastopol, CA: O'Reilly Media; 2013.

Rattenbury T, Hellerstein JM, Heer J, Kandel S. Data wrangling: techniques and concepts for agile analysts. Sebastopol, CA: O'Reilly Media; 2015.

ResearchGate. About us. 2016. https://www.researchgate.net/about. Accessed 30 Jan 2016.

Rosling H. Wealth and health of nations. 2008. http://www.gapminder.org/world. Accessed 25 Mar 2016.

Sandberg M. DataViz history: Charles Minard's flow map of Napoleon's Russian campaign of 1812. 2013, May 26. https://datavizblog.com/2013/05/26/dataviz-history-charles-minards-flow-map-of-napoleons-russian-campaign-of-1812-part-5. Accessed 25 Mar 2016.

Soubra D. The 3 Vs that define big data. 2012, Jul 5. http://www.datasciencecentral.com/forum/topics/the-3vs-that-define-big-data. Accessed 10 Jan 2016.

Stigler SM. The history of statistics: The measurement of uncertainty before 1900. Cambridge, MA: Belknap Press of Harvard University Press; 1990.

Tableau. The 5 most influential data visualizations of all time. 2016. http://www.tableau.com/top-5-most-influential-data-visualizations. Accessed 15 Jan 2016.

Trifacta. Six core data wrangling activities. 2015. https://www.trifacta.com/wp-content/uploads/2015/11/six-core-data-wrangling-activities-ebook.pdf. Accessed 10 Jan 2016.

Tufte ER. The visual display of quantitative information. Cheshire, CN: Graphcs Press; 1983.

Tufte ER. Envisioning information. Cheshire, CN: Graphics Press; 1990.

Tufte ER. Visual explanations: images and quantities, evidence and narrative. Cheshire, CN: Graphics Press; 1997.

Tukey JW. Exploratory data analysis. Boston: Addison-Wesley; 1977.

Twitter Usage Statistics. 2016. http://www.internetlivestats.com/twitter-statistics. Accessed 18 Jan 2016.

The Whitehouse. Draft consumer privacy bill of rights act. 2015. https://www.whitehouse.gov/sites/default/files/omb/legislative/letters/cpbr-act-of-2015-discussion-draft.pdf. Accessed 30 Jan 2016.

Yau N. Google's chief economist Hal Varian on statistics and data. Jan 2009. https://flowingdata.com/2009/02/25/googles-chief-economist-hal-varian-on-statistics-and-data. Accessed 5 Jan 2016.

Yau N. Data points: visualization that means something. Indianapolis, IN: Wiley; 2013.

YouTube. Statistics. 2016. https://www.youtube.com/yt/press/statistics.html. Accessed 18 Jan 2016.

Yuhanna N. The Forrester wave: in-memory database platforms, Q3. 2015, Aug 3. http://go.sap.com/docs/download/2015/08/4481ad9e-3a7c-0010-82c7-eda71af511fa.pdf

Part II
Technologies and Science of Big Data

Thomas R. Clancy

The term "big data" is often defined as data that exceeds computer processing capacity using conventional methods. Either the data is too big, in too many forms, moves too fast or does not fit the structure of the existing database system. As a result, how data is collected, stored, processed and analyzed is rapidly changing. For example, small sensors embedded in your phone, car, home and even your clothing can continuously collect data about you. Massive amounts of data from social network sites, digital images and audio signals are being stored in the cloud and through distributed file systems such as Hadoop. These large-scale data stores are managed efficiently through new software applications such as MapReduce and parallel computer processing. And the approaches used by analysts to mine and analyze data include advanced computational methods with names such as association rule learning, classification tree analysis, genetic algorithms, machine learning, regression analysis, and sentiment and social network analysis. Collectively these new methods and technologies allow analysts to process big data in ways that, in the past, were prohibitively expensive and time consuming.In many ways the new technologies and methods just described allow Big Data to speak to us. Through patterns or "knowledge value" revealed in the data, we can describe and predict important events before they happen. Examples might include calculating the risk of hospital readmissions, patient falls or pressure ulcers. Knowledge value can be represented as algorithms coded in software and then embedded in patient care technology and information systems. It is the ongoing creation of knowledge value from big data that will change how and where nurses provide care in the future. Through learning systems integrated into electronic health records and mobile technology, the next generation of clinical decision support systems will enable providers to safely shift more care from hospitals to outpatient clinic and home environments. The chapters and case studies in this part first discuss the new technologies and methods used by data scientists today, which allow the creation of knowledge value from large-scale data sets. The authors then go on to describe how knowledge value, embedded in electronic health records, mobile devices and other forms of technology are changing how and where nurses provide care through advanced clinical decision support. In the final chapter, we provide one illustration of how knowledge value is created and translated for clinical use, through the OptumLabs Research Collaborative, a unique academic and corporate partnership.

Chapter 4
A Closer Look at Enabling Technologies and Knowledge Value

Thomas R. Clancy

"We always overestimate the change that will occur in the next two years and underestimate the change that will occur in the next ten."

Bill Gates in "The Road Ahead"

Abstract Big data science has the capacity to literally transform nursing roles in the near future. By analyzing large-scale data sets, nurse data scientist can discover knowledge value, or patterns hidden in the data that can improve the health and wellbeing of patients. Such patterns can then be programmed as algorithms and embedded in software located in computers, smartphones and other mobile devices. These devices may include electronic stethoscopes, hand held ultrasound units, point of care lab testing, sleep and activity monitors and a host of other devices. New software applications can classify disease conditions, identify best practices and predict events such as sepsis, pressure ulcers and falls. Sophisticated clinical decision support applications will augment nurses' cognition and provide the capacity for an expanded scope of practice. Thus the traditional hospital centric models of the past that required patients to be physically located where advanced diagnostic and monitoring equipment was available are diminishing. And as a result nurses' roles are transitioning to that of care coordinators, health information brokers and consultants that provide care primarily in the home. This chapter takes a closer look at these enabling technologies and how they are changing nursing.

Keywords Knowledge value • Knowledge engineering • Machine learning • Intranet of Things • Quantified self-movement • Biosensors • Algorithms • Augmented cognition

T.R. Clancy, Ph.D., M.B.A., R.N., F.A.A.N.
School of Nursing, University of Minnesota, Minneapolis, MN, USA
e-mail: clanc027@umn.edu

4.1 Introduction

The year is 2025 and Hannah, a freelance, family nurse practitioner is preparing for her commute to work this morning. As one of hundreds of "freelancers" hired by the five remaining national US health systems, Hannah was recruited to her position, based on her clinical and administrative outcomes profile. Most advanced practice nurses (APN's) today are independent contractors and compete for work based on their ability to manage the health of populations they serve. Tracking software embedded in the health systems data management and analytics software continuously correlate clinical outcomes with the interventions Hannah prescribes. Her ability to optimize advanced clinical decision support systems (CDSS) to predict and prevent events that negatively impact the asthmatic population has demonstrated value to the system, and hence, ongoing job security.

As Hannah prepares for her day, she accesses the cloud through her home computer and reviews her dashboard for any alerts that may have been activated on her patients during the night. The alerts generated were triggered from patients on continuous monitoring systems in acute care settings, sensors in implanted and wearable technology, home monitoring devices (weight scales, fall carpets & beds, blood pressure monitors and refrigerators) smart phones and other mobile devices. Hannah quickly checks her predictive analytics engine and notes that two of her patients have a 65% or greater probability of being admitted to a hospital for an acute asthmatic episode within the next 48 h. The predictive model relies on a "mash-up" of data from multiple sources that include, clinical data from the patients longitudinal electronic health record (EHR), insurance claims data, genomic data, global positioning systems (GPS), streaming data from sensors in wearable, implantable and other mobile technology, and social networks. To develop the model, a nurse data scientist used a variety of machine learning techniques such as neural and Bayesian networks, support vector machines and genetic algorithms to refine the model for Hannah's population of asthmatics.

Hannah quickly reviews the recommendations created through her software's CDSS for those patients at the highest risk of readmission. The recommendations are personalized for each patient based upon evidenced based practices and the patients individual data profile. She notices that one patient will need a new genetically engineered drug based upon DNA sequencing performed on her when she was admitted to the program a year ago. Hannah contacts a nearby pharmacy to request the drug be synthesized using their 3D printer and sent to the patients address today. Hannah then puts on her wireless headset and contacts the remaining patients at high risk and follows up with appropriate interventions.

As Hannah contacts patients on her list, advanced CDSS mechanisms programmed into the software are continuously evaluating each asthmatics real time data profile. By augmenting her experiential knowledge with the software's "symptom checkers" and "prognosticators", Hannah conducts her remote patient assessments over the next hour. She reflects that, just five years ago, she would have had to make either a clinic or home visit for each of these patients. However, advances in sensor technology and artificial intelligence have allowed Hannah to augment her critical thinking skills and safely monitor these same patients in their homes.

Hannah notes that two of her patients appear to have prevented an asthma attack by using bidirectional sensors implanted in their lungs. The micro-sensors secrete bronchodilators when triggered by a combination of air quality, pollen count, geo-location (high smog area), breath nitric oxide and lung function. All of these sensors are integrated into the patients' smart phone and when a threshold is surpassed, the micro-sensors automatically secrete the bronchodilator and prevent an asthma attack. The complicated algorithm developed for this intervention was created through an "innovation ecosystem", a high-performance collaborative global web network that brought together talent, innovation, supply chains, markets, makers, capital, and experts to solve complex problems (Canton 2015). By using massive amounts of data through global networks, researchers were able to use big data science approaches to discover and program patterns hidden in the data. The ecosystem emerged through the integration of three companies, Quirky, an innovation platform that brings together inventors to present ideas, Kickstarter, an innovation community that uses crowdsourcing to fund projects and Elance, an online talent marketplace for marketing freelancers to potential clients (Canton 2015). Hannah, a member of this ecosystem, has leveraged the open source movement of innovation ecosystems to improve patient care and demonstrate her value to the health systems she contracts with. As she prepares to make her hospital, clinic and home rounds for the day, Hannah reflects on how much life has changed for her as a nurse in just the last 10 years.

4.2   Emerging Roles and the Technology Enabling Them

Although we can't predict the future, elements of the scenario just described will likely emerge over the next 10 years. Clearly, advances in information technology are transforming models of care from historic hospital centric to patient centric; and with it, the roles of clinical nurse are also changing. Nationally, acute care admissions are flat or declining, outpatient and home visits are growing and virtual visits are exploding (Avalere Health 2013). Although the total number of nurses working in hospitals has increased, the proportion of the total RN workforce in acute care has subtly declined from 67% in 1993 to 61% in 2014 (Bureau of Labor Statistics 2015). New hospital construction, typically measured by the number of beds, is downsizing in lieu of massive ambulatory care centers. For example, the new 497,000 square-foot replacement naval hospital being built at Camp Pendleton near San Diego will house only 67 beds. However, the outpatient clinics at the new facility will see over 2000 patients per day (Robeznieks 2013).

If the proportion of nurses working in hospitals is declining as a result of more care being provided on an outpatient basis or in the home, where are the remaining nurses going? Many are moving into emerging new roles enabled by advances in information technology and big data science. Some of these new roles and the technologies that support them are presented in Table 4.1:

Table 4.1 Emerging roles and the technology that enables them

Emerging role	Technology enabler	Examples	Function
Care coordination	Implanted technology	Insulin pumps, automatic defibrillators, pacemakers	Complex physiological monitoring and medication administration
	Wearable Technology	Activity, sleep, diet monitors (fitbit®, Nike® Fuel®, Garmin®), emergency alerts (Life Alert®)	Quantified self-movement (exercise, sleep and nutrition monitoring)
	Home monitoring devices	Smart scales, refrigerator monitors, fall carpets, security systems, activity motion detectors, cameras, thermostat monitors, smoke and CO_2 alarms (Nest®)	Internet of Things to monitor patients at risk in their homes (For example: seniors)
	Population Management software	Data aggregation, risk stratification and segmentation of patients with specific disease conditions (diabetes, asthma, heart failure, and other)	Decision support to manage factors that impact a specific patient population's clinical and financial outcomes (access, compliance, lifestyle and other)
Health coach	Patient portals, personal health records and patient engagement software	Email consultation, out-of-hospital care pathways, on-line support groups, test result tracking (lab tests), medication reconciliation, just-in-time education	Motivate patient engagement through on-line communication, reminders, information sharing, health education and incentive programs
	Social media	On-line support groups for specific disease conditions from Alzheimer's to weight loss	Provide emotional support, self and family disease management, and education through on-line communities
Telehealth facilitator	Telehealth equipment, software and networks	Virtual office visits (Virtuwell®, Zipnosis®), Virtual ICUs, remote telehealth assessments	Provide remote assessment and monitoring of patients using telehealth equipment
Nurse data scientist	Large-scale data repositories, parallel computer processing, distributed databases	Knowledge engineering through statistical modeling, data munging and ingestion, natural language processing (NLP), machine learning, visualization and other big data science approaches	Create knowledge value from large-scale data repositories that can be translated into improved care and treatment of patients

Table 4.1 (continued)

Emerging role	Technology enabler	Examples	Function
Nurse entrepreneur	The open source movement (software, manufacturing), crowd sourcing and funding leading to the democratization of designing and manufacturing new products and services	Innovation labs, innovation ecosystems and the Maker movement. Early adopters include Amazon®, Quirky®, Kickstarter® and other companies	Improve the care and treatment of patients through the rapid creation of new products, software applications and services

There are five key technological forces driving the shift from the traditional hospital centric model of care to a patient centric one: exponential growth in computer processing speed, the shift from analog to digital signal processing, connectivity and the network effect, consumer health and cloud computing. Over the last 40 years the number of components in a computer chip has doubled approximately every 18 months (Intel 2015). First discovered by Gordon Moore in 1965, "Moore's Law" predicts that computer processing power and storage will grow at an exponential rate. This means that next year's computers will be twice as fast, have twice the storage capacity and will cost about half the price of today's computers. And this phenomenon is predicted to continue into the foreseeable future.

Improved computer processing speed coupled with the ongoing transition to digital signal processing has significantly reduced data storage costs. For example, unlike other consumer products, electronic data costs virtually nothing to reproduce once created. This combined with improved computer chip design has accelerated advances in sensor technology. It is estimated that sensors embedded in cars, homes, mobile devices and humans through implanted and wearable technology will exceed one trillion in the next five years (Burlingame and Nelson 2014). These sensors are collecting multiple forms of data that include exercise activity, physiological data, waveforms, global positioning (GPS), sleep patterns and other data through the quantified self-movement and the "Intranet of Things".

Relentless growth of the Internet has created ubiquitous connectivity through social networks and web browsers. For example the Internet now contributes to the creation of 2.5 quintillion bytes of data daily and 90% of the stored data in the world was created in the just last two years alone (Conner 2015). The network effect generated by the Internet has been enabled through the explosive growth in mobile technology, specifically the smart phone, the new nexus of communication flow. It is now estimated that approximately six billion people in the world today have access to a cell phone, more than those who have access to a bathroom (Wang 2013).

Processing this massive inflow of new data streams would be impossible using traditional database methods of storing data on local servers or personal computers. However, the development of cloud computing, or the practice of using a network of remote servers hosted on the Internet for data management, has greatly improved our capacity to process and store data. Add to this the development of Hadoop, an open source distributed data storage framework created by Yahoo!, and MapReduce, an algorithm to query the data developed by Google, and the ability to process big data has improved exponentially (Health Catalyst 2015).

These five forces, through the technologies, described in Table 4.1, are creating new sources of data that can be used for secondary research and the creation of knowledge value. In the digital world, knowledge value is represented as patterns or correlations in the data discovered using advanced computational approaches such as data mining and machine learning. Data scientists then investigate these patterns for causation and determine if they represent new knowledge and value. These patterns can then be represented by algorithms and programmed in computer code to support clinicians in classifying disease conditions, discovering best clinical practices, predicting adverse events and a host of other things.

The approaches and methods used by data scientists to discover patterns and develop predictive models have been in existence for many years. However, the accuracy of such models was often limited by the volume of usable data that could be processed. As these barriers have been overcome in recent years, the knowledge gained from patterns discovered in the data is being embedded in software located in medical devices, electronic health records, and smart phones. For example, smart phone applications now include symptom checkers and diagnosis prognosticators, point of service lab tests, and ultrasound imaging. Electronic health records generate clinical decision support alerts that predict the probability of an acute event (such as septic shock) within the next 48 h. Traditional medical devices such as stethoscopes amplify and evaluate heart sounds and classify them by causative factors. Sensors placed in "smart homes" investigate gait patterns that may be precursors to a patient fall. In all of these examples, big data science has played a key role in discovering patterns hidden in the data outputs and then creating knowledge value that can be translated to practice.

The traditional hospital centric models of the past required patients to be physically located where advanced diagnostic and monitoring equipment was available. However, advances in pharmaco-therapeutics, evidence based practices, and technology over the last 30 years has shifted much of what was once considered acute care, to outpatient clinics and home health care. And with that shift, nursing roles also expanded. The ranks of outpatient clinic and home health care nurses grew while advanced practice nurses expanded their roles as nurse practitioners, clinical nurse specialists, nurse midwives and nurse anesthetists. With just the highest acuity patients remaining in hospitals today, we again see a transition, but this time it is from outpatient clinics to the home. And it is big data science that is driving it.

4.3   A Closer Look at Technology

It is hard to imagine, but hospitals of the future might predominantly be in patients' homes. We are in the early stages of creating technology that can extend the expertise of healthcare providers (nurses, physicians, pharmacists and others) safely outside of hospitals and clinics. Not only will we be able to assess and monitor patients remotely, we also will have the capacity to conduct sophisticated mobile lab, radiology and other forms of testing. Although home health care services have been in existence for many years, the advanced functionality of new mobile devices and telehealth monitoring tools will reshape how those services are provided. It is big data science that is discovering knowledge value in large-scale clinical data repositories and then embedding this knowledge through software in mobile devices. Let's take a closer look at some of these new technologies.

A core competency of clinical nurse training is the auscultation of heart sounds using a standard acoustic stethoscope. The ability to consistently recognize abnormal heart sounds generally requires years of practice and, in many cases, is more an art than an exact science. A disadvantage of the acoustic stethoscope is that heart sounds cannot be heard if the sound level is too low. Electronic stethoscopes, which look similar to acoustic ones, eliminate this problem by using sensors in the diaphragm to capture heart sounds and convert them into electrical signals, which can then be amplified. Studies have demonstrated that electronic stethoscopes capture heart sounds better than acoustic stethoscopes (Tourtier et al. 2011; Leng et al. 2015).

Electronic stethoscopes also have the capacity to classify abnormal heart sounds with aid from a computer or smartphone. To do so, heart sound audio waves are acquired through the stethoscopes diaphragm, then filtered and converted from an analog to a digital signal. The digital signal is then processed through an algorithm and programed in a microchip located in the stethoscope or to an attached smartphone. The algorithm was created using machine learning approaches such as neural networks and support vector machines, from thousands of data points on digitized normal and abnormal heart sounds. Studies have shown the mean (sensitivities, specificities) for diagnosing aortic regurgitation, aortic stenosis, mitral regurgitation and mitral stenosis are (89.8, 98.0%), (88.4, 98.3%), (91.0, 97.52%) and (92.2, 99.29%) (Shuang et al. 2014), respectively. The study suggests that electronic stethoscopes are potentially useful for medical application, even though they are still in the early stages of development. While the majority of electronic stethoscopes used for classification of heart sounds on the market today simply amplify, record and transfer digital signals to a computer or smartphone for analysis, they still represent a tremendous step forward toward finding an acoustic-based diagnostic tool for medical clinics as well as in bedside use (Shuang et al. 2014).

The use of electronic stethoscopes has tremendous implications for clinical nurses and how they train and practice. Will acoustic-based stethoscopes be replaced by electronic stethoscopes in nursing schools and in clinical practice settings? Will documentation of heart sounds be wirelessly transferred to electronic health records as part of the nurse's ongoing assessments? Will nurses be required to validate the classification of electronic heart sounds by acoustic based stethoscopes in their clinical documentation? These are just a few of the many questions nurses will need to answer in the next decade as their practice is transformed by big data science.

4.3.1 Handheld Ultrasound

In a 2011 Wall Street Journal article (Simon 2011), a comment made by Eric Topel, a cardiologist and author of the book, "The Patient Will See You Now", touched off a debate among providers that is still reverberating today (Topel 2015). Referring to the handheld cardiac ultrasound device he uses during his cardiac assessments, Topel remarked, "Why would I listen to the "lub dub", when I can see everything? The primary reasons why are the cost of the devices (approximately $8000) and the diagnostic quality of the images using today's technology (Barclay 2014). However, that is expected to change in the coming years.

Like the electronic stethoscope, handheld portable ultrasound devices will someday, likely transform how and where patients receive their care and treatment from providers. Yet, the use of handheld ultrasound devices won't necessarily be restricted to radiology technicians and physicians. Handheld ultrasound devices may become standard training for nurses and other healthcare providers. Big data scientists will eventually develop algorithms that in conjunction with computers and smartphones will continually improve their capacity to accurately classify abnormalities in the heart as well as other organs. And if one considers how long it takes to master heart sounds with an acoustic stethoscope, the learning curve for ultrasound training may be equivalent.

The debate on whether machines or humans can more accurately identify certain abnormalities is being tested daily. For example, a systematic review of the literature regarding the accuracy of computer interpreted electrocardiograms (ECG) found a sensitivity of 76% (95% CI, 54–89%) and a specificity of 88% (95% CI, 67–96%) for the correct interpretation of acute cardiac ischemia by a computer algorithm. For acute myocardial infarction the sensitivity was 68% (95% CI, 59–76%) and the specificity 97% (95% CI, 89–92%) (Ioannidis et al. 2001). The American College of Cardiology still recommends that all ECG's be over-read by a cardiologist. What is important to note, though, is that handheld ultrasound devices, like computer interpreted ECG, may eventually become a standard assessment tool for all providers, including nurses. At the least it will be regarded as a logical next step to enhance electronic stethoscopes, and may one day replace them.

4.3.2 Point of Care Lab Testing

Although point of care (POC) lab testing is not a new technology, in home use has been hampered by a number of factors including the need for phlebotomy, blood sample size, the number of POC tests available, pre and post-test processing, insurance reimbursement and cost. However, advances in microfluidics, a multidisciplinary field including engineering, physics, chemistry, biochemistry, nanotechnology, and biotechnology, should make self-administered, low cost, in home lab testing a reality in the coming years. Specifically, the combination of low volume blood samples processed with opto-electronic image sensors located in smart phones has the capacity to increase the accuracy and number of POC testing by an order of magnitude in the decades ahead. For example, the mega-pixel count on cell phones has been following Moore's Law and doubling approximately every 18 months. This has resulted in the smartphone becoming a general purpose microscope that now has the capability of identifying a single virus on a chip.

Advances in smartphone microscopy coupled with cloud computing should advance the availability of POC testing to include, detection of viruses, diagnosis of infectious diseases, sensing of allergens, detection of protein binding events, tumor analysis and other tests. It is estimated that because of improved access and convenience, the global POC market will grow from $17 billion in 2014 to $27 billion by 2018 (News Medical 2015). The confluence of a rapidly expanding global market for POC testing and the growing network of connected smartphones will create large-scale data warehouses of lab results. Using data science approaches, data scientists will be able to develop algorithms that quickly classify different pathologies and dynamically track the evolution of infectious outbreaks and epidemics much faster.

The implication for nurses as a result of the increased use of POC testing in the home is significant. Aside from patients no longer being required to have blood drawn in an outpatient clinic, patients can better self-manage complex disease conditions through self-administered tests. As a result, nurses' roles will transition to remote patient consultants, coaches and care coordinators. Nurse caseloads may not necessarily increase, however the overall average acuity of patients may rise as these types of patients can be more easily monitored in the home. Point of care testing of blood samples will occur in smartphones or through their connection to the cloud. Test results will be wirelessly downloaded to a patient's longitudinal electronic health record and personal health record where they can be self-monitored. This is just one example of the growing "Intranet of Things" where information is exchanged machine to machine and machine to human.

4.3.3 The Quantified Self Movement

In recent years the "quantified self-movement" has emerged as a result of improvements in sensor technology. These sensors can be located in smart phones, wristbands, and clothing as well as implanted in patients. Common sensors include accelerometers, which can detect motion and speed, ambient light sensors, and global positioning (location) to name a few. The number of health and fitness applications developed in the last five years has been enormous and include self-monitoring for vital signs, activity, diet, stress, weight, medication compliance and as well as many others. Most of these applications can send wireless, streaming data to big data repositories stored in the cloud. As a result of this, scientists utilizing data science approaches are investigating patterns that in combination with smartphones are providing both clinicians and patients insights that can improve both health and wellbeing.

4.3.4 Sleep Monitors

Sleep insufficiency has been shown to have detrimental effects on health and well-being, including physical and mental distress, activity limitations, depressive symptoms, anxiety and pain. Sleep deprived individuals are more likely to smoke, drink heavily, be physically inactive and obese (Strine and Chapman 2005). In addition conditions such as chronic sleep apnea may increase the risk of high blood pressure, heart attack, arrhythmia, stroke, obesity, and diabetes. It may also lead to or worsen heart failure (National Heart, Lung and Blood Institute 2012). The role of sleep and its impact on health has taken on increasing importance in recent years as the technology for sleep monitors in wearable technology have emerged. Most of the new, mobile sleep monitors use motion detector sensors called accelerometers to detect body movement during sleep. These types of monitors act as a surrogate for sophisticated sleep study equipment, which monitors brain wave tracings, muscle-tone evaluation and eye-movement analysis, along with a live audio/video. It is assumed that if the motion sensors in mobile sleep monitors detect movement then there is less sleep, while periods of no movement indicate sleep. This activity can be represented as waveforms on a time series graph.

The various stages of sleep such as light, deep and REM (rapid eye movement) can be classified through pattern recognition software and learning algorithms of the waveforms produced by the motion detector sensors. Although not as accurate as sophisticated sleep study equipment, many individuals claim the newer mobile units provide sufficient information to monitor the quantity and quality of their sleep (Winter 2014). The future implications of sleep monitoring for nurses are important. A core competency of clinical nurses is the assessment and ongoing monitoring of key physiological factors such as heart rate, blood pressure, respirations, temperature and pain. Because of its impact on health and the convenient access afforded by sensor technology, the quality and quantity of sleep may become a new vital sign.

4.3.5 Activity Monitors

Of the various wearable technologies on the market today, activity monitors are likely the most popular (Ferguson et al. 2015). Activity monitors use accelerometers and gyro sensors (detect changes in physical orientation) to measure the quality and quantity of physical activity. Software embedded in the device, smartphone or cloud, then calculates total energy output (often measured in calories expended). Activity monitors can be located in wristbands, smart watches, strapped to a belt and attached or embedded in clothing. Used by consumers for many years, activity monitors simply counted steps and then estimated the mileage walked: and from the mileage walked, charts approximated the calories burned and potential pounds lost. The upsurge in popularity of activity monitors can be, in part, attributed to the new microelectromechanical (MEM) accelerometers and gyros that provide more accurate estimates of physical activity (Ferguson et al. 2015). These sensors convert the analog signals of motion, velocity, and orientation and convert them to digital signals that can be processed using data science approaches. Rather than simply converting steps and estimating distance walked, machine learning techniques such as neural networks, Naïve Bayes classifiers, Markov chains, and k-means clustering, can classify the intensity of exercise (light to vigorous) and the physical orientation (standing, sitting, running) of the wearer (Mannini and Maria Sabatini 2010). However, even more intriguing is the potential to recognize patterns and correlations when data regarding diet, medications, sleep, work schedule, stress, and other elements are "mashed–up".

4.3.6   Data Mash-Ups

Mash-ups combine data collected from multiple sources and allow data scientists to discover new patterns that may lead to the creation of knowledge value, or meaningful correlations that can be represented as algorithms. These algorithms can then be codified in software programs located in computers and smartphones and made available to both healthcare providers and patients for self-management. For example, nurses could fine-tune weight loss programs for diabetics by understanding how work schedule, diet, timing of medications, and sleep impact the quality and quantity of exercise for individual patients. One company, Stonecrysus®, uses machine learning algorithms programmed with its fitness tracker to combine activity monitoring and graphical food intake to pinpoint exactly how a 20 minute jog or a chocolate bar affects fitness, metabolism and weight (Charara 2015). Rather than relying on the total steps walked per day, Mio Global's® fitness tracker measures your "personal intelligence activity score" which benchmarks heart rate patterns and pulse rate against national studies of fitness. This allows individuals to focus on a combination of activities that raise their heart rate to a level recommended by the American Heart Association (Fowler 2016) to prevent heart disease. These are two small illustrations of the emerging concept of "precision (personalized) medicine" where interventions and/or products are tailored to individual patients based on their predicted response or risk of disease (US Food and Drug Administration 2015).

4.3.7   Symptom Checkers

Like the quantified self-movement, the growth of on-line symptom checkers provides another opportunity for patient engagement through self-diagnosis. There are numerous symptom checkers on the market today and include both computer and smartphone applications. Symptom checkers use data science approaches such as branching logic, Bayesian inference and other methods to classify the likelihood of certain disease conditions based upon a user's symptoms. Most applications also provide a triage function which recommends whether or not the user should seek care and, if so where and with what urgency. Although symptom checkers are used extensively today by both patients and providers, their accuracy has come into question. A recent study (Semigran et al. 2015) demonstrated that of 23 symptom checkers evaluated, only 34% provided the correct first diagnosis and the appropriate triage advice in 57% of standardized patient evaluations.

Symptom checkers are an emerging technology and although they can only be used as guidelines today, there is no doubt that these applications will become much more accurate in the years to come. By incorporating data mash-ups (including genomics information) and data science methods, healthcare providers and patients will use symptom checkers as routine clinical decision support. In fact all of the technologies and applications previously described will form the basis for "augmented cognition".

4.3.8 Augmented Cognition

Augmented cognition represents a collaborative process between machines and the human mind that in combination creates new knowledge better than if each acted alone. This partnership is already developing as both providers and patients rely more and more on information from computers to aid them in healthcare decisions. For example, nurse practitioners utilize a variety of software applications today to aid them in diagnosing disease conditions during routine office visits (Isabel 2016). Sophisticated algorithms fed by streaming physiological data from electronic health records in acute care settings are alerting clinical nurses of the probability that a patient will progress to septic shock within the next 24 hours (Institute for Health Improvement 2015). Data mining and machine learning are identifying optimal treatment patterns and making recommendations for care plans (Ayadsi, Inc. 2016). These forms of clinical decision support will continue to improve and augment nurse cognition as data scientists and engineers discover knowledge value hidden in the data.

4.4 Big Data Science and the Evolving Role of Nurses

It is beyond the scope of this chapter to discuss all of the new diagnostic tools, monitoring devices and mobile applications that are emerging from the integration of data science and information technology. What is important to note, though, is how data science and big data will change nurses' roles in the future. Clearly, the shift toward providers safely monitoring complex patients in their homes is underway. Improvements in technology typically follow an "S" curve which, at its beginning advances slowly. But at a certain inflection point, growth increases exponentially. We currently are on the slow growth portion of the curve. But at some point in the near future, improvements and adoption of technology in the home will accelerate (Fig. 4.1).

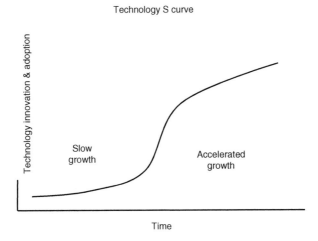

Fig. 4.1 Technology S curve (Rogers 2007)

If big data science and technology innovation continue at their current pace more care will occur in the home, patient engagement in self-care will increase, and nurses will act, in part, as healthcare information brokers. In this sense, nurses will search, gather and integrate information on behalf of patients. Much of the direct caregiving will likely be provided by nursing aides, family members, community volunteers and patients themselves but under the supervision of clinical nurses. This does not necessarily diminish the role of nurses but continues to advance them as knowledge workers.

As information brokers and consultants, nurses will need to have a deep understanding of telehealth applications used to monitor health status and patient safety in the home. This includes technologies previously described using wearable and implanted technology as well as patient safety sensors such as video, audio and motion detectors, thermostats, gait evaluation carpets, and security systems. Already technology companies, nurses and family caregivers are working collaboratively to keep patients with chronic diseases in the home for their care. For example, Joe Gaugler, PhD, a professor at the University of Minnesota School of Nursing, is evaluating the impact of a five-year, $1.25 million study to determine whether remote monitoring systems might help those with Alzheimer's disease and memory loss stay in their homes longer, reduce stress on family caregivers and potentially avoid costly emergency room visits or hospitalizations by spotting problems earlier (Crosby 2015). A key partner in the study is Healthsense®, an information technology company whose home monitoring system is being used by 23,000 people in 33 states. Through monitors placed on chairs, beds, refrigerators, walls, carpets and other areas, nurses and family members are tracking the daily habits of in-home residents with Alzheimer's disease and dementia. Data analytics tools in the software allow nurses and family caregivers to investigate patterns that, if caught early, may prevent problems later. These might include a subtle change in how often a refrigerator door is opening or how long a resident stays in a chair or bed; patterns that may indicate a developing problem with diet or mobility.

The concept of augmented cognition also has tremendous implications for nursing. If at some point in the future, clinical decision support applications are deemed equivalent or better than provider judgment, how will this impact clinical and advanced practice nurses' scope of practice? IBM's Watson, the fasted computer in the world, is already capable of storing more medical information than providers, and unlike humans, its decisions are accurate, evidence-based, consistent, and can be offered anywhere in the world (Friedman 2014). Not only that, its learning algorithms developed by data scientists can generate and evaluate hypothesis and recommend treatment. Watson is currently being used by the MD Anderson Cancer Center to recommend treatment options for leukemia patients. Wellpoint, Watson's distributor, states that Watson is better at diagnosing some diseases such as lung cancer than physicians. Although Watson still cannot match human diagnosticians consistently, it is learning more daily and will likely get there in the future.

Advance practice nurses are already using clinical decision support tools such as symptom checkers to augment their cognition in patient assessment and diagnosis. For example, Isabel, The Symptom Checker®, searches through 6000 disease conditions using sophisticated big data algorithms to rank order the probability of a specific diagnosis, based on the patient's symptoms. And although symptom checkers and diagnosis checklist systems such as Isabel are an emerging technology, they eventually will become a standard tool for clinical and advanced practice nurses.

4.5   Conclusion

The five forces previously described as driving the growth and adoption of big data science are also transitioning the future roles of nurses. Smartphones, electronic stethoscopes, hand held ultrasound devices, activity and sleep monitors, point-of-care lab testing, in-home sensors, symptom checkers, and diagnosis check lists are all tools used for augmenting our cognition. These devices classify complex patterns in big data and help providers understand and, in some cases, prevent healthcare events in patients' lives. Thus as these tools continue to improve and become accepted by practitioners, nurses' scope of practice will likely expand beyond today's standards. With so many tools at their disposal, clinical nurse roles will continue to advance as healthcare information brokers. The scenario described at the beginning of this chapter may not be as far-fetched as it seems. As more complex patients are cared for in the home, nurses will need to utilize all of the healthcare technology at their disposal to safely care for those patients.

The challenge in the future, as it has been in the past, is how to maintain the essence of caring amidst the ever-growing use of technology. A recent study (Lindquist et al. 2011) of research nurses' work using home monitoring technologies for lung transplantation patients found that nurses averaged 305 tasks/week related to 45 patients' pulmonary function data. The variation in time spent on individual tasks ranged from seconds up to 81 min (research meetings). Of note was the amount of time nurses spent on computer tasks (118.5 tasks/week). Rather than accessing data from a convenient, single source, much of the nurse's time was spent searching for data from multiple electronic files representative of the fragmented health system we live in today. Nurses were only able to spend an average of 82 min/week with the 45 patients. Thus, although the convergence of data science and new technology has created enormous opportunities for nurses, issues related to workflow, protocols, and policy will moderate those effects if not also considered. However, these barriers will likely be overcome as noted by Bill Gates in *The Road Ahead* (Gates et al. 1995), "We always overestimate the change that will occur in the next two years and underestimate the change that will occur in the next ten".

References

Avalere Health analysis of American Hospital Association Annual Survey data, 2013, for community hospitals. US Census Bureau: National and State Population Estimates. 2013, Jul 1. https://www.census.gov/popest/data/national/asrh/2013/index.html. Accessed 29 Mar 2016.

Ayadsi, Inc. The journey from volume- to value-based care starts here: understanding and managing clinical variation helps you drive high-quality care at lower costs. 2016. http://www.ayasdi.com/applications/clinical-variation/. Accessed 6 Feb 2016.

Barclay L. Could handheld ultrasound replace the stethoscope? Medscape. 2014. http://www.medscape.com/viewarticle/822956#vp_3. Accessed 12 Jan 2016.

Bureau of Labor Statistics, U.S. Department of Labor Occupational outlook handbook, (2014–15 ed) Registered Nurses. http://www.bls.gov/ooh/healthcare/registered-nurses.htm. Accessed 18 Apr 2015.

Burlingame N, Nelson L. A simple introduction to data science. 2014. [Digital Audio Recording] Audible.com http://www.audible.com/search/ref=a_mn_mt_ano_tseft__galileo/178-9919335-7111136?advsearchKeywords=a+simple+introduction+to+data+science&sprefixRefmarker=nb_sb_ss_i_0_12&sprefix=a+simple+int. Accessed 2 Dec 2015.

Canton J. Future smart: managing the game-changing trends that will transform your world, vol. 2015. Boston: De Capo Press; 2015. p. 28–33.

Charara S. How machine learning will take wearable data to the next level: because someone's got to do something with it. Wearable Tech. 2015. http://www.wareable.com/wearable-tech/machine-learning-wearable-data-sensors-2015. Accessed 24 Jan 2016.

Conner M. Data on big data. 2015. http://marciaconner.com/blog/data-on-big-data/. Accessed 1 Nov 2015.

Crosby J. New technology is aimed at keeping dementia, Alzheimer's patients at home: U nursing school is studying the effectiveness of Healthsense monitors. The Star Tribune. 2015. http://www.startribune.com/new-technology-is-aimed-at-keeping-dementia-alzheimer-s-patients-at-home/362997461/. Accessed 29 Jan 2016.

Ferguson T, Rowlands A, Olds T, Maher C. The validity of consumer-level, activity monitors in healthy adults worn in free-living conditions: a cross-sectional study. Int J Behav Nutr Phys Act. 2015;12:42. doi:10.1186/s12966-015-0201-9.

Fowler G. Stop counting 10,000 steps; Check your personal activity intelligence. Wall St J. 2016. http://www.wsj.com/articles/stop-counting-10-000-steps-check-your-personal-activity-intelligence-1453313834. Accessed 27 Jan 2016.

Friedman LF. IBM's Watson supercomputer may soon be the best doctor in the world. The Business Insider. 2014. http://www.businessinsider.com/ibms-watson-may-soon-be-the-best-doctor-in-the-world-2014-4. Accessed 31 Jan 2016.

Gates W, Myhrvold N, Rinearson P. The road ahead. New York: Viking Penguin; 1995.

Health Catalyst. Hadoop in healthcare: a no-nonsense Q and A. 2015. https://www.healthcatalyst.com/Hadoop-in-healthcare. Accessed 31 Dec 2015.

Institute for Healthcare Improvement. Early warning systems: scorecards that save lives. 2015. http://www.ihi.org/resources/Pages/ImprovementStories/Early Warning Systems Scorecards That Save Lives.aspx. Accessed 6 Feb 2016.

Intel Corporation. 50 years of Moore's law. 2015. http://www.intel.com/content/www/us/en/silicon-innovations/moores-law-technology.html. Accessed 11 Nov 2015.

Ioannidis JP, Salem D, Chew PW, Lau J. Accuracy and clinical effect of out-of-hospital electrocardiography in the diagnosis of acute cardiac ischemia: a meta-analysis. Ann Emerg Med. 2001;37(5):461–70.

Isabel, The Symptom Checker. The Isabel Healthcare Mission. 2016. http://symptomchecker.isabelhealthcare.com/home/ourmission. Accessed 31 Jan 2016.

Leng S, Tan RS, Chai KTC, Wang C, Dhanjoo G, Zhong L. The electronic stethoscope. Biomed Eng Online. 2015;14:66. doi:10.1186/s12938-015-0056-y.

Lindquist R, VanWormer A, Lindgren B, MacMahon K, Robiner W, Finkelstein S. Time-motion analysis of research nurse activities in a lung transplant home monitoring study. Prog Transplant. 2011;21(3):190–9.

Mannini A, Maria Sabatini A. Machine learning methods for classifying human physical activity from on-body accelerometers. Sensors. 2010;10:1154–75. doi:10.3390/s100201154.

National Heart, Lung and Blood. What is sleep apnea? 2012. http://www.nhlbi.nih.gov/health/health-topics/topics/sleepapnea. Accessed 20 Jan 2016.

News Medical. New report on global market for point-of-care diagnostics. 2015. http://www.news-medical.net/news/20150626/New-report-on-global-market-for-point-of-care-diagnostics.aspx. Accessed 14 Jan 2016.

Robeznieks A. Shedding beds: new hospital projects are taking ambulatory care to the extreme. Modern Healthcare. 2013. http://www.modernhealthcare.com/article/20130126/MAGAZINE/301269979. Accessed 23 Dec 2015.

Rogers E. Diffusion of innovations. 2007. https://www.researchgate.net/profile/Anja_Christinck/publication/225616414_Farmers_and_researchers_How_can_collaborative_advantages_be_created_in_participatory_research_and_technology_development/links/00b4953a92931a6fae000000.pdf#page=37. Accessed 29 Jan 2016.

Semigran HL, Linder JA, Gidengil C, Mehrotra A. Evaluation of symptom checkers for self diagnosis and triage: audit study. BMJ. 2015;351:h3480. doi:10.1136/bmj.h3480.

Shuang L, Ru San Tan K, Tshun CC, Chao W, Dhanjoo G, Zhong L. The electronic stethoscope. Biomed Eng Online. 2014;14:66.

Simon S. Medicine on the move: mobile devices help improve treatment. Wall St J. 2011. http://www.wsj.com/articles/SB10001424052748703559604576174842490398186. Accessed 12 Jan 2016.

Strine TW, Chapman DP. Associations of frequent sleep insufficiency with health-related quality of life and health behaviors. Sleep Med. 2005;6(1):23–7.

Topel E. The patient will see you now: the future of medicine is in your hands. New York: Basic Books; 2015. p. 79–103.

Tourtier JP, Libert N, Clapson P, Tazarourte K, Borne M, Grasser L, et al. Auscultation in flight: comparison of conventional and electronic stethoscopes. Air Med J. 2011;30(3):158–60. doi:10.1016/j.amj.2010.11.009.

US Food and Drug Administration. Precision (personalized) medicine. 2015. http://www.fda.gov/ScienceResearch/SpecialTopics/PersonalizedMedicine/. Accessed 24 Jan 2016.

Wang Y. More people have cell phones than toilets, U.N. study shows. Time Magazine On-line News Feed. 2013. http://newsfeed.time.com/2013/03/25/more-people-have-cell-phones-than-toilets-u-n-study-shows/. Accessed 15 Nov 2015.

Winter C. Personal sleep monitors: do they work? Huffpost Health Living. 2014. http://www.huffingtonpost.com/dr-christopher-winter/sleep-tips_b_4792760.html. Accessed 20 Jan 2016.

Chapter 5
Big Data in Healthcare: New Methods of Analysis

Sarah N. Musy and Michael Simon

Abstract With the ubiquitous availability of health-related data such as insurance claims, discharge abstracts, electronic health records, personal fitness devices or mobile phone applications, the amount of health data is increasing in size, but also in speed and in complexity. "Big data" provides new opportunities for nurse clinicians and researchers to improve patient health, health services and patient safety. Following this unprecedented amount and complexity of information available from different types of data sources, the processing and the analysis of big data challenges traditional analytical methods. For these reasons, a range of analytical approaches such as text mining and machine learning often developed in bioinformatics or engineering fields become of highest relevance to nurses wanting to work with big data. This chapter provides a brief overview of the main definitions and the analytical approaches of big data. The chapter gives two nursing research examples in the context of patient experience in cancer care and older people with dementia in nursing homes. In both cases the analytical approach (text mining and machine learning) is highly integrated into traditional research designs (a cross-sectional survey and a retrospective observational study), which highlights how traditional research designs become increasingly influenced by analytical strategies from big data or data science.

Keywords Big data • Machine learning • Text mining • Predictive modeling • Data mining and knowledge discovery • Data visualization • Natural language processing

S.N. Musy, PhD (✉) • M. Simon, PhD
Institute of Nursing Science, University of Basel, Basel, Switzerland

Inselspital, Bern University Hospital, Bern, Switzerland
e-mail: sarah.musy@unibas.ch

© Springer International Publishing AG 2017
C.W. Delaney et al. (eds.), *Big Data-Enabled Nursing*, Health Informatics,
DOI 10.1007/978-3-319-53300-1_5

79

5.1   Introduction

In the last decade the term "big data" became widely used in the literature (Gandomi and Haider 2015). The term big data refers to large and complex data sets, which are difficult to analyze with traditional data processing methods (Frost and Sullivan 2015). Because technology and also analytical methods constantly develop, the threshold of what 'difficult' constitutes is moving too, making it hard to determine, which type or size of data can be described as big data. In this context often three dimensions of big data are discussed: volume, variety and velocity (Laney 2001). *Volume* refers to the size of the data. A survey by IBM of 1144 participants found that half of respondents would consider a size of more than one terabyte as big data (see Fig. 5.1) (Schroeck et al. 2012). This answer depends also on today's storage capacities, which are constantly increasing (Gandomi and Haider 2015). *Variety* refers to the structural heterogeneity of the data, which can be structured or unstructured. Only a small part of existing data (5%) is in a structured form, referring to data stored in the traditional row-columns database or spreadsheet, such as medication data (Cukier 2010). The rest is unstructured data, including text and multimedia (i.e. pictures, audio or video files) content, such as clinical notes (Gandomi and Haider 2015). Structured data is easier to work with, but using unstructured data is particularly challenging. Finally *velocity* refers to the speed of data generation and processing. Data is produced increasingly quickly, for example the micro-blogging service twitter serves about 350,000 tweets every minute. This allows analysis of data in real time or near-real time, but also requires capacity to process analyses accordingly (Shah 2015). In addition, other characteristics have emerged, such as veracity, variability or value, which are describes elsewhere (Normandeau 2012; Katal et al. 2013; Gandomi and Haider 2015).

The rapid digitization of health and health care data is leading to a dramatic growth of information on all levels of the healthcare system (Raghupathi and Raghupathi 2014; Larson 2013). In 2012 it was estimated that worldwide size of digital healthcare data reached 500 petabytes, corresponding to more than 13 years of HD video and is expected to be multiplied by 50 in 2020 (Hersh et al. 2011).

Alongside the question of what constitutes 'big data' and where big data is generated from, a range of analytical techniques, often labeled as "data science", have

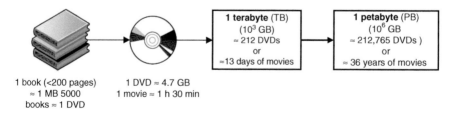

Fig. 5.1 Representation of the size of terabytes and petabytes with examples

emerged. Those techniques deal with difficulties from big data files, but also generate new opportunities with big data and data from traditional research designs. In the following sections, we will give a brief overview about the common sources of big data and typical analytical techniques and provide two examples where these techniques have been applied in nursing research. Finally, we will briefly discuss the challenges of big data and give a perspective of how data analytics from big data might influence nursing and nursing research.

5.2 Sources of Big Data

The large variety of sources and the rapid growth of healthcare data explain the interest in big data in healthcare. Figure 5.2 provides an overview of different sources. For a more detailed description, please refer to references by Weber et al. (2014) and Shah (2015). We divide data sources into two broad groups: routine data (e.g., automatically collected and readily available data) and research data (i.e., primary collected data).

Routine data is often administrative data, which is produced for financial management, such as insurance enrollment and provider claims, but might also contain medical information about the patients' health or emotional status. Research data often is not considered to be big data, but some sources like genetic data produce datasets with thousands of variables, requiring analytical approaches often used in data science.

A common source of medical data in routine data is the electronic health records (EHRs). EHRs contain patient charts (e.g., vital signs), clinical notes (from physicians and nurses), procedure reports (e.g., catheters insertion), clinical assessments

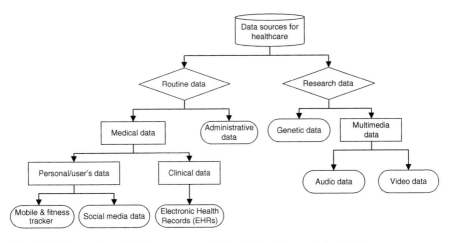

Fig. 5.2 Different types of data sources possible for the healthcare environment

(e.g., pain), care plans (e.g., nursing diagnosis), and medication information (e.g., type, dose, and time). EHRs are therefore a valuable source of information about patient demographics, diagnoses, procedures, symptoms, and medication.

Beside traditional data sources like the EHRs, new technologies coming from mobile phone applications or fitness trackers constantly collect data of users' health or activities and become an important source of medical information. Data about the heart rate, weight, height, calories intake/expenditure, distance travelled, location information through GPS, and the number of steps provide a wealth of information about healthy people and their lifestyle. Data collected in hospitals concern patients with diseases and now data from healthy people can be collected outside the hospital environment. Using this type of data is still in its infancy, but health insurance companies are becoming interested in tracker data to provide incentives for people moving more.

Another relevant source for health research is social media data. Encompassing a variety of online platforms, social media allow users to create and exchange personal or professional content and are increasingly becoming a relevant source of data for healthcare purposes (Barbier and Liu 2011; Gundecha and Liu 2012). For instance does a Facebook' friend influence you to take a certain drug? Or does the information on a blog about a drug have more impact than your physician's prescription? These are examples of questions that can be answered with data from social media. However two problems currently inhibit the use and exploitation of social media data: most social media data is not accessible to researchers because of the proprietary nature of the systems capturing this data, but also because of data protection regulations. Again this type of data is often not really used for healthcare purposes, which is not the case for marketing companies that use them to track people preference or taste in order to provide them the products of interest.

On the other side, 'big' research data is considered to be important for future developments. With lowering prices, genetic data is becoming more and more accessible. Genetic data will become a part of personalized medicine to develop and provide personalized treatments based on the patient' genome. Genomic data is in itself big data due to the large amount of variables (~3 billion base pairs in one genome).

Relatively new data sources in research are audio and video files. For instance audio analytics can support diagnosis, treatment or information about adults or children (Hirschberg et al. 2010). Patients with certain communication patterns, e.g., depression, schizophrenia, but also cancer, can benefit from audio recording support diagnosis and treatment. Another example is the analysis of infant cries, which has shown to give information about emotional status and health status (Patil 2010).

Still in its infancy compared to other data sources (Abraham and Das 2010) is real-time and pre-recorded video. Various techniques have been developed for its processing, but here big data is still a processing power challenge with one second of high-definition video being equivalent to over 2000 pages of text (Manyika et al. 2011).

5.3 Big Data Analytics

It is difficult to provide a coherent overview of the analytical techniques that are applied in the context of big data and data science because of the number of approaches, but also the disciplinary diversity ranging from informatics, engineering, statistics and others. Often these approaches are combined or overlap making it difficult to differentiate them. Hence we will introduce four areas of particular interest in the context of healthcare and nursing research: data mining, text mining, predictive modelling and machine learning.

5.3.1 Data Mining

Data mining is concerned with the detection of patterns in voluminous or complex data, where traditional methods failed to process and analyze them (Popowich 2005; Biafore 1999). Pattern recognition aims to identify potentially useful and understandable correlations in the data often to forecast or predict the likelihood of future events (Chung and Gray 1999). Predictive modeling is probably the most common application of data mining, which will be discussed in a following section. Data mining as most of the analytical tools in big data originates from database management, statistics and computer science, explaining its large panel of analytical tools.

The first step in the data mining approach is the understanding of the business at hand (Koh and Tan 2011, see Fig. 5.3). Understanding is crucial for any data mining analysis, since it aims to identify the objectives of the analysis, but also to understand how variables might be associated with each other. Understanding and preparing the data including description and visualization of the data, respectively sampling and data transformation, are important elements for any data modeling, but in particular for data mining approach. The analytical process consists of either traditional statistical methods (e.g., cluster analysis) (Copeland et al. 2009), discriminant analysis (Peterson et al. 2008), and regression analysis (Peterson et al. 2008) or non-traditional statistical methods (e.g., neural networks) (Azimi et al. 2015), decision

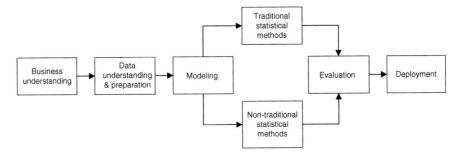

Fig. 5.3 Different steps of the data mining approach according to Koh and Tan (2011)

trees (Kang et al. 2016), link analysis (Nie et al. 2006) and machine learning (Russell et al. 1995, see Sect. 5.2.4). The evaluation stage compares models and results, and finally the deployment stage implements the data mining model.

So far data mining in health care has slowly been incorporated. But some applications do exist, such as medical insurance fraud and abuse detection. Through data mining insurers can establish norms and identify unusual claim patterns. This has led many insurers to use this information, resulting in a decrease of their losses and the costs of health care (Milley 2000).

5.3.2   Text Mining

Although great efforts have been made to underpin clinical data with structured terminologies a wide range of clinical information is still captured in unstructured format. Free-text, one possible format of data, is convenient and often used for clinical notes, procedure reports, emails, or online forums, since words contain useful information, not captured elsewhere. Text mining is a well-known method to deal with free-text. It is defined by Hearst as the process of detecting patterns and extracting knowledge from unstructured data into structured data (Hearst 1999; Popowich 2005). This transformation allows us to utilize such data for further analysis.

Figure 5.4 shows the different steps of the process. Information is retrieved to collect relevant texts, and then information is extracted (DeJong 1982). The structure of a text can be pre-processed and extracted in several ways using methods like stemming (only using the stem of a word), term document matrices (creating binary or weighted counts of words per document) or stop-word removal to make the corpora (the documents to be analyzed) accessible for further analysis (Meyer et al. 2008). Extracting information from text in a structured format is for instance useful to identify patients with certain diagnoses (Liao et al. 2015) or adverse events (Li et al. 2014) in electronic health records. The next step is a combination of semantic search (e.g., extracting sense from text and presenting it in a coherent manner) and data mining technique (e.g., by finding associations between the extracted pieces of information) (Meystre et al. 2008). Finally, in order to transform textual data into meaningful structured information, text mining is often combined with machine learning to classify text into certain categories (Gandomi and Haider 2015).

Text mining can synthesize information from many different sources and keep up-to-date with the large amounts of information. Text mining requires high levels of technical expertise and is widely used in fields like sociology, communication or bioinformatics, but so far has not gained much interest in healthcare (Raja et al. 2008).

Fig. 5.4 Different steps of the text mining approach according to Ananiadou et al. (2006)

However, some promising results for applications in healthcare research exist. For example in a single site study over a period of 6 months using the text mining approach electronic medical records from an emergency department, patients with shortness-of-breath were extracted (Cerrito and Cerrito 2006). They found that different physicians treated those complaints differently affecting care quality and costs.

5.3.3   Predictive Modelling

Predictive modelling is probably the best term representing big data analytics. Predictive modelling refers to the development of models to make accurate predictions and is often labeled as machine learning, artificial intelligence or pattern recognition depending on the discipline (Kuhn and Johnson 2013) and represents the "non-traditional" analytics part of data mining (Sect. 5.2.1). As Kuhn and Johnson (2013) describe, predictive modelling is primarily about prediction accuracy and less about the interpretation of the model. An example of this view is the identification of spam emails, where people are primarily interested in the effective trashing of spam emails as opposed to learning what features might be relevant or how an algorithm actually works. This view is based on Karl Popper's criterion for judging a theory focusing more on its predictive power and less on its ability to explain a phenomenon (Dhar 2013). This focus on prediction accuracy is probably the key ingredient driving big data analytics and dominating the analytical landscape in this area. It is difficult to provide a coherent overview of the analytical techniques that are applied in the context of big data and data science because of the number of approaches, but we will introduce an area of particular interest in the context of healthcare and nursing research: machine learning.

5.3.4   Machine Learning

Machine learning refers to analytical approaches, which allow computers to learn from data. Machine learning addresses a range of problems with supervised, unsupervised, and reinforcement machine learning being the three main types of machine learning approaches (Russell et al. 1995).

The different steps of analysis with machine learning are shown in Fig. 5.5 (Kapitanova and Son 2012). Data collection refers to the extraction of data from one database or the combining of data from multiple databases. Considerable time may be needed to preprocess the data for missing, redundant, irrelevant and outlier data. Training a model requires the split of the data into a training set and a validation set. The training set is used to train and find the right algorithms (depending on the purpose). Once an algorithm is constructed a different set is used to validate the algorithm and test its performance (evaluating the model). The last step improves the performance of the algorithm by allowing the computer to refine it, in a stepwise fashion, with new variables.

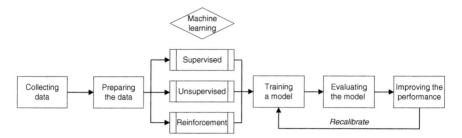

Fig. 5.5 The three types of machine learning with the associated steps of analysis according to Kononenko and Kukar (2007)

Supervised machine learning refers to analytical approaches where the computer is presented with several inputs (e.g., patient records with structured and unstructured data) and outputs (e.g., a certain patient state) in order to develop an algorithm to classify future cases according to the given characteristics. Supervised machine learning typically starts with the manual annotation or classification of a set of records (e.g., determine a certain diagnose in a patient records). The annotated records are then split into two (or more) sets: a training set and a test set. The training set is used to train the 'machine' (the computer) in order to identify records with a certain characteristic (e.g., a diagnose) through one or more algorithms. Many algorithms like support vector machines (SVM) or decision trees have been developed in order to help with the classification task. The decision which algorithm works best is determined by the complexity, sample size and noise in the training data. Often several algorithms are tested for the task at hand and the decision about which algorithm or a combination of algorithms is used can be based on the performance of the algorithm(s) in the test set.

Unsupervised machine learning refers to situations when no label, structure or classification is available and the algorithm develops its own structure to describe the data provided. The most common approaches are k-means clustering, which uses the k-mean algorithm in order to classify n observations into k clusters. For example, this approach has been used to verify the information of web-based diabetes patient education material (Thakurdesai et al. 2004). Using a list of 53 sites for the study, K-mean cluster analysis was performed to classify the web-sites into four groups based on sum of scores obtained from core educational concepts (best, medium, good, and average). The results classify 12 websites in the best category, nine in medium, 24 in good, and eight in average.

5.4 Big Data Applications in Nursing

Although big data has received much attention in the science community, the uptake in nursing seems fairly limited so far. A search with the term "big data" in the nursing core journals on Pubmed in February 2016 revealed 30 hits, with fewer than three articles showing empirical research. This is partially the case because some of the research already applying big data methods does not necessarily refer to big data or the used analytical technique is only one part of a range of methodological features. We will provide two examples of research the senior author of this chapter has been involved with, which use text mining and machine learning methodology in the context of health services research in nursing. First we will describe a study which used text mining in combination with machine learning in order to describe written patients' comments on their experiences of colorectal cancer care (Wagland et al. 2016) and, second, we will describe a study which used machine learning using a genetic algorithms to optimize covariate balance in a matched sample to explore case conferences in nursing homes (Palm et al. 2016).

Example 1 Text mining, machine learning and patient comments

Background. Surveys often contain open-ended questions providing potentially valuable and new information from survey participants. Unfortunately, data accumulated from these open-ended questions is difficult and particularly time-consuming to analyze. Traditionally this data is content analyzed in three consecutive steps, which are neither quantitative nor qualitative by nature (O'Cathain and Thomas 2004). In a first step responses are read and a coding frame is devised in order to describe the content of the comments. In a second step all comments are coded by raters, and reliability explored by double coding of a subset of comments. In the last step codes are described and the overall distribution of codes is described (O'Cathain and Thomas 2004). While this approach is sufficiently robust and reliable scaling-up to several hundreds or thousands of comments make this approach cumbersome and time-consuming task. Supervised machine learning combining text mining with machine learning allows us to train algorithms in order to detect certain types of patient comments, which then can further be analyzed.

Context of the study. The study by (Wagland et al. 2016) was challenged by the open-ended question of the national colorectal experience survey of 21,802 cancer patients in the UK, which contained 5634 responses with written comments. In a pilot study a small sample of comments was explored indicating a range of informative and relevant themes describing care experiences of colorectal cancer patients.

Methodology. In a first step a first random sample (rs1) of 400 comments was coded by three experienced qualitative researchers. The codes were developed in previous pilot study (Corner et al. 2013) and applied and adopted for this study. The framework coded comments as positive or negative experiences and whether specific forms of information to prepare patients were lacking. Cohen's Kappa between the different raters ranged from substantial (0.64) to excellent (0.87). Inconsistencies between the data and the existing framework were discussed between researchers, with disagreements jointly resolved.

In a second step another random sample (rs2) of comments was coded by the qualitative researchers. Of the overall 800 coded comments (rs1 + rs2) 50% were used to train seven different machine learning algorithms. For training the algorithms a term document matrix is created, which counts the occurrence of any used term of all included comments. In order to assess the performance of the algorithms the sample was randomly split into ten sub-samples, conducting training in nine datasets, testing in one, and repeating this process ten times (tenfold cross validation). Algorithm performance is measured as sensitivity (true positives/(true positives + false negatives)), precision (true positives/(true positives + false positives)) and by the f-score ((2 × sensitivity × precision)/(sensitivity + precision)). While sensitivity and precision are common metrics in health research the f-score describes overall performance of an algorithm, representing the harmonic mean of precision and sensitivity.

For the third step the best four algorithms to identify either positive or negative patient experiences were combined. With this approach 1688 comments were identified and finally coded by the qualitative researchers. About 81% of the comments identified by the algorithms contained positive or negative patient experiences. Figure 5.6 provides an overview about the different steps in the analytical approach of the study.

Conclusion. In summary, this study showed that combining text mining and machine learning techniques was useful and practical to identify specific free-text comments within a large dataset, facilitating resource-efficient qualitative analysis. However this is only one example requiring more experiences with other datasets in order to fully appreciate the potential and limitations of such an approach.

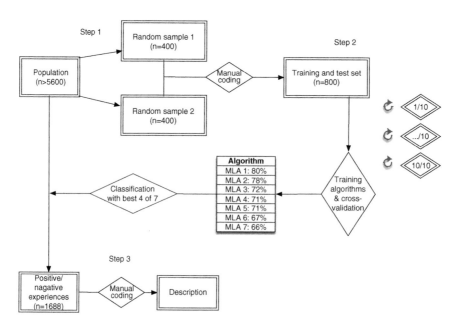

Fig. 5.6 Analytical approach of the study by Wagland et al. (2016)

Example 2 Machine learning for covariate balance using genetic algorithms

Background. Propensity score matching is a method for causal inference from observational studies (Dehejia and Wahba 2002). Causal inference in medicine is typically drawn from randomized controlled trials by assigning two groups of individuals to different treatments. The assignment can be part of regular care processes without randomization (as natural experiment) or as part of an experiment with randomization like in a randomized controlled trial to test a novel intervention. The randomization primarily serves the purpose to avoid selection bias, which occurs when individuals receiving the novel intervention are systematically different than individuals receiving usual care leading to biased estimates of the treatment effect. In a non-randomized study propensity score matching provides a mean to identify individuals in a treatment and control group sharing the same characteristics. A logistic regression model is calculated with the aim to determine the probability or 'propensity' of the participants to be in the treatment or control group given a set of observed variables. The propensity score is then used to match participants from the treatment to the control group (or vice versa). The core criterion whether the propensity score model works sufficiently well is the covariate balance between treatment and control group. For instance, whether the average age or the gender distribution in the treatment or the control group are similar, and is therefore comparable to how we would judge the randomization in a randomized controlled trial. Although conceptually the propensity score matching approach is strong, the practical application is obstructed by specifying the propensity model in a way that covariate balance is actually achieved (Diamond and Sekhon 2013).

Context. Case conferences are considered to be a key intervention to reduce challenging behavior in people with dementia in nursing homes. Dementia special care units (DSCUs) are units receiving additional funding in order to provide highly specialized care with more staff and are also expected to provide more frequent case conferences. The aim of the study was to determine whether case conferences are more frequently conducted in DSCUs than in traditional care units.

Methodology. In epidemiological terms the study is a cross-sectional observational study of 264 residents from 16 DSCUs and 48 TCUs. Information regarding case conferences was collected by the nurses using the Dementia Care Questionnaire. Several instruments were used to collect resident characteristics and challenging behavior including the Neuropsychiatric Inventory Questionnaire, the Physical Self-Maintenance Scale, Dementia Screening Scale, and sociodemographic information.

In this study propensity score matching was combined with the genetic algorithm (GenMatch). This approach solves the issue of iterative specification of the propensity score model in order to achieve covariate balance. Without the genetic algorithm, the propensity score model has to be manually specified by the researcher, which has shown to be not optimal (Austin 2009). Genetic algorithms belong to the class of evolutionary algorithms, which mimic evolution by optimizing a task (here finding covariate balance in treatment and control group) by searching for a better solution.

Conclusion. The described study provides an example of transferring a technical approach (genetic algorithm) into statistics by combining this technology with a more traditional analytical approach (propensity score matching) in order to facilitate more objective and efficient data processing and modelling. The data (264 cases and ~100 variables) would usually not considered to be big data, but the genetic algorithm used stems from the wider context of data science.

5.5 Challenges of Big Data

Big data is challenging, but some analytical approaches either developed in the field of big data, or first applied in this context now help to address some of the challenges beyond big data itself in health services and clinical research. The common denominator of the different challenges in big data analytics is the difficulties in processing or analyzing the data. The nature and the type of these difficulties can vary widely depending on the structure, the analytical aim or the contextual factors of the data. We therefore only describe some of the challenges that arise for nurse clinicians and researchers analyzing big data.

Because most big data is not collected for a specific purpose, missing data is often an issue. Therefore, approaches to "link" several data sources in order to fill gaps in the data structure is an important strategy to overcome limitations of missing data (Weber et al. 2014). For instance if standardized patient identifiers are missing, algorithms taking patient characteristics into account might allow the linkage of different data sources. Because big data is necessarily observational analyses, we are often concerned with addressing selection bias and the estimation of causal effects (Rubin 2007). Although this is not unique to big data, analytical techniques such as propensity scores and mixed models are highly relevant in this context. Finally, privacy and security concerns exist in particular in the context of big data: with more data being linked to increase the depth of information or to compensate missing data, the more difficult it becomes to de-identify individuals (Sweeney 2000; Gymrek et al. 2013). Therefore the risks for patients and the legal or ethical aspects of using big data should always carefully considered (Kohane and Altman 2005).

Despite these challenges the described examples showcase how big data analytics help to solve some of the difficulties not only in big data, but also in more traditional research. For instance genetic or textual data might contain thousands of variables, making it almost impossible to select variables manually and guided by theory, requiring new analytical approaches in order to reduce the number of variables (Shah 2015). Example 1 is such an example, the text mining generates a 1800 × 4390 matrix, which does not pose a computational problem but a conceptual—which variables of the 4390 should be used for predicting a "positive care experience"? Here supervised machine learning algorithms offer a solution, which is based on the performance of the algorithm providing a clear succinct criterion. Dealing with a different issue study 2 provides an example where the inherently difficult task of creating a sufficiently 'balanced' propensity score model is 'automatized' by the mean an evolutionary algorithm.

5.6 Conclusions

Big data coming from various sources (e.g., electronic health records, social media, audio or video data) in structured or unstructured form is growing at high speed. The processing and analysis of this mass of data is easily beyond the abilities of traditional methods. On one side, the amount of data exceeds the computing resources, and on the other side, and more importantly traditional analytical, but also conceptual means are overwhelmed by this amount of data.

Big data analytics and application in health care are still at an early stage of development and fairly slow in comparison to other industries. One reason explaining the late adoption of big data in health might be the contradiction between the population focus of big data and the practice of personalized medicine (Sacristán and Dilla 2015). On the other hand, fitness trackers being only one example are highly popular and applications using data from these devices are rapidly developing. Additionally, big data is expected to save money by preventing repeated testing and to provide better clinical decision support (Okun et al. 2013; Manyika et al. 2011; Feldman et al. 2012). Financial pressures on healthcare organizations will fuel the development and utilization of big data (Silver et al. 2001). For nursing practice, but also for nursing research big data provides great opportunities, however training and education need to keep up with this development to prepare nurses for utilizing big data for improving patient care.

References

Abraham A, Das S. Computational intelligence in power engineering. Berlin and Heidelberg: Springer-Verlag; 2010.

Ananiadou S, Kell DB, Tsujii J-I. Text mining and its potential applications in systems biology. Trends Biotechnol. 2006;24:571–9.

Austin PC. Balance diagnostics for comparing the distribution of baseline covariates between treatment groups in propensity-score matched samples. Stat Med. 2009;28:3083–107.

Azimi P, Shahzadi S, Sadeghi S. Use of artificial neural networks to predict the probability of developing new cerebral metastases after radiosurgery alone. J Neurosurg Sci; 2015.

Barbier G, Liu H. Data mining in social media. Social network data analytics. In: Aggarwal CC, editor. Social network data analytics. Boston, MA: Springer; 2011.

Biafore S. Predictive solutions bring more power to decision makers. Health Manag Technol. 1999;20:12.

Cerrito P, Cerrito JC. Data and text mining the electronic medical record to improve care and to lower costs. Proceedings of SUGI; 2006. p. 26–9.

Chung HM, Gray P. Special section: data mining. J Manag Inf Syst. 1999;16:11–6.

Copeland LA, Zeber JE, Wang CP, Parchman ML, Lawrence VA, Valenstein M, Miller AL. Patterns of primary care and mortality among patients with schizophrenia or diabetes: a cluster analysis approach to the retrospective study of healthcare utilization. BMC Health Serv Res. 2009;9:127.

Corner J, Wagland R, Glaser A, Richards SM. Qualitative analysis of patients' feedback from a PROMs survey of cancer patients in England. BMJ Open. 2013;3:e002316.

Cukier K. Data, data everywhere: a special report on managing information, Economist Newspaper. 2010. Accessed 24 Feb 2016.

Dehejia RH, Wahba S. Propensity score-matching methods for nonexperimental causal studies. Rev Econ Stat. 2002;84:151–61.

Dejong G. An overview of the FRUMP system. In: Lehnert WG, Ringle MH, editors. Strategies for natural language understanding. Hillsdale, NJ: Lawrence Erlbaum Associates; 1982.

Dhar V. Data science and prediction. Commun ACM. 2013;56:64–73.

Diamond A, Sekhon JS. Genetic matching for estimating causal effects: a general multivariate matching method for achieving balance in observational studies. Rev Econ Stat. 2013;95:932–45.

Feldman B, Martin EM, Skotnes T. Big data in healthcare hype and hope. October 2012. Dr. Bonnie 360. 2012.

Frost & Sullivan. Drowning in big data? Reducing information technology complexities and costs for healthcare organizations. White Paper. 2015.

Gandomi A, Haider M. Beyond the hype: big data concepts, methods, and analytics. Int J Inf Manag. 2015;35:137–44.

Gundecha P, Liu H. Mining social media: a brief introduction. Tutorials in Operations Research 1. 2012.

Gymrek M, Mcguire AL, Golan D, Halperin E, Erlich Y. Identifying personal genomes by surname inference. Science. 2013;339:321–4.

Hearst MA. Untangling text data mining. Proceedings of the 37th annual meeting of the Association for Computational Linguistics on Computational Linguistics, 1999. Association for Computational Linguistics; 1999. p. 3–10.

Hersh W, Jacko JA, Greenes R, Tan J, Janies D, Embi PJ, Payne PR. Health-care hit or miss? Nature. 2011;470:327–9.

Hirschberg J, Hjalmarsson A, Elhadad N. You're as sick as you sound: using computational approaches for modeling speaker state to gauge illness and recovery. In: Neustein A, editor. Advances in speech recognition: mobile environments, call centers and clinics. Boston, MA: Springer; 2010.

Kang Y, Mchugh MD, Chittams J, Bowles KH. Utilizing home healthcare electronic health records for telehomecare patients with heart failure: a decision tree approach to detect associations with rehospitalizations. Computers, Informatics, Nursing: CIN; 2016.

Kapitanova K, Son SH. Machine learning basics. In: Intelligent sensor networks: the integration of sensor networks, signal processing and machine learning, vol. 13. Boca Raton, FL: CRC Press; 2012.

Katal A, Wazid M, Goudar R. Big data: issues, challenges, tools and good practices. Contemporary Computing (IC3), 2013 Sixth International Conference on Contemporary Computing (IC3), 2013. IEEE. p. 404–9.

Koh HC, Tan G. Data mining applications in healthcare. J Healthc Inf Manag. 2011;19:65.

Kohane IS, Altman RB. Health-information altruists–a potentially critical resource. N Engl J Med. 2005;353:2074–7.

Kononenko I, Kukar M. Machine learning and data mining: introduction to principles and algorithms. Chichester: Horwood Publishing Limited; 2007.

Kuhn M, Johnson K. Applied predictive modeling. New York: Springer-Verlag; 2013.

Laney D. 3D data management: controlling data volume, velocity and variety. META Group Research Note 2001;6:70.

Larson EB. Building trust in the power of "big data" research to serve the public good. JAMA. 2013;309:2443–4.

Li Q, Melton K, Lingren T, Kirkendall ES, Hall E, Zhai H, Ni Y, Kaiser M, Stoutenborough L, Solti I. Phenotyping for patient safety: algorithm development for electronic health record based automated adverse event and medical error detection in neonatal intensive care. J Am Med Inform Assoc. 2014;21:776–84.

Liao KP, Cai T, Savova GK, Murphy SN, Karlson EW, Ananthakrishnan AN, Gainer VS, Shaw SY, Xia Z, Szolovits P, Churchill S, Kohane I. Development of phenotype algorithms using electronic medical records and incorporating natural language processing. BMJ. 2015;350:h1885.

Manyika, J., Chui, M., Brown, B., Bughin, J., Dobbs, R., Roxburgh, C. & Byers, A. H. Big data: the next frontier for innovation, competition, and productivity. 2011.

Meyer D, Hornik K, Feinerer I. Text mining infrastructure in R. J Stat Softw. 2008;25:1–54.

Meystre SM, Savova GK, Kipper-Schuler KC, Hurdle JF. Extracting information from textual documents in the electronic health record: a review of recent research. Yearb Med Inform. 2008;128-44

Milley A. Healthcare and data mining using data for clinical, customer service and financial results. Health Manag Technol. 2000;21:44–5.

Nie L, Davison BD, Qi X. Topical link analysis for web search. Proceedings of the 29th annual international ACM SIGIR conference on Research and development in information retrieval, 2006. ACM. 2006; p. 91–8.

Normandeau K. Beyond volume, variety and velocity is the issue of big data veracity. 2012. http://insidebigdata.com/2013/09/12/beyond-volume-variety-velocity-issue-big-data-veracity. Accessed 7 Jan 2016.

O'Cathain A, Thomas KJ. "Any other comments?" Open questions on questionnaires – a bane or a bonus to research? BMC Med Res Methodol. 2004;4:1–7.

Okun S, McGraw D, Stang P, Larson E, Goldman D, Kupersmith J, Filart R, Robertson RM, Grossmann C, Murray, M. Making the case for continuous learning from routinely collected data. Discussion paper. 2013. www.iom.edu/makingthecase. Accessed 31 Jan 2016.

Palm R, Trutschel D, Simon M, Bartholomeyczik S, Holle B. Differences in case conferences in dementia specific vs traditional care units in German nursing homes: results from a cross-sectional study. J Am Med Dir Assoc. 2016;17:91e9–91e13.

Patil H. "Cry baby": using spectrographic analysis to assess neonatal health status from an infant's cry. Advances in Speech Recongnition. In: Neustein A, editor. Advances in speech recognition: mobile environments, call centers and clinics. Boston, MA: Springer; 2010.

Peterson U, Demerouti E, Bergström G, Samuelsson M, Åsberg M, Nygren Å. Burnout and physical and mental health among Swedish healthcare workers. J Adv Nurs. 2008;62:84–95.

Popowich F. Using text mining and natural language processing for health care claims processing. ACM SIGKDD Explor Newsl. 2005;7:59–66.

Raghupathi W, Raghupathi V. Big data analytics in healthcare: promise and potential. Health Inf Sci Syst. 2014;2:3.

Raja U, Mitchell T, Day T, Hardin JM. Text mining in healthcare. Applications and opportunities. J Healthcare Inf Manage. 2008;22:52–6.

Rubin DB. The design versus the analysis of observational studies for causal effects: parallels with the design of randomized trials. Stat Med. 2007;26:20–36.

Russell S, Norvig P, Intelligence A. Artificial Intelligence: A modern approach. Egnlewood Cliffs: Prentice-Hall; 1995. p. 25, 27.

Sacristán JA, Dilla T. No big data without small data: learning health care systems begin and end with the individual patient. J Eval Clin Pract. 2015;21:1014–7.

Schroeck M, Shockley R, Smart J, Romero-Morales D, Tufano P. Analytics: the real-world use of big data: how innovative enterprises extract value from uncertain data. IBM Institute for

Business Value. 2012. http://www-03.ibm.com/systems/hu/resources/the_real_word_use_of_
big_data.pdf. Accessed 7 Jan 2016.

Shah NH. Using big data. Translational informatics. In: ROP P, Embi JP, editors. Translational
informatics: realizing the promise of knowledge-driven healthcare. London: Springer; 2015.

Silver M, Sakata T, Su HC, Herman C, Dolins SB, O'Shea MJ. Case study: how to apply data mining techniques in a healthcare data warehouse. J Healthc Inf Manag. 2001;15:155–64.

Sweeney L. Simple demographics often identify people uniquely. Health (San Francisco).
2000;671:1–34.

Thakurdesai PA, Kole PL, Pareek R. Evaluation of the quality and contents of diabetes mellitus
patient education on Internet. Patient Educ Couns. 2004;53:309–13.

Wagland R, Recio-Saucedo A, Simon M, Bracher M, Hunt K, Foster C, Downing A, Glaser
A,Corner J. Development and testing of a text-mining approach to analyse patients' comments
on their experiences of colorectal cancer care. BMJ Qual Saf. 2016;25:604–14.

Weber GM, Mandl KD, Kohane IS. Finding the missing link for big biomedical data. JAMA.
2014;311:2479–80.

Case Study 5.1: Value-Based Nursing Care Model Development

John M. Welton and Ellen Harper

Abstract This case study will describe the development of a national consensus model to measure patient-level nursing intensity and costs-per-patient in multiple care settings that support the continuum of care and produce objective measures of nursing value. Specifically, this case study will describe the creation of a common data dictionary that describes patient, nurse, and system-level data elements extracted from existing data sets to populate a conceptual model to measure nursing value.

Keywords Data dictionary • Data models • Value-based nursing care • Business intelligence • Nursing care quality and costs • Room and board • Diagnosis related group • Florence Nightingale

Jennifer hurried to finish her charting at the end of a busy night shift in the cardiac stepdown unit (CSU). The last item on her list was to complete her bill for each of the 4 patients she was assigned. This is a new change as her hospital recently implemented a value-based nursing care model. Billing for nursing care is a way to link individual nurses with each patient to identify the unique resources expended for each patient and use these data internally to allocate nursing time and costs. At 8:00 a.m. Jennifer attended her monthly practice council. Selected nurses from the CSU reviewed the overall patient care, adverse events, patient satisfaction and patient level nursing costs. In the past six months, the CSU was able to reduce length of stay and nursing care costs for congestive heart failure patients, their top diagnosis, by assigning a more experienced nurse on admission who often had a reduced assignment. While the first-day costs were higher than average, the nursing business intelligence analytics clearly demonstrated better outcomes by improving nursing care in the vulnerable first 24 hours after admission. Lastly, the nurses reviewed the real-time quality metrics assigned to each nurse. The CSU nursing performance metrics included medication administration delays, pain management, and glycemic control. Each nurse was rated by using time and event stamped data from the electronic health records: for example when a medication was due and the time difference for when it was administered. Jennifer smiled when she saw all her scores had improved from last month mostly from being better organized and "keeping her head in the game."

This is a fictional story. However emerging new techniques for extracting data from the electronic health record (EHR) can identify the added value of nursing care as well as the individual contribution of each nurse. Nursing care value, in its simplest

J.M. Welton, PhD, RN, FAAN (✉)
Health Systems Research, University of Colorado College of Nursing, Denver, CO, USA
e-mail: john.welton@ucdenver.edu

E. Harper, DNP, RN-BC, MBA, FAAN
School of Nursing, University of Minnesota, Minneapolis, MN, USA

form, is the relationship between quality and costs, or quality and outcomes of care (Pappas 2013; Simpson 2013). When nursing care is appropriate and optimum, adverse events such as injuries, pressure ulcers, infections, and medication errors are reduced thereby decreasing the added costs associated with morbidity and mortality (Spetz et al. 2013; Staggs and Dunton 2014; Yakusheva et al. 2014).

In the current environment, quality and outcomes of care are measured at the individual patient level and aggregated across many patients within an identifiable entity such as a hospital inpatient unit or skilled nursing facility. The actual or true costs of nursing care are not directly measured for each patient and typically averaged across many nurses and many patients (Sanford 2010). Nursing care is rolled up to daily room and board charges (Thompson and Diers 1991), which hides the added value nurses bring to the bedside. Without a direct way to measure the actual or "true" cost and resources expended by nurses for each patient, it will be impossible to measure nursing care value (Welton 2010).

5.1.1   Value-Based Nursing Care and Big Data

In attempting to arrive at the truth, I have applied everywhere for information, but in scarcely an instance have I been able to obtain hospital records fit for any purposes of comparison. If they could be obtained, they would enable us to decide many other questions besides the one alluded to. They would show subscribers how their money was being spent, what amount of good was really being done with it, or whether the money was not doing mischief rather than good; they would tell us the exact sanitary state of every hospital and of every ward in it, where to seek for causes of insalubrity and their nature; and, if wisely used, these improved statistics would tell us more of the relative value of particular operations and modes of treatment than we have any means of ascertaining at present. They would enable us, besides, to ascertain the influence of the hospital with its numerous diseased inmates, its overcrowded and possibly ill-ventilated wards, its bad site, bad drainage, impure water, and want of cleanliness—or the reverse of all these—upon the general course of operations and diseases passing through its wards; and the truth thus ascertained would enable us to save life and suffering, and to improve the treatment and management of the sick and maimed poor.

> Florence (Nightingale 1863, p. 176)

If we only had the data … Nightingale's lament may be coming close to realization. A group of nurses and other professionals began meeting in June 2013 at the University Of Minnesota School Of Nursing to address the problem of growing amounts of healthcare and nursing data (Clancy et al. 2014; Westra et al. 2015a). An action plan committed nearly 100 attendees to address a wide range of issues related to data science, informatics, and how to leverage the burgeoning amounts of information contained with the EHRs to develop new approaches, methods, and analytics that ultimately will improve patient care outcomes and decrease costs (Westra et al. 2015b).

One expert workgroup was formed to address the issue of how to measure nursing value and develop new techniques that will provide real-time metrics to monitor quality, costs, performance, effectiveness, and efficiency of nursing care (Welton

2015). During an initial one-year interaction, members of the nursing value expert workgroup identified core issues needed to explicate nursing value (Pappas and Welton 2015; Welton and Harper 2015):

1. Identify individual nurses as providers of care
2. Define nursing care as the relationship between an individual nurse and patient, family, or community
3. Link nurses directly to patients within the EHRs and measure value at each unique nurse encounter
4. Identify nurse and nursing care performance at the patient, unit and hospital or business entity unit of analysis
5. Develop patient level nursing costs based on the direct care time and other measures of resources or services provided to patients (or families, communities)

The value-based nursing care model focuses on the individual nurse rather than nursing care as the basic unit of analysis. The model is software agnostic and setting neutral. The primary analytic approach is to use events and time-stamps to link nurses and patients. For example, in the vignette, Jennifer organized her care delivered to her four patients as distinct services, interventions, assessments, etc. Each patient has unique needs and these vary across the trajectory of an illness or episode of care such as a hospitalization. The ability to discern differences in time spent with individual patients as well as the associated dollars expended across a patient population provides much greater detail and more timely and actionable information about nursing care and added value that can be used for clinical and operational decision making.

Because the value-based nursing care model is focused on individual nurses, performance can be measured for each nurse using EHR data. In the vignette, the CSU is focusing on pain management. Each nurse conducts assessments, identifies problems (e.g., acute pain), provides interventions such as administering PRN opioid medication, and reassesses a patient's response to interventions. Unit practice guidelines can be used to develop useful and objective information. For example, if the standard of care in the CSU is assessing for pain every 4 hours and follow up within 30 minutes after an intervention, extraction of nursing assessment documentation time data and pain acuity scores as well as the time and dose of PRN medications can be used to identify practice guideline adherence, patterns in using PRN opioids, and overall response to a nurse's care for patients in pain. These data can be posted or used in value-based nursing performance metrics shared within the unit.

5.1.1.1 Extracting Nursing Data from the EHR

One of the vexing problems in building new patient and nurse level analytic models is the difficulty in finding and extracting key data from the many tables in a modern EHR and developing ways to do this across multiple software platforms.

Substantial resources are needed for even simple data inquiries and reports. If hospitals or other health care settings wish to compare results and information across multiple settings, a common method and model for extracting similar data is needed.

Part of the efforts of the value-based nursing care expert workgroup is to develop a common data model that can allow multiple healthcare settings to extract similar nursing related data. The model is a roadmap for information technology and business analyst professionals to develop extraction code and pull data into a common repository for planned or ad hoc analysis. A preliminary model has been proposed by the value-based nursing care expert workgroup (Welton & Harper 2016). The common data model allows extraction and collection of complex nursing related data across many different software platforms and settings (Fig. 5.1.1). Ultimately, this common data model provides a framework for using and analyzing data about nursing care in many different settings and across many different nurses.

5.1.1.2   Nursing Business Intelligence and Analytics (NBIA)

The data model provides a template as well as the key data that can be used in complex analysis and business intelligence efforts. For example, future systems will be able to monitor performance of nurses administering medications by deriving the time between when a medication was due and when administered using bar code technology (Welton 2013). Pattern detection algorithms will be able to detect when medication administration is becoming increasingly delayed due to high workload, which may be a precursor to late medication doses. Medication administration times can be used to analyze the relationship between unit churn, patient acuity and medication administration performance. Individual nurse performance can be monitored and specific questions addressed that may indicate difficulty in meeting clinical needs due to high complexity and the amount of drugs administered (Kalisch et al. 2011, 2014; Ausserhofer et al. 2014). Focused examination of high-risk drugs such as aminoglycoside antibiotics can be used to link operational aspects of nursing care such as staffing and assignment, with short term clinical goals such as avoidance of nephrotoxicity.

5.1.2   The Cost of Nursing Care

In the value-based nursing care model, the actual services delivered as well as the associated time are allocated to each patient. This overcomes a longstanding problem of using average time to identify costs of nursing care. What is an "average" patient? The ability to link nurses to patients and apply different nursing care resources accurately to each patient provides a means to detect differences in nursing intensity and costs across an episode of care such as a hospitalization as well as compare a similar patient within a specific diagnosis located in a Diagnosis Related Group (DRG). Having the actual cost of nursing care and overhead costs such as management, benefits, and so forth, provides a way to estimate actual dollars expended for each patient and links to the billing and reimbursement system. In Fig. 5.1.1, components of the data model link patients, nurses, and charges.

This nursing value common data model provides a way to extract similar data across different EHRs and link nurses directly to patients. For example, if we were interested in examining the effects of young nurses (new graduates with less than

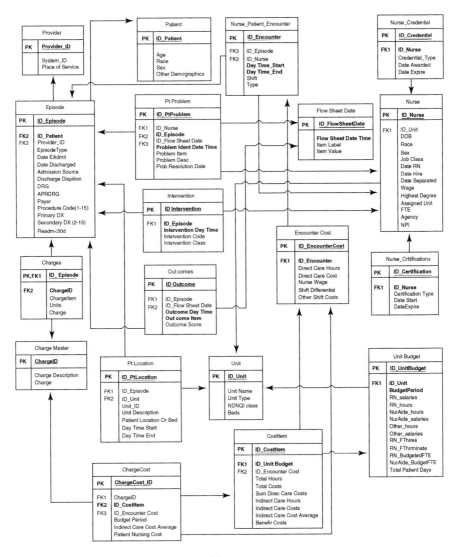

Fig. 5.1.1 Nursing value common data model

1 year experience), the DateRN field in the Nurse table would be used as the date the nurse was first licensed and then calculate the difference in time in months for the current patient that nurse was caring for and then identify the cost of care (Wage) multiplied by the hours of care given to a specific patient. Data on nursing costs and experience could be summed for an individual patient across a hospitalization, and compared to other patients based on similar or different outcomes such as length of stay, total nursing time and costs, etc.

Patient-level costing for nursing care achieves the goal of better understanding cost drivers within the health care system (Kaplan and Porter 2011). Using nursing business intelligence and analytic tools previously described, nursing and health-care finance leaders could identify nursing cost drivers and possible nursing intensity outlier patients. Using this new information, nursing care could be adjusted to achieve better matching of nurses to patients, and optimize assignments to achieve best clinical outcomes at the lowest costs of nursing care.

5.1.3   Summary

Value-based nursing care represents a new approach to using data to identify the added value nurses bring to patient care. It is a way to realize Nightingale's vision. Collecting and analyzing data at the individual nurse-patient encounter provides greater granularity for identifying clinical and operational intelligence that can inform providers in near real-time to a range of important clinical and operational information needs.

In the vignette, Jennifer was an active participant in interpreting complex data derived from the clinical assessments, interventions, and outcomes identified in the nursing documentation of the EHR. In this fictional setting, nurses are individually and collectively accountable for their care. The goal is to achieve high value outcomes by optimizing nursing care at all the "touchpoints" where nurses interact with patients.

New analytic techniques will inform clinical and operational decision making in the here and now rather than waiting weeks or months. These new data tools will decrease the time between information and action. A value driven data environment will create new information to measure the nurse-patient encounter and share and compare data across the broad spectrum of healthcare that ultimately will achieve Nightingale's vision to seek and realize excellence in everything we do.

References

Ausserhofer D, Zander B, Busse R, Schubert M, De Geest S, Rafferty AM, et al. Prevalence, patterns and predictors of nursing care left undone in European hospitals: results from the multi-country cross-sectional RN4CAST study. BMJ Qual Saf. 2014;23(2):126–35.

Clancy TR, Bowles KH, Gelinas L, Androwich I, Delaney C, Matney S, et al. A call to action: engage in big data science. Nurs Outlook. 2014;62(1):64–5.

Kalisch BJ, Tschannen D, Lee H, Friese CR. Hospital variation in missed nursing care. Am J Med Qual. 2011;26(4):291–9.

Kalisch BJ, Xie B, Dabney BW. Patient-reported missed nursing care correlated with adverse events. Am J Med Qual. 2014;29(5):415–22.

Kaplan RS, Porter ME. How to solve the cost crisis in health care. Harv Bus Rev. 2011;89(9):3–18.

Nightingale F. Notes on hospitals. Third ed. London: Longman, Green, Longman, Roberts & Green; 1863.

Pappas SH. Value, a nursing outcome. Nurs Adm Q. 2013;37(2):122–8.

Pappas SH, Welton JM. Nursing: essential to health care value. Nurse Leader. 2015;13(3):26–9. 38

Sanford KD. Nurse staffing. Finding the right number and mix. Healthc Financ Manage. 2010;64(9):38–9.

Simpson RL. Guest editorial. What is the value of nursing care? Nurs Adm Q. 2013;37(2):92–4.

Spetz J, Brown DS, Aydin C, Donaldson N. The value of reducing hospital-acquired pressure ulcer prevalence: an illustrative analysis. J Nurs Adm. 2013;43(4):235–41.

Staggs VS, Dunton N. Associations between rates of unassisted inpatient falls and levels of registered and non-registered nurse staffing. Int J Qual Health Care. 2014;26(1):87–92.

Thompson JD, Diers D. Nursing resources. In: Fetter RB, Brand DF, Gamache D, editors. DRGs their design and development. Ann Arbor: Health Administration Press; 1991. p. 121–83.

Welton JM. Value-based nursing care. J Nurs Adm. 2010;40(10):399–401.

Welton JM. Nursing and the value proposition: how information can help transform the healthcare system. Minneapolis, MN, August 12–13: University of Minnesota, School of Nursing, Center for Nursing Informatics; 2013. Accessed 1 May 2016

Welton JM. Conference report: big data in nursing 2015. Voice of Nursing Leadership. 2015;13(5):10–1.

Welton JM, Harper EM. Nursing care value-based financial models. Nurs Econ. 2015;33(1):14–9.

Welton JM, Harper EM. Measuring nursing care value. Nurs Econ. 2016;34(1):7–14.

Westra BL, Clancy TR, Sensmeier J, Warren JJ, Weaver C, Delaney CW. Nursing knowledge: big data science-implications for nurse leaders. Nurs Adm Q. 2015a;39(4):304–10.

Westra BL, Latimer GE, Matney SA, Park JI, Sensmeier J, Simpson RL, et al. A national action plan for sharable and comparable nursing data to support practice and translational research for transforming health care. J Am Med Inform Assoc. 2015b;22(3):600–7.

Yakusheva O, Lindrooth R, Weiss M. Nurse value-added and patient outcomes in acute care. Health Serv Res. 2014;49(6):1767–86.

Chapter 6
Generating the Data for Analyzing the Effects of Interprofessional Teams for Improving Triple Aim Outcomes

May Nawal Lutfiyya, Teresa Schicker, Amy Jarabek, Judith Pechacek, Barbara Brandt, and Frank Cerra

Abstract In this chapter we describe the creation of a data repository, the National Center Data Repository (NCDR) and a national network of performance sites generating that data, the National Innovation Network (NIN) for the National Center for Interprofessional Practice and Education (hereafter the National Center). We describe the raison d'être, characteristics, and ecosystem of the NIN-NCDR. The need for rigorously produced, scientifically sound evidence regarding interprofessional collaborative practice and education (IPE) and whether or not it has the capacity to positively affect the patient experience of care, the health of populations, and the per capita cost of healthcare (Triple Aim) underscored the need for the creation of the NIN-NCDR. Among other things, the NIN-NCDR is an attempt to realize a higher level of analysis of healthcare outcomes as these relate to IPE by creating the capacity to search, aggregate, and cross-reference large data sets focused on IPE related interventions. The endeavor of creating the NIN-NCDR was infused with interprofessional and cross-disciplinary energy and influences from a number of health professions. The principal creators represented medicine, education, nursing, informatics, health economics, program evaluation, public health, and epidemiology. The multiple perspectives and mix of input from many different health professions has proven to be invaluable.

Keywords NCDR • Data repository • IPE • Interprofessional practice and education

M.N. Lutfiyya, Ph.D., F.A.C.E. (✉) • T. Schicker, M.P.A. • A. Jarabek, M.S.A., M.Ed.
B. Brandt, Ph.D. • F. Cerra, M.D.
National Center for Interprofessional Practice and Education, University of Minnesota, Minneapolis, MN, USA
e-mail: nlutfiyy@umn.edu

J. Pechacek, D.N.P.
School of Nursing, University of Minnesota, Minneapolis, MN, USA

© Springer International Publishing AG 2017
C.W. Delaney et al. (eds.), *Big Data-Enabled Nursing*, Health Informatics,
DOI 10.1007/978-3-319-53300-1_6

6.1 Introduction

Even though we know we need big data to demonstrate the effects of interprofessional teams on Triple Aim outcomes, at present there is no commonly accepted definition of big data (BD) (Berwick et al. 2008; Boyd and Crawford 2012; Brandt et al. 2014; Cerra et al. 2015). For some it is a method for realizing "… a higher level of analysis of healthcare outcomes … (Berwick et al. 2008)." For others BD is "… about a capacity to search, aggregate, and cross-reference large data sets (Cerra et al. 2015)." Despite the absence of a commonly accepted definition of BD, there are some shared characteristics and contextual descriptions that frequently surface in BD scholarship (Berwick et al. 2008; Boyd and Crawford 2012; Brandt et al. 2014; Cerra et al. 2015; Cooper 2014; D'Amour and Oandasan 2005; Gordon 2013). The concepts of volume (amount), velocity (timeliness), variety (multiple data types and sources), and value (benefit) are mentioned in varying degrees of depth as the major characteristics of BD (Boyd and Crawford 2012) that are integrated by and into the BD ecosystem of a particular or specific initiative. BD ecosystems typically are comprised of infrastructure (process of collecting and storing data that can be analyzed), analytics (capacity to create information and eventually knowledge), and applications (use for created/developed knowledge) (D'Amour and Oandasan 2005; Gordon 2013; Graves and Corcoran 1989).

In this chapter we describe the creation of a data generating and data repository (the NIN-NCDR) for the National Center for Interprofessional Practice and Education (hereafter the National Center). Earlier publications from the National Center describe the research questions (Guyer and Green 2014), comparative effectiveness research strategies including the development of a national research network (Hansen et al. 2014), and the proof of concept for the informatics platform developed (Institute of Medicine 2015). Herein, we describe the raison d'être, characteristics, and ecosystem of the NCDR. Among other things, the NCDR is an attempt to realize a higher level of analysis of healthcare outcomes as they relate to interprofessional education and collaborative practice (IPE) by creating the capacity to search, aggregate, and cross-reference large datasets focused on IPE-related interventions.

The endeavor of creating the NCDR was infused with interprofessional and cross- disciplinary energy and influences from a number of health professions. The principal creators represented medicine, education, nursing, informatics, health economics, evaluation science, public health, and epidemiology. Furthermore, to assist with all aspects of the NCDR creation process an advisory committee comprised of national experts with health professions related informatics, database management, and analytics backgrounds was convened and began meeting in the very early stages of NCDR development. The multiple perspectives and mix of and input from many different health professions has proven to be invaluable, with resulting proof of concept with the demonstration of producing, managing, analyzing and reporting the data in the creation of new knowledge.

6.2 Raison D'être for the NCDR

The National Center, funded through a public–private partnership, was created from a competitive process to provide leadership, scholarship, evidence, and national visibility advancing IPE as a viable and efficient healthcare delivery model. The healthcare delivery model referenced is one grounded in the Triple Aim outcomes delineated by Berwick et al., focusing on improved patient experience of care, improved population health, and reduction in per capita cost of healthcare (Lutfiyya et al. 2016).

IPE entails the spectrum of interprofessional education and collaborative practice as these occur within a nexus. Defined as the alignment of health-professions education and collaborative practice, an IPE nexus informs change at the micro, meso and macro levels and informs and facilitates the redesign of health care delivery and education (Guyer and Green 2014; Hansen et al. 2014; Institute of Medicine 2015; Matney and Brewster 2011). Table 6.1 describes micro, meso, and macro level outcomes. This nexus is central to the work of the National Center (Fig. 6.1).

Table 6.1 Outcomes at multiple levels (micro, meso and macro)

Change level	Clinical	Education	Nexus
Micro	Provides care of patients and operates within its own environment and ecology, participants are committed to working together	Teaching environment to educate learners	Intentionally create relationship at the practice and education microsystem level to achieve Triple Aim
Meso	Senior leadership and governing structures in clinical systems, corporate offices, governing boards	University/college presidents, provosts, deans, and senior administration; governing boards and trustees; regents	Greater understanding of synergies between health system transformation and meeting higher education needs; support IPE implementation at micro level
Macro	Political, financial, accreditation and policy environment; state, regional and/or national level; increasingly complex	Political, financial, accreditation and policy environment; state, regional and/or national level; increasingly complex	Political, financial, accreditation and policy environment; state, regional and/or national level; increasingly complex

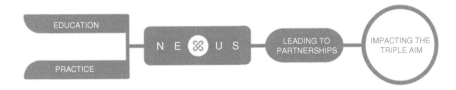

Fig. 6.1 The National Center's Nexus

In 2014, National Center staff published a scoping review of IPE literature through the lens of the Triple Aim (Merelli et al. 2014) observing that despite being an area of inquiry for more than four decades, IPE research and analysis has mostly focused on: (1) immediate or short-term changes on learner knowledge, skills and attitudes, (2) clinical practice-based processes—but not outcomes, and (3) health-related organizational-level policy changes. Very little of the literature reviewed focused on either population health or patient health outcomes, and none on healthcare cost reduction. This scoping review concluded that: "Moving forward requires asking questions about the impact of IPE in new ways, which calls for the collection and generation of data allowing examination of as yet untested causal pathways between and among the domains of IPE, [clinical] practice and healthcare delivery, health outcomes, and healthcare costs [14]." Generalizable findings were called out as an important focus going forward, with the authors arguing that rigorous research and thoughtful, careful data analysis using multiple methods were essential to producing such findings. Ultimately, the need for rigorously produced, scientifically sound evidence regarding IPE and whether or not it has the capacity to positively affect the patient experience of care, the health of populations, and the per capita cost of healthcare underscored the need for the creation of the NCDR. A recent Institute of Medicine (IOM) report (Pechacek et al. 2015) concurred with many of the findings just discussed reinforcing the need for a data repository such as the NCDR and its efficient and effective use in producing meaningful and actionable information and evidence regarding the nexus and its effects on Triple Aim outcomes.

6.2.1   Characteristics of the NCDR

To actually create the NCDR, the National Center had to develop and nurture a research network (Guyer and Green 2014; Hansen et al. 2014; Institute of Medicine 2015) that was scalable (could grow) and that would produce analyzable data that could generate useable knowledge by answering multi-level (micro, meso, and macro) questions (Guyer and Green 2014; Hansen et al. 2014; Institute of Medicine 2015; Matney and Brewster 2011) about the impact of IPE. This network—named the Nexus Innovation Network (NIN)—is comprised of multiple intervention research projects spanning the US. Early on, a decision was made to aim for the generation of large multifaceted data sets to ensure generalizability, answer established research questions, test a priori hypotheses, as well as discover possible serendipitous findings. Table 6.2 displays the research questions and associated hypotheses. BD was, in part, the informatics model (Institute of Medicine 2015). The characteristics of BD mentioned in the introduction can be used to describe crucial aspects of the NCDR.

Table 6.2 National Center's research questions related to Triple Aim outcomes and associated hypotheses (directional and non-directional)

Research questions	Hypotheses	Variables[a]
In patients and/or learner populations what impact does IPE have on patient reported satisfaction with healthcare quality?	Team-based care grounded in collaborative practice *improves* patient reported satisfaction with healthcare quality	DV = patient reported satisfaction with healthcare quality IV = team-based care grounded in collaborative practice
In patients and/or learner populations what impact does IPE have on have on reducing healthcare cost?	Team-based care grounded in collaborative practice *reduces* healthcare costs	DV = healthcare costs IV = team-based care grounded in collaborative practice
In patients and/or learner populations what impact does IPE have on improving population health?	Team-based care grounded in collaborative practice *improves* population health outcomes	DV = population health outcomes IV = team-based care grounded in collaborative practice
In patients and/or learner populations what impact does IPE have on the effect sizes of various ecological factors associated with Triple Aim outcomes?	Ecological/environmental factors essential for achieving improved health outcomes can be identified through data analyses of IPE intervention covariates	DV = improved health outcomes IV = ecological/environmental factors (IPE intervention covariates)
In patients and/or learner populations what impact does IPE have on the transformation of the process of care?	Factors essential for transformation of the process of care can be identified through data analyses of IPE intervention covariates	DV = for transformation of the process of care IV = IPE intervention covariates
In patients and/or learner populations what impact does IPE have on the transformation of health professions education?	Team-based care grounded in collaborative practice *improves* health professions education	DV = health professions education IV = team-based care grounded in collaborative practice

[a]DV = Dependent Variable(s); IV = Independent Variable(s)
Both types of variables can and will be computed from multiple and mutually exclusive variables collected by either electronic health records, standardized and validated instruments, standardized surveys, and/or project developed surveys

6.2.2   Data Volume

One aim was/is to facilitate the generation, collection, and storage of what will eventually be a large volume of analyzable data that are nationally representative of IPE interventions and combined education and clinical practice settings (nexi). Since ongoing plans are for the NIN to continually grow, data volume will accordingly continue to increase. The data volume began with data collected from eight pioneer sites in 2013 and has grown to data from 24 current fully on-boarded projects with 27 projects in earlier stages of development. Research intervention projects are currently in process in 25 States (sites). Figure 6.2 displays the NIN by State and stage.

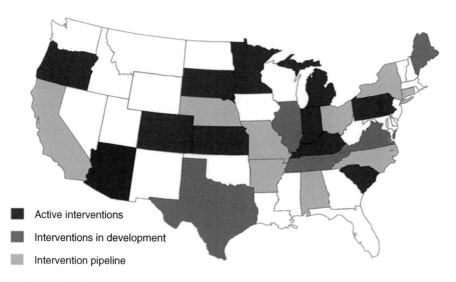

Fig. 6.2 Nexus innovations network data variety

6.2.3 Data Velocity

With big data there needs to be a balance struck between the time needed to collect valid and reliable data to the time data are analyzed and relevant useful knowledge generated. Data need to be sound yet not stale. Hence, collecting data that can be analyzed to assess the impact of IPE project interventions in as timely a manner as possible is an imperative aim.

In addition to specific NCDR data collection surveys, data from electronic health records (EHRs), site specific data collection tools/surveys, and validated data collection tools (e.g., ACE-15, PHQ-9, SF-16, HEDIS, etc.) are also provided and/or generated by intervention projects and are collected and stored in the NCDR. All intervention projects provide data (collected by NCDR specific surveys) on IPE-related education, network users (educators, clinicians, and clinical staff), student users (from multiple health professions programs), costs (project and delivered health care), intervention-specific outcomes, project format and structure, and critical incidents. Table 6.3 displays the standardized NCDR specific surveys by respondent and question content.

6.2.4 Data Value

The benefit accrued from the NCDR is related directly to the demonstration of the possible positive impact (actionable value) of IPE as a viable and efficient health-care delivery model. Heretofore, the data have not been sufficient or appropriate

Table 6.3 National Center data repository surveys

Survey	Respondents	Questions
Demographics	All participants	6 questions creating a personal profile
Education survey	Project lead with input from associated educational unit(s)	24 questions about the IPE program (one survey per unique facility or site)
Costs survey	Project lead with consultation from all relevant others	25 questions related to general finances (one survey per unique facility or site)
Network user survey	All clinical and educational participants in the intervention (e.g., clinicians, faculty)	32 questions related to IPE and CP at the intervention clinical performance site
Student user survey	All students participating in the intervention	16 questions related to related to IPE and CP at the intervention clinical performance site
Intervention specific survey	Project lead with consultation from all relevant others	80 questions related to the care processes of the specific project
Intervention outcome survey	Project lead with consultation from all relevant others	Delineation of all outcomes being measured as well as how and when they are being measured
Critical incidents survey	Any clinical or educational participant in the intervention (e.g., clinicians, faculty)	5 questions asking the who, what, where, when, how of the incident and your subsequent actions and completed only when a "critical incident" occurs

to make a definitive case for IPE. With the development of the NCDR scientifically sound and appropriate evidence regarding IPE will be collected at a sufficient volume (power) for analysis to answer questions about whether or not IPE positively affects the patient experience of care, the health of populations, and the per capita cost of healthcare.

6.2.5 Ecosystem of the NCDR

Ecosystems are any system or network of interconnecting and interacting parts. They are dynamic entities or enterprises that vary in size and number of elements that comprise them. Ecosystems are impacted by both internal and external factors. When references are made to BD ecosystems different descriptions are offered depending on the immediate purpose being discussed. In this section we describe the NCDR ecosystem in terms of infrastructure, analytics, and applications. Not only are these domains inter-related and interactive but so are the elements within each domain.

6.2.6   Infrastructure

This is the underlying foundation or basic framework of a system or organization. There are both critical and less critical elements to every infrastructure. The critical infrastructure elements of the NCDR are: projects, project management, scientific review, scientific support, data repository structure, and data repository function.

The first critical elements of the NCDR infrastructure are the intervention research *projects* that constitute the NIN. Without these there would be no data generated focused on assessing the impact of IPE on healthcare outcomes, the health status of populations, or healthcare costs. Because of the critical importance of the intervention research projects to the NCDR, the National Center recruits IPE projects from across the country and has developed a clear and detailed onboarding procedure/process. Figure 6.3 depicts the onboarding process for the intervention research projects of the NIN. As part of the onboarding process, each intervention research project must develop a scientific workplan and provide a diagram of the fundamental components (e.g., situation needing change, research question, dependent variables, independent variables, assumptions being made about change, study design, etc.) of the project. This diagram is the project's logic model. Figure 6.4 provides an illustration of an intervention research project logic model.

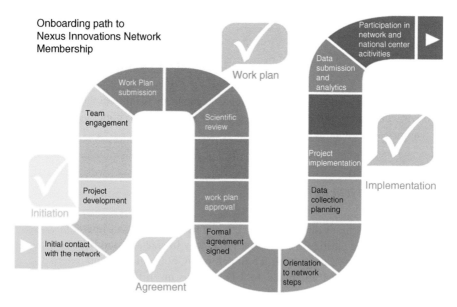

Fig. 6.3 NIN onboarding process

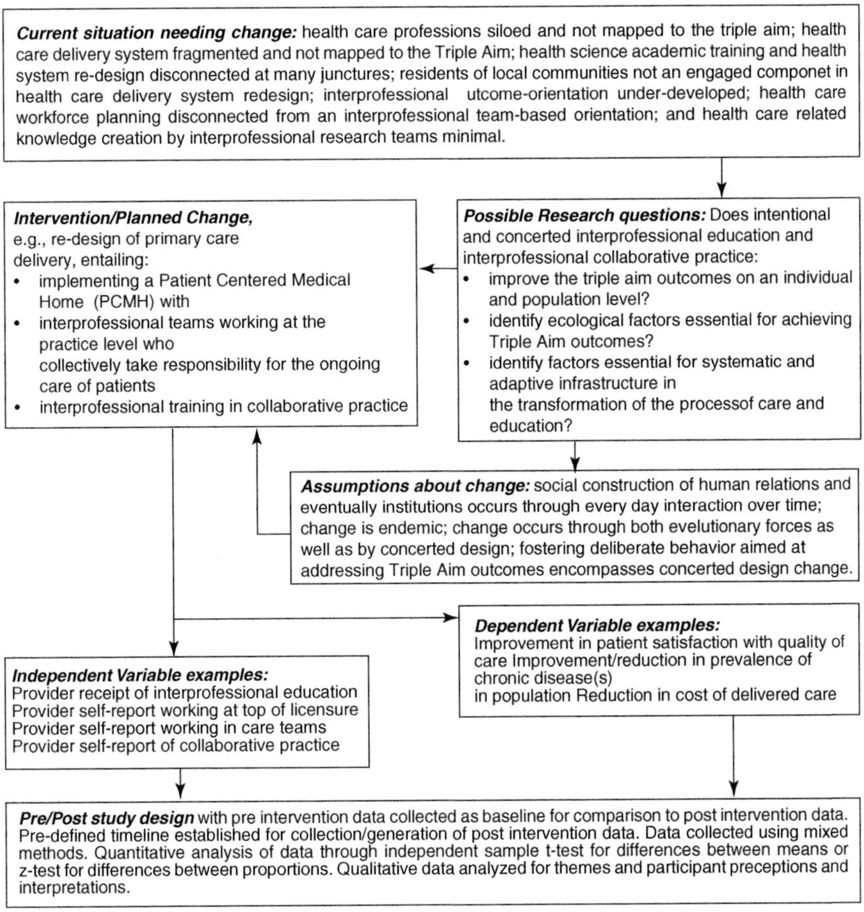

Current situation needing change: health care professions siloed and not mapped to the triple aim; health care delivery system fragmented and not mapped to the Triple Aim; health science academic training and health system re-design disconnected at many junctures; residents of local communities not an engaged componet in health care delivery system redesign; interprofessional   utcome-orientation under-developed; health care workforce planning disconnected from an interprofessional team-based orientation; and health care related knowledge creation by interprofessional research teams minimal.

Intervention/Planned Change,
e.g., re-design of primary care
delivery, entailing:
- implementing a Patient Centered Medical Home (PCMH) with
- interprofessional teams working at the practice level who collectively take responsibility for the ongoing care of patients
- interprofessional training in collaborative practice

Possible Research questions: Does intentional and concerted interprofessional education and interprofessional collaborative practice:
- improve the triple aim outcomes on an individual and population level?
- identify ecological factors essential for achieving Triple Aim outcomes?
- identify factors essential for systematic and adaptive infrastructure in the transformation of the processof care and education?

Assumptions about change: social construction of human relations and eventually institutions occurs through every day interaction over time; change is endemic; change occurs through both evelutionary forces as well as by concerted design; fostering deliberate behavior aimed at addressing Triple Aim outcomes encompasses concerted design change.

Dependent Variable examples:
Improvement in patient satisfaction with quality of care Improvement/reduction in prevalence of chronic disease(s)
in population Reduction in cost of delivered care

Independent Variable examples:
Provider receipt of interprofessional education
Provider self-report working at top of licensure
Provider self-report working in care teams
Provider self-report of collaborative practice

Pre/Post study design with pre intervention data collected as baseline for comparison to post intervention data. Pre-defined timeline established for collection/generation of post intervention data. Data collected using mixed methods. Quantitative analysis of data through independent sample t-test for differences between means or z-test for differences between proportions. Qualitative data analyzed for themes and participant preceptions and interpretations.

Fig. 6.4 Intervention research project example

6.2.6.1   Project Management

Project management based at the National Center is also a crucial/critical element of the NCDR infrastructure. Multiple project managers (presently two) work closely with each intervention project in the NIN in order to maintain a consistent point of contact, ensure that each project is able to enter required data into the NCDR specific surveys, and that project specific data are entered into the NCDR in a timely and problem-free manner. Project managers also coordinate meetings between National Center science staff and project-specific staff as these are needed.

6.2.6.2 Scientific Review

Scientific review entails the assessment of and approval for each intervention research project that eventually becomes part of the NIN. This review is completed by an interprofessional team comprising two physicians (a surgeon and a family physician) and a chronic disease epidemiologist. Scientific review is crucial to ensuring that the intervention research project designs are scientifically sound and that the data the project plans on collecting is valid and reliable in light of the research question(s) being asked.

6.2.6.3 Scientific Support

Scientific support goes hand in glove with the scientific review, although not all of those involved in the scientific review continue to provide scientific support. The type of support provided involves assisting with identifying and defining meaningful and appropriate intervention-related outcomes, metrics, and measurement tools/approaches. Scientific support is essential to ensure that the data collected by each intervention project in the NIN will be able to yield analyses that answer not only the project specific research questions(s) but also the National Center's key research questions and adequately test the established a priori hypotheses of the National Center (see Table 6.2).

6.2.6.4 Data Repository Structure and Function

Data repository structure and function is a critical and vital element of the NCDR's infrastructure since it warrants that NCDR surveys can be appropriately linked for analyses, that intervention-specific outcome data imported into the NCDR can be linked with other NCDR data for analyses, and that data leveraging for volume across intervention research projects can be achieved. Ease of data entry is also a critical part of the structure and function of the data repository.

6.2.6.5 Analytics

In the context of the NCDR this domain encompasses the capacity to create information and eventually knowledge. Information is processed data and is essential to knowledge creation (Institute of Medicine 2015; Porche 2014; Ross et al. 2014; Weinberg et al. 2013). Information is contextualized rather than raw data. Knowledge is derived from the modeled analyses of information leading to the discovery or identification of patterns and relationships between types of information. Knowledge is the synthesis of information or the analysis of information identifying and formalizing relationships (Institute of Medicine 2015; Porche 2014; Ross et al. 2014; Weinberg et al. 2013). Knowledge results from hypothesis testing or model development of some sort. Ultimately, based on the analytics of the NCDR data producing knowledge, evidence-based action plans for the transformation of healthcare delivery and the process of care becomes possible.

6.2.6.6 Applications

In the NCDR this domain references the use of the created/developed/generated knowledge. While a critical goal of the NCDR is to create actionable knowledge regarding IPE and its impact on health outcomes, healthcare delivery, and healthcare costs we are in the very early stages of building this aspect of the NCDR ecosystem. If the knowledge produced from the data in the NCDR bear the fruit we believe it will—in other words that there is a positive causal relationship between IPE and the Triple Aim outcomes, (Guyer and Green 2014; Hansen et al. 2014; Institute of Medicine 2015) then this knowledge could strongly influence the concerted transformation of healthcare delivery in the US. Moreover, this knowledge would give credence to the importance of the nexus and IPE integrated across health professions education with an eye toward improving clinical practice.

6.3 Conclusions

It is one of the founding beliefs of the National Center that high-functioning interprofessional healthcare teams can improve the experience, outcomes and costs of healthcare. Little research, however, has been conducted to generate evidence providing unequivocal support for this belief—leaving it an open question. To rectify this and essentially call the question, the National Center has articulated a research agenda, created and is growing a national IPE intervention research network, and has (most importantly) developed a data repository using BD informatics as its model. The NCDR has the potential of becoming the jewel in the crown of the IPE field of inquiry. This data repository has been a long time in coming and its importance to the field of IPE cannot and should not be understated.

References

Berwick DM, Nolan TW, Whittington J. The triple aim: care, health, and cost. Health Aff. 2008;27:759–69.

Boyd D, Crawford K. Critical questions for big data. Inform Commun Soc. 2012;15:662–79. doi: 10.1080/1369118X.2012.678878.

Brandt BF, Lutfiyya MN, King JA, Chioreso C. A scoping review of interprofessional collaborative practice and education using the lens of the triple aim. J Interprof Care. 2014;28:393–9.

Cerra F, Pacala J, Brandt BF, Lutfiyya MN. The application of informatics in delineating the proof of concept for creating knowledge of the value added by interprofessional practice and education. Healthcare. 2015;3:1158–73. doi:10.3390/healthcare3041158.

Cooper P. Data, information, knowledge and wisdom. Anaesth Intensive Care Med. 2014;15:44–5.

D'Amour D, Oandasan I. Interprofessionality as the field of interprofessional practice and interprofessional education: An emerging concept. J Interprof Care. 2005;19:8–20.

Gordon K. What is big data? ITNOW. 2013;55(3):12–3. doi:10.1093/itnow/bwt037.

Graves JR, Corcoran S. The study of nursing informatics. J Nurs Sch. 1989;21:227–31.

Guyer M, Green ED. The National Institute of Health's Big Data to Knowledge (BD2K) initiative: capitalizing on biomedical big data. J Am Med Inform Assoc. 2014;21:957–8.

Hansen MM, Miron-Shatz T, Lau AYS, Paton C. Big data in science and healthcare: a review of recent literature and perspectives contribution of the IMIA Social Media Working Group. Yearb Med Inform. 2014;9:21–6. doi:10.15265/IY-2014-0004. Accessed 30 Mar 2016.

Institute of Medicine. Measuring the impact of interprofessional education on collaborative practice and patient outcomes. Washington, DC: The National Academies of Science; 2015.

Lutfiyya MN, Brandt BF, Delany C, Pechacek J, Cerra F. Setting a research agenda for interprofessional education and collaborative practice in the context of US health system reform. J Interprof Care. 2016;30(1):7–14. doi:10.3109/13561820.2015.1040875.

Matney S, Brewster PJ. Philosophical approaches to nursing informatics data-information-knowledge-wisdom framework. Adv Nurs Sci. 2011;34:6–16.

Merelli I, Pérez-Sánchez H, Gesing S, D'Agostino D. Managing, analyzing, and integrating big data in medical bioinformatics: open problems and future perspectives. Biomed Res Int. 2014;134023. doi:10.1155/2014/134023. Accessed 30 Mar 2016.

Pechacek J, Cerra F, Brandt B, Lutfiyya MN, Delaney C. Creating the evidence through comparative effectiveness research for interprofessional education and collaborative practice by developing a national intervention network and a national data repository. Healthcare. 2015;3:146–61.

Porche DJ. Men's health big data. Am J Mens Health. 2014;8:189.

Ross MK, Wei W, Ohno-Machado L. "Big data" and the electronic health record. Yearb Med Inform. 2014;9:97–104. doi:10.15265/IY-2014-0003. Accessed 30 Mar 2016.

Weinberg BD, Davis L, Berger PD. Perspectives on big data. J Marketing Analytics. 2013;1:187–201. doi:10.1057/jma.2013.20.

Chapter 7
Wrestling with Big Data: How Nurse Leaders Can Engage

Jane Englebright and Edmund Jackson

Abstract The opportunities arising in health systems from the emergence of large-scale clinical data repositories are immense. The application of data science approaches to "big data" is transformational and has the potential to dramatically improve the health and wellbeing of individuals on a national scale. However, although big data science offers great rewards, it has its challenges too. This is no more evident than in nursing where the sheer amount of data flowing through health systems can be overwhelming. Clinical documentation in electronic health records, publically reported quality measures and business intelligence reports are just a few of the many data requirements nurse leaders encounter daily. This chapter describes the challenges nurse leaders face today and discusses strategies that nurse leaders can use to leverage big data to meet the Triple Aim of improving quality, improving the patient experience while reducing cost.

Keywords Nurse executives and big data • Data models • Standardized nursing languages • Nursing and data analytics • Electronic health records • Interoperability

7.1 Introduction

The emergence of large-scale clinical and administrative data repository's, or "big data", has provided nurse leaders tremendous opportunities but in the face of enormous challenges. Today's nurse leaders are literally inundated with data. From clinical documentation in the electronic health record (EHR), to publicly reported outcomes, to business intelligence reports, nurse leaders are awash in a tsunami of

J. Englebright, Ph.D., R.N., C.E.N.P., F.A.A.N. (✉) • E. Jackson, Ph.D.
Hospital Corporation of America, Nashville, TN, USA
e-mail: jane.englebright@hcahealthcare.com

© Springer International Publishing AG 2017
C.W. Delaney et al. (eds.), *Big Data-Enabled Nursing*, Health Informatics,
DOI 10.1007/978-3-319-53300-1_7

data. Clearly the success of nurse executives in the future will depend on their ability to manage this tidal wave of data in a way that meets the Triple Aim of improving quality, improving the patient's experience and reducing cost.

To be successful, nurse administrators of the future will need to collaborate with experts in the field of data science, and familiarize themselves with different approaches to analyzing big data. These methods include knowledge discovery through data mining and machine learning: techniques that can identify and categorize patterns in the data and predict events such as stroke, heart failure, falls, pressure ulcers and readmission to the hospital. They can also predict which staffing model is most likely to yield the best outcomes for a specific patient population.

Equally important will be nurse administrator's ability to communicate real time system performance through data visualization dashboards used to report clinical analytics and business intelligence. Fortunately today's health systems are starting to realize the vast opportunities that big data and data science can unlock. This chapter focuses on the state of big data and data science and what is possible related to nursing leadership.

7.2 Defining Big Data and Data Science

The terms big data and data science are often used interchangeably. However, there is a fundamental difference between the two areas. Data Science is an interdisciplinary field that seeks to capture the underlying patterns of large complex data sets and then program these patterns (or algorithms) into computer applications. An example of a computer application using a programmed algorithm is the Modified Early Warning (MEW) system which predicts how quickly a patient experiencing a sudden decline receives clinical care. The algorithm uses six factors to predict a MEWs score: respiratory rate, heart rate, systolic blood pressure, conscious level, temperature, hourly urine output (for previous 2 h) (AHRQ Innovations Exchange Webpage 2016).

Big data refers to data sets too large or complex for traditional data management tools or applications (Big Data 2015). Said another way, big data encompasses a volume, variety, and velocity of data that requires advanced analytic approaches. So, in this chapter, when we refer to big data, we are referring to the combination of structured, semi-structured and unstructured data stores and the special analytic approaches required to harness, to analyze, and to communicate insights from these data. Big data in contrast to data science looks to collect and manage large amounts of varied data to serve large-scale web applications and vast sensor networks.

7.3 Nursing Leader Accountabilities and Challenges

The chief nursing officer is a member of the executive leadership team of the organization. As such, this leader has accountabilities to both the organization and to the profession. At each level of nursing leadership, nurse leaders balance fiduciary and ethical accountabilities to the employer with accountabilities for the professional practice of nursing. This balancing act inevitably leads to some difficult decisions about deployment of scarce or expensive resources, continuing or discontinuing low volume services, or optimizing the model of care delivery to achieve best outcomes at the lowest cost.

7.4 Systems Interoperability

To meet these accountabilities, nurse leaders use data, lots of data, from many different types of data systems. Generally these data systems can be classified as financial, human resources, operational and clinical. Figure 7.1 summarizes the kinds of data nurse leaders frequently use from these different types of data systems.

These systems have varying levels of sophistication within a given organization but typically the systems are not interoperable. This means nurse leaders are often analyzing financial data first, followed by human resource data, then operational data, and finally clinical data. Nurse leaders then manually synthesize these separate analyses to make decisions. Figure 7.1 depicts this process.

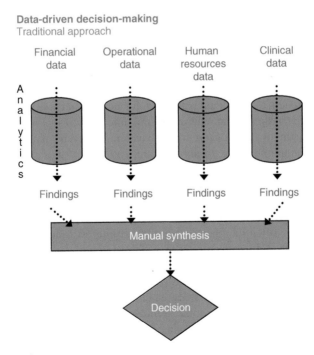

Fig. 7.1 Data-drive decision making: Traditional Approach

7.5 Non-Standardization

To further complicate this traditional approach to data analysis, the data systems may have different data definitions and different time periods that hinder cross-functional analysis. For example, financial systems may be tied to a fiscal calendar with well-defined month-end close activities. Human resource systems may be based on bi-weekly pay periods that do not coincide with the monthly fiscal calendar. Allocating external agency staff hours and costs to the department, shift, and patients across these two data systems for operational or clinical analyses is difficult and may result in different values and conclusions depending on which data source is used. For example, a "day" in the financial system is usually a 24-h period that begins at midnight and ends at 2359 (using the 24-h clock). A "day" in an operational scheduling system is usually defined by shift start and end times. So, the number of patient days or patient encounters in a given department on a specific day may vary, depending on which definition of "day" is used.

7.6 The Invisibility of Nursing

Financial data poses yet another challenge for nursing leaders. While financial information systems are usually the most mature information systems in the organization, nursing financial information is frequently not cataloged in the systems in a manner that enables analysis in relation to patient outcomes, operational changes or specific nurses. Lags in accounting processes for contract labor and lack of alignment with shift times were previously discussed. The inability to assign individual nursing practitioners to individual patients in financial systems is a significant hindrance to understanding the amount of nursing time each patient is consuming and therefore estimating the cost of care for each patient. This type of analysis is important to understanding the value of nursing care.

In addition to these technical challenges, traditional approaches to data analysis have some significant limitations for nurse leaders. Most importantly, these analyses often lack variables that are important to nurse leaders. Variables such as patient care requirements, nurse competence levels, departmental structure and care processes are often the phenomena of most interest to nurse leaders but are not reflected in the traditional data sources in most organizations. Secondly, traditional data sources may not be timely enough to inform operational decision-making. Many traditional data systems produce metrics on biweekly, monthly or quarterly bases. Cross-functional analyses that include clinical outcomes, such as infection rates, may be several months removed from the implementation of the novel clinical process that is being evaluated. Thirdly, many of the variables in the traditional systems are not actionable for the nurse leader. Clinical and human resource metrics calculated at the organizational level, such as pressure ulcer and turnover rates, do not identify the specific departments that need leadership intervention. Finally, the manual nature of the cross-functional analysis may systematically under- or over-estimate the interactions between the variables in the different systems because it relies on the judgment of the nurse leader.

Nurse leaders also expend significant effort helping non-nursing leaders within an organization understand the value of nursing to the organization, in terms of both patient outcomes and financial outcomes. The ability to measure the value of nursing in relation to the outcomes produced and the cost of that production is critical in an era of value-based health care (Pappas 2013). Pappas cites a need for a reporting framework that combines cost and quality as an important first step in documenting the value of nursing to outcomes and error avoidance. Pappas also identifies a need for data that makes tangible some of the intangible work of nursing, such as surveillance. Big data and data science offer techniques for quantifying or datafying and communicating (Mayer-Schonberger and Cukier 2013) concepts that have traditionally been intangible or subjective, giving nurse leaders a new language for expressing the value of nursing to decision-makers at all levels within the organization. These include concepts such as 'complexity of care' when discussing patients or 'novice to expert' when discussing nurses.

7.7   A Common Data Repository Across the System

The chief nursing officer (CNO) role has traditionally been specific to an organization. Thus the data used by the CNO has been confined to data from a single facility. The growth of health systems, and system CNOs (Englebright and Perlin 2008), has created an avenue for aggregating multi-facility data and engaging in big data. The desire to harness the power of big data to answer nursing questions will require nurse leaders within and across organizations to adopt national data standards and to explicitly define data models to generate sharable and comparable data. This is the first for step toward realizing the promise of big data for nursing.

7.8   The Value of Big Data for Nurse Leaders

Clancy and Reed (2015) discuss the data tsunami that is facing nurse executives. Traditional data sources are being supplemented with new data from sensors on equipment, patients and staff within the care environment, with social media interactions, and with retail information. Finding patterns and correlations among these diverse data sources is the province of big data and data science.

Big data and data science promise to answer some of the most perplexing questions facing nursing leaders today. What is the optimal way to organize and staff a patient care service? Big data provides a way to combine large amounts of different types of data to answer complex questions. Traditionally nurse leaders have had to ask financial questions of financial systems, human resource questions of human resource systems, clinical questions of clinical systems, and operational questions of operational systems. Answering these questions for an emergency department would require analyzing volume, payer mix, revenue and cost in the financial system; payroll costs, turnover, skill mix and staff satisfaction in the human resources systems; patient outcomes in clinical systems; and patient arrival patterns and throughput times in operational systems. Nurse leaders are then forced to use their knowledge, experience and intuition to make a judgment on the best way to organize and staff the department. Data science approaches can help by combining large amounts of data from multiple data sources into a coherent analysis and creating predictive models that provide the foundation for evidence based executive decision-making.

Big data and data science would embrace all these traditional data sources but might also include new data sources such as website hits searching for information on flu symptoms during flu season to anticipate a spike in volumes or police reports of accidents near the emergency department. Big data can search for correlations that might not be apparent to nurse leaders, such as the operating hours of nearby businesses. From this data, a model could be derived that predicts patient arrival patterns and care needs. This model could be used for long range planning, but can also be adjusted in near-real time to help the front line manager adjust staffing to be ready to meet the needs of patients that are about to present to the emergency department. This is big data and data science at its best, combining a large volume of highly variable data at high velocity.

7.9 The Journey to Sharable and Comparable Data in Nursing

Sounds almost too good to be true. So, what's the catch? Big data is a group activity and healthcare is still largely an industry of individual or small providers who take great pride in their individuality. To be most effective, big data should rest on a foundation of standardized, coded data elements. SNOMED, LOINC, and ICD10 are well-known data standards in healthcare. However, these taxonomies do not address many of the concepts important to nursing (Rutherford 2008).

Attempts to map the domain of nursing knowledge have been underway for many years. The American Nurses' Association currently recognizes twelve standardized nursing languages, which are summarized in Table 7.1 (American Nurses' Association 2012). The Health Information Technology Standards Panel (HITSP)

Table 7.1 Currently recognized nursing languages

Language	Recognized by ANA
NANDA-Nursing Diagnoses, Definitions, and Classification	1992
Nursing Interventions Classification system (NIC)	1992
Clinical Care Classification system (CCC) formerly Home Health Care Classification system (HHCC)[a]	1992
Omaha system[b]	1992
Nursing Outcomes Classification (NOC)	1997
Nursing Management Minimum Data Set (NMMDS)	1998
PeriOperative Nursing Data Set (PNDS)	1999
SNOMED CT	1999
Nursing Minimum Data Set (NMDS)	1999
International Classification for Nursing Practice (ICNP®)	2000
ABC codes	2000
Logical Observation Identifiers Names and Codes (LOINC®)	2002

Note Adapted from American Nurses' Association (2012). ANA recognized terminologies that support nursing practice. Retrieved March 12, 2016 at: http://www.nursingworld.org/npii/terminologies.htm
[a]Recognized by HITSP in 2007 (Alliance for Nursing Informatics 2007)
[b]Recognized by HITSP in 2008 (personal communication Luann Whittenburg)

recognized two of these terminologies, Clinical Care Classification System (CCC) and the Omaha System, as meeting the criteria required for the harmonization efforts undertaken as a component of the American Recovery and Reinvestment Act and Meaningful Use Electronic Health Record (EHR) Incentive Program (HITSP 2009; CMS 2016).

The goal of the significant investment in health information technology funded by the Health Information Technology for Economic and Clinical Health (HITECH) Act was to create the infrastructure for sharable comparable clinical data that promises to improve care processes and care outcomes (Payne et al. 2015). The detailed data now available in the electronic health record offers the promise for more detailed, more specific and more actionable indicators of nursing effectiveness. However, the ability to apply big data techniques to nursing clinical data is limited by the lack of adoption of standard terminologies that are essential for comparable and sharable data. Currently, there is no requirement that healthcare providers or health information technology vendors use these standardized nursing languages. Adoption is voluntary and uptake has been slow across the industry.

The nursing profession has been working toward shareable and comparable data to describe nursing practice for many years. In 2004, the National Quality Forum endorsed National Voluntary Consensus Standards for Nursing-Sensitive Care; see Table 7.2 (National Quality Forum 2004). This set of 15 voluntary metrics included patient-centered outcome measures, nursing-centered intervention measures, and

Table 7.2 National Quality Forum nurse-sensitive measures

Category	Measure
Patient-centered outcome measures	1. Death among surgical inpatients with treatable serious complications (failure to rescue) 2. Pressure ulcer prevalence 3. Falls prevalence 4. Falls with injury 5. Restraint prevalence (vest and limb only) 6. Urinary catheter-associated urinary tract infection for intensive care unit (ICU) patients 7. Central line catheter-associated blood stream infection rate for ICU and high-risk nursery (HRN) patients 8. Ventilator-associated pneumonia for ICU and HRN patients
Nursing-centered intervention measures	9. Smoking cessation counseling for acute myocardial infarction 10. Smoking cessation counseling for heart failure 11. Smoking cessation counseling for pneumonia
System-centered measures	12. Skill mix (RN, LVN/LPN, UAP and contract) 13. Nursing care hours per patient day (RN, LPN and UAP) 14. Practice Environment Scale—Nursing Work Index 15. Voluntary turnover

Note Adapted from National Quality Forum (2004). National voluntary consensus standards for nursing-sensitive care: an initial performance measure set. Retrieved March 29, 2016 from: https://www.qualityforum.org/Projects/n-r/Nursing-Sensitive_Care_Initial_Measures/Nursing_Sensitive_Care__Initial_Measures.aspx

system-centered measures. These metrics are incorporated into the National Database of Nursing Quality Indicators™ (NDNQI®) (Montalvo 2007) and are considered by the American Nurses Credentialing Corporation (ANCC) in the Magnet Recognition Program®. Over 2000 hospitals participate in NDNQI (Press Ganey 2015), submitting standard data elements, generating benchmarks, and stimulating learning and improvement.

Nurse leaders may be able to learn much about nursing practice from analyzing data from electronic health records. Data generated by nurses comprises a significant portion of the information in the EHR. Yet, most of this data has not been entered into the EHR in a standard format that allows it to be reused, shared, analyzed and compared. Kaiser Permanente and the US Department of Veterans Affairs worked together to compare patient data from the electronic health record related to pressure ulcers (Chow et al. 2015). They created a prototype of a common nursing information model with standard terms that allowed the two organizations to share and compare patient data to improve care coordination and quality of care. The process they used to standardize nursing concepts and assign codes using Clinical LOINC and SNOMED CT was both rigorous and onerous. Chow et al. (2015) cite the challenges of heterogeneous EHR systems, architectural limitations, and lack of data harmonization that were encountered in sharing one clinical data set between two organizations.

Clancy and Reed (2015) highlight the importance of data models in organizing data and standardizing how they relate to one another, how they comport to computer fields within the EHR and how they map to standard nursing terminologies. Data models are the "Rosetta stone" that allows data from distinct systems to be combined for analysis without losing the meaning or context of the original intent or meaning. Data models are particularly important for data originating from clinical documentation due to the variability that exists in both technology and philosophy.

7.10   Gaining Insight from Data in Real Time

While there are amazing insights to gain from EHR data and from clinical measures of nursing performance, the real value of big data comes from sourcing it from multiple systems to address complex questions such as how to operate a nursing service that generates the best patient and staff outcomes at the lowest cost. Big data and data science provide the tools and techniques to bring together large volumes of data from different types of information systems and to analyze it in novel ways that provide insights and support the evidence based decision-making that has been lacking. When you consider the velocity of big data, you create more than understanding or insight; you can create tools that nurse leaders use to manage nursing services in new ways. No longer do nurse administrators need to wait until the end of the month or the end of the quarter to receive time sensitive reports. Today CNO's are receiving data in near-real time, enabling point of care application of data insights. In other words, nurse leaders at all levels of the organization can use this data, including the charge nurse in the middle of a busy shift. The trick is how to deliver the output to the user within their workflow so that it adds value and does not detract from the important work of taking care of patients.

7.11   Strategies for Moving Forward

In 2013, the University of Minnesota School of Nursing began convening an annual meeting of nurses interested in advancing big data and data science applications within the profession. The participants work collaboratively throughout the year to advance key initiatives related to big data (University of Minnesota 2015). One output of the group in 2015 was The Chief Nurse Executive Big Data Checklist, see Table 7.3 (Englebright and Caspers 2016). The checklist identifies three arenas in which nurse leaders can begin to advance the adoption of big data and data science in nursing.

Table 7.3 The chief nurse executive big data checklist

Create a data culture
Incorporate data-driven decision-making into clinical and operational processes at all levels of the organization
Create continuous, timely feedback loops from data to decision-maker within clinical and operational technologies and processes
Drive access to data to all levels of the organization
Define the nursing terms needed in the existing financial, human resources, operations and clinical data systems
Adopt standard nursing taxonomies to structure and codify nursing terms in each of these data systems to enable internal analysis and external benchmarking • Nursing Minimum Data Set (NMDS) • Nursing Management Minimum Data Set (NMMDS) • American Nurses Association (ANA) recognized clinical terminologies
Collaborate with health care information technology (HIT) professionals to assemble nursing terms from the existing financial, human resources, operational and clinical data systems into a common platform for analysis
Develop data competencies
Secure access to critical informatics and data analytics skills, including statistical analysis, benchmarking, dash boarding and data visualization to realize full benefit of investment in Big Data
Establish the data and informatics competencies required for each nursing role and create a path to achieving those competencies
Create a data infrastructure
Establish a governance structure that includes direct care nurses and is charged with approving changes to the data sources and nursing terms, balancing the benefits and costs of expanding data collection activities
Align the nursing data agenda with the overall informatics strategy for the organization by integrating nursing and informatics professionals into the decision-making committees at all levels
Enforce use of big data, nursing business intelligence and clinical data and, establish data security
Assure that information is integrated across departments/silos, and that data are reliable and consistently used across the enterprise

Note Adapted from Englebright J and Caspers B (2016). The role of the chief nurse executive in the big data revolution. *Nurse Leader.* In Press

7.12   Instilling a Data-Driven Culture Through Team Science

The first arena identified on The Chief Nurse Executive Big Data Checklist is instilling a data-driven culture in the organization (Englebright and Caspers 2016). Nurse leaders do this by embedding data into all decision-making processes in the organization, assuring all levels of leaders have access to data, and that data feedback loops to leaders and staff are timely and transparent. The second arena identified on the checklist is developing competencies for understanding and using big data in self and others. This is a rapidly moving target as big data and data science evolves. The final arena on the checklist is creating the organizational infrastructure that ensures nursing and nursing informatics is an important component of data analytics within the organization.

Nurse leaders must connect to a data team in order to engage in big data and data science. In today's healthcare environment data science is an emerging discipline and hence the question frequently arises as how to form the data team to bring the analytics and technical aspects of this work to fruition. Data scientists are, to read the literature, unicorns, and competition for such talent is fierce. We have found that instead of trying to hire data scientists, it is more fruitful to grow a team.

A critical aspect is to realize that data science or advanced data analytics is distinct from informatics or information technology, it is increasingly part of the operations of the organization. The complexity of the big data and data science questions asked by the leaders of the organization require frequent iteration with the data science team. The necessary players on the data science team starts with a leader who can interact credibly and communicate clearly with both the organizational leadership and the analytical team. That team is composed of nursing domain experts, software developers, user interface specialists, business intelligence developers, statisticians and machine learning experts. They interact with a technical team in information technology (IT) comprised of database architects and engineers, big data architects and engineers, software developers, extract-transform-load (ETL) engineers, testers, and product support. The key for success amid this complexity is communication and teamwork: and like nursing, data science is a team sport.

7.13 Putting It All Together: An Example

This example illustrates how big data and data science can be applied to drive improved performance on inpatient nursing units. This example shows a progression from descriptive analytics (what happened), through diagnostic analytics (why did it happen), to predictive analytics (what will happen next) and finally to prescriptive analytics (what should I do about it).

A nursing unit is a highly dynamic and complex environment and managing this requires a holistic picture encompassing the most important variables from financial, human resources, operational and clinical domains of performance. The question is which are the most important variables?

7.13.1 Step 1: Diagnostic Analytics

This project began with a key user analysis in the patient experience domain, identifying those variables correlated, and causative, with high "Willingness to Recommend" and "Overall Patient Satisfaction" scores on patient experience surveys. One theme that emerged was the management of pain. Diagnostic analytics revealed that patients who reported better experiences with pain management had the highest scores on Willingness to Recommend and Overall Satisfaction. This led to the development of indicators of effective pain management practices that could be incorporated into the unit-based dashboard or data portal. A subset of patient experience scores includes patients with catheter associated urinary tract infections.

7.13.2   Step 2: Diagnostic Analytics

Catheter associated urinary tract infections (CAUTI) are an important, nurse sensitive, measure of clinical quality. In creating a dashboard one might be tempted to display the CAUTI rate in the clinical domain. This is valuable, but is not only in arrears, and *a-definitio,* unmanageable (the event has already happened) it is also difficult to attribute to a specific unit or nurse as a stimulus for action or improvement. Diagnostic analytics fuses many variables to find the leading variables. Unsurprisingly, the mean indwelling time of urinary catheters on a unit is highly predictive of the CAUTI rate on that unit. So, on a management dashboard we display that mean indwelling time (with navigation down to the details of each catheter and patient) rather than the CAUTI rate. The display of average hours on catheter is both meaningful and actionable for the manager and the nurse and directly correlated to the outcome they are trying to impact, infection rate.

7.13.3   Step 3: Predictive Analytics

The diagnostic analytics revealed that that urinary catheter indwelling time is predictive of CAUTI.

7.13.4   Step 4: Prescriptive Analytics

Now that we know that having a urinary catheter indwelling for more than 24 hours is highly predictive of CAUTI, we can simply create a list each shift, ranking patients with catheters by the length of time the catheter has been present and suggesting which need to be cleaned or removed in the present shift. This is actionable for both the manager and the nurse, providing them with tangible steps to take this shift to improve outcomes by driving down CAUTI rates.

This four-step process was repeated for each domain of performance to create a nursing unit dashboard. Figure 7.2 illustrates the first iteration of the dashboard with three domains of performance, clinical, patient experience, and productivity. Still in development are human resources and financial as well as additional clinical indica-

Financial	Human resources	Operational	Clinical
Volume	Turnover	Nursing hours/patient day	Pressure ulcer prevalence
Revenue	Skill mix	Discharges before noon	Patient falls with injuries
Supply cost	Percent BSN	Patient wait time	Cather-associated urinary tract infection rate
Labor cost	RN engagement	Follow-up phone call completions	Patient experience

Fig. 7.2 Examples of nursing metrics from each type of data system

tors. By taking the same approach though all domains areas of nursing performance, identifying key driver variables, displaying them in a consistent way, and combining that with action-enabling tools, we create a truly potent data-driven management approach.

Big data and data science provide nurse leaders with data that is timely, consistent and relevant and presented in ways that prompt insight and action. This is the realm of clinical decision support and is the effector arm of big data in the clinical setting. The technical components necessary to acquire, process and deliver analytics in support of Clinical Decision Support are significant, most especially when integrating disparate non-clinical sources. Substantial preparation is necessary to ensure success.

7.14 Conclusions

Big data and big data science are an exciting new frontier for nurse leaders. They offer new tools and techniques for simplifying data-driven decision-making. Nurse leaders operate within a tsunami of data and big data and data science offer strategies for finding the meaningful signals and patterns in the data, allowing the leader to provide focus and direction to the organization with confidence.

References

AHRQ Innovations Exchange Webpage. 2016. https://innovations.ahrq.gov/qualitytools/modified-early-warning-system-mews. Accessed 21 Feb 2016.

Alliance for Nursing Informatics. Nationwide health information technology standard for nursing. 2007. http://www.allianceni.org/docs/news012007.pdf. Accessed 28 Mar 2016.

American Nurses' Association. ANA recognized terminologies that support nursing practice. 2012. http://www.nursingworld.org/npii/terminologies.htm. Accessed 28 Mar 2016.

Big Data. Merriam-Webster's online dictionary, 11th ed. 2015. http://www.merriam-webster.com/dictionary/big%20data. Accessed 16 Mar 2016.

Centers for Medicare and Medicaid Services (CMS). Electronic Health Records (EHR) incentive programs. 2016. https://www.cms.gov/Regulations-and-Guidance/Legislation/EHRIncentivePrograms/index.html. Accessed 1 Mar 2016.

Chow M, Beene M, O'Brien A, Greim P, Cromwell T, DuLong D, Bedecarre D. A nursing information model process for interoperability. J Am Med Inform Assoc. Advance online publication. 2015;22:608–14. doi:10.1093/jamia/ocu026. Accessed 28 Mar 2016.

Clancy T, Reed L. Big data, big challenges: implications for chief nurse executives. J Nurs Adm. 2015;46(3):113–5.

Englebright J, Caspers B. The role of the chief nurse executive in the big data revolution. Nurse Leader. 2016;14(4):280–284. doi:10.1016.

Englebright J, Perlin J. The chief nurse executive role in large healthcare systems. Nurs Adm Q. 2008;32(3):188–94. doi:10.1097/01.NAQ.0000325175.30923.ff.

Health Information Technology Standards Panel. HITSP EHR-Centric Interoperability Specification. Publication No. IS 107. 2009. http://www.hitsp.org/InteroperabilitySet_Details.aspx?MasterIS=true&InteroperabilityId=513&PrefixAlpha=1&APrefix=IS&PrefixNumeric=107. Accessed 18 Mar 2016.

Mayer-Schonberger V, Cukier K. Big data: a revolution that will transform how we live, work and think. Boston: Houghton Mifflin Harcourt; 2013.

Montalvo I. The National Database of Nursing Quality Indicators®(NDNQI®). Online J Issues Nurs. 2007;12(3):2. doi:10.3912/OJIN.Vol12No03Man02.

National Quality Forum. National Voluntary Consensus Standards for Nursing-Sensitive Care: an initial performance measure set. 2004. https://www.qualityforum.org/Projects/n-r/Nursing-Sensitive_Care_Initial_Measures/Nursing_Sensitive_Care__Initial_Measures.aspx. Accessed 27 Mar 2016.

Pappas SH. Value, a nursing outcome. Nurs Adm Q. 2013;37(2):122–8. doi:10.1097/NAQ.0b013e3182869dd9.

Payne TH, Corley A, Cullen TA, Gandhi TK, Harrington L, Kuperman GJ, et al. Report of the AMIA EHR 2020 task force on the status and future direction of EHRs. J Am Med Inform Assoc. 2015;22(1):1102–10. doi:10.1093/jamia/ocv066.

Press Ganey. Nursing Quality (NDNQI). 2015. http://www.pressganey.com/solutions/clinical-quality/nursing-quality. Accessed 10 Mar 2016.

Rutherford M. Standardized nursing language: what does it mean for nursing practice? Online J Issues Nurs. 2008;13(1). doi:10.3912/OJIN.Vol13No01PPT05.

University of Minnesota School of Nursing. Improved patient care through sharable, comparable nursing data. Proceedings of the conference: nursing knowledge: big data and science for transforming health care. University of Minnesota, Minneapolis 4–5 June 2015. 2015. http://www.nursing.umn.edu/prod/groups/nurs/@pub/@nurs/documents/content/nurs_content_504042.pdf. Accessed 14 Mar 2016.

Case Study 7.1: Improving Nursing Care Through the Trinity Health System Data Warehouse

Nora Triola, Miriam Halimi, and Melanie Dreher

Abstract This case study describes how Trinity Health, a national health system, uses their data repository to discover how clinical nursing actions contribute to knowledge value and meet the Triple Aim. The design and results of four projects using the Trinity Health Data Repository are discussed and include the following exemplars: Decreasing mortality using Interdisciplinary Plans of Care (IPOC's); Identifying factors that contribute to pressure ulcers; Preventing venous thromboembolism (VTE) with pharmaceutical prophylaxis: Predicting malpractice claims for missed diagnosis in the emergency room.

Keywords Big data • Plans of care • Venous thromboembolism (VTE) • Pressure ulcers • Nursing informatics • Data science • Data warehouse • Knowledge value • Predictive modeling • Algorithms

7.1.1 Introduction

In the acute care setting, nurses have the most direct contact with patients. As a result, nurses provide the majority of the documentation and data regarding patient care within the electronic health record (EHR). The explosion of "big data" tools and concepts allows for the manipulation of electronically stored information to create new nursing knowledge. The generation of new knowledge requires the use of scientific inquiry through the EHR and public health data and emerging methods that are revealed as we understand the multiple uses of big data (Brennan and Bakken 2015). While it is apparent that the primary use of EHRs is clinical documentation, the secondary use of stored EHR data for research has taken on new meaning for transforming the way nursing observations and assessments are translated into large data sets to improve patient care.

Recently, Trinity Health assembled a summit of internal interprofessional thought leaders from around the country to discern what "big data" means to the organization. The overarching findings of the summit were that big data means, "Data working for us and not us working for the data". The group discussed the opportunities to leverage the Trinity Health data warehouse in new ways, regardless of the heterogeneity of its initial form. For example, there are multiple ways blood pressure can be entered into an electronic flowsheet due to the many EHRs and data models within the Trinity Health system. Blood pressure can be represented as a discrete entry for systolic and diastolic blood pressure, or concatenated into one data element such as arterial pressure. Despite the lack of homogeneity, there are numerous questions to be asked. The summit concluded that the value of putting data to work in more meaningful ways would benefit both clinical care and financial outcomes.

N. Triola, Ph.D., R.N., N.E.A.-B.C. • M. Halimi, D.N.P., M.B.A., R.N.-B.C.
M. Dreher, Ph.D., R.N., F.A.A.N. (✉)
Trinity Health System, Livonia, MI, USA
e-mail: ntriola@trinity-health.org

Trinity Health's clinical leadership had a vision to assemble data in a single location to better manage duplication and standardize inputs. Data outputs support the organization when they can be used (with minimal transformation or supplementation) to discover new knowledge about nursing practice, operations and patients. In these case studies, the focus is on the relationship between the role of big data and its impact on patients, clinicians, business and Trinity Health's approach for using data to inform nursing through secondary data analysis.

7.1.2 Trinity Health

Trinity Health was formed in May 2013 when Catholic Health East (CHE) and Trinity Health consolidated to form the second largest Catholic healthcare organization in the United States (Brokel et al. 2006). This consolidation and several recent acquisitions have formed an organization of 91 hospitals in 21 states, 51 home care agencies and hospice locations, 14 PACE locations, and 61 continuing care facilities. Trinity Health staffs 3900 employed physicians, 23,900 affiliated physicians and greater than 35,000 registered nurses (see Fig. 7.1.1).

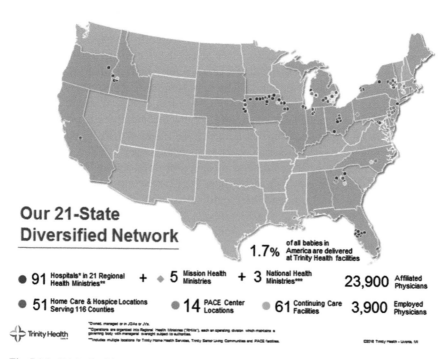

Fig. 7.1.1 Trinity health

Trinity Health's strategic plan, titled "People Centered Care 2020", focuses on three core areas: (1) Episodic care management for individuals with a focus on efficient and effective episode delivery initiatives; (2) Population health management with a focus on efficient care management initiatives; and (3) Community health and well-being with a focus on serving those who are poor, other populations, and impacting the social determinants of health. The Trinity Health strategy is designed to meet the requisites of the Triple Aim: better health, better care, and lower costs.

The integration of the two organizations has led to a complex data environment with multiple source applications in use across the various continuum environments. Although complex, Trinity Health has one of the largest data warehouses within the nation with more than 13 million unique patient records providing a data rich environment for Trinity Health to leverage for big data analytics and nursing research.

Nursing at Trinity Health, over the last 10 years, with the implementation of the electronic health record (EHR), has focused on capturing structured and standardized information electronically across facilities and care venues. This focus now allows for the secondary use of data to study outcomes measurements, practice level improvements, surveillance, population health, clinical research and decision support (Brokel et al. 2012).

7.1.2.1 Trinity Health Data Warehouse: A Cross-Continuum Data Environment

The Trinity Health data environment is comprised of five data flow layers that allow for normalizing, indexing and mastering patient and provider level data for clinical, operational and financial dimensions. The first layer, the source systems layer, represents the core systems across all Trinity Health venues of care. These systems are comprised of both internal systems, housed by Trinity Health, and external systems, hosted by outside organizations in the cloud or subscription based applications. The second layer, the discovery layer, is where the data from the source systems sit into a persistent state. The data is loaded in its natural state into the discovery layer with little to no quality control. The goal of the discovery layer is to load source data quickly so it can be analyzed for business intelligence. The third layer, the integrated data layer, is where data from multiple source systems are integrated into one model. The model conforms to the third normal form (3NF) data model, a model that is used to normalize a database design to reduce the duplication of data and allow for flexible data integration. The fourth layer, the semantic layer, abstracts and simplifies the complexity of the integrated data model, providing multiple business friendly views of data assets depending on needs. The last layer, the analytics layer, is where the business users interact with the data. This last layer represents the big data ecosystem within the vast data environment (see Fig. 7.1.2).

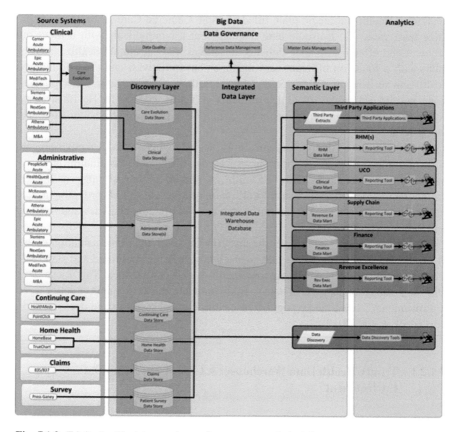

Fig. 7.1.2 Trinity health data warehouse-future state analytic infrastructure (courtesy of Rob Katofiasc and Marian Moran)

7.1.3 Case Studies

Since its big data summit in 2014, Trinity Health has launched or completed several analyses that put big data to work making new discoveries for nursing care. These new discoveries identified correlations between nursing interventions or actions and patient outcomes and in return have informed clinical practice requirements.

7.1.3.1 Interdisciplinary Plans of Care (IPOC) Case Study

Trinity Health developed Interdisciplinary Plans of Care (IPOCs) to guide the delivery and documentation of care, using condition-specific guidance distilled from published evidence for nursing and other professionals including respiratory, spiritual care, rehabilitation care, and clinical nutrition. The goal of IPOC development and usage was to assure consistent care delivery across all Trinity Health hospitals by embedding IPOCs within the clinical workflows and the EHR (Gardiner et al. 2014; HIMSS. HIMSS CNO-CNIO vendor roundtable 2015). IPOCs prompt nurses to make specific observations and document the progress a patient makes towards a condition-specific goal. Trinity Health has 141 IPOCs, with 497 patient care problems for nurses and ancillary providers to select based on the patient's specific condition. In addition to the use of disease specific condition IPOCs, the Adult Core IPOC, consisting of sub-phases for *Deficient Knowledge, Risk of Infection, Difficulty Coping Related to Hospital Stay, Pain, Risk of Deep Vein Thrombosis, Fall Prevention,* and *Risk of Pressure Ulcer* are initiated on all adult patients in medical/surgical and ICU units.

Trinity Health hypothesized that IPOC usage might affect quality outcomes such as mortality. To test this hypothesis, inpatient mortality rates with and without IPOCS were compared using data from the Trinity Health Data Warehouse. Administrative medical records of 165,203 adult medical/surgical discharges for a 12 month period for 18 hospitals across the United States were abstracted. It was noted that IPOC'S guided care in most discharges (see Table 7.1.1). Among the IPOC discharges, the risk of mortality was about half of the mortality in discharges without IPOC guidance. However, the relative risk of mortality remained unchanged when adjusted for acuity factors, despite the confounding influence of higher IPOC usage among higher acuity patients. The study demonstrated that structured documentation, supported by condition specific IPOC's, resulted in lower mortality rates irrespective of acuity. Trinity Health has now incorporated the use of both the Adult Core IPOC and a disease specific IPOC into its nursing practice. It is unclear from the data sets whether the observed decline in mortality rates was causal or associative with IPOC use. Further analysis is needed to understand the causation.

In looking at the use of the Adult Care IPOC's and a disease specific IPOC's, Trinity Health studied data from patients with a primary or secondary diagnosis of congestive heart failure (CHF). The study demonstrated lower mortality risk in patients with disease specific IPOCS, such as Cardiovascular Management or CHF, when compared individually to just the Adult Core IPOC. Hence, the data for congestive heart failure shows that any IPOC is better than no IPOC. Further, condition-specific IPOCs are better than the Adult Core IPOC alone. For primary CHF, the specific CHF IPOC appears slightly better than the general Cardiovascular Management IPOC, but the best performance was obtained when all three (Adult Core, CHF and Cardiovascular Management IPOCs) were used. Patients with a secondary CHF diagnosis, who had the Adult Core with the Cardiovascular Management IPOCs initiated, performed slightly better compared to patient with the Adult Core and CHF IPOC's. Because most of the secondary CHF patients have very acute primary diagnoses, patients fared better with the IPOC specific to their primary condition (see Table 7.1.2).

The case study demonstrates how system-wide, structured documentation supported by IPOC's does improve the quality of care to patients. This also illustrates the value of big data to inform consumers, government agencies, and other health professionals how nursing actions deliver quality care and advance the Triple Aim.

Table 7.1.1 IPOC data-number of inpatient records by health ministry (Trinity legacy) with IPOC information integrated at the patient record level

	2011	2012	2013	2014	2015	Total
Total	52,669	189,745	232,833	267,310	293,510	1,036,092
Holy Cross	27,713	31,226	27,990	31,115	34,612	152,656
Oakland	17,441	18,428	17,216	18,073	20,002	91,160
Dubuque	7515	8148	7049	7868	8580	39,160
Boise		15,299	14,428	14,259	15,733	59,725
Mishawaka		12,903	11,501	13,456	14,746	52,611
Sioux City		10,437	8483	8910	9254	37,088
Livonia		13,906	11,234	14,283	15,637	55,062
Pt. Huron		3553	1535	2752	3274	11,115
Ann Arbor		30,393	27,738	29,915	32,168	120,220
Gr Rapids		16,048	14,446	15,770	18,227	64,492
Mason City		10,771	8670	9027	9410	37,878
Muskegon		8452	6172			14,667
Clinton		4552	3879	4116	4261	16,808
St. Anne's		2537	12,843	15,879	18,894	50,153
Mt. Carmel West		2277	12,589	13,940	15,396	44,202
New Albany		815	4453	4362	4301	13,931
Fresno			15,776	20,830	21,770	58,376
Mt. Carmel East			13,510	19,185	20,558	53,253
Hackley			6527	7411	9427	23,365
Chelse			3522	3644	3694	10,860
Livingston			3045	2749	2759	8553
Lakeshore			277			277
Nampa				4717	5150	9867
Ontario				2437	2722	5159
Plymouth				1754	1989	3743
Baker City				858	903	1761

7.1.3.2 Pressure Ulcer Case Study

Trinity Health engaged with a local university to analyze gaps in care. The group settled on pressure ulcers as the first condition to study. A review of the literature showed an increase in pressure ulcer risk in African Americans compared with European Americans. Using primarily nursing observation data stored in the Trinity Health data warehouse, the group discovered that the risk of pressure ulcers was equivalent between races. Further, the group discovered that smaller body mass index was a much more important factor than race as previously thought (Pidgeon 2013). The results of the study were used to inform clinical nurses that careful documentation of body mass is important in assessing pressure ulcer risk.

Table 7.1.2 Mortality performance by IPOC usage-patients with primary or secondary diagnosis of CHF

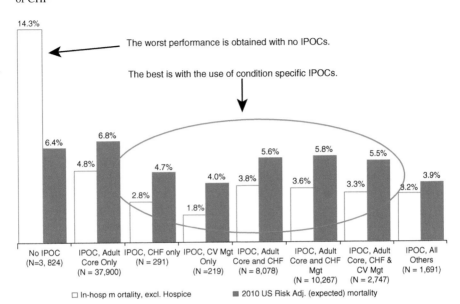

7.1.3.3 Venous Thromboembolus (VTE) Advisory Case Study

As part of the general EHR deployment, Trinity Health has a VTE advisor algorithm embedded within the EHR which is used for all admissions to assess VTE risk and offer prophylactic nursing and pharmacological options. For clinicians, the VTE advisor offers insight and a double-check mechanism for assessing risk and steps to prevent VTEs in their patients. The use of structured nursing documentation such as allergies, laboratory results, weight, documentation in the medication administration record and other information relevant to the VTE risk assessment enables the advisor to give precise prophylaxis recommendations.

As part of ongoing EHR value (return on investment) assessments, Trinity Health analyzed the effectiveness of the VTE advisor as a suggested preventive step. This study demonstrated that prevention of VTE was higher among providers who used the VTE advisor compared to those that did not. Further, it was found that earlier use of the VTE advisor was associated with shorter length of stay, fewer hospital acquired VTE's, lower in-hospital mortality and lower direct variable cost per case. This study demonstrated how observation and documentation by nurses helps drive a highly important preventive prophylaxis regimen. When documentation and observation is guided through the VTE advisor, the precision of the prophylaxis is superior to the prophylaxis alone.

7.1.3.4 General to Specific and Failure to Diagnose Case Study

As part of an ongoing malpractice risk analysis, Trinity Health discovered that failure to diagnose within an emergency department (ED) setting contributed greatly to overall malpractice claims. However, actually observing a failed diagnosis is difficult to capture within an EHR. The use of a proxy allows for a figure to approximate in place. Choosing a proxy for this uncaptured phenomenon, the proxy stated that a patient returning to the ED within 10 days of a prior ED visit for the same clinical condition is at high risk of becoming a failure to diagnose claimant. Using the Center for Medicare/Medicaid Services (CMS) Clinical Classification System (CCS), Trinity Health developed an algorithm to classify each ED admissions chief complaint to a general clinical category. These clinical categorizations ranged from general (abdominal pain) to specific conditions (appendicitis). With the general to specific categorizations, it was discovered that patients who present with general abdominal complaints and leave with a specific diagnosis are half as likely to return within 10 days. Further, when expending no fewer than the median number of diagnostic resources, patients are twice as likely to leave the ED with a specific diagnosis. This finding will lead to new ED discharge advisors or possibly to different follow-up care for those leaving without a specific diagnosis.

This case study demonstrated that the delivery and documentation of diagnostic activities, primarily facilitated by nurses, helped transition general patient complaints to a specific diagnosis. The act of moving from general to specific delivered higher quality patient care and lower malpractice claims.

7.1.4   Conclusion

Modern nursing care requires copious amounts of data from observations and assessments. However, no longer is this data disregarded as valuable for use in secondary data analysis. As shown in the case studies described, liberating nursing data allows for the development of new patient care paradigms. Efforts over the years to transition from free text notes to structured documentation has contributed to new discoveries and guides us toward the Triple Aim of better health, better care at lower costs. Big data science plays a key role in the future of nursing as it offers key clinical insights to inform and transform evidence-based practices that can truly deliver people centered care.

Acknowledgements

For the data warehouse infrastructure graphic, we thank Rob Katofiasc and Marian Moran. For their contributions to generation and abstraction of case study data, we thank Robert Sloan, Charles Bowling and Eric Hartz.

References

Brennan PF, Bakken S. Nursing needs big data and big data needs nursing. J Nurs Scholarsh. 2015;47(5):477–84.

Brokel JM, Shaw MG, Nicholson C. Expert clinical rules automate steps in delivering evidence-based care in the electronic health record. Comput Inform Nurs. 2006;24(4):196–205; quiz 206–197.

Brokel JM, Ward MM, Wakefield DS, Ludwig A, Schwichtenberg T, Atherton D. Changing patient care orders from paper to computerized provider order entry-based process. Comput Inform Nurs. 2012;30(8):417–25.

Gardiner JC, Reed PL, Bonner JD, Haggerty DK, Hale DG. Incidence of hospital-acquired pressure ulcers–a population-based cohort study. Int Wound J. 2014; doi:10.1111/iwj.12386.

HIMSS. HIMSS CNO-CNIO vendor roundtable. 2015. www.himss.org/cno-cnio-vendor-roundtable. Accessed 23 Apr 2016.

Pidgeon E. Trinity health and catholic health East mark first day as consolidated system. 2013. www.trinity-health.org/body.cfm?id=196&action=detail&ref=35. Accessed 1 Feb 2016.

Chapter 8
Inclusion of Flowsheets from Electronic Health Records to Extend Data for Clinical and Translational Science Awards (CTSA) Research

Bonnie L. Westra, Beverly Christie, Grace Gao, Steven G. Johnson, Lisiane Pruinelli, Anne LaFlamme, Jung In Park, Suzan G. Sherman, Piper A. Ranallo, Stuart Speedie, and Connie W. Delaney

Abstract Clinical data repositories increasingly are used for big data science; flowsheet data can extend current CDRs with rich, highly granular data documented by nursing and other healthcare professionals. Standardization of the data, however, is required for it to be useful for big data science. In this chapter, an example of one CDR funded by NIH's CTSA demonstrates how flowsheet data can add data

B.L. Westra, Ph.D., R.N., F.A.A.N., F.A.C.M.I. (✉)
Center for Nursing Informatics, School of Nursing, University of Minnesota,
Minneapolis, MN, USA
e-mail: Westr006@umn.edu

B. Christie, D.N.P., R.N.
Integrated Quality & Patient Safety, Fairview Health Services,
Minneapolis, MN, USA

G. Gao, D.N.P., R.N.-B.C. • L. Pruinelli, Ph.D.-C., R.N. • J. In Park, Ph.D.-C., R.N.
C.W. Delaney, Ph.D., R.N., F.A.A.N., F.A.C.M.I.
School of Nursing, University of Minnesota, Minneapolis, MN, USA

S.G. Johnson, Ph.D.-C., Ph.D. • S. Speedie, Ph.D., F.A.C.M.I.
Institute for Health Informatics, University of Minnesota, Minneapolis, MN, USA

A. LaFlamme, D.N.P., R.N.
University of Minnesota Medical Center—Fairview Health Services,
Minneapolis, MN, USA

S.G. Sherman, Ph.D., R.N.
Nursing and Allied Health, Fairview Health Services, Minneapolis, MN, USA

P.A. Ranallo, Ph.D.
Six Aims for Behavioral Health, Minneapolis, MN, USA

© Springer International Publishing AG 2017
C.W. Delaney et al. (eds.), *Big Data-Enabled Nursing*, Health Informatics,
DOI 10.1007/978-3-319-53300-1_8

repositories for big data science. A specific example of pressure ulcers demonstrates the strengths of flowsheet data and also the challenges of using this data. Through standardization of this highly granular data documented by nurses, a more precise understanding about patient characteristics and tailoring of interventions provided by the health team and patient conditions and states can be achieved. Additional efforts by national workgroups to create information models from flowsheets and standardize assessment terms are described to support big data science.

Keywords Flowsheet data • Clinical data • Repository • Data science • Normalizing data • Informatics

8.1   Introduction

Data stored in electronic health records (EHRs) and other information systems are increasingly used for meeting the quadruple aim of better health and patient experience, cost reduction, and better work life of health care providers (Bodenheimer and Sinsky 2014). These data are extracted and organized into clinical data repositories (CDRs) using common data (information) models and data standards for big data science. There is a need to expand the data typically represented in CDRs with nursing and interprofessional patient assessments and interventions. Flowsheet data from EHRs is a major source of nursing documentation and include time-based structured and semi-structured data essential to represent nursing and interprofessional care. If flowsheet data is included in CDRs, a more complete representation of patients with highly granular and longitudinal tracking of clinical states, care, and outcomes is possible. These additional data enhances our ability to conduct big data science projects that effectively target interventions based on individual patient characteristics and circumstances. One such model is the Extended Clinical Data initiative at the University of Minnesota, funded by the National Institutes of Health (NIH) through Clinical Translational Science Awards (CTSA).

8.2   CTSAs to Support Big Data Science

The coalescence of the complexity of today's pressing health issues and diseases, availability of advanced technology and increased analytics capacity, and challenging resources demands collaboration, teamwork, and networking of health care providers, stakeholders, and data expertise. Coupled with the interdependence, joint ownership, and collective responsibility between and among scientists trained in a plethora of different fields, the National Institutes of Health committed to reforming and transforming the way biomedical research is conducted, overcoming specific hurdles, and filling the gaps needed to explore and tackle these complex health and biomedical problems (https://ombudsman.nih.gov/collaborationTS.html). In 2003 Elias Zerhouni, NIH Director, announced the 2004 program addressing the NIH Roadmap: programming that would focus on new pathways to discovery, research teams of the future, and reengineering of the clinical research enterprise.

In 2006, the national CTSA program was launched to address the mission: Accelerating Discoveries Toward Better Health. The CTSA program created a definable academic home for clinical and translational research. Institutions funded by the CTSA program work to transform the local, regional, and national environments to increase the efficiency and speed of clinical and translational research across the country. Currently, 62 medical research institutions in 31 states and the District of Columbia are active members of the CTSA consortium. These institutions work together to accelerate discoveries toward better health (https://ctsacentral.org/consortium/institutions/). Moreover, several CTSA institutions have self-assembled into regional consortia, often including non-CTSA members. Figure 8.1 is a listing of these consortia, each one comprised of at least one CTSA institution (https://ctsacentral.org/consortium/regional-consortia/).

The consortia have established best practices for education and training, regulatory knowledge including human subjects' advocacy, research networking, biostatistics, community engagement, comparative effectiveness research, and more. They have established data sharing agreements and networked data access. These resources in part enable large data science.

Nursing professionals and engagement have been visible since the inception of the CTSA program in 2006. NIH National Institute of Research (NINR) collaborates with

- Appalachian Translational Research Network
- SPIRiT Consortium
- CONCERT Consortium: CTSAs Addressing COPD
- Midwest Consortium
- Chicago Consortium for Community Engagement
- New York and Connecticut Consortium - NYCON Consortium
- Upstate New York Translational Research Network
- Southeast CTSA Consortium
- Texas Regional CTSA Consortium
- Western Emergency Services Translational Research Network
- The Massachusetts Research Subject Advocacy Group
- The Ohio Consortium
- Western States Consortium
- UC BRAID Consortium
- Wisconsin Network for Health Research
- Midwest Area Research Consortium for Health
- Greater Plains Collaborative Clinical Data Research Network
- Greater Los Angeles CTSA Consortium

Fig. 8.1 Regional CTSA consortia

the CTSA to contribute in addressing the NINR mission (https://www.ninr.nih.gov/) to lead nursing research, which develops knowledge to:

- Build the scientific foundation for clinical practice
- Prevent disease and disability
- Manage and eliminate symptoms caused by illness
- Enhance end-of-life and palliative care

The NINR Nurse Scientist Translational Research Interest Group fosters collaboration to discuss and implement processes in which clinical and translational nursing science investigators can be represented, recognized, and garner resources in the CTSAs. Non-CTSA affiliated nurse scientists have participated since inception with the mutual goal of accelerating the speed of scientific discovery translation to benefit and improve health. It is obvious that resources, teams, and collaboration mandate permeate the CTSAs program. Questions that can be addressed when nursing data are incorporated into CDRs include: What interventions provided by nurses prevent adverse events such as pressure ulcers? What is the likelihood of hospital readmission for patients with severe sepsis if the Surviving Sepsis Campaign guidelines are followed? How does risk adjustment using nursing assessments compare in predicting hospital readmission with other risk assessments such as co-morbidity indices?

8.3   Clinical Data Repositories (CDRs)

As healthcare data becomes increasingly available in electronic form, there is a growing need to integrate data about a patient derived from various resources in an organized way to conduct quality improvement and research activities that evaluate and extend evidence-based practice guidelines to make decisions to improve outcomes. For instance, a data repository (a.k.a. data warehouse) is a database that centralizes and integrates information from many data sources into a single location for easier reporting and analysis. The data from each source is normalized; the meaning of the information is consistent across different sources.

Data repositories exist for different purposes. For example, in business, consumer information can be combined for sales and marketing purposes. In health care, a CDR is a data warehouse that consists primarily of demographic and clinical information about a patient. It can be a repository that stores a longitudinal record of all information about a patient (such as demographics, family history, laboratories, medications, procedures and flowsheets). It can be very specialized. Examples of specialized CDRs (often labelled Registries) exist for specific diseases and conditions such as the United Network for Organ Sharing/Organ Procurement and Transplant Network (UNOS/OPTN) (http://optn.transplant.hrsa.gov/converge/latestData/step2.asp), the Diabetes Collaborative Registry, the National Cardiovascular Data Registry (https://www.ncdr.com/). Other CDRs are designed specifically for research collaboration such as the HMO Research Network, the National Patient-Centered Clinical Research Network

(PCORnet), and MiniSentinel to research adverse drug reactions, and Observational Medical Outcomes Partnership (OMOP) to study the effects of medical products (Johnson et al. 2015). However, currently very few CDRs include data by or about nursing to demonstrate the influence of nursing care on patient outcomes (Goossen et al. 2013).

One hospital-based CDR that contains nursing data is the National Database of Nursing Quality Indicators (NDNQI), which contains data designed to evaluate indicators of nursing quality, processes, and outcomes (Montalvo 2007). Developed by the American Nursing Association (ANA), the NDNQI contains information from more than 1000 organizations. Member organizations contribute nursing data on a periodic basis and in a standard way. In exchange, they receive benchmarking data—reports comparing their quality measures with those of their peers. Every month each organization sends encounter level information about the patient, facility/unit, skin assessments and pressure ulcer attributes. Data is required to conform to rigid guidelines to ensure the data have the same meaning across every reporting organization. For example, the Pressure Ulcer Survey includes eight risk assessment items and six items for Pressure Ulcer interventions as well as the rates of pressure ulcers by stage (Bergquist-Beringer et al. 2013). Standardization of data and processes enables cross-organizational comparisons to be valid. The ability to compare information in a CDR across organizations is important in order to measure and compare patient outcomes across institutions, establish best practices, determine how different organizations document care differently and how those differences may affect outcomes. Organizations can exchange this information to improve patient outcomes, decrease costs, and the patient experience.

8.3.1 CDR Structure and Querying Data

Clinical data repositories (CDRs) support aggregating data into a single database instance (centralized) or data can be stored in multiple databases often at different locations. In a centralized CDR, all participating systems copy their data to a single, central data repository where data are organized, integrated, and stored using a common data standard. In a federated CDR, individual source systems maintain control over their data, but allow querying of this information by other participating systems upon request. NDNQI is an example of a centralized CDR. The advantages of the centralized model are that queries are executed more quickly and the data has better consistency due to enforcement of normalization rules and data quality checks. The HMO Research Network is an example of a federated CDR and is composed of 19 sites that share data through a virtual data warehouse (federated) to conduct research such as cancer, drug safety, cardiovascular disease and mental health (http://www.hcsrn.org/). The virtual data warehouse means that each site maintains their own data using standard definitions and formats that then can be queried by researchers across sites to conduct research. In a federated search, a query is distributed to the databases or organizations participating in the federation.

The federated search then aggregates the results that are received from the search engines. The federated approach enables each organization to maintain control and transparency on exactly when and how data for its patients is being used.

8.3.2   Standardizing Patient Data

In order to have a CDR that is useful for research and benchmarking, it is important to capture robust and complete patient health information. Patients may see many providers who use different EHR systems, which generally are not linked together. It is difficult to bring all of the patient's records from all providers together to obtain a complete patient representation. Interoperability is the term that describes the capability to exchange and share electronic information across different settings and providers. Integration of this data requires standardized data and organization of the data in standardized models. The Office of the National Coordinator and informatics organizations such as Health Level 7 (HL7) have developed a number of consensus-based information models and recommended data standards to make health information exchange possible (https://www.healthit.gov/). One such model is the Consolidated Clinical Data Architecture (CCDA), which includes critical information to support patient transitions across health settings. The newest method to standardize data is Fast Healthcare Interoperability Resources (FHIR), which uses building blocks of information that can be assembled for exchanging information or organizing data in a data warehouse to solve real world clinical and administrative problems. FHIR is suitable for use in a wide variety of contexts—mobile phone apps, cloud communications, EHR-based data sharing, server communication in large institutional healthcare providers, and much more. Clinical data repositories can benefit from FHIR standardization, making it easier to bring in clinical data from different sources.

There has also been work in nursing to standardize nursing information. Terminologies and classifications systems can be used to code and structure nursing diagnosis, interventions and outcomes. For example, Harris and colleagues have been defining information models and standard terminology for pressure ulcers (Harris et al. 2015). However, health systems design their pressure ulcer documentation in ways that may not be consistent with this pressure ulcer model, resulting in the need to map EHR data after the fact to the model to enable comparison of data across systems. Additionally, there currently are no FHIR models for the exchange of nursing data which would allow CDRs to more easily incorporate nursing information to support nursing research.

Other efforts are also underway to build information models, following standardized processes and subsequently linking standardized terminologies for extending CDRs to include data of interest to nurse researchers. An example of a CDR that contains nursing information is maintained by the University of Minnesota. Data from Fairview Health Services, University of Minnesota Physicians, and the

Fig. 8.2 CDR clinical observations by dimension

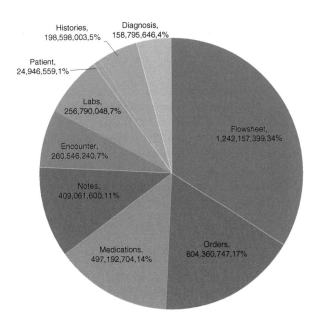

Minnesota Death Index are included. This is a centralized CDR that incorporates information from a single EHR instance from more than 40 clinics and 8 hospitals. The CDR contains over 3.6 billion clinical observations across 9 patient dimensions including patient history and demographics, encounters, procedures, diagnosis, notes, laboratories, medications, and flowsheets. The data are stored in 153 tables in the data warehouse (Johnson et al. 2015). The distribution of observations by each dimension is shown in Fig. 8.2; note that flowsheet data represent the largest portion of data in the CDR.

8.4   What Are Flowsheets?

Flowsheets are time-based templated documentation by clinicians (nurses and other health professionals) that can be highly granular with longitudinal tracking of patient clinical states and status for achieving and measuring patient outcomes. Flowsheets include structured or semi-structured data for rapid documentation and visualization of assessments and interventions reported by clinicians. They are organized similar to spreadsheets where the questions or observation types identify the rows with documentation of individual answers in the columns organized by date and time (Waitman et al. 2011). Each cell on the flowsheet represents the documentation of a particular assessment or intervention at a certain point in time. Data can be manually entered as free text, alphanumeric value, or chosen from a defined pick list. Data can also be entered automatically through interfaces with patient monitoring equipment. Each cell has an audit trail (time stamp) that identifies the person documenting, the data documented, and when the documentation was saved.

Flowsheets used by nursing are important because of the focus on important aspects of the patient documented over time that indicate changes in status as well as the interventions provided. "Nurse flowsheets" are a document type in the **Logical Observation Identifiers Names and Codes (LOINC)** clinical documentation ontology. Assessments can include the patient's physical, emotional and/or psychosocial condition, personal and family history, vital signs, response to treatment, and education provided. Examples of pressure ulcer interventions include risk assessments, activity type and frequency, positioning, pressure ulcer staging, and wound healing.

8.4.1 How Do Organizations Decide What to Record on Flowsheets?

Flowsheets are multifaceted and are created in a systematic way so clinicians can document findings using a systematic approach (e.g., following a head to toe assessment); subsequently, clinicians can quickly locate the data to understand patient trends and outcomes of care (e.g., temperature or intake over time). During EHR implementation, the system vendor may provide standard flowsheet templates that can be customized by the organization to meet their unique patient needs; organizations can also completely customize flowsheets without vendor assistance. Electronic flowsheets can facilitate the creation of clinical decision support (CDS) tools in the form of reminders, additional documentation options and cascading of questions if certain criteria are met. Flowsheets can provide applicable documentation tools for regulatory requirements, quality reporting, and to support evidence-based practices and best practices.

Decisions regarding what content to include on flowsheets (and where to include that content) differ between health care organizations. Often, these decisions are based on clinical workflow, patient populations, and efficiency. For example, a flowsheet template (screen view in an EHR) for vital signs can be used to focus on a specific group of questions for intensive care patients which can vary compared to an outpatient setting. In addition, flowsheets can be built as a form of clinical decision support—helping the clinician to determine what is expected and what is not through defined normal assessments, links to policies/procedures, and practice expectations. Practice guidelines and procedures can be internally generated or externally supplied by outside content vendors. For example, evidence-based chronic disease management and patient education content may be supplied by a vendor that is integrated with the flowsheets.

8.4.2 Strengths and Challenges of Flowsheet Data

There are strengths and challenges in using flowsheet data for big data science. Strengths of flowsheet data are numerous. First, flowsheets provide a history and account of the patient's care over time, offering a view of the patient's story and serves as a permanent record for purposes of look back (legal or otherwise). Second, flowsheet data can be tailored to the unique circumstances and needs of the patient and in a variety of care settings (e.g. inpatient, procedural areas, ambulatory care, home care, and long-term care). Third, flowsheets provide a useful communication tool for exchange of information with other providers over time and in other settings. Finally, when data entered into flowsheets are structured, they can be used in a variety of ways including sharing information across care settings and for patient or population level reporting and analytics. This allows trending within patient visits, detecting patterns in populations, and the ability to automate quality and regulatory reporting.

While the data within flowsheets are a rich source of information, there is limited standardization of flowsheet content and how staff documents that content. Available data may be incomplete because flowsheet documentation can be time consuming when entering discreet data. Also, some organizations "chart by exception" (Smeltzer et al. 1996), resulting in missing data if staff do not to understand the rules for this type of charting policy and philosophy. The learning curve to become proficient in documenting using flowsheets can be challenging; novice nurses can be reluctant to add data fields or rows for important information about their patients, potentially resulting in documentation gaps. Redundant or repetitive data entry occurs among various disciplines (e.g. nurses and respiratory therapy may both chart lung sounds). Changes, updates, and optimization of the EHR that affect flowsheets can occur frequently (e.g. monthly, quarterly, annually) and thus, communication regarding these changes needs to be robust and well understood, so that staff use the flowsheets as intended with each new version of the EHR. Standardized documentation is required to support data sharing, comparing, and reuse both within and across locations and entities (Kim et al. 2008). It is important for health care organizations to have a robust process for decision making for standardization across settings and disciplines; lack of standardization results in the inability to share or collate information across those settings.

Hospitals are complex adaptive systems requiring robust support for accurate and timely clinical documentation (Clancy et al. 2006). As such, flowsheet documentation is not the only tool to fulfill this need, but is one of the primary tools for nursing documentation. As a result, understanding the overall patient story may not be represented by flowsheet documentation alone; additional information from the medication administration record (MAR), care plan, provider progress notes, and other sources are required. Even so, the ever expanding amount and complexity of flowsheet data recorded over time necessitates structure and standardization to perform reporting and analytics on these data (Waitman et al. 2011).

Given all of these challenges, one might struggle to understand how flowsheets could be used to the benefit of patients and the nursing profession. A real benefit is the ability to use documentation tools such as the Braden skin assessment across settings. Use of such a standard tool allows comparable assessments over time and across settings. Clearly, standardized information models that include essential and generalizable concepts of interest for analytics and research are needed. This is where definition of data sets and standardization of documentation for nursing care become critically important factors for the nursing profession to collect, share, and compare flowsheet data across organizations.

8.4.3   Example of Pressure Ulcer

Pressure ulcers reduce quality of life, increase mortality, and lead to unnecessary costs (Bergquist-Beringer et al. 2013); they are considered adverse events as a result of potentially poor care. Patients at risk, if identified early, can avoid pressure ulcers. This has an impact on the bottom line for hospitals financially. Since nurses are the front-line contacts, frequent patient assessments identify patients at risk and interventions provided by nurses and others can reduce the incidence of pressure ulcers. Typically, these frequent assessments and interventions are documented in EHR flowsheets. Therefore, flowsheet data are valuable for a precise understanding of patient care to extend knowledge for comparative effectiveness research across institutions and knowledge discovery for tailoring evidence-based practice interventions to patient risk factors.

In a study from the University of Minnesota's AHC-IE (Westra et al. 2016), a random subset of de-identified data documented in the EHR between October 20, 2010 and December 26, 2013 was extracted containing 66,660 patients with 199,665 encounters from inpatient and outpatient settings. The flowsheet data represented 562 templates (screen views), with 2693 groups of observations (assessments and interventions), 14,450 unique flowsheet measures (observations), and 153,049,704 data points or observations. From this subset, there were 84 unique concepts for pressure ulcer assessments representing 96 unique flowsheet observations. A frequently used standardized assessment, the Braden Scale, was documented 475,091 times for 52,104 encounters representing 41,181 hospitalized patients. On average, the Braden Scale was documented 9.1 times per patient encounter. The frequent assessment, captured as normal documentation, provides an unprecedented amount of data to track patient risk for pressure ulcers over time. Approximately 10,000 patients had more than one encounter indicating that risk of pressure ulcers can be tracked across admissions. The cost of reusing EHR data versus original data collection is a tremendous savings for conducting big data research. There are several challenges with the use of flowsheet data that researchers need to be aware of. One is that the Braden Scale occurred on 34 different templates (screen views) such as adult medical-surgical units, obstetrics, behavioral health, observation units, and other settings. Non-standardized observations, however, can be a challenge. For instance, additional assessments of the wound base lacked specificity, such as "unable to visualize" (n = 825) and 363 "other" (n = 363) times. Similarly, for documentation of treatment interventions, the most documented treatment was "per plan of care" (n = 151). As each patient's plan of care differs, this gives no discrete information about what type of treatment the patient received for their pressure ulcer.

These strengths and challenges are demonstrated using the example of pressure ulcer flowsheets from the AHC-IE. The documentation of pressure ulcers is part of skin assessments and includes specialized risk assessments, descriptions, and interventions for prevention or treatment. One problem that is evident is that duplicate documentation is present in the pressure ulcer data set. There are two fields listed on the flowsheet for staging—one for the nurse and one for a trained Wound Ostomy Continence (WOC) nurse. They have similar, but different pick lists based on scope of practice, as can be seen in Fig. 8.3.

Due to the large number of data points and the variation in how they are documented in flowsheets make it difficult to extract exact data for pressure ulcer reporting. For instance, users can document multiple answers from a pick list in a single

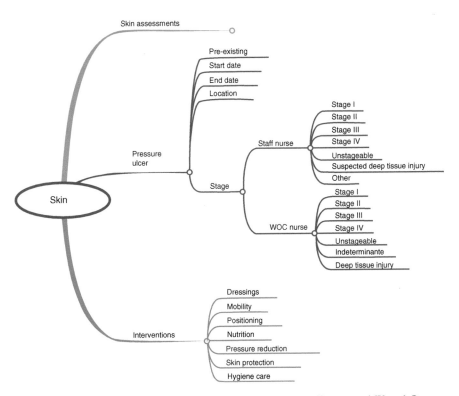

Fig. 8.3 Documented assessments for pressure ulcer stage by staff nurse and Wound Ostomy Continence (WOC) nurse

date/time field, making these data difficult to discern. In the case of pressure ulcers wound base assessment, the pick list includes 19 choices of descriptors ranging from color (e.g. pink, gray, black), painful, blanchable, non-blanchable, tunneling present, unable to visualize, and other. This means there are numerous possible combinations of documentation that could occur for this one assessment. There is a standard information model that includes the most common questions and answers that can provide a way to map flowsheet data from EHRs for comparison across systems (Harris et al. 2015). The information model can also be useful for optimizing systems to decrease options for standardization of data and save time when reusing the data for big data research.

Flowsheet data are important to tell the patient story and capture essential data for big data research. Flowsheets are multifaceted and are created in a systematic way so clinicians can document findings using a systematic approach (e.g. following a head to toe assessment); subsequently, clinicians can quickly locate the data to understand patient trends and outcomes of care (e.g. temperature or intake over time). In addition, flowsheets can be built as a form of clinical decision support—helping the clinician to determine what is expected and what is not through defined normal assessments, links to policies/procedures, and practice expectations. Practice guidelines and procedures can be internally generated or externally supplied by outside content vendors. They provide a visual of patient status over time, quickly provide communication to the healthcare team, and can provide trending data for health trajectory data. The challenge is the lack of standardization of data collected, potential for incomplete data, and duplication of data across time, disciplines, and locations. Standardized information models that include essential and generalizable concepts of interest for analytics and research are needed either for design of EHRs or to normalize data in CDRs.

8.5 Standardization Essential for Big Data Science

As emphasized previously, big data science requires common data (information) models and standardized data to aggregate the data within and across EHRs. As evidenced from the previous pressure ulcer example, implementation of EHRs requires an understanding of documentation requirements for multiple providers and multiple requirements for use of information. Certainly high quality EHR documentation is critical to ensure optimal patient care, support communication between providers, and coordinate care across time and settings. If EHRs are designed by attending to both primary and secondary use of data, considerable effort can be saved normalizing the data for research. Secondary use of EHR data is facilitated by having an information model for consistent capture and representation of data in standardized ways (processes) using standardized data elements and terminologies. If process and data standardization is not implemented from the beginning, then considerable effort is required to rework the data post-hoc (Goossen and Goossen-Baremans 2013)

8.5.1 Nursing Information Models

Big data science with flowsheet data will only become a reality with a commonly accepted information model representing clinical knowledge across settings. Multiple terms can be used to describe the result of this modeling process such as common data model, information model, and domain analysis model (Goossen et al. 2010). Even though different nuances are represented by these terms, the point is that these models identify content and organization in the database in ways that standardize the data and allow for application of data standards to generate and communicate flowsheet data in EHR's across providers and systems. An information model is constructed to represent standardized and reusable clinical concepts, their context and requirements (Tao et al. 2013). With such modeling, common data elements and terminologies are required to connect clinical information over time and across location irrespective of technology or different EHRs. In this sense, a healthcare information model enables the construction and specification of concepts and data elements representing clinical knowledge in computable and interoperable environments. The benefits of information modeling are that it standardizes ways for (1) analyzing, sorting, formalizing and structuring data elements for clinical use; (2) conceptual modeling of data elements, structures, and relationships in CDRs; (3) deploying data element standards in different information systems; and, (4) ensuring quality control for quality improvement, clinical purposes and research. Flowsheet data can be mapped to the concepts within an information model. Once mapped and standardized, flowsheet data built on an information model can facilitate conducting comparative effectiveness research across topics and care settings. Information searches within structured flowsheet data will also become more accurate and efficient. The vision is to build a global repository to store and share nursing and interprofessional information models (Goossen 2014).

8.5.2 Example Nursing Information Models and Processes

A data-driven process was used to generate a standardized information model for skin and wound assessment across several acute care organizations (Harris et al. 2015). Six organizations participated, by sharing data from EHRs to create a pressure ulcer model from flowsheet data and map concepts in the information model to standardized data. The goal was to create a precise and unambiguous model that could be generalized across information systems and organizations that included a generalized skin assessment, pressure ulcer risk assessment, and skin alteration/ wound assessment. Terms were compared from each of the organizations' EHRs. A consensus-based approach was used to determine essential content and definitions. Subsequently, concepts were compared to data standards using LOINC for assessment concepts and Systematized Nomenclature of Medicine Clinical Terms (SNOMED-CT) for answers or choices in picklists. The coding standards are consistent with Federal standards for health information exchange. There were 419 unique concepts identified; of these, 70% could be mapped to existing data standards. The remaining concepts were submitted for inclusion in LOINC and SNOMED-CT. The pressure ulcer information model can be found at http://www.fhims.org/press_ulcer.html.

8.5.3   National Collaborative to Standardize Nursing Data

The University of Minnesota initiated an effort to create a National Action Plan for sharable and comparable nursing data for big data science through annual conferences and ongoing virtual workgroups (Westra et al. 2015). The Nursing Knowledge: Big Data Science conference was first convened in 2013 to develop the National Action Plan with virtual working groups focused on educational issues, health policy, data standardization, streamlining of workflow, and data science. The intent is to have deliverables that are shared openly for consistency in building and optimizing EHRs and other information systems to create and use sharable and comparable nursing data. Two groups have specifically focused on development of information models and mapping concepts from information models to national data standards.

The development of flowsheet information models was initiated by the University of Minnesota and partially funded by their CTSA. Subsequently, the "Clinical Data Set and Analytics" workgroup from the Nursing Knowledge: Big Data Science initiative is collaborating across health systems to validate assessments and interventions in 14 information models. Data were derived from practice to create and validate information models. Priorities for information model development first focused on flowsheet data to support five clinical quality measures related to: pain, falls, pressure ulcers, catheter associated urinary tract infection (CAUTI), and venous thrombosis embolism (VTE). The processes for information model development and validation build on previous efforts to create information models for nursing data (Goossen 2014; Harman et al. 2012; Harris et al. 2015; Warren et al. 2015) After developing information models for the five clinical quality measure related flowsheet data, additional physiological and behavior health information models subsequently were developed and ready for validation. The information models include assessments, goals, interventions, and outcomes. The resulting 14 information models developed are shown in Table 8.1.

A second workgroup focuses on mapping concepts from a minimum physiological nursing assessment to LOINC and SNOMED-CT. The assessment framework for mapping nursing assessments is shown in Fig. 8.4.

Table 8.1 Flowsheet information models to data in CDRs

Information model name	Definition
Review of systems/clinical quality measures	
Cardiovascular system	The heart and the blood vessels by which blood is pumped and circulated through the body
Gastrointestinal system	The digestive structures stretching from the mouth to anus, but does not include the accessory glandular organs (liver, biliary ducts, pancreas)
Genitourinary system (including CAUTI)	All organs and their functions involved in the formation and release of urine including external genitalia. Included are data related to catheter associated urinary tract infections (CAUTI)
Muscoloskeletal system	The combination of muscles and nerves working together to permit movement. Musculoskeletal conditions range from fine to gross motor functioning
Pain	An unpleasant sensation induced by noxious stimuli which are detected by nerve endings of nociceptive neurons
Peripheral neurovascular system (including VTE)	The combination of circulation, sensation, or motion of an extremity to assess, prevent, or treat venous thromboembolism (VTE), the formation of blood clots in the vein (s)
Respiratory system	The tubular and cavernous organs and structures, by means of which pulmonary ventilation and gas exchange between ambient air and the blood are brought about
Pressure ulcers	Injuries to skin and underlying tissue resulting from prolonged **pressure** on the skin
Falls (safety)	An unplanned event which results in a person coming to rest inadvertently on the ground or floor or other lower level (with or without injury)
Vital signs, height/weight	The measurement of basic body functions that may be monitored or measured, namely pulse rate, respiratory rate, body temperature, and blood pressure as well as height and weight
Psychosocial/behavioral health	
Aggression and interpersonal violence	The intentional, non-consensual infliction of physical, verbal and/or psychological harm on one or more persons by another person
Psychiatric mental status exam	A person's mental state, with a primary focus on evaluating psychiatric signs and symptoms and ruling out known physiological etiologies
Suicide and self harm	The intentional infliction of physical injury or death on oneself without a motive of sexual arousal or gratification
Substance abuse	Use of substances either in quantities, or for purposes, other than those indicated for effective treatment of a specific physical or mental condition

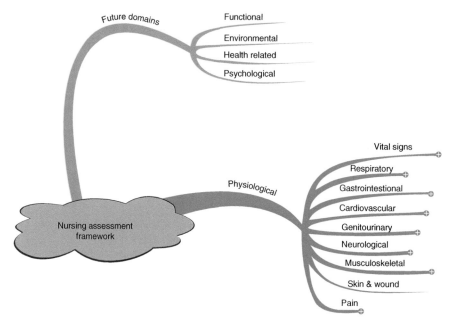

Fig. 8.4 Nursing assessment framework

The framework and subsequent concepts within it were developed through a collaboration across six organizations, using a data-driven process, identified common nursing assessment items that constitute 106 observations (50% new LOINC), and 348 observation values (20% New SNOMED CT) organized into fifteen panels (86% new LOINC) (Matney et al. 2016) (https://search.loinc.org/search.zul?query=nursing+physiologic).

8.6   Conclusion

Clinical data repositories increasingly are used for big data science, and flowsheet data can extend current CDRs with rich, highly granular data documented by nursing and other healthcare professionals. Standardization of the data, however, is required for it to be useful for big data science. In this chapter, an example of one CDR funded by NIH's CTSA demonstrates how flowsheet data can add data repositories for big data science. A specific example of pressure ulcers demonstrates the strengths of flowsheet data and also the challenges of using this data. Through standardization of this highly granular data documented by nurses, a more precise understanding about patient characteristics and tailoring of interventions provided by the health team and patient conditions and states can be achieved. Additional efforts by national workgroups to create information models from flowsheets and standardize assessment terms are described to support big data science.

References

Bergquist-Beringer S, Dong L, He J, Dunton N. Pressure ulcers and prevention among acute care hospitals in the united states. Jt Comm J Qual Patient Saf. 2013;39(9):404–14.

Bodenheimer T, Sinsky C. From triple to quadruple aim: care of the patient requires care of the provider. Ann Fam Med. 2014;12(6):573–6. doi:10.1370/afm.1713.

Clancy TR, Delaney CW, Morrison B, Gunn JK. The benefits of standardized nursing languages in complex adaptive systems such as hospitals. J Nurs Adm. 2006;36(9):426–34. doi:00005110-200609000-00009 [pii].

Goossen WT. Detailed clinical models: representing knowledge, data and semantics in healthcare information technology. Healthc Inform Res. 2014;20(3):163–72. doi:10.4258/hir.2014.20.3.163.

Goossen WT, Goossen-Baremans A. Clinical professional governance for detailed clinical models. In: Hovenga EJS, Grain H, editors. Health information governance in a digital environment. Amsterdam: IOS Press; 2013. p. 231–59.

Goossen W, Goossen-Baremans A, van der Zel M. Detailed clinical models: a review. Healthc Inform Res. 2010;16(4):201–14. doi:10.4258/hir.2010.16.4.201.

Goossen W, Boterenbrood F, Krediet I. Exchanging nursing oncology care data with use of a clinical data ware house. ETELEMED 2013: The Fifth International Conference on eHealth, Telemedicine, and Social Medicine; 2013 Feb 5. p. 183–7.

Harman TL, Seeley RA, Oliveira IM, Sheide A, Kartchner T, Woolstenhulme RD, et al. Standardized mapping of nursing assessments across 59 U.S. military treatment facilities. AMIA Annu Symp Proc. 2012;2012:331–9.

Harris MR, Langford LH, Miller H, Hook M, Dykes PC, Matney SA. Harmonizing and extending standards from a domain-specific and bottom-up approach: an example from development through use in clinical applications. J Am Med Inform Assoc. 2015;22(3):545–52. doi:10.1093/jamia/ocu020.

Johnson SG, Byrne MD, Christie B, Delaney CW, LaFlamme A, Park JI, et al. Modeling flowsheet data for clinical research. AMIA Joint Summits on Translational Science Proceedings AMIA Summit on Translational Science; 2015. p. 77–81.

Kim H, Harris MR, Savova GK, Chute CG. The first step toward data reuse: disambiguating concept representation of the locally developed ICU nursing flowsheets. Comput Inform Nurs. 2008;26(5):282–9. doi:10.1097/01.NCN.0000304839.59831.28.

Matney SA, Settergren TT, Carrington JM, Richesson RL, Sheide A, Westra BL. Standardizing physiologic assessment data to enable big data analytics. West J Nur Research. 2016:63–77.

Montalvo I. The national database of nursing quality indicators™ (NDNQI®). OJIN. 2007;12(3):Manuscript 2. doi:10.3912/OJIN.Vol12No03Man02.

Smeltzer CH, Hines PA, Beebe H, Keller B. Streamlining documentation: an opportunity to reduce costs and increase nurse clinicians' time with patients. J Nurs Care Qual. 1996;10(4):66–77.

Tao C, Jiang G, Oniki TA, Freimuth RR, Zhu Q, Sharma D, Pathak J, Huff SM, Chute CG. A semantic-web oriented representation of the clinical element model for secondary use of electronic health records data. J Am Med Info Association: JAMIA. 2013;20(3)554–62.

Waitman LR, Warren JJ, Manos EL, Connolly DW. Expressing observations from electronic medical record flowsheets in an i2b2 based clinical data repository to support research and quality improvement. AMIA Annu Symp Proc. 2011;2011:1454–63.

Warren JJ, Matney SA, Foster ED, Auld VA, Roy SL. Toward interoperability: a new resource to support nursing terminology standards. Comput Inform Nurs. 2015;33(12):515–9. doi:10.1097/CIN.0000000000000210.

Westra BL, Latimer GE, Matney SA, Park JI, Sensmeier J, Simpson RL, et al. A national action plan for sharable and comparable nursing data to support practice and translational research for transforming health care. J Am Med Inform Assoc. 2015;22(3):600–7. doi:10.1093/jamia/ocu011.

Westra BL, Christie B, Johnson SG, Pruinelli L, LaFlamme A, Park JI, Sherman SG, Byrne MD, Svenssen-Renallo P, Speedie S. Expanding Interprofessional EHR Data in i2b2. AMIA Jt Summits Transl Sci Proc. 2016;20:260–8. eCollection 2016.

Chapter 9
Working in the New Big Data World: Academic/Corporate Partnership Model

William Crown and Thomas R. Clancy

Abstract The emergence of national and regional research networks consisting of providers and consumers of data are generating large-scale data sets that can be used for secondary research purposes. Many research collaboratives develop by building on past associations with one or two constituencies—typically academic institutions and integrated healthcare delivery systems. An emerging model is the academic/corporate model whose research partners include life science companies, professional organizations and academic institutions. This chapter will focus on the benefits and challenges of academic/corporate research collaboratives using OptumLabs as an exemplar.

Keywords Academic/corporate partnerships • Big data • Claims data • Intellectual property • Electronic health records • Natural language processing • Machine learning • OptumLabs

W. Crown, Ph.D., M.A. (✉)
Chief Scientific Officer, OptumLabs™, Cambridge, MA, USA
e-mail: William.crown@optum.com

T.R. Clancy, Ph.D., M.B.A., R.N., F.A.A.N.
University of Minnesota School of Nursing,
Minneapolis, MN, USA

© Springer International Publishing AG 2017
C.W. Delaney et al. (eds.), *Big Data-Enabled Nursing*, Health Informatics,
DOI 10.1007/978-3-319-53300-1_9

9.1 The Evolving Healthcare Data Landscape

The availability of data to support healthcare research, clinical and policy translation, and innovation is expanding rapidly. This expanding data infrastructure offers the potential to accelerate the generation of evidence about the value of clinical and policy interventions, hasten the adoption of such evidence, and encourage innovation. However, the specific objectives of large-scale data aggregation efforts differ considerably and have implications for who has access to the data and for what purpose.

Much of the healthcare evidence from retrospective claims-based studies has been conducted by academics using Medicare data. Notable examples include the Dartmouth Atlas studies on variation in care (Manning et al. 1987; Weeks 2016). However, an Institute of Medicine study demonstrated that measures of care variation and value differed substantially between Medicare and commercially insured patients (Institute of Medicine 2013a, b). This highlights the issues of generalizability of conclusions drawn from observational studies using data reflecting the healthcare utilization patterns of different patient populations. Retrospective database studies of healthcare interventions, policies, variation in care and many other topics have generated a massive literature but there are relatively few retrospective database studies of nursing care (Dunn et al. 2015; Aponte 2010; Smaldone and Connor 2003). The reasons for this are not immediately obvious. Dunn et al. (2015) suggest that the relative dearth of nursing research using retrospective data may be partly due to lack of knowledge about available research databases on the part of nursing researchers. Another issue may be lack of knowledge about the resources and analytic skills needed to successfully analyze retrospective data. The medical claims and electronic medical record data assets available to study the value of nursing services are substantial and improving rapidly. But the data management and statistical skills needed to analyze such large and complex datasets can be a significant barrier to initiating and completing nursing research projects.

Several groups have begun to aggregate very large healthcare databases (Curtis et al. 2014; Wallace et al. 2014; Oye et al. 2015). These databases differ in terms of the types of research they are intended to support and who has access to the data. Prominent examples include:

1. The HealthCare Cost Institute (HCCI) is a non-profit established in 2011 to facilitate access to commercial healthcare claims data for research. HCCI pools de-identified medical claims, drug claims, and enrollment data for several large insurers (approximately 40 million lives). Because the claims reflect reimbursed amounts from several insurers, HCCI is particularly valuable for conducting studies of the cost of healthcare interventions (www.healthcostinstitute.org).

2. PCORnet is a network of clinical sites collectively reflecting electronic medical record data and patient reported outcomes on approximately 47 million patients. PCORnet has been constructed with funding from the Patient Centered Outcomes Research Institute to proactively create a research infrastructure to support patient centered outcomes research (PCOR). The PCORnet infrastructure is

being used to conduct a number of large-scale clinical trials and observational studies (www.pcornet.org).

3. Mini-Sentinel is a network of medical claims and electronic medical records from multiple insurers. It was established under funding from the U.S. Food and Drug Administration (FDA) in 2008 to facilitate comparative safety studies for pharmaceutical products. The database currently contains data on over 193 million individuals with 39 million individuals currently enrolled and accumulating data. Access to these data is provided to researchers conducting studies requested by the FDA (www.mini-sentinel.org).

4. OptumLabs was established in 2013 in a partnership between Optum and the Mayo Clinic. Its core linked data assets include de-identified claims data for privately insured and Medicare Advantage enrollees, de-identified electronic medical record (EMR) data from a nationwide network of provider groups, and a de-identified database of consumer lifestyle information. The database contains longitudinal health information on more than 150 million enrollees, representing a diverse mixture of ages, ethnicities and geographical regions across the United States. The EMR data is sourced from provider groups and reflects all payers, including uninsured patients (www.optum.com/optumlabs).

Despite the proliferation of data and data infrastructure, most research databases do not contain linkage of clinical information with information on healthcare utilization for the same patients. Where such data exists, it tends to be for special populations (e.g., the Veterans Administration), which are often not broadly accessible to the research community, or smaller databases from specific provider organizations such as Geisinger Health System (www.medmining.com). These data are often extremely rich but lack generalizability. Ideally, researchers would like to be able to link EMR, claims, and other data for patients representing a broad group of different patient populations, payer types, and care coordination models.

9.2   The Promise and Complexity of Working with Multiple Sources of Data

Researchers have been working with claims data for twenty-plus years. We are quick to disparage claims data because we know them so well. However, claims data have many strengths from a statistical and analytic standpoint. In particular, claims data captures patient treatment across settings in a fairly comprehensive way. Among other things, this is very valuable for measuring the comorbidities profiles of patients. Claims data also, of course, has information on healthcare utilization and cost, which are very important in the context of today's environment as we think about value in healthcare.

The availability of EHR data in the United States has expanded exponentially—primarily due to the meaningful use provisions of the Health Information Technology for Economic and Clinical Health Act (Public Law 111–5 2009). This increased availability, in combination with the well-known limitations of claims data with respect to clinical outcomes and severity measures, has spurred tremendous interest in conducting research with EHR data. Unfortunately, the state of knowledge regarding the use of EMR data for research is similar to that of the analysis of medical claims 20 years ago. This can lead to frustration when researchers attempt to use EHR data for research for the first time.

There are several characteristics of EHR data that make it challenging to use for research and which also raise specific methodological issues:

- EHR data has its origins with healthcare providers and tends to be site-specific. As a result, the ability to understand patient comorbidity profiles and the breadth of interaction of patients with the healthcare system can be very limited. Medical claims are very good at capturing the breadth of healthcare utilization across care settings but may be very poor with respect to clinical data content and quality.
- Many large clinical organizations such as integrated delivery networks (IDNs) have multiple EHRs for their different clinical sites. Moreover, these clinics may use completely different EHR vendors (e.g., EPIC, Cerner, AllScripts) and these are not typically linked, nor are the data reported in a common data format.
- Although EHRs usually have structured fields for variables such as height, weight, blood pressure, common laboratory results, etc., these fields are often empty and the data remains in unstructured notes. This makes the data very difficult to use for research.
- Data in EHRs reflects patient encounters with the healthcare system but do not capture information needed to determine denominators (e.g., health plan enrollment) needed for many research questions such as calculating incidence or prevalence rates.
- EMRs are good at capturing initial prescription orders but notoriously bad at capturing refill behavior. Claims, on the other hand, are very good at capturing refills but may miss prescriptions altogether if the patient never takes the prescription to the pharmacy to get it filled.

9.3   Implications of Linked Claims and EHR Data for Nursing Studies

There are relatively few retrospective database studies of nursing care using claims data. In addition to the issues of lack of knowledge about the availability of such data for research, and experience with conducting statistical analyses on such large data files, it may also be difficult to tease out the impact of nursing interventions as they are often not coded in the claims. This is an area where electronic health records may be helpful. Unfortunately, 80% of EHR data is unstructured text and natural language processing (NLP) can only do so much to isolate nursing interventions and their outcomes. It is important that health systems mandate the use of EHR data models that incorporate a standardized nursing language such as SNOMED-CT or LOINC that will allow healthcare data warehouses to be searched for the link between nursing interventions and clinical and financial outcomes.

For example, claims data contain at least two sets of codes that are useful for identifying nursing services. Provider specialty codes identify licensed practical nurses as a separate category. Similarly, provider category codes identify nurses, nurse practitioners, physician assistants, and registered nurses as separate categories. In general, claims data are quite good for capturing the breadth of the experience of patients, their medical comorbidities, the drugs they take, their visits, etc., but they are not very good for measuring disease severity, cancer stage, biomarkers, etc. On the other hand, electronic health records data are much stronger for capturing clinical detail but they can often be confined to particular sites such as hospitals, oncology clinics, etc. In comparison with claims data, much of the knowledge about comorbidities may be missing. As a consequence statistical models estimated with EHR data alone are likely to be biased because they will be missing important information on comorbidities. Similarly, models estimated with claims data alone are likely to be biased because they will lack important controls for clinical severity. Historically, this has been one of the fundamental challenges with the analysis of observational data; researchers are often relegated to working with subsets of data that are not complete enough to enable the derivation of reliable statistical inferences. The linkage of datasets should help to address many of these issues and improve the ability of machine learning or traditional statistical methods to generate more reliable models.

Although linkage of claims and EHR data may go a long way toward addressing the limitations of each data type analyzed in isolation, it is important to consider the potential biases or implications for generalizability that linkage may imply. For example, EHR data, which is sourced from provider organizations, typically includes patients will all types of insurance, including no insurance at all. However, claims data is associated with a particular payer. Some claims datasets contain information on multiple payers but any observed claim refers to insurance coverage provided by a particular payer. Thus, linked claims and EHR data implies that the sample represents only patients with the insurance type represented by the payer, rather than the broader set of patients in the provider setting. In other words, data linkage may reduce bias due to missing clinical variables but the results of the analysis will apply to the subpopulation with the insurance coverage represented by the claims data.

There are, of course, many other types of data beyond medical claims and electronic medical records. For example, patient reported outcomes (PRO) data is extremely limited in big administrative data sets. This is a potentially major limitation in evaluating nursing outcomes because a major component of nursing value may reside in patient satisfaction measures, pain management, and a variety of other metrics that are not captured in claims data. Nevertheless, some PRO data may reside in unstructured notes in EHR data. A good example is pain scores, which are an important patient-reported outcome. Figure 9.1 shows the distribution of pain scores in the EHR data from the OptumLabs Data Warehouse.

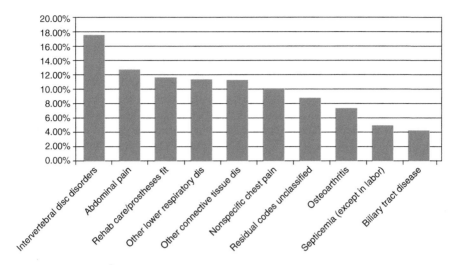

Fig. 9.1 Percent of days with a pain score by top 10 diagnoses. *Source*: Tabulations from OptumLabs data warehouse, December 2015

About 28% of the patients in the database had a valid pain score in their record. However, it was necessary to use natural language processing to search for the scores in the unstructured notes. In general, natural language processing is necessary even for items where there is a field in the structured data as clinicians often just record such values in the notes rather than click through the EHR menus to store the data in a structured database. On a positive note, several recent policy initiatives, such as the Medicare pilot programs of bundled payment for total joint replacement and the Oncology Care Model of the Centers for Medicare and Medicaid Services (CMS), include patient-reported outcomes for calculating payments and measuring quality (Lavallee et al. 2016). Such initiatives may drive uptake in the use of PROs much as the meaningful use provisions of the 2009 Health Information Technology for Economic and Clinical Heatlh Act drove adoption of electronic health records.

9.4 Big Data Methods

The growing volume, variety, and velocity of data (Laney 2001) have significant implications for the analytical methods used to draw meaningful inferences from healthcare data. Data linkages, such as EHR and medical claims, offer the potential to improve the depth of data available for analysis but also create gaps in the data where such linkages are not possible. For some types of information, such as data feeds from personal fitness devices, the data may be available for only a small proportion of the overall sample. The complexity of the data structures from these different linkages poses challenges for traditional statistical methods such as multivariate regression analysis.

The term "machine learning" refers to a family of mathematical and statistical methods that have historically been focused on prediction (Hastie et al. 2009). Machine learning methods consist of a large number of alternative methods including classification trees, random forests, neural networks, support vector machines, and lasso and ridge regression to name a few. We are often interested in prediction in health care. What strain of flu is likely to be prevalent in the coming flu season? How many vials of flu vaccination must be prepared to meet treatment demand? But prediction is not the same thing as estimating treatment effects. In clinical settings the challenge is to isolate the effect of a treatment on patient outcomes so that the correct treatment can be selected. Policy evaluations face the same statistical challenges. Some machine learning methods have the ability to estimate treatment effects and some do not. But it is fair to say that relatively little emphasis has been placed on treatment-effect estimation in the machine learning literature.

The basic approach with all machine learning is to segment the data into learning and validation data sets to develop highly accurate classification algorithms (Crown 2015). Once the algorithms have been developed they are applied to the full dataset to do the prediction. The idea is that one should be able to perform these classifications without human intervention, and the methods should also be able to operate on very large data sets and be very fast. In the machine-learning literature, this process of using learning and training datasets to develop prediction algorithms is known as K-fold cross validation. The approach is fairly straightforward. The idea is to take the initial data set and randomly split it into several (typically 5 or 10) sub-samples. For each sub-sample that is held aside, the classification algorithms are built on each of the other remaining sub-samples. Once the algorithms have been built, each is used to predict the membership prediction error that is associated with each one of the sub-samples. Finally, a sum of the prediction errors is calculated over all the subsamples. Using this approach, one can evaluate different machine learning methods simultaneously and then compare the average errors associated with each model to determine which method performs the best. The process can be completely automated. The best algorithm is applied to the entire data set—typically to do a prediction.

Some machine learning approaches use regression-based methods for prediction. For example, Lasso methods utilize a correction factor to reduce the risk of over fitting (Tibshirani 1996). Because the Lasso method can also force the coefficients of some variables to zero, it is also useful for variable selection. Most importantly, since Lasso regression involves the estimation of coefficients in a multivariate model, it is a short step to thinking about the use of machine learning to obtain estimates of treatment effects. Many researchers would feel uncomfortable letting computers choose the specification of the final model. This is understandable. However, researchers can certainly evaluate the final model for its theoretical or clinical plausibility, as well as subject it to the usual battery of specification tests. Moreover, the risk of ending up with an implausible model can be broadly managed by selection of the set of starting variables from which the model is constructed. Machine learning methods enable the starting set of variables to be much larger than is normal practice in health services research but it is not necessary to completely throw out the concept of a theoretical or clinical model. Finally, the K-fold cross validation approach used in machine learning can be thought of as a more sophisticated and systematic version of the best practice of splitting one's sample into two—one for model development and the other for final model estimation.

Unfortunately, there is nothing magical about machine learning that protects against the usual challenges encountered in observational data analysis. In particular, just because machine-learning methods are operating on big data does not protect against bias. Increasing sample size—for example, getting more and more and more claims data—is not going to correct the problem of bias if the dataset is lacking in key clinical severity measures such as cancer stage in a model of breast cancer outcomes. Large sample sizes can be helpful in estimating instrumental variables models which tend to have large standard errors.

9.5 Beyond Research—Accelerating Clinical/Policy Translation and Innovation

An important challenge with healthcare research is to accelerate the translation of research findings into meaningful clinical and policy change. Although it is often cited that it takes 17 years for new research findings to become the standard of care (Morris et al. 2011) this depends on many factors such as whether the clinician is a primary care doctor or a specialist, whether the new evidence requires a change to physician workflow. Pachman et al. (2015) found that new evidence on the use of calcium and magnesium in treating chemotherapy-induced neuropathy resulted in immediate changes in treatment.

The application of data science methods to large, complex health care datasets creates the opportunities for algorithm development and product innovation that promise to dramatically accelerate clinical translation. It is easy to imagine that in a few years, EHR systems will automatically place patients in clinical sub clusters and display, for physicians, treatment alternatives that have worked well for patients like the one being seen. Machine learning methods are already being used to identify sub clusters for patients with conditions like heart failure. When such algorithms are built into the tools being used by clinicians in the process of care, clinical translation will take place independently of the knowledge diffusion process currently associated with the peer reviewed literature.

Similarly, data science methods can be used to build predictive models that predict the risk of events before they occur. Such tools are already in widespread use but will become more robust when developed using large, complex healthcare datasets and big data methods. An example is shown in Fig. 9.2, which displays a screen shot from a model to predict diabetes (McCoy et al. 2016). This particular model was estimated using machine-learning methods on claims data alone. Machine learning algorithms can be continuously applied to claims or EHR data in healthcare settings—flagging patients at risk and triggering interventions such as nurse contacts with patients to check on diet, medication adherence, etc.

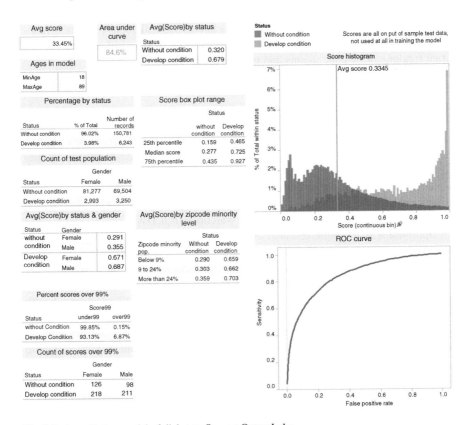

Fig. 9.2 A predictive model of diabetes. *Source*: OptumLabs

9.6   Innovation and Management of Intellectual Property in Academic/Corporate Partnerships

Data science talent is scarce and in great demand. The infrastructure required to assemble large-scale healthcare datasets and support data science analytic methods is daunting. For these reasons and more, the analysis of big healthcare data requires teamwork. Nurse scientists will need to be prepared to work with a diverse team of experts. For example, simply accessing large-scale data files may require data engineers with expertise in hardware and database structures to map raw data to different formats for preprocessing. Domain experts with a deep understanding of data dictionaries may be needed to cleanse the data, assist with feature extraction, and standardize it. Data scientists, experienced in predictive modeling using machine learning and data mining approaches will be necessary to search for and analyze patterns in the data. Statisticians may be called upon to validate study results, infer conclusions from the data and present findings using sophisticated visualization software applications. Subject matter experts in nursing will act as primary investigators to advance hypotheses for testing, develop the research design, apply for grant funding, draw conclusions and publish the results.

As can be seen, the infrastructure required to conduct big data science research can be significant. Providing both the expertise and resources available to manage big data projects within one organization requires a substantial investment. As a result, organizations will increasingly partner. An emerging model is the academic/corporate partnership, which can bridge gaps in resources and expertise needed to conduct big data research. Academic and corporate partners working together provide the opportunity to advance big data research better than if each acted alone. One exemplar is the OptumLabs research collaborative where academic health centers, life science companies and professional organizations have formed a partnership to conduct research and innovation using the OptumLabs Data Warehouse (OLDW).

OptumLabs is a unique research collaborative initially developed in a partnership between OptumLabs and the Mayo Clinic in 2013. The collaborative is made up of over 20 partners whose research teams conduct secondary data analysis using the OLDW. The OLDW contains over 150 million lives of de-identified health insurance claims data spanning a period of 20 years, linked with electronic health records. Approximately 25% of the patient lives are linked across insurance claims and electronic health records, which allows study of the entire health continuum. The use of health insurance claims data provides one example of how corporate "primary" use data has value as "secondary" use data for healthcare research.

The University of Minnesota is a partner in the OptumLabs research collaborative through its School of Nursing (UMSON). The UMSON facilitates access to the OLDW for any University of Minnesota researcher and currently is managing 11 studies from faculty in various Schools and Colleges including: Nursing, Medicine, Dentistry and Public Health. The data access model for the OLDW (See Fig. 9.3.) is unique in that researchers access the OLDW in a secure "sandbox" hosted by OptumLabs. For security of confidential and de-identified health information, data cannot be removed from the sandbox. However, partners, with approval, can install their own statistical software programs into the sandbox for data analysis.

Partners have the option to explore the OLDW prior to obtaining a data license using an OptumLabs software application named the Natural History of Disease (NHD). The NHD provides investigators the opportunity to explore OLDW data fields, determine frequency counts, create control groups, review odds ratios and determine the feasibility of conducting a study. All research applications using the OLDW are reviewed by a research committee, comprised of representatives from across the partnership, to assess feasibility and alignment with the OptumLabs mission. Partners using data from the OLDW for funded research can build the cost of data licenses into the grant budget. Intellectual property rights and commercialization of new discoveries, as a result of OLDW studies, are clarified in a "Statement of Work" agreement between the partner and OptumLabs prior to the project. Partners are encouraged to publish results of their studies and share results at partner ideas exchange meetings, which occur on a semi-weekly basis.

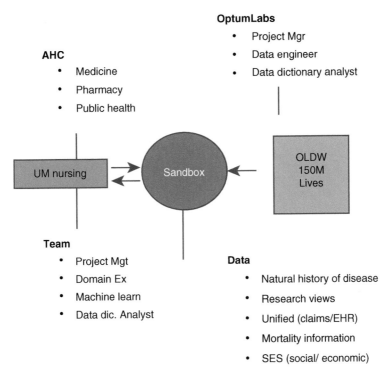

Fig. 9.3 OptumLabs research collaborative structure. University of Minnesota School of Nursing

The team science benefits of academic/corporate partnerships such as the OptumLabs research collaborative are readily apparent. Figure 9.4 presents the emerging roles needed to conduct big data research. For example, OptumLabs provides the hardware/software and data engineers to manage the OLDW and update it, extract data and prepare research views in the sandbox for partners, and consult on the data dictionary and various methods of analysis. The research partner contributes domain experts, project managers, data scientists and analysts to pre-process the data and apply statistical and computational methods in the sandbox. The semi-weekly partner ideas exchange meetings allow research teams from each of the partners to offer recommendations on study design, analysis and evaluation methods to other members. These roles can vary depending on the type and scope of the study being conducted.

Although there are benefits to academic/corporate partnerships, there also are challenges. New knowledge, such as algorithms or predictive models, created from studies using the OLDW may have commercial value. Intellectual property, licensing fees and royalties may need to be clarified up front. Data license and technical support fees may be incurred by research partners. It is important to discuss these items in advance and document them in the Statement of Work to avoid problems at a later date.

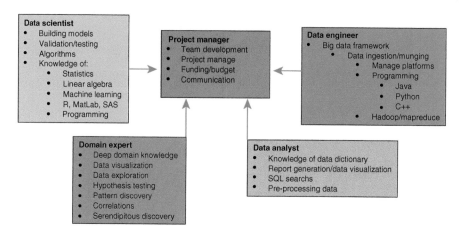

Fig. 9.4 Emerging research roles: team science and secondary analysis of big data

9.7 The Ongoing Debate About the Merits of RCTs Versus Observational Studies

A significant barrier to the adoption of new evidence from observational studies is the degree of confidence that clinicians and policy makers have in the conclusions drawn from non-randomized studies. Few topics engender stronger responses from researchers than their opinions about whether randomization is necessary in order to draw causal inferences from observational data. In one camp are those who argue that, due to the many potential sources of bias in observational studies, the best one can hope for is to draw inferences about correlation, rather than causation. Few would disagree with their conclusion that because randomization distributes both observed and unobserved characteristics of patients across treatment groups (assuming the sample size is sufficient) it is the strongest research design for the unbiased estimation of treatment effects.

On the other hand, even though researchers in the observational data camp would readily agree that randomization is the strongest research design, they assert that it does not necessarily follow that inferences about causality are impossible with observational studies. Such studies offer the opportunity to generate timely evidence on many research questions provided the data are sufficiently complete and well-measured, and appropriate research design and statistical methods are used.

Despite general perceptions to the contrary, a recent review of the literature concluded that observational studies generally generate similar average treatment effects to clinical trials when care has been taken to mirror the inclusion and exclusion criteria of the RCTs (Anglemyer et al. 2014). Earlier studies have reached similar conclusions (Concato et al. 2000; Benson and Hartz 2000).

Nevertheless, this is no reason to be complacent. Observational studies are subject to the risk of bias from a host of sources. In particular, bias can be introduced by a variety of different types of measurement error—the most prevalent probably being missing or poorly measured variables. A common example of this is studies using medical claims databases to estimate treatment effects for conditions where controls for clinical severity influence both treatment selection and outcomes. For instance, lack of data on HER-2 status, metastasis status, and cancer stage for studies of breast cancer treatment and outcomes would likely result in biased estimates of treatment effects. It is important to make the distinction, however, between data availability and observational versus randomized designs. As data availability continues to improve, the ability to include relevant control variables will also improve. From a statistical standpoint, an observational model containing a complete set of well-measured control variables should result in similar treatment effect estimates to those of an RCT designed to evaluate the same treatment in a similar patient population.

Most of the multivariate methods for estimating treatment effects focus upon the observed variables but economists, in particular, have developed methods that attempt to test for missing variables or other sources of measurement error. The method of instrumental variables is one of the best known of these methods but it is often difficult to implement because it requires the identification of variables that are correlated with treatment but uncorrelated with outcomes. This generally turns out to be a difficult task—leading to the conclusion that the researcher can sometimes make the bias problem worse by trying to repair it with instrumental variables versus acknowledging the problem and leaving it uncorrected (Crown et al. 2011).

Healthcare stakeholders need evidence to make decisions that range from coverage and reimbursement of new drugs on the part of payers to point-of-care clinical decisions on the part of physicians. Meta analyses of well-designed, large, randomized clinical trials provide the highest level of evidence for the questions they are intended to answer. But they are not very informative about how a therapy may work in real world patient populations that are usually very different from those studied in registration trials. Moreover, we simply do not have the time or the money to answer all questions with clinical trials. By necessity, observational research will be the source of much of the evidence with respect to the effectiveness, safety, and value of alternative therapies. As the quality of observational data continues to improve, so too, will the quality of observational studies themselves. Ongoing comparisons of results from observational and randomized studies evaluating comparable treatment effects in similar patient populations will continue to be helpful in informing where we can productively rely upon evidence from observational studies versus where we should use randomization to address the remaining challenges arising from confounding.

9.8 Conclusions

The availability of healthcare data is expanding rapidly. A number of large-scale claims and electronic medical record databases have been assembled to support healthcare research of different types. The missions of these large-scale data aggregation efforts vary including a focus on safety studies, health economics, patient centered outcomes, and more broad-based collaborative research that spans questions of clinical outcomes, value, and innovation.

Retrospective database studies of nursing are relatively rare. This is attributable to a variety of factors including lack of knowledge on the part of nursing researchers of available retrospective claims and EMR databases, lack of expertise in data management and statistical analysis of very large data files, and the poor coding of nursing services in claims databases. However, with the availability of electronic medical record data and, particularly, linked claims and EMR data the feasibility of conducting database studies of nursing should improve.

As both the size and complexity of healthcare databases grows, traditional statistical methods will suffer some serious drawbacks—especially as a result of missing data and the increasing non-representativeness of samples with complete information. Big data methods such as random forests, neural networks, lasso models, etc. offer new opportunities to extract insights from observational where traditional multivariate statistical models often break down.

Big data methods also have an advantage of being focused on prediction. Prediction isn't always the objective in the analysis of healthcare data but it is often an important component in improving care. Algorithms developed from observational data analysis can be built into tools used in delivery systems of care—dramatically accelerating the dissemination of knowledge into clinical practice.

For clinicians and policy makers to act on results generated by the analysis of observational data they have to believe that the evidence is credible. This requires moving beyond randomized clinical trials as the only credible source of evidence. There are valid scientific reasons for randomization being the strongest research design. However, this doesn't mean that all other designs are incapable of generating reliable evidence. Good data, coupled with strong research design and appropriate statistical methods, has the ability to generate credible evidence as well. Moreover, the evidence can be generated and made available nearly real-time on relevant patient populations.

References

Anglemyer A, Horvath HT, Bero L. Healthcare outcomes assessed with observational study designs compared with those assessed in randomized trials (review). Cochrane Database Syst Rev. 2014;4:MR000034. doi:10.1002/14651858.MR000034.pub2.

Aponte J. Key elements of large survey data sets. Nurs Econ. 2010;28(1):27–36.

Benson K, Hartz AJ. A comparison of observational studies and randomized, controlled trials. N Engl J Med. 2000;342(25):1878–86.

Concato J, Shah N, Horwitz RI. Randomized controlled trials, observational studies, and the hierarchy of research designs. N Engl J Med. 2000;342(25):1887–92.

Crown W. Potential application of machine learning in health outcomes research—some statistical cautions. Value Health. 2015;18(2):137–40.

Crown W, Henk H, VanNess D. Some cautions on the use of instrumental variables (IV) estimators in outcomes research: how bias in IV estimators is affected by instrument strength, instrument contamination, and sample size. Value Health. 2011;14:1078–84.

Curtis L, Brown J, Platt R. Four health data networks illustrate the potential for a shared national multipurpose big-data network. Health Aff. 2014;33(7):1178–86. doi:10.1377/hlthaff.2014.0121.

Dunn SL, Arslanian-Engoren C, Dekoekkoek T, Jadack R, Scott L. Secondary data analysis as an efficient and effective approach to nursing research. West J Nurs Res. 2015;37(10):1295–307.

Hastie T, Tibshirani R, Friedman J. The elements of statistical learning: data mining, inference and prediction. New York: Springer Verlag; 2009.

Institute of Medicine. Large simple trials and knowledge generation in a learning health system: workshop summary. Washington, DC: The National Academies Press; 2013a.

Institute of Medicine. Variation in healthcare spending: target decision making, not geography. Washington, DC: The National Academies Press; 2013b. Released July 24.

Laney D. 3D management: controlling data volume, velocity, and variety. Stamford: Meta Group Inc.; 2001.

Lavallee D, Chenok K, Love R, Petersen C, Holve E, Segal C, Franklin P. Incorporating patient-reported outcomes into health care to engage patients and enhance care. Health Aff. 2016;35(4):575–82.

Manning WG, Newhouse JP, Duan N, Keeler EB, Leibowitz A. Health insurance and the demand for medical care: evidence from a randomized experiment. Am Econ Rev. 1987;77:251–77.

McCoy R, Smith S, Nori V. Development and validation of HealthImpact: an incident diabetes prediction model based on administrative data. Health Serv Res. 2016; doi:10.1111/1475-6773.12461. [Epub ahead of print]

Morris Z, Wooding S, Grant J. The answer is 17 years, what is the question: understanding time lags in translational research. J R Soc Med. 2011;104(12):510–20. doi:10.1258/jrsm.2011.110180.

Oye K, Jain G, Amador M, Arnaout R, Brown J, Crown W, Ferguson J, Pezalla E, Rassen J, Selker H, Trusheim M, Hirsch G. The next frontier: fostering innovation by improving health data access and innovation. Clin Pharmacol Ther. 2015;98(5):514–21.

Pachman DR, Ruddy K, Sangarlingham LR, et al. Calcium and magnesium use for oxaliplatin-induced neuropathy: a case study to assess how quickly evidence translates into practice. J Natl Compr Canc Netw. 2015;13(9):1097–101.

Public Law 111–5. Health Information Technology for Economic and Clinical Health Act. Title XIII of the American Recovery and Reinvestment Act of 2009. 110th United States Congress. Washington, DC: United States Government Printing Office; 2009.

Smaldone AM, Connor JA. The use of large administrative data sets in nursing research. Appl Nurs Res. 2003;16:205–7. doi:10.1016/s0897-1897(03)00040-5.

Tibshirani R. Regression shrinkage and selection via the lasso. J R Stat Soc. 1996;Series B:58:267–88.

Wallace P, Shah N, Dennon T, Bleicher P, Crown W. OptumLabs: building a novel node in the learning health care system. Health Aff. 2014;33(7):1187–94.

Weeks W. Geographic variation in medicare expenditures, 2003–2012. JAMA Intern Med. 2016;176(3):405–7. doi:10.1001/jamainternmed.2015.7814.

Case Study 9.1: Academic/Corporate Partnerships: Development of a Model to Predict Adverse Events in Patients Prescribed Statins Using the OptumLabs Data Warehouse

Chih-Lin Chi and Jin Wang

Abstract This case study provides an illustration of the roles and methods data science team members use to collaboratively complete an academic/corporate big data project. The project focuses on creation of a machine-learning model, using the OptumLabs Data Warehouse that predicts adverse events in patients prescribed statin drug therapy. The benefits and challenges of using large complex health insurance claims data are discussed as well as the various machine-learning approaches that can be applied.

Keywords OptumLabs • Machine • learning • Academic/corporate partnerships • Statin therapy • Data science • Big data • Cardiovascular disease • Data-mining • Nursing informatics

9.1.1 Introduction: Research Objective

This case study discusses an example of an academic corporate partnership, using the OptumLabs Data Warehouse (OLDW), aimed at reducing the frequency of adverse events in patients prescribed statin drugs. The OLDW includes de-identified administrative claims data on commercially insured and Medicare Advantage enrollees, as well as de-identified electronic medical record (EMR) data from a nationwide network of provider groups. The OLDW contains longitudinal health information on more than 150 million enrollees, representing a diverse mixture of ages, ethnicities and geographical regions across the United States. A detailed description of the OLDW can be found in Chap. 5.

C.-L. Chi, Ph.D., M.B.A. (✉)
Optum Labs™, Cambridge, MA, USA

School of Nursing, University of Minnesota, Minneapolis, MN, USA

Institute for Health Informatics, University of Minnesota, Minneapolis, MN, USA
e-mail: cchi@umn.edu

J. Wang, M.S., M.H.I. (✉)
Optum Labs™, Cambridge, MA, USA

Institute for Health Informatics, University of Minnesota, Minneapolis, MN, USA

Statins are highly effective drugs for lowering low-density lipoprotein cholesterol, non-high-density lipoprotein cholesterol, and apolipoprotein B levels in plasma, all of which are key contributors to the leading cause of death worldwide, atherosclerotic cardiovascular disease (ASCVD) (Cohen et al. 2012; Taylor et al. 2013). Extensive clinical trials and studies support the use of this class of drugs for the treatment, primary prevention, and secondary prevention of ASCVD (Stone et al. 2014). A study from the U.S. Department of Health and Human Services demonstrated that, during the 2007–2010 time period, approximately 47% of Americans older than 65 years were taking statins or other cholesterol-lowering drugs (National Center for Health Statistics 2014). This usage rate of cholesterol-lowering drugs has increased approximately seven-fold since 1988–1994, due in part to the introduction and acceptance of statin drugs. However, it has been reported that more than 50% of patients discontinue statin medication within 1 year after treatment initiation (Cohen et al. 2012). Adverse events (AEs) associated with statin use is the primary reason for such discontinuation (Mancini et al. 2011; Maningat et al. 2013).

Statin-related AEs include muscle symptoms and more uncommonly, myopathy, rhabdomyolysis, renal events, hepatic events, increased risks for hyperglycemia and cognitive effects (U.S. Food and Drug Administration 2014). The consequence of AEs results in the discontinuance of statin treatment and hence the clinical benefits of lower plasma cholesterol levels and decreased risk of primary and secondary cardiovascular events.

Typical strategies to manage and control AEs include reducing dosage, changing statin agents, using a statin holiday, or trying alternative cholesterol-lowering therapy (Brewer 2003; Fernandez et al. 2011). In practice settings, these traditional approaches control the AE problem only after the event, leading many people to discontinue statin treatment. However, these strategies can also be problematic. Switching to a lower statin intensity can reduce AE risk, but also lessens the reduction in ASCVD risk (Fernandez et al. 2011). Switching to non-statin therapy is problematic as well, given that many non-statins have worse AE profiles and poorer ASCVD risk reduction than statins.

This project will use machine-learning, predictive modeling, and big-data approaches to develop a proactive strategy for patients at risk for statin AEs. By identifying patients at risk for statin AEs, providers can prescribe a personalized statin treatment plan that minimizes the individual's risk of AEs. The idea of 'personalization' and 'proactiveness' is similar to the decision-making process of a courtroom judge who typically considers previously similar cases to help conclude a decision. In a similar manner, when developing a proactive strategy, we use a machine learning algorithm to identify previously similar patients and their treatment plans from the data. We then generalize the treatment plan and use it to predict the personalized treatment plan that maximizes the reduction of AE risk to support the prescription. Big data plays an important role when developing this proactive strategy because of the large number of previously similar patients, treatment plans, and AE outcomes.

9.1.2 Resources Needed for Big-Data Analysis in the OptumLabs Project

9.1.2.1 Multi-disciplinary Teamwork

Big data research studies are a team science, and collaboration across organizations and among domain experts is important to successfully conduct projects. The statin project is an exemplar for academic/corporate partnerships that involve interdisciplinary collaboration among OptumLabs, nursing, medicine, pharmacy, and computer science. The interdisciplinary teamwork system structure is shown in Fig. 9.1.1. Each expert collectively contributes domain knowledge in a specific area for the project.

The benefits of an academic/corporate partnership are often not apparent. For instance, OptumLabs not only provides data, but also brings valuable resources to the partnership. OptumLabs analysts provide technical support for database questions and how to navigate the OLDW data dictionary. Specific questions or concerns are answered by either OptumLabs support analysts, domain experts from the University of Minnesota, or both. In addition, supporting documents, training, and monthly OLDW user group meetings are great resources to help investigators familiarize themselves with how to best use the OLDW for research. OptumLabs also provides an annual Research and Translation forum and semi-weekly Partner Ideas Exchange meetings aimed at sharing research ideas, encouraging research collaboration, presenting future, ongoing or completed projects, and reporting research results using the OLDW.

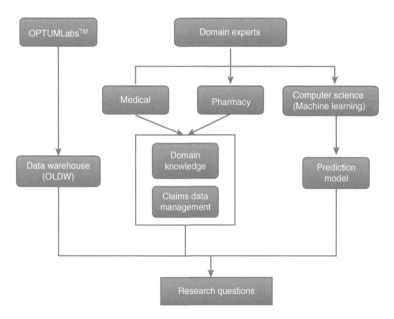

Fig. 9.1.1 The interdisciplinary teamwork system structure

The cohort for this study included patients diagnosed with cardiovascular disease (CVD) who were prescribed statins. To select CVD patients using claims data, clinician expertise was needed to determine which CVD billing codes (diagnosis and procedure) to include, and then to identify those codes in medical claims. Each OLDW claim line included both primary and secondary ICD-9 diagnosis codes (not to exceed nine) and six CPT/HCPCS procedure codes. To determine whether or not all codes, or a portion them, should be included, clinical domain experts in medicine, nursing and pharmacy were consulted. This was especially important for identifying statin users in pharmacy claims, identifying outcomes of interest, medication usage and patient behaviors in the selected data.

In big data projects where machine learning is deployed, experts play a critical role in conducting data analyses and supporting technical issues that may arise. These include determining a machine learning method, choosing a research design, selecting an algorithm, and identifying methods to best fit the data. In the statin project, these experts helped to develop and adapt existing algorithms to specific research questions and maximized the productivity and feasibility of the developed/adapted algorithms. OptumLabs played an integral part of the interdisciplinary team by providing data as well as database knowledge while experts from the fields of medicine, pharmacy, nursing and machine learning brought forward research questions, provided domain expertise, and designed research approaches.

9.1.2.2   Remote Data Analysis

For reasons of patient privacy and data security, the OLDW can only be accessed remotely from a secure sandbox hosted by OptumLabs. Within the sandbox, Aqua Data Studio® (ADS) is provided as a structured query language (SQL) tool that allows researchers to access data in the sandbox. SQL is a programing language developed to communicate with relational databases. It is used to design, maintain, and update databases, and most importantly for research purposes, is used for data retrieval and extraction. The ability to run rapid queries makes SQL a very useful tool in big-data projects. The OLDW contains hundreds of related tables and by using SQL, different tables can easily be linked by using "join clauses" to extract data. For example, there are several tables related to medication prescriptions including National Drug Codes (NDC), and pharmacy claims (including Part D and cost tables). Pharmacy claims and pharmacy claim Part D tables represent prescription claims and include fill dates, days' supply on hand, prescribing provider, and other data. However, detailed drug information such as product information (generic name, brand name, strength, and drug form), labels, and package size are contained in the NDC table. Thus the unique NDC key(s), which exist(s) in both tables, are used to link these tables.

In addition to data extraction, SQL can also perform certain types of data analysis such as descriptive statistics and data distributions. The remote computing environment (sandbox) also provides the statistical software program, such as "R" for advanced data analysis. Depending on project needs and user preference, researchers may also use SAS®, Matlab®, or other data analysis software for research. For purposes of the statin project, the statistical program Matlab® was utilized.

9.1.2.3 Hardware and Software

An important consideration in big-data analysis is computational speed and runtime. When analyzing large, complex, datasets analytic tools can encounter software crashes and errors or long processing times to produce results. Appropriate hardware and software approaches are necessary to minimize these problems. Consideration of the computer's CPU (central processing unit) speed, memory size, data transfer rate, and the number of cores is essential. Together, these elements determine the speed and runtime feasibility for generating analytic results. In many cases, a single computer is insufficient and a cluster of several computers (local and/or remote) is needed to conduct data analyses. The use of multiple computers to simultaneously process data, or parallel computing, benefits big data analysis by running many calculations together. For example, in a previous clinical trial simulation consisting of data from more than one-million virtual patients, 50 computers (each having 30 cores with CPU speeds of 2GHz) were used in parallel to conduct one run of a clinical trial simulation and required about 24 hours of processing time. Hence parallel computing is especially important when sophisticated computing approaches, such as machine learning algorithms are needed.

The choice of software used to facilitate data analysis is also crucial. Depending on the project type and if parallel computing and large computational power is required, the relevant software choice to support a large-scale computing environment is very important. Some users take advantage of graphics processing units (GPU), which are embedded in today's video cards. Graphic processing units are very efficient for processing large-scale and complex data used in computer graphics and electronic images. The highly parallel structure of GPU's makes them more efficient than general-purpose CPUs when processing data through software encoded algorithms. To generate computing power from the GPU, specialized software, such as Compute Unified Device Architecture (CUDA) technology is often employed.

In some cases, the dataset is so large that it cannot be fitted to the computers' memory. In such cases, MapReduce (software for parallel processing of large data sets) can be used to conduct analysis on several chunks of data and then the results are combined on a Hadoop server. Hadoop is a Java-based programming framework that supports the processing of large data sets in a distributed computing environment. The software application Matlab® was selected as the analytic tool for this project because its relevant toolboxes can be used to conduct machine learning work, support parallel computing, GPU computing, and MapReduce functions. To reduce project complexity, data extraction (conducted in ADS) and feature selection (conducted in Matlab®) methods were used to remove non-essential data from the dataset. Although this resulted in a smaller sample size and reduced number of variables, it improved the relevance of the data used for the study.

9.1.3   Research Process

9.1.3.1   Extracting Relevant Data for Research

Before data analysis and model construction are initiated two important functions must first take place to create an accurate and desirable dataset: data extraction and data cleansing. Although the OLDW includes over 150 million lives from over a 20 year period, only a subset of the data records were relevant to this research project. Appropriate data-extraction decisions not only help to select reasonable data to support the research scope and questions, but also help to conduct it in a realistic time frame using suitable processing power.

When extracting data for this study it was decided to include records after the year 2010 for several reasons. These included:

1. Medicare Part D went into effect in 2006 which resulted in reduced pharmaceutical prices and increased utilization of prescription medications. Thus, it was better to choose claims after 2006.
2. Seven statins were included in the study including Pitavastatin, the newest statin in the US, which received FDA approval in 2009 and was brought to the market in 2010. Therefore, the study included all completed prescriptions statins since 2010.
3. The Affordable Care Act (ACA), enacted in 2010, changed healthcare in many aspects, including alterations to insurance coverage, standards, premiums, and healthcare costs.

9.1.3.2   Preprocessing Data to Support Machine-Learning Work

The next step after data extraction is data preprocessing. Medical claims data are messy and can take significant time to preprocess before machine-learning methods are employed. Systematic coding and human errors, outliers, missing values, irrelevant and redundant information, changes in individual names, address locations, and health insurance are just a few of the problems encountered. This makes the work of data cleaning extremely important.

Outliers can be an issue in preprocessing data. Outliers (Barnett and Lewis 1994) are extreme observations that are significantly different from the other values in a dataset. There are some approaches used for outlier detection. The simplest way is to determine if any value falls outside a normal range: as an abstract but easily comprehensible example, if an individual's age is stored as 200 years old or someone's height is 30 feet. It is important to note that outliers sometimes are errors but may also indicate important information. Consultation from domain experts in healthcare and informatics to determine the best strategy to handle and analyze outliers is critical.

Missing values occur when there is no data in an input field for a variable. There are three types of missing data (Nakagawa and Freckleton 2008): data missing completely at random (MCAR), data missing at random (MAR), and data missing not at random (MNAR). Careful consideration is needed to decide appropriate strategies needed to discover missing values. In general, if many input fields have missing values then eliminating this specific variable should be considered. If an algorithm cannot handle a large number of missing values, a more appropriate way to impute them may be needed (Royston 2004).

Depending on the project, feature transformation (Kusiak 2001) may be needed as some algorithms can only process certain types of data and may require consultation from data science experts. For example, discretization may be used to convert continuous variables into categorical variables and binarization to convert continuous and/or categorical variables into binary variables to facilitate learning from certain machine-learning algorithms. Data scientists can provide consultation on ways to prevent the loss of valuable information that existed in the continuous data before transformation.

Normalization is a method of rescaling variable values to a specific range, such as [0,1]. It is also an important step before model construction. Variables measured in different scales are not comparable and do not have equal contributions to prediction. For instance, large value variables have more weights than smaller ones and can dominate the results of calculations thus biasing the prediction model or cluster analysis. Normalization transforms all of the variables to the same scale, and hence reduces the chance that some variables will dominate the model.

Feature selection (Guyon and Elisseeff 2003), a method used to select a subset of the most useful and relevant features for research questions, is also an important task in data preprocessing. Feature selection can reduce the number of variables and the complexity of the model, improve computational speed, and most importantly, improve prediction of the algorithm. Although the OLDW has more than 1300 potential variables it is impractical to use all of them in a prediction model. Strategies typically used for feature selection include the:

1. Filter Method: A subset of features is selected independently based on the characteristics of data without considering any classifier. Examples of the filter method include Rough Sets, Relief and its extension ReliefF.
2. Wrapper Method: A subset of features is selected based on a specific classifier for evaluation. Stepwise regression (backward/forward selection) is a well-known wrapper-based model.
3. Embedded Feature Selection: An embedded mechanism that selects feature subsets during model construction.
4. Literature review and consultation from domain experts.

9.1.3.3 Selecting Appropriate Machine-Learning Approaches for Data Analysis

Once data preprocessing has been completed data analysis can begin: the first step being the selection of an appropriate analytic approach. In general, statistical methods are used in population observation, explanation, and confirmation studies. In contrast, machine learning is typically used in prediction and exploratory studies. Machine learning is a powerful tool to discover hidden patterns from large-scale datasets. Supervised (Kotsiantis 2007) and unsupervised machine learning (Gentleman and Carey 2008) are two popular methods used to group data. They are distinguished by whether or not the training data has known outcomes. Supervised learning includes classification and regression approaches, while unsupervised learning includes association rule learning and clustering analysis. To choose the best-fit algorithm, it is first necessary to know the characteristic of the data and relevant model choices. In the statin study, a supervised learning model was selected to predict AEs for patients prescribed statins using the training data's input (patient characteristics and statin treatment plans). The outputs (AEs or not) were then used to predict new patients' AEs. If the outcome is continuous, then machine-learning approaches such as regression can be used. If the outcome is binary, or multi-class, classification analysis is recommended.

There are a number of different algorithms used in machine learning approaches. For example, decision trees, support vector machines (SVM), K-nearest neighbor (k-NN), and random forest are algorithms commonly used for classification problems. K-means, density based spatial clustering of applications with noise (DBSCAN), and agglomerative hierarchical clustering is commonly used for clustering problems. How to choose an optimal algorithm that can best fit your research question and data is complicated and requires deep knowledge and experience of the problem domain.

9.1.4 Conclusion

This project's primary aim was to develop a proactive strategy to reduce adverse effects in patients prescribed statins. Use of the OLDW illustrated the benefits of big data and its capacity to develop prediction models for individual treatment plans. A key lesson learned from the project was that big-data analysis typically involves diverse expertise from several disciplines. In many cases, domain experts from both healthcare and data science are required. Based on this project's objectives, team members with expertise in database management, machine learning, pharmacy, nursing, and cardiovascular disease prevention were included. Appropriate choices of hardware, computation speed, memory, data transfer rate, parallel computing, software, and the approaches to facilitate big-data analysis are key considerations. Finally, reducing the number of irrelevant variables through data extraction and feature selection approaches reduces data complexity and improves the accuracy of the dataset.

References

Barnett V, Lewis T. Outliers in statistical data. 3rd ed. New York: John Wiley; 1994.
Brewer HB. Benefit-risk assessment of Rosuvastatin 10 to 40 milligrams. Am J Cardiol. 2003;92:23–9.

Cohen JD, Brinton EA, Ito MK, Jacobson TA. Understanding statin use in America and gaps in patient education (USAGE): an internet-based survey of 10,138 current and former statin users. J Clin Lipidol. 2012;6:208–15. doi:10.1016/j.jacl.2012.03.003.

Fernandez G, Spatz ES, Jablecki C, Phillips PS. Statin myopathy: a common dilemma not reflected in clinical trials. Cleve Clin J Med. 2011;78:393–403.

Gentleman R, Carey VJ. Unsupervised machine learning. In: Bioconductor case studies. New York: Springer; 2008. p. 137–57.

Guyon I, Elisseeff A. An introduction to variable and feature selection. J Mach Learn Res. 2003;3:1157–82.

Kotsiantis SB. Supervised machine learning: a review of classification techniques. Informatica. 2007;31:249–68.

Kusiak A. Feature transformation methods in data mining. IEEE Trans Electron Packag Manuf. 2001;24:214–21. doi:10.1109/6104.956807.

Mancini GBJ, Baker S, Bergeron J, et al. Diagnosis, prevention, and management of statin adverse effects and intolerance: proceedings of a Canadian working group consensus conference. Can J Cardiol. 2011;27:635–62.

Maningat P, Gordon BR, Breslow JL. How do we improve patient compliance and adherence to long-term statin therapy? Curr Atheroscler Rep. 2013;15(1):291.

Nakagawa S, Freckleton RP. Missing inaction: the dangers of ignoring missing data. Trends Ecol Evol. 2008;23:592–6. doi:10.1016/j.tree.2008.06.014.

National Center for Health Statistics. Health, United States 2013: with special feature on prescription drugs. 2014.

Royston P. Multiple imputation of missing values. Stata J. 2004;4:227–41.

Stone NJ, Robinson JG, Lichtenstein AH, et al. 2013 ACC/AHA guideline on the treatment of blood cholesterol to reduce atherosclerotic cardiovascular risk in adults: a report of the American College of Cardiology/American Heart Association Task Force on practice guidelines. Circulation. 2014;129:S1–S45. doi:10.1161/01.cir.0000437738.63853.7a.

Taylor F, Huffman MD, Macedo AF, et al. Statins for the primary prevention of cardiovascular disease. Cochrane Database Syst Rev. 2013;1:CD004816. doi:10.1002/14651858.CD004816. pub4.

U.S. Food and Drug Administration. FDA expands advice on statin risks. 2014. http://www.fda.gov/downloads/ForConsumers/ConsumerUpdates/UCM293705.pdf. Accessed 31 Jan 2014.

Part III
Revolution of Knowledge Discovery, Dissemination, Translation Through Data Science

Connie W. Delaney

Big data and data science challenge our views of the universe, health care, nursing, and humanity. Data science challenges how we define, build and organize knowledge, re-examine knowledge transfer, and embrace systems that support local-to-global health and individual/family/community well being. These challenges and opportunities for redesign, re-imagination and inventiveness permeate our society's vision of the core pillar of research/scholarship. These challenges and opportunities ask us to re-examine our policies and human, environmental, and financial investments. In this part the national transformation of research by the National Institutes of Health initiatives (Hardy and Bourne) is described, with a detailed nursing exemplar related to genomics and epigenetics (Daack-Hirsch). Kim and Selby discuss big data, networks and people-centric enhanced science empowered through the Patient-Centered Outcomes Research Institute (PCORI) initiate. New federal data resources that specifically empower big data-supported discoveries related to disparities are illustrated by Correa-de-Araujo. Translation of research discoveries and knowledge to practice and achieving clinical care excellence are exemplified by Troseth and team, while Landstrom presents a health system transformation supporting macro-level research-supported care. Finally, big data and data science demand a new look at knowledge discovery methods. Aliferis provides a succinct description of data analytics essential to realizing the benefits of big data in changing the outcomes, quality, costs, and patient and provider experiences in health care.

Chapter 10
Data Science: Transformation of Research and Scholarship

Lynda R. Hardy and Philip E. Bourne

Abstract The emergence of data science as a practice and discipline is revolutionizing research potential in all disciplines, but healthcare science has the potential to affect the health of individual lives. The use of existing data provides fertile ground for healthcare professionals to conduct research that will maximize quality outcomes, develop algorithms of care to increase efficiency and safety, and create predictive models that have the ability to prevent illness events and reduce healthcare expenditures. Data science can change practice through using existing and growing amounts of data to conduct research and build scholarship. Clinical trials, in some cases, may no longer be required to examine interventions. The pragmatic and efficient use of existing large cohort datasets has the ability to generate sample and control groups to determine efficacy. The collection of data from electronic medical records can provide substantial data to determine trends, construct algorithms, and consider disease and health behaviors modeling that alter patient care. *Digital research* incorporating vast amounts of data and new analytics has the ability to influence global healthcare.

Keywords Big data • Data science • Translational research • Schema-on-write Schema-on-read

L.R. Hardy, Ph.D., R.N., F.A.A.N. (✉)
College of Nursing, The Ohio State University, Columbus, OH, USA
e-mail: unclyn@gmail.com; hardy.305@osu.edu

P.E. Bourne, Ph.D.
Data Science Institute, University of Virginia, Charlottesville, VA, USA

© Springer International Publishing AG 2017
C.W. Delaney et al. (eds.), *Big Data-Enabled Nursing*, Health Informatics,
DOI 10.1007/978-3-319-53300-1_10

10.1   Introduction to Nursing Research

The purpose of this chapter is to understand how research in the digital age and big data are transforming health-related research and scholarship suggesting a paradigm shift and new epistemology.

Health science and the translation of research findings are not new. Florence Nightingale, a social reformer and statistician who laid the foundation for nursing education, conducted early health-related research during the Crimean War by collecting data on causality of death in soldiers. Nightingale was a data collector, statistician, and concerned with data visualization as indicated by her Rose Diagram, a topic of much research. The diagram originally published in *Notes on Matters Affecting the Health, Efficiency, and Hospital Administration of the British Army. Founded Chiefly on the Experience of the Late War. Presented by Request to the Secretary of State for War* graphically presented data indicating that more soldiers died because of disease than of their battle-related injuries. Nightingale was a pioneer at discovery and data-based rationale underpinning the practice of nursing and health-related implications. Informed by philosophy, she was a systematic thinker, who understood the need for systematic data collection (Fig. 10.1). Nightingale designed survey instruments and determined their validity through vetting with experts in the field. She was a statistician basing her findings on mathematics. Through using the findings of her work to change practice and policy, Nightingale reformed conditions for workhouse poor, patient care standards and the right to a meaningful death (McDonald 2001, Selanders and Crane 2012). Imagine what she might have done in the digital age, with computers, and big data.

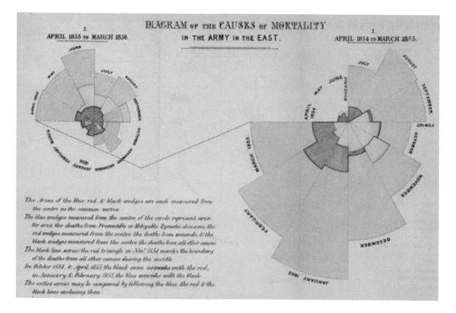

Fig. 10.1 Nightingale rose diagram. *Source* History of Information.com. (Accessed 5 Dec 2015)

Nursing research, as defined by the National Institute of Nursing Research, is knowledge development to "build the scientific foundation for clinical practice; prevent disease and disability; manage and eliminate symptoms caused by illness, and enhance end-of-life and palliative care" (NINR 2015). Grady and Gough (2015) further suggest that nursing science provides a bridge from basic to translational research via an interdisciplinary or team science approach to increasing the understanding of prevention and care of individuals and families through personalized approaches across the lifespan (Grady and Gough 2015).

10.1.1 Big Data and Nursing

The profession of nursing has been intricately involved with healthcare data since the beginning of nurse's notes that documented patient care and outcomes. Notes and plans of care are reviewed, shared and modified for future care and better outcomes. The magnitude of healthcare data is complex thereby requiring nontraditional methods of analysis. The interweaving of multiple data streams to visualize, analyze and understand the entirety of human health demands a powerful method of data management, association, and aggregation. The digitization of healthcare data has consequently increased the ability to aggregate and analyze these data focusing on historical, current and predictive possibilities for health improvement.

The foundation of nursing research is the integration of a hypothesis-driven research question supported by an appropriate theoretical framework. The fourth paradigm, a phrase originally coined by Jim Gray (Hey et al. 2009), or data science, is considered by some to be the end of theory-based research, thus creating a reborn empiricism whereby knowledge is derived from sense experience. This view is not without its detractors. There were big data skeptics before the explosion took place. Dr. Melvin Kranzberg, professor of technology history, in his 1986 Presidential Address, commented that "Technology is neither good nor bad: nor is it neutral ... technology's interaction with the social ecology is such that technical developments frequently have environmental, social, and human consequences that go far beyond the immediate purposes of the technical devices and practices themselves" (Kranzberg 1986, p. 545). Kitchin (2014) further suggested that "big data is a representation and a sample shaped by technology and platform used, the data ontology employed and the regulatory environment" (p. 4). Kitchin's statement reinforces the idea that data cannot explain itself but requires a lens (e.g., theoretical framework) through which to interpret the data. Data is *raw/without interpretation* and cannot interpret outliers or aberrancies suggesting bias. It provides a bulk of information where a specific analytic process is applied—but at the end, the data must be interpreted. Use of a theoretical framework provides a pathway of understanding that can support statistical findings of data analysis.

Bell (2009) suggests that "data comes in all scales and sizes ... data science consists of three basic activities, capture, curation, and analysis" (p. xiii). He also comments on Jim Gray's proposal that scientific inquiry is based on four paradigms:

Table 10.1 Research paradigms

Paradigm	Gray	General research	Research type
1	Experimental science	Positivism	Quantitative
2	Theoretical science	Anti-positivism	Qualitative
3	Computational science	Critical theory	Critical and action oriented
4	Data science		

experimental, theoretical, computational, and data science (Hey et al. 2009). Table 10.1 provides an integration of Gray's paradigms with general research paradigms. Gray's fourth paradigm supports the integration of the first three

The explosion of big data (defined by volume, velocity and veracity) in healthcare provides opportunities and challenges. Healthcare providers, researchers and academics have the ability to visualize individual participant data from multiple sources (hospital, clinic, urgent care and school settings, claims data, research data) and in many forms (laboratory, imaging, provider notes, pharmacy, and demographics). The challenge of aggregating and analyzing these data streams is possessing usable standardized data and the right analytic tools with the power to aggregate and understand data types. The outcome of having appropriate data and tools is the ability to improve health care outcomes and develop predictive models for prevention and management of illness. Other advantages include those related to clinical operations, research, public health, evidence-based medicine/care, genomic analytics, device (wearable and static) monitoring, patient awareness, and fraud analysis (Manyika et al. 2011). Platforms have been developed to assist with the analyses of the various data streams, e.g., Hadoop, Cloudera CDH, Hortonworks, Microsoft HDInsight, IBM Big Data Platform, and Pivotal Big Data Suite. These platforms frequently use cloud computing—ubiquitous elastic compute and large data storage engines from the likes of Google, Amazon and Microsoft. Thus not only is the scientific paradigm changing, but also the compute paradigm from local processing to distributed processing. These changes are accompanied by a new software industry focusing on such areas as data compression, integration, visualization, provenance and more.

More specifically analytical methods, other than using a theoretical pathway to determine what data should be collected and the method of analysis should be considered to allow the data and not the theory to provide the pathway. Many methods are becoming available as a means to analyze and visualize big data. One method, point cloud, uses a set of data points in a three dimensional system for data visualization (Brennan et al. 2015). Other methods use various forms of data clustering such as cluster analysis (groups a set of objects/data points into similar clusters) (Eisen et al. 1998) and progeny clustering (applies cluster analysis determining the optimal number of clusters required for analysis) (Hu et al. 2015). A variety of methods exist to examine and analyze healthcare data providing rich data for improving patient outcomes, predictive modeling and publishing the results.

Theory-based research uses the method of *schema-on-write* (Deutsch 2013) which is usually a clean and consistent dataset but the dataset is more limited or narrow. This method, where data were pre-applied to a plan, requires less work initially but also may result in a more limited result. Research opportunities today provide the ability to broaden the scope and magnitude of the data by allowing for the expansion and use of multiple types/streams of data using the method of *schema-on-read*. *Schema-on-read* identifies pathways and themes at the end of the process, allows the researcher to cast a wide data net incorporating many types of structured and unstructured data, and finally applies the theoretical pathway to allow the analysis to 'make sense' of the data. The data is generally not standardized or well organized but becomes more organized as it is used. The data has the ability to be more flexible thus providing more information (Pasqua 2014). In summary, *schema-on-read* provides the ability to create a dataset with a multidimensional view; these traits magnifying the usability of the dataset. This expands the nurse researcher's ability to explain the research question leading to development of preventive or treatment interventions.

The digital environment and diversity of data have created the need for interdisciplinary collaboration using scientific inquiry and employing a team science approach. This approach provides an environment to maximize self-management of illness, increase, maintain a level of individual independence, and predict usefulness of interventions within and external to professional health environments. Individual empowerment allows individuals to participate, compare outcomes, and analyze their own data. This is accomplished using simple, smart-phone applications and wearable devices. It is the epitome of self-management and participatory research.

10.1.2 Nursing and Data

The Health Information Technology for Economic and Clinical Health (HITECH) Act, an initiative passed in 2009, provides financial support for electronic health records (EHRs) to promote meaningful use in medical records thereby expanding EHR use and healthcare information. Nurse scientists previously gathered data from smaller data sets that were more narrowly focused such as individual small research studies and access to data points within the EHR. This approach provided a limited or constricted view of health-related issues. The unstructured nature of the data made data extraction difficult due to non-standardization. Using the narrow focus of the data inhibited the ability to generalize findings to a larger population often resulting in the need for additional studies. Moreover, EHR data focused on physician diagnosis and related data, failing to capture the unstructured but more descriptive data, e.g., nursing data (Wang and Krishnan 2014).

The big data tsunami allows nurse scientists access to multiple data streams and thus expands insight into EHRs, environmental data that provides an exposome or human environment approach, genetic and genomic data allowing for individuality of treatment and technology driven data such as wearable technology and biosensors allowing nearly real-time physiologic data analysis. Moreover, data sharing provides power that has not previously existed to detect differences in health dis-

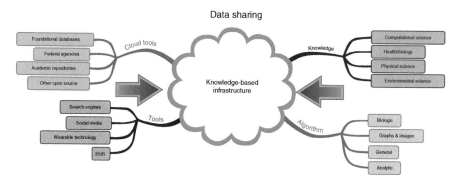

Fig. 10.2 Challenges of big data. Adapted from the high-level research design in the fourth discipline (p. 169)

parities. Data aggregation generates a large participant pool for use in pragmatic studies to understand the depth and breadth of health disparities (Fig. 10.2).

For example, technology-driven data that impacts elder care and can prevent adult injury is fall-related data collected by retrieving data from hospital or home systems (Rantz et al. 2014). This data widen the scope of evidence-based healthcare by providing a multidimensional interpretation of the data. Larger, more inclusive datasets such as claims data from the Centers for Medicare and Medicaid Services (CMS) add to the ability to create a more complete view of health and healthcare. However, this data is constrained by what claims form information is captured and available for research purposes. Data digitization and open access publication provide a rich environment for all disciplines to access, correlate, analyze and predict healthcare outcomes. Future data sharing and reuse will likely capture more of the research continuum and process.

10.2   The New World of Data Science

Data science and resulting data use are not new and are growing at warp speed. It is estimated that 2.7 zettabytes of data were generated daily in 2012; 35 zettabytes of data per day are anticipated by 2020; 90% of current data were collected in the past 2 years and 5 exabytes are generated daily (Karr 2012). A brief history of data science shows that Tukey (1962) in describing his transformation from an interest in statistics to one in data analysis initiated the thought that there was a difference between the two. It was not until the mid-1990s that a more formalized approach to data (analysis) science was developed that began to look at the increased accumulation of data and data analytics (Tukey 1962).

The advent of data science is compared to Fordism. The world changed when Henry Ford discovered new methods of building automobiles. Manufacturing processes were modernized, modifying knowledge and altering methods of understanding the world. Fordism changed society and behaviors impacting everyone (Baca 2004). The big data explosion follows the same trajectory. New methods of capturing, storing and analyzing data have and will continue to have an impact on society. Today's data has exploded into multiple data streams, structured and unstructured.

Rapid growth of the big data or data science ecosystem emphasizes the need for interdisciplinary approaches to interpreting and understanding health and healthcare data. The data explosion provides an environment with various data types require the same breadth of scientists to interpret the data accurately. Just as generation of data is from a plurality of sources so must the composition of the team assigned to its analysis. The data explosion provides nurse scientists, as well as other disciplines, the ability to work within highly diverse teams to provide deep knowledge integration and comprehensive analysis of the data. This expansive inclusion of expertise extends to employing the skills of citizen scientists.

The data explosion also provides a new world data alchemy allowing for transformation, creation and combination of data types to benefit healthcare outcomes through accurate decision-making and predictability. Data standards are becoming more prevalent. One key example of this prevalence is the NIH's Big Data to Knowledge (BD2K) program recently establishing a Standards Coordinating Center (SCC). One example of standards work is Westra and Delaney success in having the Nursing Management Minimum Data Set (NMMDS) incorporated into Logical Observation Identifiers Names and Codes (LOINC) a universal system for test, observations and measurement (Westra et al. 2008). Computers are becoming ever more powerful and due to the ubiquitous nature of data and data-driven algorithms allow deeper and more complex analyses resulting in greater accuracy in patient care-related decision making (Provost and Fawcett 2013). Making use of existing and future data necessitates training of data scientists. The need for data scientists is growing with an anticipated shortage of between 140,000 and 150,000 people (Violino 2014).The combination of these elements—standards, data, analytics and a trained workforce—increases the accuracy and predictive use of data with numerous opportunities for scholarship, including publication and analytic developments.

10.3 The Impact of Data Proliferation on Scholarship

Scholarship (FreeDictionary: academic achievement; erudition; Oxford Dictionary: learning and academic study or achievement) was once solely a paper journal publication (p-journal) and an academic requirement for tenure. The era of big data and data science increases the realm of scholarship by adding a variety of publication/ dissemination forms such as electronic journal publication (e-journal), web-based formal or informal documentation, reference data sets, and analytics in the form of software and database resources. Digitalization of online information and data provide fertile ground for new scholarship. The technological provision of shared data, cloud computing and dissemination of publications places scholarship in the fast lane for nursing and other disciplines. The information superhighway is clearly the next generation infrastructure for scholarship even as the academic establishment's adoption of such change is behind that pace of change. Such a gap and the migration of a skilled workforce of data scientists from academic research to the private sector are concerns.

> "Scholarship represents invaluable intellectual capital, but the value of that capital lies in its effective dissemination to present and future audiences."
> AAU, ARL, CNI (2009)

Much of academic scholarship is based on Boyer's model espousing that original research is centered on discovery, teaching, knowledge and integration (Boyer 1990). The American Association of Colleges of Nursing (AACN) adopted Boyer's model defining nursing research as: "… those activities that systematically advance the teaching, research, and practice of nursing through rigorous inquiry that (1) is significant to the profession, (2) is creative, (3) can be documented, (4) can be replicated or elaborated, and (5) can be peer-reviewed through various methods" (2006). A hallmark of scholarship dissemination continues to be a process of peer-reviewed publications in refereed journals. The Association of American Universities (AAU), the Association of Research Libraries (ARL) and the Coalition for Network Information (CNI) published a report emphasizing the need to disseminate scholarship. Big data now questions if the scholarship model requires updating to be more inclusive of the sea change in information (culturally, socially, and philosophically) technology has introduced (Boyd and Crawford 2012).

Today's digital environment and the need for dissemination of scholarly work suggest expanding the definition to allow for other methods of scholarship. Borgman and colleagues, in their 1996 report to the National Science Foundation (NSF), developed the information life cycle model as a description of activities in creating, searching and using information (Borgman et al. 1996). The outer ring denotes life cycle stages (active, semi-active, inactive) with four phases (creation, social context, searching and utilization), where creation is the most active. The model includes six stages, which further assist the context of scholarship utilization (Fig. 10.3).

The incorporation of the information life cycle model into the AACN scholarship definition adds the need for dissemination and the inclusion of sources outside the normal process of p-journal publications. This incorporation would highlight that publication is a multi-dimensional continuum requiring three main criteria. First, the information must be publicly available via sources such as subscriptions, abstracts and databases/datasets allowing for awareness and accessibility of the work. Second, the scholarly work should be trustworthy; this is generally conducted through peer review and identified institutional affiliation. Finally, dissemination and accessibility are the third criterion that allows visualization of the scholarly work by others (Kling 2004).

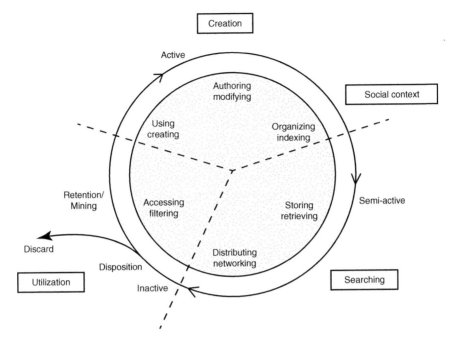

Fig. 10.3 Information life cycle. *Source* Borgman et al. (1996) report to NSF. *Note* The outer ring indicates the life cycle stages (active, semi-active, and inactive) for a given type of information artifact (such as business records, artworks, documents, or scientific data). The stages are superimposed on six types of information uses or processes (*shaded circle*). The cycle has three major phases: information creation, searching, and utilization. The alignment of the cycle stages with the steps of information handling and process phases may vary according to the particular social or institutional context

The digital enterprise is no longer relegated to p-journals but has increased to include e-journals, data sets, repositories (created to collect, annotate, curate and store data) and publicly shared scholarly presentations. Citations of data sets now receive credibility and validity through this new scholarship type, in part through the work of the National Institutes of Health Big Data to Knowledge (BD2K) initiative. The rapid ability to access, analyze either through cloud computing or novel software designed for big data, and disseminate through multiple methods provides nursing and all disciplines a more rapid ability to publicize and legitimize their scholarly work.

10.4   Initiatives Supporting Data Science and Research

Many government agencies have initiated work designed to build processes within the digital ecosystem to assist teams focusing on data science. These initiatives have been developed as a means of assisting in faculty and student training, collaboration with centers of excellence, development of software designed to facilitate the analysis of large datasets, and the ability to share data and information through a cloud-based ecosystem that maximizes the use of existing multi-dimensional data to better understand and predict better patient outcomes. Further, multiple agencies have open funding sources consistent with nursing science. Examples follow.

10.4.1   National Institutes of Health

The National Institutes of Health (NIH) spearheaded the big data program with the creation of the Big Data to Knowledge or BD2K initiative in 2012 when an advisory committee convened by Dr. Francis Collins, the NIH Director, investigated the depth and breadth of big data potential. Dr. Collins and key members of his leadership team reviewed the committee's findings and committed to providing a 'data czar' to facilitate data science that would span the 27 Institutes that comprised NIH. The BD2K team, led by Dr. Phil Bourne (co-author of this work), created a data science ecosystem incorporating (1) **training** for all levels of data scientists, (2) centers that would work independently and in concert with all BD2K **centers** to build a knowledge base, (3) a **software** development team focused on creating and subsequently maintaining new methods for big data analysis, and (4) a **data indexing** team focused on creating methods for indexing and referencing datasets. (https:// datascience.nih.gov). Taken together the intent is to make data FAIR—Findable, Accessible (aka usable), Interoperable and Reusable.

10.4.1.1   Training

Training focused on establishing an effective and diverse biomedical data science workforce using multiple methods across educational and career levels—students through scientists. Training focused on the continuum of scientists who see biomedical data science as their primary occupation to those that see biomedical data science as a supplement to their skill set. Development and funding of a training coordination center that ensures all NIH training materials are discoverable is paramount.

10.4.1.2   Centers

Centers included the establishment of 11 Centers of Excellence for Big Data Computing and two Centers that are collaborative projects with the NIH Common Fund LINCS program (the LINCS-BD2K Perturbation Data Coordination and Integration Center, and the Broad Institute LINCS Center for Transcriptomics). Centers are located throughout the United States providing training to advance big data science in the context of biomedical research across a variety of domains and datatypes.

10.4.1.3 Software

Software focus included targeted Software Development awards to fund software and methods for the development of tools addressing data management, transformation, and analysis challenges in areas of high need to the biomedical research community.

10.4.1.4 Commons

Commons addressed the development of a scalable, cost effective electronic infrastructure simplifying, locating, accessing and sharing of digital research objects such as data, software, metadata and workflows in accordance with the FAIR principles (https://www.force11.org/group/fairgroup/fairprinciples).

10.4.1.5 Data Index

Data Index is a data discovery index (DDI) prototype (https://biocaddie.org/) that indexes data that are stored elsewhere. The DDI will increasingly play an important role in promoting data integration through the adoption of content standards and alignment to common data elements and high-level schema.

10.4.2 *National Science Foundation*

The National Science Foundation (NSF) is a United States government agency supporting research and education in non-medical fields of science and engineering. NSF's mission is "to promote the progress of science; to advance the national health, prosperity, and welfare; to secure the national defense…" (http://www.nsf.gov accessed 12/23/2015). The annual NSF budget of $7.3 billion (FY 2015) is the funding source for approximately 24% of all federally supported basic research conducted by U.S. colleges and universities.

NSF created the Directorate for Computer & Information Science & Engineering (CISE) with four goals related to data science to:

- Uphold the U.S. position of world leadership in computing, communications, and information science and engineering;
- Promote advanced computing, communications and information systems understanding;
- Support and provide advanced cyberinfrastructure for the acceleration of discovery and innovation across all disciplines; and
- Contribute to universal, transparent and affordable participation in an information-based society.

CISE consists of four divisions, each organized into smaller programs, responsible for managing research and education. These four divisions (the Division of Advanced Cyberinfrastructure; the Division of Computing & Communication Foundations; the Division of Computer and Network Systems; and the Division of Information and Intelligent Systems) incorporate program directors acting as the point of contact for sub-disciplines that work across each division and between divisions and directorates. NSF CISE provides funding in the areas of research infrastructure, advancing women in academic science and engineering, cybersecurity, big data hub and spoke designs to advance big data applications, and computational and data science solicitations to enable science and engineering (http://www.nsf.gov/cise/about.jsp accessed 12/23/2015).

10.4.3   U.S. Department of Energy

The U.S. Department of Energy's (DOE) mission is to "ensure America's security and prosperity by addressing its energy, environmental and nuclear challenges through transformative science and technology solutions" (http://energy.gov/mission accessed 12/23/2015). A prime focus of the DOE is to understand open energy data through the use and access to solar technologies. The DOE collaborated with its National Laboratories to harness data to analyze new information from these large data sets (Pacific Northwest National Lab), train researchers to think about big data, and to focus on issues of health-related data (Oak Ridge National Lab). As an example, the Oak Ridge National Laboratory's Health Data Sciences Institute, in concert with the National Library of Medicine (NLM), developed a new and more rapid process to accelerate medical research and discovery. The process, Oak Ridge Graph Analytics for Medical Innovation (ORiGAMI), is an advanced tool for literature-based discovery.

10.4.4   U.S. Department of Defense

The U.S. Department of Defense (DOD) focuses on cyberspace to enable military, intelligence, business operations and personnel management and movement. The DOD focuses on protection from cyber vulnerability that could undermine U.S. governmental security. The four DOD foci include (1) resilient cyber defense, (2) transformation of cyber defense operations, (3) enhanced cyber situational awareness, and (4) survivability from sophisticated cyber-attacks (http://www.defense.gov/accessed 12/23/2015).

The Defense Advanced Research Projects Agency (DARPA) is an agency within the DOD that deals with military technologies. DARPA focuses on a 'new war' they call a network war for cyber security. Understanding that nearly everything has a computer, including phone, television, watches, and military weapons systems, DARPA is utilizing a net-centric data strategy to develop mechanisms to thwart potential or actual cyber-attacks.

10.5 Summary

The big data tsunami created fertile ground for the conduct of research and related scholarship. It opened doors to healthcare research previously unimagined; it provided large data sets containing massive amounts of information with the potential of increasing knowledge and providing a proactive approach to healthcare. It also created a firestorm of change reflecting how access to data is accomplished. Big data is the automation of research. It also is an epistemological change that questions certain ethical morés; for example, just because we can access the data does not mean we should. Fordism changed the manufacturing world with a profound societal impact; big data is the new Fordism impacting society.

The societal and ethical impact of big data requires the attention of all disciplines. The impact, while requiring a priori decisions, will provide an unprecedented opportunity to influence healthcare and add to the global knowledge base and scholarly work. Big data has opened an abyss of opportunity to explore what is, hypothesize what could be, and provide methods to change practice, research and scholarship.

Big data is central to all areas of nursing research. Areas with the most prominent interface with other major healthcare initiatives, from an NIH perspective, include the precision medicine initiative seeking to further personalize a patient's health profile; the U.S. cancer moonshot, which has at its core the greater sharing and standardization of data supporting cancer research; and the Environmental influences of Child Health Outcomes (ECHO). We are truly entering the era of data-driven healthcare discovery and intervention.

References

AACN. Position statement on nursing research. Washington, DC: American Association of Colleges of Nursing; 2006. http://www.aacn.nche.edu/publications/position/nursing-research

Baca G. Legends of fordism: between myth, history, and foregone conclusions. Soc Anal. 2004;48(3):169–78. doi:10.3167/015597704782342393.

Borgman CL, Bates MJ, Cloonan MV, Efthimiadis EN, et al. Social aspects of digital libraries. Final report to the National Science Foundation; 1996.

Boyd D, Crawford K. Critical questions for big data: provocations for a cultural, technological, and scholarly phenomenon. Inform Commun Soc. 2012;15(5):662–79. doi:10.1080/13691 18X.2012.678878.

Boyer EL. Scholarship reconsidered: priorities of the professoriate. Carnegie Foundation for the Advancement of Teaching; 1990.

Brennan PF, Ponto K, Casper G, Tredinnick R, Broecker M. Virtualizing living and working spaces: proof of concept for a biomedical space-replication methodology. J Biomed Inform. 2015; doi:10.1016/j.jbi.2015.07.007.

Deutsch T. Why is schema-on-read so useful? In: IBM big data and analytics hub. 2013; http://www.ibmbigdatahub.com/blog/why-schema-read-so-useful. Accessed 31 Jan 2016.

Eisen MB, Spellman PT, Brown PO, Botstein D. Cluster analysis and display of genomic-wide expression patterns. Proc Natl Acad Sci. 1998;95:14863–8.

Grady PA, Gough LL. Nursing science: claiming the future. J Nurs Scholarsh. 2015;47(6):512–21. doi:10.1111/jnu.12170.

Hey T, Tanslery S, Tolle K. The fourth paradigm: data-intensive scientific discovery. Redmond, Washington: Microsoft Research; 2009.

Hu CW, Kornblau SM, Slater, JH, Outub AA. Progeny clustering: a method to identify biological phenotypes nature; 2015. doi:10.1038/srep12894.

Karr D. The flood of big data; 2012. https://www.marketingtechblog.com/ibm-big-data-marketing. Accessed 22 Dec 2015.

Kitchin R (2014). The Data Revolution: Big Data, Open Data, Data Infrastructures and their Consequences. Sage: London

Kling R. The internet and unrefereed scholarly publishing. Annu Rev Inform Sci Technol. 2004;38(1):591–631. doi:10.1002/aris.1440380113.

Kranzberg M. Technology and history: Kransberg's laws. Technol Cult. 1986;27(3):544–60.

Manyika J, Chui M, Brown B, Bughin, J, Dobbs R, Roxburgh C, Byers A. Big data: the next frontier for innovation, competition, and productivity. McKinsey & Company, The McKinsey Institute; 2011.

McDonald L. Florence Nightingale and the early origins of evidence-based nursing. Evid Based Nurs. 2001;4(3):68–9. doi:10.1136/ebn.4.3.68.

National Institute of Nursing Research. 2015. http://www.ninr.nih.gov/. Accessed 17 Dec 2015.

Pasqua J. Schema-on-read vs schema-on-write. In: MarkLogic; 2014. http://www.marklogic.com/blog/schema-on-read-vs-schema-on-write/. Accessed 17 Dec 2015.

Provost F, Fawcett T. Data science and its relationship to big data and data-driven decision making. Big Data. 2013;1(1):BD51–9. doi:10.1089/big.2013.1508.

Rantz MJ, Banerjee TS, Cattoor E, Scott SD, Skubic S, Popescu M. Automated fall detection with quality improvement rewind to reduce falls in hospital rooms. J Gerontol Nurs. 2014;40(1):13–7. doi:10.3928/00989134-20131126-01.

Selanders LC, Crane PC. The voice of florence nightingale on advocacy. Online J Issues Nurs. 2012;17(1):1. doi:10.3912/OJIN.Vol17No01Man01.

Tukey JW. The future of data analysis. Ann Math Stat. 1962;33(1):1–67. doi:10.1214/aoms/1177704604.

Violino B. The hottest jobs in IT: training tomorrow's data scientists. Forbes/Transformational Tech; 2014.

Wang W, Krishnan E. Big data and clinicians: a review on the state of the science. JMIR Med Inform. 2014;2(1):e1. doi:10.2196/medinform.2913.

Westra BL, Delaney CW, Konicek D, Keenan G. Nursing standards to support the electronic health record. Nurs Outlook. 2008;56(5):258–266. e251. doi:10.1016/j.outlook.2008.06.005.

Case Study 10.1: Complexity of Common Disease and Big Data

Sandra Daack-Hirsch and Lisa Shah

Abstract Human development, health, and disease processes are the culmination of complex interactions among DNA sequences, gene regulation—epigenetics, and the environment. To truly create individualized interventions to address prevention and treatment, complex data systems that are integrated are needed. As an exemplar, this case study will explore the complexity of information and the vast sources of big data and related analytics needed to better understand causes of Type 2 Diabetes (T2D), which in turn will drive individualized interventions to reduce risk and better treat individuals. Personalized healthcare (also called precision medicine) is becoming a reality. President Obama announced the Precision Medicine Initiative in February 2015. Conceptually, precision medicine has been defined as prevention and treatment strategies that take individual differences into account to generate knowledge applicable to the continuum of health and disease. To that end, this case study describes current initiatives to assemble and analyze the vast and complex phenotypic, genetic, epigenetic, and exposure information generated or that will be generated by researchers and clinicians on individuals, within a specific clinical example of T2D.

Keywords Type 2 diabetes • Genetics/genomics • Exposome • Epigenetics • Omics • Big data • Personalized health care

10.1.1   Type 2 Diabetes (T2D) as a Significant Health Problem

Diabetes is a significant public health problem and its prevalence is increasing. As of 2012 an astounding 29.1 (9.3%) million Americans of all ages and racial/ethnic groups were affected by either Type 1 or Type 2 diabetes although, roughly 90–95% are Type 2. An estimated 8.1 million of those who have diabetes are undiagnosed (National Diabetes Information Clearinghouse [NDIC] 2014; Valdez et al. 2007) which potentiates the likelihood that they will also develop secondary health complications related to untreated diabetes (Klein Woolthuis et al. 2007). Diabetes is the leading cause of kidney failure, nontraumatic lower-limb amputation, heart disease, stroke, and new cases of blindness in the United States. In 2010 it was the seventh leading cause of death. As of 2012 medical expenses for people with diabetes (any form) were more than twice those for people without diabetes, costing Americans an estimated $245 billion in direct and indirect costs (NDIC 2014).

S. Daack-Hirsch, Ph.D., R.N. (✉) • L. Shah, M.S.N., R.N.
The University of Iowa, Iowa city, IA, USA
e-mail: sandra-daack-hirsch@uiowa.edu

Type 2 diabetes (T2D) is a complex metabolic disease characterized by persistently elevated blood glucose caused by insulin resistance coupled with insulin deficiency. The substantial genetic component is thought to interact with environmental risk factors (plentiful diets and limited physical activity) to produce T2D. Several genetic variants have been associated with an increased risk to develop T2D (Wolfs et al. 2009), but the clinical validity of genetic variants alone to estimate diabetes risk remains limited, and genetic variants explain only a small portion of total risk variation (Burgio et al. 2015). While the staggering increase in prevalence of T2D is well documented, until recently the sharp increase in prevalence was largely thought to be driven by environmental factors experienced during adulthood (mainly an imbalance between energy intake and energy expenditure). However, there is also a sharp increase in T2D and obesity among persons under the age of twenty that challenges our understanding of risk factors (NDIC 2014). Just how genetic and environmental factors work in concert to produce the T2D phenotype remains unclear. Moreover, the environmental contribution to T2D is far more complex than an imbalance between energy intake and energy expenditure.

10.1.2 Factors Contributing to T2D

10.1.2.1 Genetics/Genomics

The genetic contribution to T2D has been established through research involving family history, twins studies, and genetic analysis. The lifetime risk for developing T2D in the Western world is reported to be between 7 and 10% (Burgio et al. 2015; Wolfs et al. 2009). Narayan et al. (2003) estimated that for individuals born in the United States in 2000, the lifetime risk for T2D is 1 in 3 for males and 2 in 5 for females. Family studies reveal that unaffected first-degree relatives of individuals with T2D have a two- to almost six-fold increase in risk to develop T2D over the course of a lifetime compared to people without a family history of T2D (Harrison et al. 2003; Valdez et al. 2007). The concordance rate for T2D among identical twins is high and is consistently reported to be greater than 50% in many populations (Medici et al. 1999; Newman et al. 1987; Poulsen et al. 2009). The fact that there are monogenetic (single gene) forms of diabetes (e.g., Maturity Onset Diabetes of the Young [MODY] and Permanent Neonatal Diabetes Mellitus [PNDM]) provides further evidence for a genetic role in the diabetes phenotype. Nevertheless, the majority of cases of diabetes are not one of the monogenetic forms; rather, T2D is a genetically complex disorder in which any number of genetic variants predispose an individual to develop the disease. Advances in genotyping technology have led to large-scale, population-based genetic studies to identify genetic variants (single nucleotide polymorphisms [SNPs]) associated with T2D. For example, the most recent studies have verified up to 65 SNPs to be associated with T2D (Morris et al. 2012; Talmud et al. 2015).

Discounting the rare monogenetic forms of diabetes and given the genetic and phenotypic heterogeneity of T2D[1] drawing firm conclusions about genotype—phenotype correlation using standard statistical analyses is difficult. To address statistical limitations, several efforts are underway to develop new algorithms to generate genetic risk scores (GRS) that predict T2D (Keating 2015; Talmud et al. 2015). As knowledge of the underlying genetic contribution grows, the prediction models improve. Mounting evidence shows that combining GRS and clinical risk factors (e.g., BMI, age, and sex) further improves the ability to detect incident cases (Keating 2015). However, the clinical utility of GRS remains problematic (Lyssenko and Laakso 2013). Developing risk prediction algorithms that combine GRS and phenotypic data for T2D is challenging in part due to the high heterogeneity in both the genetic factors and phenotypic elements of the disease. Moreover, genes and genetic variants at different locations in the genome (polygenic loci) that are associated with T2D are involved in multiple physiologic processes such as gluconeogenesis, glucose transport, and insulin homeostasis and many are also implicated in obesity (Burgio et al. 2015; Keating 2015; Slomko et al. 2012; Wolfs et al. 2009). To date, GRS use common genetic variants that have the strongest main effects. Other sources of genetic variance include rarer higher-penetrant variants, epigenetics, gene-gene and gene-environment interactions, and sex-specific genetic signals (Keating 2015; Lyssenko and Laakso 2013). The phenotype is also highly complex with patients presenting in various combinations of body type, age, family history, gestational diabetes, drug treatments, and comorbidities including obesity and metabolic syndrome. Detecting genetic differences is difficult when they are rare. Combining complex phenotype, interaction (gene x gene and or gene x environment), and gene variant information requires new data science approaches in order to leverage the complexity and create information that is clinically useful.

10.1.2.2 The Environment

In 2005 Wild coined the term "exposome" to describe the complementary environmental component of the gene-environment interaction indicative of complex traits and diseases (Wild 2005). As with the genetic component of T2D, the environmental component is also complex and plays a major role in the diabetes phenotype. Most of our knowledge of the T2D exposome is limited to the behavioral or modern living environment (Slomko et al. 2012). The modern living environment is characterized by increased access to low-cost, calorie-dense foods and increased sedentary lifestyle. The modern living environment is most amenable to intervention, and in fact interventions targeting diet and exercise are known to be effective in preventing or delaying the onset of T2D (Diabetes Prevention Program [DPP] Research Group 2002; Lindström et al. 2003; Venditti 2007). In the context of the modern living environment there is an emerging awareness of "unavoidable exposures" and their

---

[1] Over 200 variants associated with Type 2 diabetes are recorded in the database of Genotypes and Phenotypes ([dbGaP] found at http://www.ncbi.nlm.nih.gov/gap).

connection to T2D (Slomko et al. 2012). These are exposures to man-made chemicals through ambient particles, water, food, and use of consumer or personal care products—some are found in plastics and resins. These chemicals are ubiquitous in the everyday environment at levels below standards set by the Environmental Protection Agency and other regulatory agencies. While a single exposure is not likely to cause harm, little is known about chronic low-level exposure and risk for disease. Burgio et al. (2015) summarized the growing evidence that suggests endocrine-disrupting chemicals such as brominated flame retardants and organochlorine pesticides, heavy metals, and pharmaceuticals (e.g., corticosteroids, antipsychotics, beta-blockers, statins, thiazide diuretics) may interfere with β-cell function and induce insulin resistance (Burgio et al. 2015, p.809; Diabetes.co.uk 2015).

There is also emerging evidence that the gut microbiota composition could affect risk for T2D. Gut microbiota are important for intestinal permeability, host metabolism, host energy homeostasis, and human toxicodyamics (how chemicals affect the body). Changes in microbiota composition that interfere with these functions can lead to increased activation of inflammatory pathways which in turn interferes with insulin signaling, increase in energy harvesting and fat storage in adipose tissue, and potentiate the effect of chemical exposure—all potential pathways to increase risk for metabolic syndrome, obesity and or T2D. For a more in depth review of gut microbiota and T2D refer to Burgio et al. (2015) and Slomko et al. (2012).

10.1.3   Epigenetics

10.1.3.1   Overview of Epigenetics

Epigenetics may explain how genetic and environmental factors work in concert to produce T2D. "Epigenetics is the study of heritable changes (either mitotically or meiotically) that alter gene expression and phenotypes, but are independent from the underlying DNA sequence …"(Loi et al. 2013, p. 143). The epigenome is a series of chemical modifications (often referred to as tags or marks) that are superimposed on to the genome. In humans epigenetic modifications can either affect the proteins that are involved in the packaging of DNA into chromatin (known as chromatin modification), or directly attach to the DNA (e.g., DNA methylation). Epigenetic modification regulates gene expression by either activating (turning on) or deactiving (turning off) genes or segments of the DNA at given times (Genetics Learning Center 2014). Chromatin modification and DNA methylation are functionally linked to transcription and likely provide the mechanisms by which cells are programmed from one generation to the next. In other words, the epigenome activates genome in what is manifested as the phenotype.

During the pre-genomic era it was thought that disease and specific human traits were the direct result of variants in the DNA sequence (e.g., direct mutation of a single gene). However, very few diseases/traits are associated with only gene variants. To varying degrees, other factors such as poverty, nutrition, stress, and environmental toxin exposures can also contribute to health or lack thereof; yet none fully explain susceptibility to disease or variations in human traits. Environmental and social signals such as diet and stress can trigger changes in gene expression without changing the sequence of the DNA (Heijmans et al. 2008; McGowan et al. 2009; Mathers et al. 2010; Radtke et al. 2011; Weaver et al. 2004). Some of these epigenetic tags are cell specific and differentiate phenotype at a cellular level with respect to cell type and function. In a differentiated cell, only 10 to 20% of the genes are active (Genetic Science Learning Center 2014). Some epigenetic tags are acquired and lost over the life course of an individual, and some tags are passed on from generation to generation and may take several generations to change.

10.1.3.2 Examples of Epigenetic Modification and T2D

While it is widely known and accepted that maternal nutrition is of paramount importance to the health and development of the offspring, the precise biologic mechanisms linking maternal nutrition to offsprings' wellbeing are just beginning to be understood. Epigenetic mechanisms may provide one such link. Evidence for epigenetic modification in the form of fetal programing can be found among individuals who were prenatally exposed to famine during the Dutch Hunger Winter in 1944–45. These individuals had less DNA methylation (hypomethylation) of the insulin-like growth factor 2 (IGF2) gene compared to their unexposed, same-sex siblings. IGF2 is a key factor in human growth and development. This epigenetic modification acquired in utero persisted throughout the children's lifetime (Heijmans et al. 2008) and has been associated with higher rates of T2D, obesity, altered lipid profiles, and cardiovascular disease (Schulz 2010) among these children (Burgio et al. 2015).

A number of recent studies report changes in methylation patterns of specific genes associated with T2D (Rönn et al. 2013; Zhang et al. 2013; Ling et al. 2008; Yang et al. 2011; Kulkarni et al. 2012; Yang et al. 2012; Hall et al. 2013; Ribel-Madsen et al. 2012). Studies also reveal differential methylation patterns in genes associated with T2D among those affected by T2D (Zhang et al. 2013; Ling et al. 2008; Yang et al. 2011, 2012; Kulkarni et al. 2012; Hall et al. 2013) and in tissue specific samples (pancreases and mitochondria). These types of studies provide evidence that genotypes (DNA sequences) and their regulation (epigenetic modifications) are important factors contributing to T2D and that the epigenome is modifiable providing targets for interventions.

The environmental exposures described above ("unavoidable exposures") could also lead to changes in gut microbiota composition. In fact, changes in gut microbiota composition have been shown to interfere with epigenetic regulation of *FFAR3* gene in patients with T2D (Remely et al. 2014). *FFAR3* is normally expressed in the pancreatic β-cells and mediates an inhibition of insulin secretion by coupling with other proteins (National Center for Biotechnology Information [NCBI] 2015).

Interfering with the epigenetic regulation of FFAR3 would in turn lead to an inability to regulate insulin secretion appropriately.

10.1.3.3 Summary of Factors Contributing to T2D

T2D is the combination of biological contributing factors (genetics), environmental contributing factors (exposome), and the synthesis of biology and environment (epigenetics). Evolving epigenetic evidence suggests that epigenetic modifications could be important biomarkers for predicting risk, monitoring effectiveness of interventions, and targeting for therapy development to both prevent and treat T2D. Epigenetic patterns may serve as biomarkers connecting the exposome and genome (Fig. 10.1.1), thereby providing more comprehensive risk information for T2D. Unfavorable epigenetic modification may be reversed by lifestyle interventions, such as by modifying diet, increasing physical activity, and enriching the in utero environment. The rapid advances in genetic, exposome, and epigenetic sciences offer exciting possibilities for future discovery that will deepen our understanding of the complex balance between the environment and the genome, and how that balance influences health. Clearly an in-depth understanding T2D is largely dependent on big data and related advanced analytics.

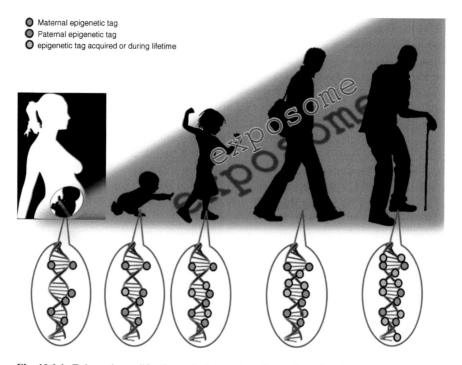

Fig. 10.1.1 Epigenetic modifications: the interaction of genetic and environmental risk factors over the life course

10.1.4 Current Initiatives to Leverage the Power of Big Data for Common Disease

10.1.4.1 Omics

Omics is the application of powerful high through-put molecular techniques to generate a comprehensive understanding of DNA, RNA, proteins, intermediary metabolites, micronutrients, and microbiota involved in biological pathways resulting in phenotypes. Scientists and informaticians are working on ways to integrate the layers of omic sciences and the exposome to better quantify an individual's susceptibility to diseases such as T2D and to capitalize on his or her inherent protections against disease (Slomko et al. 2012). These techniques would allow for massive amounts of genomic, epigenomic, exposure, and phenotypic data to be analyzed in concert in order to build more powerful prediction models and provide targets for the development of prevention and treatment modalities.

10.1.4.2 Clinical Genomic Resources

Several initiatives are underway to assemble the vast and complex phenotypic, genetic, epigenetic and exposure information pertaining to wellness and disease states. These initiatives will leverage health information that is currently generated or will be generated by researchers and clinicians on individuals.

ClinGen. (http://clinicalgenome.org/) is a project to develop standard approaches for sharing genomic and phenotypic data provided by clinicians, researchers, and patients through centralized databases, such as ClinVar—a National Database of Clinically Relevant Genetic Variants (CRGV). ClinGen investigators are working to standardize the clinical annotation and interpretation of genomic variants. Goals of ClinGen include:

- **Share** genomic and phenotypic data through centralized databases for clinical and research use
- **Standardize** clinical annotation and interpretation of variants
- Improve understanding of variation in **diverse populations**
- Develop **machine-learning algorithms** to improve the throughput of variant interpretation
- Implement **evidence-based expert consensus** for curation of clinical validity
- Assess the **'medical actionability'** of genes and variants to support their use in clinical care systems
- **Disseminate** the collective knowledge/resources and ensure EHR interoperability (http://www.genome.gov/27558993)

Currently ClinGen efforts are focused on cardiovascular disease, pharmacogenomics, hereditary (germline) cancer, somatic cancer, and inborn errors of metabolism. However, knowledge generated on structure and process will serve as a template for approaching other diseases.

eMERGE. The Electronic Medical Records and Genomics (eMERGE) Network is a National Institutes of Health (NIH)-organized and funded consortium of U.S. medical research institutions. The eMERGE Network brings together researchers from leading medical research institutions across the country to conduct research in genomics, including discovery, clinical implementation and public resources. eMERGE was announced in September 2007 and began its third and final phase in September 2015 (http://www.genome.gov/27558993). The Network is comprised of six workgroups (see Table 10.1.1).

The primary goal of the eMERGE Network is to develop, disseminate, and apply approaches to research that combine biorepositories with electronic medical record (EMR) systems for genomic discovery and genomic medicine implementation research. In addition, the consortium includes a focus on social and ethical issues such as privacy, confidentiality, and interactions with the broader community (eMERGE https://emerge.mc.vanderbilt.edu/).

PhenX. One of the limitations in being able to interpret findings from genome-wide association (GWA) studies is lack of uniform phenotypic descriptions and measures. For example, hundreds of associations between genetic variants and diabetes have been identified. However, most GWA studies have had relatively few phenotypic and exposure measures in common. Development and adoption of standard phenotypic and exposure measures could facilitate the creation of larger and more comprehensive datasets with a variety of phenotype and exposure data for cross-study analysis, thus increasing statistical power and the ability to detect associations of modest effect sizes and gene-gene and gene-environment interactions (http://www.genome.gov/27558993). PhenX was developed in recognition of the need for standard phenotypic and exposure measures, particu-

Table 10.1.1 eMerge workgroup summary. eMERGE https://emerge.mc.vanderbilt.edu/

Workgroup	Primary function of the workgroup
Phenotyping workgroup	Creation, validation, and execution of phenotype algorithms across the network and beyond
Genomics workgroup	Create a single and uniform data set for all individuals genotyped across the network
Return of Results (RoR)/ Ethical, Legal, and Social Implications (ELSI) workgroup	Define an initial set of variants that are potentially useful in clinical practice for purposes such as assessment of genetic risk for complex disorders or selection or dosing of drugs, focusing on common disease risk variants and pharmacogenetic variants for which we expect to have data; assess the levels of evidence supporting these variants and consider the cost and benefit of incorporating them into patient care
EHR Integration (EHRI) workgroup	Develop standards and methods for incorporating genomic data into the EHR and optimizing the utilization of that data for patients and physicians
Clinical annotation workgroup	Work with Central Sequencing and Genotyping Centers to make pathogenic/non-pathogenic calls on samples
Outcomes workgroup	Study economic and non-economic outcomes of genomic research integrated into healthcare

larly as related to GWA studies. The National Human Genome Research Institute (NHGRI) initiated the PhenX Toolkit in 2006 with the goal of identifying and cataloguing 15 high-quality, well-established, and broadly applicable measures for each of 21 research domains (diabetes is one of these) for use in GWA studies and other large-scale genomic research (www.phenxtoolkit.org).

Roadmap Epigenetics Mapping Consortium. The National Institute of Health (NIH) Roadmap Epigenetics Mapping Consortium was created in an effort to understand epigenetic modifications and how these interact with underlying DNA sequences to contribute to health and disease. The project will provide publically available epigenetic maps on normal human tissues, support technology development, and provide funding in epigenetics research (National Institutes of Health 2015; Slomko et al. 2012).

Precision Medicine/Personalized Healthcare. Precision Medicine/personalized healthcare is a medical model that proposes to customize healthcare by incorporating medical decisions, practices, and products that are based on individual variability in genes, environment, and lifestyle. The potential ability of applying this concept more broadly has been dramatically improved by the recent development of large-scale biologic databases described above. The Precision Medicine Initiative Cohort Program proposes to:

- Identify genomic variants that affect drug response
- Assess clinical validity of genomic variants associated with disease
- Identify biomarkers that are early indicators of disease
- Understand chronic diseases and best management strategies
- Understand genes/pathways/factors that protect from disease
- Assess how well novel cellphone-based monitors of health work
- Evaluate the ability of EHRs to integrate research data
- Learn and apply new ways of engaging participants in research
- Develop methodology for data mining and statistical analysis (https://www.nih.gov/precision-medicine-initiative-cohort-program)

10.1.5 Scope and Practice of Genetics/Genomics Nursing

The American Nurses Association in collaboration with the International Society of Nurses in Genetics provides an excellent resource for nurses interested in clinical genetics and nursing, the Genetics/Genomics Nursing: Scope and Standards of Practice, 2nd Edition (2016). This resource summarizes the role of nurses in genetics/genomics, which focuses on the actual and potential impact of genetic/genomic influences on health. Genetics/genomics nurses educate clients and families on genetic/genomic influences that might impact their health and intervene with the goals of optimizing health, reducing health risks, treating disease, and promoting wellness. This practice depends upon research and evidence-based practice, interprofessional collegiality and collaboration with genetics/genomics professionals and other healthcare professionals to provide quality patient care.

10.1.6 Conclusion

In conclusion, T2D is an increasingly common and complex disorder with genome, exposome, and epigenome factors contributing to the widely variable phenotype. Initiatives in precision medicine propose to customize healthcare by integrating data and information pertaining to individual variability in genes, environment, and lifestyle and interpreting this information to inform medical decisions, practices, and products that prevent, delay, and more effectively treat individuals who are at risk or have T2D. While many of our current initiatives build the evidence base needed to guide clinical practice for the individual, society also needs to be mindful of the social inequalities of opportunity including education, environmental quality, and access, not only to health care but to nutritious food, recreation, and community supports that contrite health and disease. These social determinants are part of the individual's exposome, and yet are often beyond the control of the individual. Finally, motivating individuals at higher risk to engage in lifestyle changes to reduce their risk for T2D remains challenging. Communicating risk information about T2D is further complicated by how a person personalizes and rationalizes his or her risk to develop it (Shah et al. in press; Walter and Emory 2005). Knowing about genetic risk is not enough to motivate people to change behaviors (Grant et al. 2013). An important knowledge gap to fill is our understanding of how people at increased risk for T2D come to understand and manage behaviors to reduce their risk for disease. Understanding a person's beliefs may facilitate effective collaboration with healthcare providers, and improve risk reduction education using a truly comprehensive personalized approach.

References

American Nurses Association. Genetics/genomics nursing: scope and standards of practice. 2nd ed. Washington, DC: ANA; 2016.

Bell G (2009). Forward in *The Fourth Paradigm: Data-Intensive Scientific Discovery* (p.xv). Redmond, WA, Microsoft.

Burgio E, Lopomo A, Migliore L. Obesity and diabetes: from genetics to epigenetics. Mol Biol Rep. 2015;42:799–818. doi:10.1007/s11033-014-3751-z.

Diabetes.co.uk the Global Diabetes Community. Drug induced diabetes. 2015. http://www.diabetes.co.uk/drug-induced-diabetes.html. Accessed 30 Jan 2016.

Knowler WC, Barrett-Connor E, Fowler SE, Hamman RF, Lachin JM, Walker EA, Nathan DM. Diabetes prevention program research group 2002. Reduction in the incidence of Type 2 diabetes with lifestyle intervention or Metformin. N Engl J Med. 2002;346(6):393–403. doi:10.1056/NEJMoa012512.

Diabetes Prevention Program [DPP] Research Group 2002. The diabetes prevention program (DPP): description of lifestyle intervention. Diabetes Care. 2002;25(12):2165–71. doi:10.2337/diacare.25.12.2165.

Genetic Science Learning Center. Epigenetics. 2014. http://learn.genetics.utah.edu/content/epigenetics/. Accessed 30 Jan 2016.

Grant RW, O'Brien KE, Waxler JL, Vassy JL, Delahanty LM, Bissett LG, Green RC, Stember KG, Guiducci C, Park ER, Florez JC, Meigs JB. Personalized genetic risk counseling to motivate diabetes prevention: a randomized trial. Diabetes Care. 2013;36(1):13–9. doi:10.2337/dc12-0884.

Hall E, Dayeh T, Kirkpatrick CL, Wollheim CB, Dekker Nitert M, Ling C. DNA methylation of the glucagon-like peptide 1 receptor (GLP1R) in human pancreatic islet. BMC Med Genet. 2013;14:76. doi:10.1186/1471-2350-14-76.

Harrison TA, Hindorff LA, Kim H, Wines RC, Bowen DJ, McGrath BB, Edwards KL. Family history of diabetes as a potential public health tool. Am J Prev Med. 2003;24(2):152–9. doi:10.1016/S0749-3797(02)00588-3.

Heijmans BT, Tobi EW, Stein AD, Putter H, Blauw GJ, Susser ES, Slagboom PE, Lumey LH. Persistent epigenetic differences associated with prenatal exposure to famine in humans. Proceedings from the National Academy of Science. 2008;105(44):17046–9. doi:10.1073/pnas.0806560105.

Jaenisch R, Bird A. Epigenetic regulation of gene expression: how the genome integrates intrinsic and environmental signals. Nat Genet. 2003;33(Suppl):245–54. doi:10.1038/ng1089.

Keating BJ. Advances in risk prediction of type 2 diabetes: integrating genetic scores with Framingham risk models. Diabetes. 2015;64(5):1495–7. doi:10.2337/db15-0033.

Klein Woolthuis EP, de Grauw WJ, van Gerwen WH, van den Hoogen HJ, van de Lisdonk EH, Metsemakers JF, van Weel C. Identifying people at risk for undiagnosed Type 2 diabetes using the GP's electronic medical record. Fam Pract. 2007;24(3):230–6. doi:10.1093/fampra/cmm018.

Kulis M, Esteller M. DNA methylation and cancer. Adv Genet. 2010;70:27–56. doi:10.1016/B978-0-12-380866-0.60002-2.

Kulkarni SS, Salehzadeh F, Fritz T, Zierath JR, Krook A, Osler ME. Mitochondrial regulators of fatty acid metabolism reflect metabolic dysfunction in Type 2 diabetes mellitus. Metabolism. 2012;6(2):175–85. doi:10.1016/j.metabol.2011.06.014.

Lindström J, Louheranta A, Mannelin M, Rastas M, Salminen V, Eriksson J, Uusitupa M, Tuomilehto J. Finnish Diabetes Prevention Study group 2003. The Finnish Diabetes Prevention Study (DPS): lifestyle intervention and 3-year results on diet and physical activity. Diabetes Care. 2003;26(12):3230–6. doi:10.2337/diacare.26.12.3230.

Ling C, Del Guerra S, Lupi R, Rönn T, Granhall C, Luthman H, Masiello P, Marchetti P, Groop L, Del Prato S. Epigenetic regulation of PPARGC1A in human Type 2 diabetic islets and effect on insulin secretion. Diabetologia. 2008;51(2):615–22. doi:10.1007/s00125-007-0916-5.

Lyssenko V, Laakso M. Genetics screening and the risk for Type 2 diabetes: worthless or valuable? Diabetes Care. 2013;36(Suppl(2)):S120–6. doi:10.2337/dcS13-2009.

Loi M, Del Savio L, Stupka E. Social epigenetics and equality of opportunity. Public Health Ethics. 2013;6(2):142–53. doi:10.1093/phe/pht019.

McGowan PO, Sasaki A, D'Alessio AC, Dymov S, Labonté B, Szyf M, Turecki G, Meaney MJ. Epigenetic regulation of the glucocorticoid receptor in human brain associates with childhood abuse. Nat Neurosci. 2009;12(3):342–8. doi:10.1038/nn.2270.

Mathers JC, Strathdee G, Relton CL. Induction of epigenetic alterations by dietary and other environmental factors. Adv Genet. 2010;71:3–39. doi:10.1016/B978-0-12-380864-6.00001-8.

Medici F, Hawa M, Ianari A, Pyke DA, Leslie RD. Concordance rate for type II diabetes mellitus in monozygotic twins: actuarial analysis. Diabetologia. 1999;42(2):146–50. doi:10.1007/s001250051132.

Morris AP, Voight BF, Teslovich TM, et al. Wellcome Trust Case Control Consortium; Meta-Analyses of Glucose and Insulin-related traits Consortium (MAGIC) Investigators; Genetic Investigation of ANthropometric Traits (GIANT) Consortium; Asian Genetic Epidemiology Network–Type 2 Diabetes (AGEN-T2D) Consortium; South Asian Type 2 Diabetes (SAT2D) Consortium; DIAbetes Genetics Replication and Meta-analysis (DIAGRAM) Consortium 2012. Large-scale association analysis provides insights into the genetic architecture and pathophysiology of type 2 diabetes. Nat Genet. 2012;44:981–90. doi:10.1038/ng.2383.

National Diabetes Information Clearinghouse [NDIC]. National diabetes statistics. 2014. http://diabetes.niddk.nih.gov/dm/pubs/statistics/#Prevention. Accessed 27 Jan 2016.

National Center for Biotechnology Information (NCBI). FFAR3 free fatty acid receptor 3 [Homo sapiens (human)]. 2015. http://www.ncbi.nlm.nih.gov/gene/2865. Accessed 30 Jan 2016.

National Institutes of Health. The National Institutes of Health (NIH) roadmap epigenomics mapping consortium. 2015. http://www.roadmapepigenomics.org. Accessed 30 Jan 2016.

Narayan KM, Boyle JP, Thompson TJ, Sorenson SW, Williamson DF. Lifetime risk for diabetes mellitus in the United States. JAMA. 2003;290(14):1884–90. doi:10.1001/jama.290.14.1884.

Newman B, Selby JV, King MC, Slemenda C, Fabsitz R, Friedman GD. Concordance for type 2 (non-insulin-dependent) diabetes mellitus in male twins. Diabetologia. 1987;30(10):763–8. doi:10.1007/BF00275741.

Poulsen P, Grunnet LG, Pilgaard K, Storgaard H, Alibegovic A, Sonne MP, Carstensen B, Beck-Nielsen H, Vaag A. Increased risk of Type 2 diabetes in elderly twins. Diabetes. 2009;58(6):1350–5. doi:10.2337/db08-1714.

Puumala SE, Hoyme HE. Epigenetics in pediatrics. Pediatr Rev. 2015;36:14–21. doi:10.1542/pir.36-1-14.

Radtke KM, Ruf M, Gunter HM, Dohrmann K, Schauer M, Meyer A, Elbert T. Transgenerational impact of intimate partner violence on methylation in the promoter of the glucocorticoid receptor. Transl Psychiatry. 2011;1:21. doi:10.1038/tp.2011.21.

Remely M, Aumueller E, Merold C, Dworzak S, Hippe B, Zanner J, Pointner A, Brath H, Haslberger AG. Effects of short chain fatty acid producing bacteria on epigenetic regulation of FFAR3 in type 2 diabetes and obesity. Gene. 2014;537:85–92. doi:10.1016/j.gene.2013.11.081.

Ribel-Madsen R, Fraga MF, Jacobsen S, Bork-Jensen J, Lara E, Calvanese V, Fernandez AF, Friedrichsen M, Vind BF, Højlund K, Beck-Nielsen H, Esteller M, Vaag A, Poulsen P. Genome-wide analysis of DNA methylation differences in muscle and fat from monozygotic twins discordant for type 2 diabetes. PLoS One. 2012;7(12):e51302. doi:10.1371/journal.pone.0051302.

Rönn T, Volkov P, Davegårdh C, Dayeh T, Hall E, Olsson AH, Nilsson E, Tornberg A, Dekker Nitert M, Eriksson KF, Jones HA, Groop L, Ling C. A six months exercise intervention influences the genome-wide DNA methylation pattern in human adipose tissue. PLoS Genet. 2013;9(6):e31003572. doi:10.1371/journal.pgen.1003572.

Schulz LC. The Dutch hunger winter and the developmental origins of health and disease. Proceedings from the National Academy of Science. 2010;107(39):16757–8. doi:10.1073/pnas.1012911107.

Shah L, Perkhounkova Y, Daack-Hirsch S. Evaluation of the perception of risk factors for Type 2 diabetes instrument (PRF-T2DM) in an at-risk, non-diabetic population. J Nurs Meas. In press.

Slomko H, Heo HJ, Einstein FH. Minireview: epigenetics of obesity and diabetes in humans. Endocrinology. 2012;153(3):1025–30. doi:10.210/en.2011-1759.

Strahl BD, Allis CD. The language of covalent histone modifications. Nature. 2000;403(6765):41–5. doi:10.1038/47412.

Talmud PJ, Cooper JA, Morris RW, Dudbridge F, Shah T, Engmann J, Dale C, White J, McLachlan S, Zabaneh D, Wong A, Ong KK, Gaunt T, Holmes MV, Lawlor DA, Richards M, Hardy R, Kuh D, Wareham N, Langenberg C, Ben-Shlomo Y, Wannamethee SG, Strachan MW, Kumari M, Whittaker JC, Drenos F, Kivimaki M, Hingorani AD, Price JF, Humphries SE. Sixty-five common genetic variants and prediction of Type 2 diabetes. Diabetes. 2015;64(5):1830–40. doi:10.2337/db14-1504.

Valdez R, Yoon PW, Liu T, Khoury MJ. Family history and prevalence of diabetes in the U.S. population: the 6-year results from the National Health and Nutrition Examination Survey (1999–2004). Diabetes Care. 2007;30(10):2517–22. doi:10.2337/dc07-0720.

Venditti EM. Efficacy of lifestyle behavior change programs in diabetes. Curr Diab Rep. 2007;7(2):123–7. doi:10.1007/s11892-007-0021-7.

Walter FM, Emery J. Coming down the line—patients' understanding of their family history of common chronic disease. Ann Fam Med. 2005;3(5):405–14. doi:10.1370/afm.368.

Weaver IC, Cervoni N, Champagne FA, D'Alessio AC, Sharma S, Seckl JR, Dymov S, Szyf M, Meaney MJ. Epigenetic programming by maternal behavior. Nat Neurosci. 2004;8:847–54. doi:10.1038/nn1276.

Wild CP. Complementing the genome with an 'exposome': the outstanding challenge of environmental exposure measurement in molecular epidemiology. Cancer Epidemiol Biomark Prev. 2005;14(8):1847–50. doi:10.1093/mutage/gen061.

Wolfs MG, Hofker MH, Wijmenga C, van Haefte TW. Type 2 diabetes mellitus: new genetic insights will lead to new therapeutics. Curr Genomics. 2009;10(2):110–8. doi:10.2174/138920209787847023.

Yang BT, Dayeh TA, Kirkpatrick CL, Taneera J, Kumar R, Groop L, Wollheim CB, Nitert MD, Ling C. Insulin promoter DNA methylation correlates negatively with insulin gene expression and positively with HbA(1c) levels in human pancreatic islets. Diabetologia. 2011;54(2):360–7. doi:10.1007/s00125-010-1967-6.

Yang BT, Dayeh TA, Volkov PA, Kirkpatrick CL, Malmgren S, Jing X, Renström E, Wollheim CB, Nitert MD, Ling C. Increased DNA methylation and decreased expression of PDX-1 in pancreatic islets from patients with type 2 diabetes. Mol Endocrinol. 2012;26(7):1203–12. doi:10.1210/me.2012-1004.

Zhang Y, Kent 2nd JW, Lee A, Cerjak D, Ali O, Diasio R, Olivier M, Blangero J, Carless MA, Kissebah AH. Fatty acid binding protein 3 (fabp3) is associated with insulin, lipids and cardiovascular phenotypes of the metabolic syndrome through epigenetic modifications in a Northern European family population. BMC Med Genet. 2013;6:9. doi:10.1186/1755-8794-6-9.

Chapter 11
Answering Research Questions with National Clinical Research Networks

Katherine K. Kim, Satish M. Mahajan, Julie A. Miller, and Joe V. Selby

Abstract In recent years, the healthcare community has recognized that our current clinical research system, for all of the great advances it produces, is in need of improvement. Research—especially large clinical trials—is currently not only expensive, but also slow in both the setup and conduct of a study. Expanding the nation's capacity to conduct clinical studies quickly and economically requires new infrastructure that takes advantage of data gathered in clinics, hospitals, and other sites where patients receive care—as well as patient registries. PCORnet, the National Patient-Centered Clinical Research Network, works with patients, clinicians, health systems leaders, informaticians, and clinical researchers to connect 29 individual networks of patients and healthcare systems into a large interoperable, secure national network that turns millions of patient encounters into valuable data points. This chapter includes a case study of the Patient-centered SCAlable National Network for Effectiveness Research (pSCANNER), a stakeholder-governed, distributed clinical data research network that is part of PCORnet. pSCANNER leverages data from its clinical sites—over 30 million patients in all 50 states—for comparative effectiveness and patient-centered outcomes research to improve care of conditions such as obesity, heart failure, and Kawasaki disease. Implications for nursing research and practice are also offered.

Keywords Clinical data research network • Distributed data network • Patient centered outcomes research • Comparative effectiveness research • Patient engagement • Secondary use of health data

K.K. Kim, Ph.D., M.P.H., M.B.A. (✉) • S.M. Mahajan, Ph.D., M.Stat., M.Eng., R.N.
Betty Irene Moore School of Nursing, University of California—Davis, Sacramento, CA, USA
e-mail: kathykim@ucdavis.edu

J.A. Miller, Ph.D. • J.V. Selby, M.D., M.P.H.
Patient-Centered Outcomes Research Institute (PCORI), Washington, DC, USA

© Springer International Publishing AG 2017
C.W. Delaney et al. (eds.), *Big Data-Enabled Nursing*, Health Informatics,
DOI 10.1007/978-3-319-53300-1_11

11.1 The Vision

In recent years, the healthcare community has recognized that our current clinical research system, for all of the great advances it produces, is in need of improvement. Research—especially large clinical trials—is currently not only expensive, but also slow in both the setup and conduct of a study. Often, completed studies turn out not to be large enough or properly designed to answer the questions most critical to patients. Findings may fail to be relevant to actual clinical practice because they don't involve patients who are representative of important population groups or the interventions tested aren't practical in typical healthcare settings.

A national clinical research network would capitalize on the vast amounts of valuable health information created every day during routine patient visits. Today, opportunities to use this information for research are often missed because the systems that hold this information can't easily communicate or collaborate with each other. Expanding the nation's capacity to conduct clinical studies quickly and economically requires new infrastructure that takes advantage of data gathered in clinics, hospitals, and other sites where patients receive care—as well as patient registries. The NIH Roadmap for Medical Research, issued in 2003, envisioned that by 2013 electronic health records from everyone in the United States would provide data to a national clinical research system.

While we are far from that destination, limited research networks have been established in a variety of agencies, including the Food and Drug Administration, National Institutes of Health, Agency for Health Research and Quality, and Veterans Health Administration. Although the exact structures of the networks vary, each links sites and investigators to pool resources and conduct multiple collaborative research protocols. An ideal national data network would cover large, diverse, defined populations in routine care settings and capture longitudinal standardized data. Networks can realize efficiencies that facilitate the rapid conduct of trials to answer important research questions. Patient organizations, as well as academic centers, pharmaceutical companies, and health systems, have expressed interest in participating—and having a voice—in clinical research networks.

11.2 Electronic Data

The availability of health information in electronic form underlies the feasibility of any national clinical research network. Three important types of information are electronic health records, insurance claims data, and patient-reported outcomes.

Electronic health records (EHRs) are digital records that contain not only data collected in a provider's office but also a more comprehensive patient history. Multiple healthcare organizations can contribute to a patient's EHR. A single EHR can contain information about a patient's medical history, diagnoses, allergies, medications, immunizations, and imaging and lab results from current and past care providers, emergency facilities, school and workplace clinics, pharmacies, laboratories, and medical imaging facilities. Providers, including hospitals, large ambulatory care systems and individual care providers, are increasingly using EHRs, thanks in large part to the Center for Medicare and Medicaid Services' (CMS) Meaningful Use initiatives. EHRs may improve patient care by providing more accurate and complete information, enabling better crisis care, assisting providers to coordinate care, and giving patients and their families useful information to help them share in decisions.

Insurance claims data consists of the coded information that physicians, pharmacies, hospitals, and other healthcare providers submit to payers (e.g., insurance companies, Medicare). These data describe specific diagnoses, procedures, and drugs. Claims data provides abundant, standardized patient information because a claim results from almost every patient encounter with the medical system. The data also may include information on medication compliance and services provided (e.g., eye exam) that may not show up in EHRs. Claims data also provides a population framework needed for longitudinal outcomes studies, and captures many outcomes, regardless of where they occur.

Patient-reported outcomes (PROs) are those that are best reported by patients themselves. They include symptoms, such as pain, that cannot be reliably or accurately assessed by other means. They can also be used to report quality of life and activities of daily living. PROs can be collected electronically through computers and touch-screen devices and telephones with an interactive voice response system. Validated methods, including questionnaires, are available for measuring some PROs. However, researchers may need to work with patients to identify new measures that reflect what is significant to them. Caregiver reports may be appropriate if the patient cannot self-report outcomes of interest.

11.3 Distributed Data Networks

A distributed data network enables researchers to collaborate in the use of electronic health data. It provides investigators with access to a pool of data larger than that of a single site. This type of network has no central repository. Data is owned and maintained by each original data holder behind its firewall. Each data holder remains in control of data security, patient privacy, and proprietary interests for its information. This type of network also safeguards protected health information and proprietary data. With its larger information pool, a distributed network can achieve greater statistical power. It can study rare outcomes and specific populations.

Within the network, each data holder standardizes its data to a common model, allowing indirect analysis by partners. A researcher sends a query, in the form of a computer program, to the data holder. The partner executes the program and returns the results. Access to the local data is allowed through controlled network functions and an agreed-upon governance framework, rather than requiring direct integration between systems or export of data sets. This approach encourages development of novel analytic and statistical methods and reusable, flexible, and scalable programs.

It also reduces legal, regulatory, privacy, proprietary, and technical barriers typically associated with data sharing.

11.3.1 The Mini-Sentinel Distributed Database

The 2007 Food and Drug Administration Amendments Act (FDAAA) required the FDA to create a national electronic system for monitoring the safety of the medical products it regulates. These include drugs, vaccines, other biologics (such as blood products), and medical devices. The agency was directed to obtain the information on these products from existing electronic healthcare data.

A pilot project, the Mini-Sentinel Distributed Database, contains quality-checked data on almost 200,000 individuals. That data was collected by partner organizations—which include insurance companies and hospitals—in the course of its normal operations. Most Mini-Sentinel data comes from health claims submitted to payers. Neither clinicians nor patients report directly to Mini-Sentinel, and EHR data is available for only a small minority of the records.

Each data partner retains control over its data and executes standardized computer programs to respond to FDA's questions regarding the safety and use of regulated medical products. Usually, the data partners provide aggregated results, which become publicly available. A goal of the network is to respond to many questions within days or weeks.

Institutions collaborating with Mini-Sentinel include academic partners as well as data partners. Together these organizations provide scientific, technical, methodologic, and organizational expertise. Representatives of the collaborating institutions participate in planning, operations, and other Mini-Sentinel activities.

Some of the projects to date assess exposures to medical products or health outcomes among individuals exposed to medical products. For example, one recently completed project found no evidence of elevated risk of febrile seizures among young children who received a trivalent influenza vaccine during the 2010–2011 season.

11.4 PCORnet, the National Patient-Centered Clinical Research Network

Large-scale comparative clinical effectiveness research (CER), like medical product safety assessment, requires access to data held by multiple organizations. It often requires greater detail than is found in claims data (e.g. severity of illness, indications, and treatment exposures). 2 years ago, the Patient-Centered Outcomes Research Institute (PCORI) began an ambitious journey to find a faster, less expensive, more powerful way to conduct CER to improve the nation's health and health care. The idea was that a "network of networks" could harness the power of large amounts of health information, including data from EHRs, while taking advantage of strong partnerships between patients, clinicians, and health systems. Such a national network, guided by the wide range of healthcare stakeholders, would enable researchers to efficiently address the real-world needs of patients and those who care for them.

In December 2013, PCORI launched PCORnet, the National Patient-Centered Clinical Research Network. The PCORI Board of Governors approved about $100 million in initial funding to work with patients, clinicians, health systems leaders, informaticians, and clinical researchers to connect 29 individual networks of patients and healthcare systems into a large interoperable, secure national network that would turn millions of patient encounters into valuable data points. The goal: a large electronic data infrastructure that could support researchers in learning from clinical and patient-reported outcomes in large observational outcomes studies and could also support the conduct of large, pragmatic clinical trials embedded within participating delivery systems. Rigorous confidentiality and privacy protocols were developed under the guidance of patients and other healthcare stakeholders to enhance trust in this network, and streamlined contracting and IRB review protocols were implemented to increase the efficiency of the network for conducting large studies.

PCORnet is using data from EHRs, claims, and patient-reported data shared by patients as they manage their conditions in their daily lives. By working closely with healthcare delivery systems, clinicians, and patients, PCORnet will identify approaches to conducting high-quality randomized and observational studies, while integrating research activities into clinical practice without interrupting the flow of care between patient and healthcare providers. When fully operational, PCORnet will be a national resource used for research projects conducted by organizations within the partner networks as well as projects by other research institutions supported by other funders.

11.4.1 The Partner Networks

PCORnet is made up of two types of networks. Clinical Data Research Networks (CDRNs) are based in healthcare systems such as hospitals, integrated delivery systems, and federally qualified health centers. CDRNs hold clinical, patient-level data on diagnoses, treatments, bio-physiologic measures and clinical outcomes from multiple healthcare delivery systems. Data may be archived from EHRs, claims, referrals, and other computerized information sources. In each CDRN, patients serve as representatives in individual network and PCORnet governance, representing patient interests at that level.

Patient-Powered Research Networks (PPRNs), operated and governed by groups of patients and their research partners, are focused on one or more specific conditions or communities, and are interested in sharing health information and participating in research. In each PPRN, a patient serves as principal investigator or co-principal investigator. PPRNs originally assembled virtually or through some other means to share participants' patient-reported data and facilitate patient-generated healthcare research. Through PPRNs, patients may share their health-related experiences, address questions, and participate in formal research studies.

11.4.2 Governance

PCORnet is the first large distributed data network to involve patients and those who care for them in a substantive way in the governance of the network and in determining what questions will be studied. Rules governing the use of PCORnet data will be transparent.

During Phase I, PCORnet's decision-making and policy-making activities were overseen by a governance structure that included PCORI senior leadership and representatives of PCORnet's various stakeholder communities. That initial group quickly began working to refine the governance structure based on stakeholder feedback and an assessment of the various operational challenges and opportunities.

New governance policies, approved August 2015, put more decision-making authority, including strategic planning, into the hands of CDRN and PPRN representatives, operating through PCORnet's Council and its Executive Committee, rather than the Coordinating Center or PCORI. This new approach grew out of a commitment by PCORI and PCORnet leadership to apply the lessons learned as the network developed and to move toward independence and long-term sustainability. Voting members of the council include one representative from each individual network, the coordinating center, and PCORI. Its Executive Committee, elected by Council members, also includes a representative from PCORI and the coordinating center.

CDRNs and PPRNs have individually adopted governance structures and processes. Each participating network meets the following requirements:

- Patients are partners in the research process.
- Patients are involved in network leadership positions.
- Patients are involved in decision making about network participation in proposed studies.
- Patients are involved in formulating research questions, defining essential characteristics of each study, and designing and suggesting plans for dissemination and implementation activities.
- Patients who contribute to PCORnet at large and to network leadership are compensated.

11.4.3 Data Handling

The PCORnet Common Data Model (CDM) is based on the Mini-Sentinel Common Data Model and has been informed by other distributed initiatives such as the HMO Research Network, the Vaccine Safety Datalink, various AHRQ projects, and the ONC Standards and Interoperability Framework Query Health Initiative. The PCORnet CDM is positioned within healthcare standard terminologies to enable interoperability with and responsiveness to evolving data standards.

Attention to data security and data privacy is critical to establishing and sustaining patient trust and encouraging participation in research. A critical feature of this effort is that patients will be deeply involved in creating these policies. In addition to general policies that will apply to PCORnet as a whole, each individual network will develop security measures that are appropriate for its members. The networks will benefit from comparing approaches, so that the most successful policies and procedures can be shared and refined.

11.5 Current State

PCORnet recently completed its 18-month developmental phase. During this period, PCORnet consisted of 29 individual networks plus a coordinating center that provided logistical assistance. All the network partners are committed to learning from one another and working together to construct a research network with high value and utility to all stakeholders in healthcare—researchers, patients, and health systems. Among the networks were 11 Clinical Data Research Networks (CDRNs), based in large healthcare systems, and 18 Patient-Powered Research Networks (PPRNs), based in organizations led by patients and caregivers and focused on particular conditions.

During this first phase, the partner networks built out PCORnet's basic structure, created models for engaging patients and other stakeholders to play a central and meaningful role, laid the foundation for making their health data "research ready" while protecting privacy and security, and established productive partnerships. Toward the end of the developmental phase, PCORnet launched three demonstration studies to answer important questions while testing the technical capacity and function of the health data networks that comprise PCORnet.

PCORnet's Early Demonstration Studies

The first demonstration study compares the benefits and harms of two doses of aspirin taken daily to prevent heart attacks and strokes among patients diagnosed with heart disease. The trial will randomly assign 20,000 patients to take a daily low-dose or regular-strength aspirin. This project involves researchers, clinicians, and patients from six CDRNs and one PPRN, which will provide a secure patient portal for data collection.

The next two demonstration studies are observational studies designed to answer key questions about obesity. One compares benefits and risks of the three most commonly used types of weight loss surgery. Involving 10 CDRNs, representing 53 healthcare organizations, this project will review records from 60,000 patients who had one of these procedures in the past 10 years. The other study will investigate effects of antibiotic use during a child's first 2 years of life on growth and risk for obesity. The study will use clinical records on 1.6 million children maintained by nine CDRNs representing 42 healthcare systems.

In July 2015, PCORI's Board approved a second investment of $142 million for PCORnet's next 3 years. These awards will fund the 13 CDRNs and 20 PPRNs. These individual networks include most of those that participated in Phase I. The six additional established networks, capable of quickly integrating into PCORnet, include two focused on specific populations rather than conditions.

11.6 Future Plans

In a short time, PCORnet has grown from a theory about how to improve patient-centered CER to a maturing enterprise that will provide a powerful national resource for faster, more efficient, and more patient-centered studies. Phase II will further develop the network so that researchers from both within and outside PCORnet can tap into the resource to conduct a wide range of important studies. During this period, PCORnet will also make plans for its long-term sustainability.

PCORnet plans to fund additional studies to further test PCORnet's capabilities and answer key research questions. It will also implement and test a range of tools and procedures to streamline and speed the process of conducting clinical research.

PCORnet has plans to increase collaborations within PCORnet and with the wider research community. Finally, PCORnet will continue to expand its common data model with additional data resources. The partner networks will continue to add data and data categories after agreeing on standards for data representation and storage. In the interest of developing joint research proposals, PCORnet is engaging with America's Health Insurance Plans, FDA's Sentinel program, and the Centers for Medicare and Medicaid Services. In the future, PCORnet will attract research funding from multiple sources and share the research resources developed with the research community.

There is confidence that the national research community and research funders will see PCORnet as a vital resource—and use and invest in it. More importantly, it seems that the culture and practices PCORnet is building can offer a model of research integrated into care delivery and communities that others will adopt. This will lead to more efficient studies and more useful results from our national investments in improving health and health care.

11.7 PCORnet in Practice: pSCANNER

Patient-centered SCAlable National Network for Effectiveness Research (pSCANNER) is one of the clinical data research networks (CDRNs) that make up PCORnet. pSCANNER is a stakeholder-governed, distributed clinical data research network with access to over 30 million patients in all 50 states (Ohno-Machado et al. 2014). The network includes academic medical centers, community hospitals, public health systems, and community health centers. (See http://pscanner.ucsd.edu for a complete list of participating institutions.) pSCANNER uses a distributed

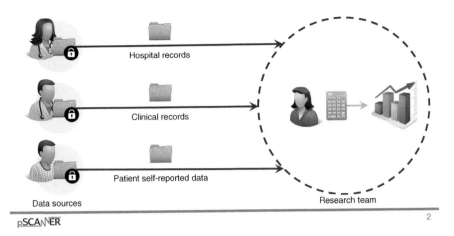

Fig. 11.1 Traditional federated data mechanism: bringing data to the computation

architecture with a common data model and federated query system to enable comparative effectiveness and patient-centered outcomes research. The initial areas of research focus are weight management and obesity, heart failure, and Kawasaki disease. All CDRNs in PCORnet contribute to weight management and obesity studies. On the basis of the clinical and research expertise at its participating institutions, pSCANNER chose to also study heart failure, as a second common chronic condition, and Kawasaki disease, as a rare condition.

The institutions standardize their data model to Observational Medical Outcomes Partnership (OMOP) and to the PCORnet CDM. In addition, the institutions install a pSCANNER node that supports a privacy-preserving distributed computation model (Meeker et al. 2015). A traditional federated data mechanism relies on the transfer of data to a central repository or the ability to access local data sources. Data must be delivered to the researcher to conduct the analysis, with potential issues related to privacy and security (Fig. 11.1). In contrast, pSCANNER distributes queries that include multivariate analytical modules, thereby avoiding the need to transfer data sets to a central repository and the attendant privacy and security challenges. This distributed query model brings computation to the data and returns results to the researcher (Fig. 11.2). pSCANNER demonstrated implementation of its network operational capabilities in PCORnet's 18-month long Phase I.

11.7.1   Stakeholder Engagement

pSCANNER is stakeholder-governed; stakeholders including patients, caregivers, clinicians, and researchers are engaged in all aspects of governance during PCORnet phase I. Four stakeholder advisory boards (SABs) with 49 members were convened. The SAB-governance provides input to policy development for the network.

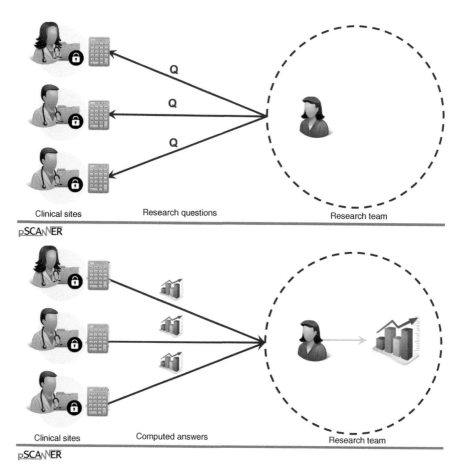

Fig. 11.2 pSCANNER federated query mechanism: bringing computation to the data

Recognizing that governance policies may vary depending on the purpose, type of data shared, and the flow of data, a flexible policy framework was used to guide discussions of research data governance (Kim et al. 2013, 2014). The framework applied the Office of the National Coordinator (ONC) for Health Information Technology's Nationwide Privacy and Security Framework for Electronic Exchange of Individually Identifiable Health Information, which in turn was adapted from the US Fair Information Practice Principles (FIPPs) for stewardship of sensitive personal information (National Coordinator for Health Information 2008). The FIPPs include stating the purpose for collecting information; limiting the collection and use of the information to the minimum necessary; being open and transparent about the information collected about individuals; adopting reasonable security protections; and creating a system of accountability for abiding by laws and policies governing data use and disclosure (McGraw 2013).

An SAB-education board was convened for each of the three conditions of interest—weight management and obesity, heart failure, and Kawasaki disease. These

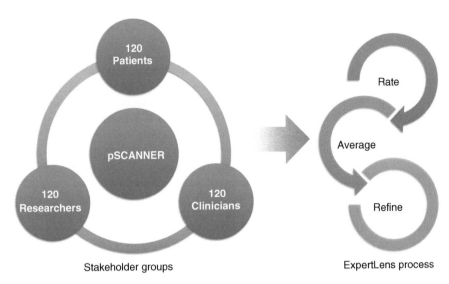

Stakeholder groups   ExpertLens process

Fig. 11.3 Scalable stakeholder engagement via online Delphi consensus process

boards advise on recruitment of diverse advisors and participants, propose initial research topics, suggest appropriate educational materials, and dissemination.

One of the objectives of pSCANNER was to develop and assess an innovative method for engaging patients and other stakeholders in generating patient-centered outcomes research priorities, with a particular focus on scalability of the method to engage large groups of stakeholders. The SABs contributed to the design (Fig. 11.3) of online stakeholder research prioritization panels for each of the three cohorts of interest. The panels were conducted using a modified-Delphi method, a deliberative and iterative approach to attaining consensus with discussion and statistical feedback. The panel interactions were entirely online via ExpertLens™—an online modified-Delphi platform that combines rounds of questions with a round of statistical feedback and online discussion and automatically analyzes group responses (Dalal et al. 2011).

11.7.2 Research in pSCANNER

pSCANNER is now beginning PCORnet phase II, which includes implementation of research and development of network sustainability strategies. During this phase, pSCANNER will both respond to PCORnet data queries and national studies that may encompass a variety of health conditions and topics and propose its own research focusing on its three conditions of interest. Deepening engagement of stakeholders via participatory design teams made up of patients and caregivers, clinicians, and researchers was demonstrated. The teams will design research studies based on the priorities selected by the Delphi stakeholder research prioritization panels in phase I.

One project that is leveraging pSCANNER data is prediction of 30-day readmission risk for heart failure. Heart failure is a debilitating chronic disease affecting

5.7–6.6 million adults in the United States (Fitzgerald et al. 2012), with annual reports of 550,000 new cases (Delgado-Passler and McCaffrey 2006). Although improvements in mortality after diagnosis have been noted, hospitalizations have increased by 175% from 1979 to nearly 1.1 million in 2005 (Dunlay et al. 2009). The heart failure disease trajectory is characterized by frequent rehospitalizations (Howie-Esquivel and Dracup 2008). Heart failure represents the most expensive Diagnosis Related Groups (DRG) diagnosis for hospitalizations in general, and the most frequent diagnosis for 30-day readmissions (Fida and Pina 2012). The American Heart Association forecasts 215% increase in direct cost and 80% increase in indirect cost, and a corresponding increase in prevalence by 25% for heart failure by 2030 (Heidenreich et al. 2011). Thus, reducing the readmission rate, especially for heart failure, is important for quality conscious and fiscally responsible health systems. CMS has established the Readmissions Reduction Program for hospitals as one of the initiatives to reduce prospective payments to hospitals with excess readmissions (Readmissions Reduction Program 2014).

A data-based risk assessment approach would facilitate identification of patients at potentially high risk for heart failure readmissions (O'Connor et al. 2008). Many researchers have proposed such predictive models for worsening symptoms, mortality, and/or readmission but there is no consistency across the models in terms of the readmission timeframe, data sources and algorithms used, and they exhibit modest predictive power. Some models have used either registry or trial-based clinical data (O'Connor et al. 2008, 2012) while others have used hospital-based or Medicare claims-based administrative data (Keenan et al. 2008; Krumholz et al. 2000; Philbin and DiSalvo 1999). This fragmented approach has hindered our progress toward systematic risk prediction and stratification of patient cohorts for optimal interventions. Recently, researchers have suggested examining data surrounding patients' overall health (Chin and Goldman 1997; Hammill et al. 2011) to increase the accuracy of such models predictions.

Satish Mahajan's work in progress involves a prototype risk prediction model based on a comprehensive and systematic combination of structured data from clinical, administrative, and psychosocial domains to predict the risk of 30-day readmission for heart failure. This prototype is implemented using a dataset from one Veteran's Health Administration hospital and its results are being cross-validated using the pSCANNER dataset from the University of California Davis Health System (UC Davis). The cross-validation effort at UC Davis is based on the OMOP/PCORnet Common Data Model and therefore the dataset extraction is universal across all the pSCANNER nodes. This effort is informed by first querying patients with a heart failure diagnosis for a certain timeframe (typically 5–8 years) and then gathering data for all the encounters for the identified cohort of patients for predictors in all domains of interest. Scaling of the model to the five UC sites has the potential to investigate relevant facility-level and aggregate-level factors of significance.

Since heart failure is a progressively degenerative chronic disease, care coordination and patient/caregiver engagement are critically important to maintenance of health and prevention of complications. In fact, pSCANNER's 116-member heart failure research prioritization panel (composed of patients, clinicians, and researchers) determined that the top priority for research was to understand whether and how care coordination activities impact patient outcomes care coordination. This priority is supported by a large body of research that suggests that heart failure readmissions to hospitals are strongly correlated to psychosocial and behavioral factors surrounding patients (Chin and Goldman 1997; Hammill et al. 2011). Patient communities could also be engaged to understand unique patient perspectives and narratives that could be studied using qualitative research techniques (Retrum et al. 2013). A collaboration with PCORnet PPRNs such as Health eHeart (https://www.health-eheartstudy.org/community) might be initiated to understand functional status, validate patient-reported outcomes instruments, develop relevant nutrition education, or offer broad dissemination of research findings.

11.7.3 UC Davis Betty Irene Moore School of Nursing's Role in pSCANNER

The Betty Irene Moore School of Nursing (SON) at UC Davis advances health and ignites leadership through innovative education, transformative research and bold system change. The SON team has led interdisciplinary investigations of the governance model and stakeholder engagement in pSCANNER. The pSCANNER stakeholder engagement team involved collaborators from nursing, medicine, research, health communication, patient communities, and students, providing a rich and productive environment for contributing to the science of engagement which is described in this chapter. In addition, the SON contributed these perspectives to PCORnet through the Patient and Consumer Engagement Task Force and participation in PCORI convenings.

11.8 Role of Nursing Science in and with PCORnet

11.8.1 Nursing Data

With federal programs to promote the adoption of Electronic Health Record (EHR) systems, electronic care documentation has become an integral part of care provision (American Recovery and Reinvestment Act 2009). In practice, nursing documentation is critical to care provision in institutional settings such as hospitals, clinic visits, and nursing homes and in distributed environments such as discharge planning, care management, home health, and telehealth. These data in EHRs along with administrative and claims data serve as the foundational data sets necessary for large-scale comparative effectiveness research. Nursing data such as assessments of social determinants of health, nursing care plans and shared care plans, flowsheets, and nursing notes are often the source of person-centered health processes and person-relevant outcomes.

A project recently implemented at one of the Department of Veterans Affairs hospitals makes this point relevant. According to infection-control data from the hospital's quality management department, one unit experienced high rates of decubitus ulcers (bedsores) for three consecutive months. Utilization data from nursing management showed that the unit was experiencing a shortage of nursing assistants. Nursing notes and discussion with the nursing staff indicated that the problem resulted from an increased number of patients during the winter months. Further analysis of nursing supervisors' dataset validated the seasonal increase in patients for this unit. When the number of nursing assistants was increased, the rate of bedsores decreased. That analysis required multiple data sources and a systematic scientific approach.

This work could be further extended to address patient education by discharge planners regarding infection and wound care and resources necessary for such care. Currently, activities related to in-hospital stay do not necessarily carry over to home care. PCORNet-style data networks would allow researchers to move seamlessly through large clinical, administrative, operational, and patient-centered datasets and increase our statistical power to test hypotheses.

These types of datasets are also necessary for generating evidence regarding the most effective approaches to collaborating with patients, caregivers, and communities regarding such activities as care coordination and lifestyle self-management. However, EHRs currently do not routinely capture potentially useful nursing data in structured and standardized forms that can be used computationally. Hence, the nursing community has an important role to play in informing the development of strategies to collect relevant and usable data.

As the largest segment of the healthcare workforce, nurses are principal contributors to translation of research findings into effective person-centered health programs and incorporation of guidelines and decision support into practice without hindering clinical workflows. Nurses are also powerful allies in communication and dissemination of appropriate evidence-based recommendations to patients and caregivers. As functional designers, implementers, and end users, nurses thus contribute to every stage of data-driven healthcare systems (Barrett 2002).

As PCORNet becomes operational, nurse researchers can examine many compelling questions that are particularly relevant to nursing and require access to large, representative data sets. Among the possibilities are studies related to patient safety concerns that may be nurse-modifiable, such as prevention of bedsores; infections due to invasive lines, catheters, and ventilators; and medication administration error. Nursing can and should contribute to performance and outcomes measurement related to these topics as well as others of importance to learning health systems (Lamb and Donaldson 2011).

Nurse informaticians are contributing to the development and refinement of clinical data standards, particularly semantic standards. These standards will be critical to PCORnet's effectiveness as a national resource for CER. Practical applications, such as risk-prediction models, could be developed based on such standard datasets. The nursing workforce will be pivotal in prospective validation of such models before large-scale deployment for clinical guidance.

Finally, an emerging area that would benefit from active nursing participation is the science of engagement. A long history of nursing research and expertise in theoretical models and methodologies for engagement of patients, families, caregivers, and other stakeholders would be an important contribution to the operationalization of PCORnet infrastructure and its capacity to generate timely and relevant data for health decision making (Barello et al. 2012; Pelletier and Stichler 2013).

Acknowledgement The pSCANNER Team (http://pscanner.ucsd.edu/people). pSCANNER is supported by the Patient-Centered Outcomes Research Institute (PCORI), contract CDRN-1306-04819.

References

Barello S, Graffigna G, Vegni E. Patient engagement as an emerging challenge for healthcare services: mapping the literature. Nurs Res Pract. 2012;2012:7. doi:10.1155/2012/905934.

Barrett EAM. What is nursing science? Nurs Sci Q. 2002;15(1):51–60. doi:10.1177/089431840201500109.

Chin MH, Goldman L. Correlates of early hospital readmission or death in patients with congestive heart failure. Am J Cardiol. 1997;79:1640–4. doi:10.1016/S0002-9149(97)00214-2.

Dalal S, Khodyakov D, Srinivasan R, Straus S, Adams J. ExpertLens: a system for eliciting opinions from a large pool of non-collocated experts with diverse knowledge. Technol Forecast Soc. 2011;78(8):1426–44. doi:10.1016/j.techfore.2011.03.021.

Delgado-Passler P, McCaffrey R. The influences of postdischarge management by nurse practitioners on hospital readmission for heart failure. J Am Acad Nurse Pract. 2006;18(4):154–60. doi:10.1111/j.1745-7599.2006.00113.

Dunlay SM, Redfield MM, Weston SA, et al. Hospitalizations after heart failure diagnosis a community perspective. J Am Coll Cardiol. 2009;54(18):1695–702. doi:10.1016/j.jacc.2009.08.019.

Fida N, Pina IL. Trends in heart failure hospitalizations. Curr Heart Fail Rep. 2012;9(4):346–53. doi:10.1007/s11897-012-0117-5.

Fitzgerald AA, Allen LA, Masoudi FA. The evolving landscape of quality measurement for heart failure. Evolving Challenges in Promoting Cardiovascular Health. Ann N Y Acad Sci. 2012 Apr;1254:131–9. doi:10.1111/j.1749-6632.2012.06483.x.

Hammill BG, Curtis LH, Fonarow GC, et al. Incremental value of clinical data beyond claims data in predicting 30-day outcomes after heart failure hospitalization. Circ Cardiovasc Qual Outcomes. 2011;4(1):60–7. doi:10.1161/CIRCOUTCOMES.110.954693.

Health Information Technology for Economic and Clinical Health Act. American Recovery and Reinvestment Act of 2009 Division A Title XIII 2009.

Heidenreich PA, Trogdon JG, Khavjou OA, et al. Forecasting the future of cardiovascular disease in the United States: a policy statement from the American Heart Association. Circulation. 2011;123(8):933–44. doi:10.1161/CIR.0b013e31820a55f5.

Howie-Esquivel J, Dracup K. Does oxygen saturation or distance walked predict rehospitalization in heart failure? J Cardiovasc Nurs. 2008;23(4):349–56. doi:10.1097/01.JCN.0000317434.29339.14.

Keenan PS, Normand SL, Lin Z, et al. An administrative claims measure suitable for profiling hospital performance on the basis of 30-day all-cause readmission rates among patients with heart failure. Circ Cardiovasc Qual Outcomes. 2008;1(1):29–37. doi:10.1161/CIRCOUTCOMES.108.802686.

Kim KK, McGraw D, Mamo L, Ohno-Machado L. Development of a privacy and security policy framework for a multistate comparative effectiveness research network. Med Care. 2013;51(8 Suppl 3):S66–72. doi:10.1097/MLR.0b013e31829b1d9f.

Kim KK, Browe DK, Logan HC, Holm R, Hack L, Ohno-Machado L. Data governance requirements for distributed clinical research networks: triangulating perspectives of diverse stakeholders. J Am Med Inform Assoc. 2014;21(4):714–9. doi:10.1136/amiajnl-2013-002308.

Krumholz HM, Chen YT, Wang Y, Vaccarino V, Radford MJ, Horwitz RI. Predictors of readmission among elderly survivors of admission with heart failure. Am Heart J. 2000;139(1):72–7.

Lamb G, Donaldson N. Performance measurement—a strategic imperative and a call to action: an overview of forthcoming papers from the American Academy of Nursing Expert Panel on Quality. Nurs Outlook. 2011;59(6):336–8. doi:10.1016/j.outlook.2011.09.002.

McGraw D. Building public trust in uses of Health Insurance Portability and Accountability Act de-identified data. J Am Med Inform Assoc. 2013;20(1):29–34. doi:10.1136/amiajnl-2012-000936.

Meeker D, Jiang X, Matheny ME, et al. A system to build distributed multivariate models and manage disparate data sharing policies: implementation in the scalable national network for effectiveness research. J Am Med Inform Assoc. 2015;22(6):1187–95. doi:10.1093/jamia/ocv017.

O'Connor CM, Abraham WT, Albert NM, et al. Predictors of mortality after discharge in patients hospitalized with heart failure: an analysis from the Organized Program to Initiate Lifesaving Treatment in Hospitalized Patients with Heart Failure (OPTIMIZE-HF). Am Heart J. 2008;156(4):662–73. doi:10.1016/j.ahj.2008.04.030.

O'Connor CM, Mentz RJ, Cotter G, et al. The PROTECT in-hospital risk model: 7-day outcome in patients hospitalized with acute heart failure and renal dysfunction. Eur J Heart Fail. 2012;14(6):605–12. doi:10.1093/eurjhf/hfs029.

Office of the National Coordinator for Health Information. Nationwide privacy and security framework for electronic exchange of individually identifiable health information. 2008. http://healthit.hhs.gov/portal/server.pt/community/healthit_hhs_gov__privacy___security_framework/1173. Accessed 10 Jan 2013.

Ohno-Machado L, Agha Z, Bell DS, et al. pSCANNER: patient-centered Scalable National Network for Effectiveness Research. J Am Med Inform Assoc. 2014;21(4):621–6. doi:10.1136/amiajnl-2014-002751.

Pelletier L, Stichler J. Action brief: patient engagement and activation: a health reform imperative and improvement opportunity for nursing. Am Acad Nurs Policy. 2013; doi:10.1016/j.outlook.2012.11.003.

Philbin EF, DiSalvo TG. Prediction of hospital readmission for heart failure: development of a simple risk score based on administrative data. JACC. 1999;33(6):1560–6. doi:10.1016/S0735-1097(99)00059-5.

Readmissions Reduction Program. 2014. http://www.cms.gov/Medicare/Medicare-Fee-for-Service-Payment/AcuteInpatientPPS/Readmissions-Reduction-Program.html. Accessed 24 Mar 2015.

Retrum JH, Boggs J, Hersh A, et al. Patient-identified factors related to heart failure readmissions. Circ Cardiovasc Qual Outcomes. 2013;6(2):171–7. doi:10.1161/CIRCOUTCOMES.112.967356.

Chapter 12
Enhancing Data Access and Utilization: Federal Big Data Initiative and Relevance to Health Disparities Research

Rosaly Correa-de-Araujo

Abstract High value health and healthcare data are available from both the U.S. Federal Government and the private sector. Advancements in technology and networking capabilities, as well as the rise in wearable devices and consumer health care applications, have led to a considerable increase in the volume and variety of these data. While very large databases are being combined within electronic health records systems (EHRs), information reflecting nursing and potentially other relevant interprofessional notes is still to be integrated. Numerous federal initiatives are ongoing to improve access to and use of big data, and enhance and develop new analytical tools to maximize and accelerate the integration of big data and data science into biomedical research. Nurse practitioners and researchers have the opportunity to use big data to support clinical decisions, increase understanding of patient self-reported data, and expand the emerging research field of symptoms management to ultimately contribute to elimination of health and health care disparities and promote health equity.

Keywords Big data • Data science • Databases

The views expressed in this chapter are those of the author and do not necessarily represent the views of the National Institutes of Health-National Institute on Aging, The U.S. Department of Health and Human Services, or the U.S. Federal Government.

R. Correa-de-Araujo, MD, MSc, PhD
U.S. Department of Health and Human Services, Division of Geriatrics and Clinical Gerontology, National Institute on Aging, National Institutes of Health, Bethesda, MD, USA
e-mail: Rosaly.correa-de-araujo@nih.gov

© Springer International Publishing AG 2017
C.W. Delaney et al. (eds.), *Big Data-Enabled Nursing*, Health Informatics,
DOI 10.1007/978-3-319-53300-1_12

High value health and healthcare data are available from both the U.S. Federal Government and the private sector to help us better understand health, disease, access, utilization, quality, and outcomes of care. Advancements in technology and networking capabilities, as well as the rise in wearable devices and consumer health care applications have led to a considerable increase in the volume and variety of these data (Tan et al. 2015; Holland et al. 2015). As a result, healthcare data presented in many different formats is generated from multiple sources including clinical data from electronic health records (physicians notes and prescriptions, imaging, laboratory, pharmacy, insurance, administrative data); machines/sensors (vital signs, wearable devices); social media posts (Twitter, blogs, Facebook and other platforms, web pages); emergency care data, news feeds, and scientific literature (Raghupathi and Raghupathi 2014).

While very large databases are being combined within electronic health records systems and related data repositories, information reflecting nursing and potentially other relevant interprofessional notes is lacking (Clancy et al. 2014; Keenan 2014). This makes it difficult to use the data to ascertain all factors affecting patient outcomes. Efforts, however, are ongoing to integrate nursing data and information into electronic health records systems, and implement standardized language to represent nursing diagnoses, interventions, and outcomes of care. Other general concerns with the use of big data to advance research and knowledge also exist. Kahn et al. (2014) recommend that: "approaches to informed consent must be reconceived for research in the social-computing environment to take advantage of the technologies available and develop creative solutions that will empower users who participate in research, yield better results, and foster greater trust." The loss of confidentiality and commodification of patient/consumer-generated data are possible challenges. Methodological rigor (selecting appropriate data streams; understanding data derivation; data extraction techniques; calibration/recalibration of algorithms designed/selected for application to the data sources; handling of confounders/biases; approaches for pattern recognition, uncertainty modeling, predictive analytics) remains critical to big data analyses.

The current trends toward the use of larger integrated datasets in health care benefited from prior experiences with big data indicating that potential multiple costs savings and improvements in business processes, customer services, and forecasting can be achieved. Health care stakeholders continue to be eager with the availability of such high quality big datasets to particularly enable: (1) Comparative effectiveness research to identify and disseminate best practices that can reduce the over and under treatment; (2) Research and development efforts to study predictive modeling and improvements in efficiencies and analyses of clinical trial data; (3) Personalized medicine taking into account the individual's health care history and genomic profile to support style practices, prevention and early detection of diseases; (4) Development of innovative business models (e.g., online platforms for the communities of patients, administrators, clinicians) derived from the aggregation and synthesis of clinical and claims data; and (5) Payment pricing strategies that are balanced and based on fraud detection, application of health economics principles, and outcomes research (McKinsey Global Institute 2011; Garcia 2015).

Despite the challenges with data complexity and the inability to use traditional data processing tools and techniques to analyze the data, the availability of real-time big data is a very attractive and necessary feature of big data (Gartner 2011; Baro et al. 2015). Information technology (IT) infrastructure including data lakes and cloud data storage and management solutions make big data analytics possible. Since most IT systems still rely on data warehouse structures, the benefits big data may provide are limited. However, new strategies, analytic tools and techniques are needed for researchers, health care administrators, clinicians, and policymakers to deal with such data. New tools and techniques are essential to better process, analyze, interpret and utilize research findings. Further success in creating big data value in the health care sector depends on policy changes to balance the potential societal benefits of big data approaches and the protection of patient's confidentiality. The many current practices and policies related to data use, access, sharing, privacy, and stewardship will also need to be revised to allow promoting data sharing with the appropriate protections, and advance collaboration on a common goal of delivering better outcomes at lower costs (Roski et al. 2014).

Analyzing big data may support the identification of gaps, challenges and opportunities in health and health care; the design and implementation of strategies to improve the delivery of health services; and, the management and development of new policies. Big data also provides valuable information on predicting epidemics, healing disease, improving quality of life, and avoiding potentially preventable deaths. Zhu et al. (2015) describe a new conditional logistic regression model derived from inpatient data analyses that increased by 20% the accuracy of predicting 30-day hospital readmission for heart failure. Disseminating innovative research and practices derived from the analyses of big data can ultimately benefit patients through improved health outcomes and reduction of health care costs (Shmueli and Koppius 2011; Cortada et al. 2012; Marr 2015; Tan et al. 2015).

12.1 The U.S. Department of Health and Human Services and the Health Data Initiative

The establishment of the Health Data Initiative by the U.S. Department of Health and Human Services (HHS) led to the development, in 2012, of a very resourceful website—*healthdata.gov*. More than 1000 datasets are made available at this site, increasing the accessibility to high value health data. The data can be particularly helpful to those seeking potential solutions to ongoing or emerging challenges in their health care organizations. Examples of information made available include clinical care provider quality information, nationwide health service provider directories, databases of the latest medical and scientific knowledge, consumer product data, community health performance information, and government spending data.

A Health Data Consortium based in Washington, D.C. also functions as a public-private partnership that advocates for the availability and use of health data, in particular government health datasets, and for the improvement of health and health care through increased accessibility to patient data and the innovative use of such data.

The Health Data Initiative Forum (a.k.a. "Health Datapalooza") has been hosted annually since 2011. The Health Datapalooza convenes stakeholders to continuously educate and inform them about the value of data, and new and ongoing initiatives using health data to develop programs and applications. This forum has showcased a considerable number of the best products and services developed by companies that have harnessed HHS data to help consumers get the information they need; health care providers to deliver better care; employers to promote health and wellness; and policymakers to make more informed decisions. (See http://healthdatapalooza.org/.)

The ultimate goal of the HHS Health Data Initiative is to improve health, health care, and the delivery of human services by coupling the power of data with a culture of innovative data use by institutions in the public and private domains, communities, research and policymaking groups. By making existing data easier to use, HHS is also committed to making the data machine readable (i.e. data presented in a format that can be read by computers; while all machine readable format is also a digital format, not all digital format is machine readable), downloadable, and accessible via applications such as programming interfaces, while continuing to protect data privacy and confidentiality.

The current limitations of appropriate tools to facilitate data use and analyses, restricted data accessibility, and the insufficient training of health care professionals and researchers in data science methods including translation into policies and practices, remain significant barriers to the rapid improvement in data use and translational impact. In an effort to address these issues and make health data openly available, several policy developments have contributed to the thriving Health Data Initiative. The Open Government Initiative launched in 2009 by President Obama has provided significant support for public access to government data. With the Patient Protection and Affordable Care Act (ACA) signed by the President in March 2010, new data resources that advance transparency in the health care provider and health insurance markets became available. The ACA also authorizes HHS Centers for Medicare and Medicaid Services (CMS) to evolve how it pays care providers from "pay for volume of services" to "pay for health." This shift is creating strong incentives for health care providers to leverage the power of data and technology to help minimize errors, decrease preventable hospital admissions and readmissions, enhance coordination of care, and ensure patient engagement. Innovative big data techniques and approaches are being used by CMS to generate actionable information from CMS data more effectively, i.e., the use of real-time analytics for program monitoring and detecting of fraud and abuse, and the increased provision of data to providers, researchers, beneficiaries, and other stakeholders (Brennan et al. 2014).

Section 4302 of the ACA (Understanding Health Disparities: Data Collection and Analysis) focuses on standardization, collection, analysis, and reporting of health disparities data. It requires that data collection standards for race, ethnicity, sex, primary language and disability status be established and included on all population health surveys, to the extent practicable. These standards were announced by the U.S. Secretary of Health in October 2011. (See http://minorityhealth.hhs. gov/omh/browse.aspx?lvl=3&lvlid=53.) In addition, Section 4203 of the ACA deals specifically with accessibility standards for the design of medical diagnostic equipment for people with disabilities. This means that new data may become available on how health care organizations are doing as relates to improving access to safe care and enhancing health outcomes for this population. (See https://www.gpo.gov/ fdsys/pkg/PLAW-111publ148/pdf/PLAW-111publ148.pdf;   https://www.access-board.gov/guidelines-and-standards/health-care/about-this-rulemaking/proposed-standards/introduction-and-background.)

The Agency for Healthcare Research and Quality (AHRQ) has supported collaborations of informatics experts, researchers, and clinicians to develop better ways to link electronic health records with administrative claims data, pharmacy data, and diagnostic data, and to collect patient-specific information (such as medication side effects, pain after surgery, disease symptoms, and experience of care) in order to create a comprehensive picture of the patient experience, while protecting patient confidentiality. The Agency has also supported the development of new analytic tools to rapidly analyze the data to produce useful information that can improve decisions and clinical care. For example, AHRQ's registry user's guide is the gold standard for best practices in designing and analyzing registry data. The resulting increase in the quality of registry data has enabled new initiatives, such as a CMS rule allowing participation in a qualified clinical data registry as a proxy for care quality reporting. The guide is part of a larger initiative—Registry of Patient Registries (RoPR)—which is a database of more than 107 clinical registries including disease registries focused on treatment effectiveness, quality improvement, public health, postmarketing surveillance of drugs or devices already on the market, and other issues.

Both CMS and the Food and Drug Administration (FDA) are using registry data for policy decisions. CMS uses registry data to make decisions about which procedures or medical devices to cover for a specific condition and to determine which procedure/device works for patients older than 65. The FDA uses registry data for evidence evaluation throughout the product life cycle, including premarketing evaluation, postmarketing surveillance, labeling extension, and device tracking. For example, the Total Joint Replacement registry provides the FDA with early postmarketing surveillance data on different types of artificial knee joints. The Function and Outcomes Research for Comparative Effectiveness in Total Joint Replacement (FORCE-TJR) project surveillance includes post-TJR implant complications and patient-reported pain, both events that precede revisions. FORCE-TJR is also testing novel methods for monitoring implant performance using direct-to-patient strategies, including a pilot study of an FDA-developed app that patients use to report adverse events (AHRQ 2015).

The OpenFDA Initiative launched in 2014 provides application programming interfaces (APIs) and is investigating access/download source/original data from various high-value structured datasets to a number of high-value structured datasets, including adverse events, drug product labeling, and recall enforcement reports. Its platform is currently in public beta, available at https://open.fda.gov/#sthash.VNxLMnmm.dpuf.

The Data Science Initiative at the National Institutes of Health (NIH) involves not only very large data or large numbers of data sources but also the complexity of the biomedical data (imaging, phenotypic, molecular, exposure, health, behavioral, and other fields) that can be used toward new drug discoveries or to determine the genetic and environmental causes of disease. A considerable array of opportunities exist to maximize the potential of existent data and support new directions for research including improved data accuracy to sustain the development of precision methods for health care. In addition to the challenges associated with the considerable amount of information, the lack of organization and access to appropriate analytic data tools, these opportunities are also affected by insufficient workforce trained in data science methods.

The NIH Data Science Initiative has acknowledged the role of big data to advance science by launching a set of funding opportunities designed to develop resources and increase the biomedical workforce's competency in data science related to computer science, biostatistics, and biomedical science. The main focus of the NIH Big Data to Knowledge (BD2K) Program is to support research and development of innovative and transforming approaches and tools to maximize and accelerate integration of big data and data science into biomedical research. Several active funding opportunity announcements (FOAs) became available. (See https://datascience.nih.gov/bd2k/announcements.) In these funding opportunities, the terminology "Big Data Science" is meant to capture the opportunities and address the challenges facing all biomedical researchers in releasing, accessing, managing, analyzing, and integrating datasets of diverse data types (e.g., imaging, phenotypic, molecular [including–omics], physiological, anatomical, clinical, behavioral, environmental, many other biological and biomedical data types, and data generated for other purposes such as social media, search histories, and cell phone data). Specifically, active funding opportunities currently seek research applications looking into the early stage development of technologies (new software, tools and related resources) as well as the fundamental research (methodologies and approaches) in biomedical computing, informatics and big data science (http://grants.nih.gov/grants/guide/pa-files/PA-14-155.html). Another opportunity targets the extended development, maintenance, testing, evaluation, hardening and dissemination of existing biomedical software in the context of the same research areas mentioned above (http://grants.nih.gov/grants/guide/pa-files/PA-14-156.html). The need for contemporary software that is easy to modify and extend is highlighted as a key feature to support the open-ended and ever-changing scientific discovery process. The issues of interoperability and portability are also important. Interoperability among different software packages or among software and existing databases is a major challenge for which extended interoperability remains a target. Portability to different types of hardware is important to allow software to operate on a variety of platforms employing different operating systems. The McKinsey (2011) report also states that the most basic first step necessary to create analyzable "big data" systems is to digitize and structure the data so that it is interoperable (e.g., collected, represented, measured, and stored in the same way across electronic health records and organizations). Electronic health records within organizations have generally been assembled and tailored to meet the unique needs of each organization. This practice severely compromises the ability to compare data collected within one organization to data collected across organizations. This ability to use data from diverse organizations is needed to create research opportunity and benefit from "big data."

The main vision of the big data initiative at NIH is to achieve efficiency in biomedical research by making it easier for researchers to locate and manipulate data and software. According to the NIH data leader 5% improvement in efficiency in research means more than $150 million/year in NIH budget that can be spent on funding additional people/research. However, challenges to overcome include: sustaining support of data resources under the current budget environment; training of experimental researchers to use a wide variety of analysis tools, software carpentry, working with big data, and taking skills to the labs.

Under the big data initiative, NIH is building the COMMONS, a virtual space initiative where researchers can share, locate, use, and cite datasets. Examples of impact from the big data initiative include Vanderbilt University's a virtual cohort built by linking phenotypic information from de-identified electronic health records to a DNA repository (200,000 samples). It provides real-world clinical data that can be used to study disease and drug response under the precision medicine initiative. A major component of this initiative is the establishment of a large longitudinal epidemiology cohort of a million or more participants in whom genetic and environmental determinants of a wide variety of human diseases can be studied. (See http://www.newswise.com/articles/vanderbilt-s-biovu-databank-now-world-s-largest-human-dna-repository-linked-to-searchable-electronic-health-information.)

The University of Arizona College of Engineering is currently working on a Data Compression Software to Make Biomedical Big Data Universally Available. Digitalized biomedical data has been assembled in a format and size that allows professionals, including pathologists with limited resources in remote areas, to access, analyze and store the data. This capability is very relevant because it provides the opportunity for quicker second opinions and diagnoses. (See https://projectreporter.nih.gov/project_description.cfm?project number=1U01CA198945-01.)

Over the past many years major discussions and plans related to the future of epidemiology have occurred. The National Cancer Institute (NCI) and National Heart, Lung and Blood Institute (NHLBI) have launched recommendations including transforming the field and its funding strategies within the context of big data science, technological developments, and resource constraints. The Centers for Disease Control and Prevention (CDC) has particular interest in the field as it expands to include sharing of resources, data, and metadata; evaluation of new methods and technologies to measure exposures, susceptibility, and outcomes; and identification of new ways of collecting personal (e.g., mobile health or "m-health") and macro-level data. In particular, shared resources such as whole genome sequencing of study participants will be valuable information in epidemiologic studies across age and disease spectra (Khoury 2015). One of CDC's major contributions to big data is CDC Public Health Genomics Knowledge, an online, searchable database of published scientific literature, CDC resources, and other material that addresses the translation of genomic discoveries into improved health care and disease prevention.

The HuGE Navigator provides access to a continuously updated knowledge base in human genome epidemiology, including information on population prevalence of genetic variants, gene-disease associations, gene-gene and gene- environment inter- actions, and evaluation of genetic tests. (See https://phgkb.cdc.gov/GAPPKB/phg-Home.do?action=about.)

The National Institute on Nursing (NINR) includes data science as a major com- ponent of its research programs. The institute has particularly focused on the devel- opment of common data elements to allow comparisons of variables across studies and patient populations. A 2011 Request for Information called for public input on an initiative to identify the best common measures for symptom assessment. (See http://grants.nih.gov/grants/guide/notice-files/NOT-NR-11-010.html.)   Common data elements for symptom science were also the subject of discussion sessions during the regional nursing research conferences in 2015, resulting in the publica- tion of an article that identifies common data elements to be used across NINR- supported Centers to measure pain. (See http://grants.nih.gov/grants/guide/ notice-files/NOT-NR-11-010.html.), sleep, fatigue, and affective and cognitive symptoms (Redeker et al. 2015). A more recent funding opportunity announcement calls for applications on self-management of symptoms to include common data elements in the research plan. (See http://grants.nih.gov/grants/guide/rfa-files/ RFA-NR-16-002.html.)   (See   http://grants.nih.gov/grants/guide/notice-files/ NOT-NR-11-010.html.)

Other NINR-supported research of impact included a large study employing big data in which a group of scientists investigated patient safety, decision-making, and nurse staffing characteristics. Discharge data from nine European countries was studied. It was found that an increased workload of one patient per nurse was asso- ciated with a 7% increase in the odds of surgical inpatient mortality within 30 days of admission. Researchers then suggested that nurse staffing cuts to save money might adversely affect patient outcomes and that an increased emphasis on bache- lor's degree education for nurses could reduce preventable hospital deaths (Aiken et al. 2014).

Students enrolled in a nurse practitioner program routinely used a handheld mobile device to document their clinical encounters; this system has been shown to help assess student performance and to strengthen evidence-based nursing practice. In a randomized controlled trial, decision support software was integrated into the devices that provided clinical practice guidelines for the screening and management of obesity and overweight, tobacco use, and depression in adults and children. The trial data included over 34,000 unique clinical encounters; data analysis of these encounters found that screening increased, but varied across conditions in the devel- opment of management plans for the conditions diagnosed. By improving health care processes, nursing science is improving health care outcomes (Bakken et al. 2014).

At the P30 Center of Excellence for Biobehavioral Approaches to Symptom Management, nurse scientists pooled symptom data from research projects across diverse patient populations and different diseases and conditions to examine rela- tionships between psychological states and nerve signaling chemicals. By using common data elements, these researchers were able to determine strong correlations among symptoms and perceived stress, but found weaker relationships between nerve signaling chemicals and symptoms, indicating the need for further research in the taxonomy of symptoms and biomarkers (Lyon et al. 2014).

12.1.1 Integrating Nursing Data into Big Data

Generating high quality interoperable big nursing data is needed to contribute to big healthcare data analyses. Because most nursing and other clinical data are not standardized, they are not interoperable (Staggers 2013). As stated previously in this chapter, efforts are ongoing to integrate nursing information into electronic health records systems, implement standardized language to represent nursing diagnoses, interventions, and outcomes of care; modify and standardize nursing informatics education to build understanding and competencies; and, influence policy and standards for documenting and coding nursing information in healthcare knowledge systems. The Nursing Knowledge Conference sponsored by the University of Minnesota School of Nursing pioneers these efforts through a broad action plan to support the development of interoperable nursing data for big data healthcare research. (See http://www.nursing.umn.edu/prod/groups/nurs/@pub/@nurs/documents/content/ nurs_content_482402.pdf.) The following are some key recommendations from that plan: (1) Adopt standardized nursing terminologies (SNTs) (ANA recognized) at the point of care and Nursing Information Dataset Evaluation Center (NIDSEC) database standards; (2) Promote institutional and public policies that support the use, refinement, and expansion of IT standards to enable the documentation and exchange of key nursing data across systems including empowering IT nurses to advocate for the integration of SNTs into electronic health records; (3) Create and implement a strategy to educate nursing students, nurses, faculty, nurse executives, IT nurses, and the inter-professional care disciplines on key aspects of nursing informatics. In addition, the Healthcare Information and Management Systems Society CNO-CNIO Vendor Roundtable Big Data Principles Workgroup provides the following recommendations to guide the ability to capture and use big data in nursing (See http://www.himss.org/cno-cnio-vendor-roundtable.): (1) Promote standards and interoperability in all healthcare delivery settings to implement an American Nurses Association (ANA) recognized nursing terminology mapped to national standards, such as Systematized Nomenclature of Medicine—Clinical Terms (also known as SNOMED CT) or Logical Observation Identifiers Names and Codes (also known as LOINC); (2) Advance quality eMeasures including meaningful use clinical quality measures and nursing-sensitive performance indicators. The data needed to populate these measures come from multiple sources, some of which are not currently available in electronic health records. Clinical quality eMeasures will become essential for enabling analytics and big data initiatives to generate new evidence and knowledge; (3) Leverage nursing informatics experts to better identify, define, manage, and communicate data, information, knowledge, and wisdom in nursing practice.

To make informed clinical decisions in any healthcare setting, nurses need access to accurate, real-time information. Structuring data in standard ways allows sharing and comparing information. Interoperable systems, advances in electronic health records, and alignment of standards and terminologies can demonstrate the value of consistent and accurate care in clinical practice.

12.2 Eliminating Health Disparities and Building Health Equity with Big Data

12.2.1 The Social Determinants of Health

The World Health Organization (WHO) states that a person's health is shaped by the conditions in which they are "born, grow, live, work and age, including the health system," and "distribution of resources at global, national and local levels." These conditions are the social determinants of health, which are mostly responsible for health inequities and consequently the potentially avoidable differences in health status seen among population groups within and between countries. Poverty is known to limit access to healthy foods and safe neighborhoods. More education is a predictor of better health. Differences in health are striking in communities with poor social determinants of health such as unstable housing, low income, unsafe neighborhoods, or substandard education. The knowledge about the impact of social determinants of health can support the development of strategies to improve health at the individual and population levels, decrease health disparities and advance health equity. (See http://www.who.int/social_determinants/en/.)

Both health equity and social determinants are acknowledged as critical components of the post-2015 sustainable development global agenda and of the push towards progressive achievement of universal health coverage. The WHO states that in order to reduce health inequities, both social determinants of health and universal health coverage must be addressed in an integrated and systematic manner.

Despite the impact of social determinants of health, translating them into policies and practice is significantly challenging. The reason lies in the difficulty in linking health policy with public health practice. Generally, health data are reported as individual indicators, instead of representing a comprehensive geographic area within the context of non-medical indicators affecting health. To properly assess the impact of policies and programs on people's health, a clear relationship between health outcomes and social determinants should be established. Moreover, many past years of data collection have not led to policies and public health programs consistent with the principles surrounding the social determinants of health. Therefore, a considerable amount of work is still pending to properly use, interpret and implement quality improvement strategies and interventions based on the impact of these determinants on health outcomes. Improving and sustaining policy changes at the system level and empowering communities should consequently result in long-term impact in population health (Russell et al. 2013).

Table 12.1 displays information on selected data sources containing variables related to the social determinants of health. The data sources are relevant to support the development of research agendas and improvements of programs targeting elimination or reduction of health disparities and advances in health equity.

Table 12.1 Selected data sources with social determinants of health (See http://www.cdc.gov/socialdeterminants/data/index.htm.)

Data source	Description
Chronic disease indicators http://www.cdc.gov/cdi/index.html	State and selected metropolitan-level data for chronic diseases and risk factors, particularly relevant to public health professionals and policymakers.
Community health status indicators http://wwwn.cdc.gov/CommunityHealth/	Health profiles of every state and county and the District of Columbia. A set of primary and associated indicators that affect health outcomes and population health in various domains (e.g., health care access and quality, health behaviors, social factors, and physical environment) is accessible along with county demographics. Comparisons of the value for each indicator with those of demographically similar "peer counties" in the United States and with *Healthy People 2020* targets can be obtained.
Health indicators warehouse http://www.healthindicators.gov/	National, state, and local health indicators. Multiple dimensions of population health, health care, and health determinants can be queried, data can be viewed in map, chart, or table format. Indicator data can be downloaded. This resource, however, began a retirement phase on June 30, 2016. The site remained available for use in its current state until April 15, 2017, but no longer updated. To provide continued access to the content of this resource, different segments of its database are planned to be made available through Healthdata.gov. Much of the data and indicators can also be located through Health, United States (http://www.cdc.gov/nchs/hus/index.htm), Healthy People (http://www.healthypeople.gov/), County Health Indicators (http://wwwn.cdc.gov/communityhealth/), and other federal and nongovernmental data and indicator access tools and sites.
Interactive atlas of heart disease and stroke http://nccd.cdc.gov/dhdspatlas/	Online county-level mapping of heart disease and stroke by race/ethnicity, gender, and age group. Maps can show social and economic factors and health services for the United States, specific states, or territories.
National center for HIV/AIDS, viral hepatitis, STD, and TB prevention atlas http://www.cdc.gov/nchhstp/atlas/index.htm	Interactive maps, graphs, tables, and figures showing geographic patterns and time trends of reported occurrence of HIV, AIDS, viral hepatitis, tuberculosis, chlamydia, gonorrhea, and primary and secondary syphilis. Data are based on nationally notifiable infectious diseases in the United States and can be used to examine disparities.

(continued)

Table 12.1 (continued)

Data source	Description
National environmental public health tracking network http://ephtracking.cdc.gov/showHome.action	A system of integrated health, exposure, and hazard information and data from a variety of national, state, and city sources. Maps, tables, and charts with data about environmental indicators (e.g., particulate matter in the air) are available.
Vulnerable populations footprint tool http://www.communitycommons.org/chna/	Maps and reports that identify geographic areas with high poverty rates and low education levels—two key social determinant indicators of population health, can be created. Thresholds for target areas are adjustable, allowing the tool to be used in geographic areas where regional rates may be higher or lower than the national average.
Sortable risk factors and health indicators http://sortablestats.cdc.gov/#/	Sortable Stats is an interactive data set comprised of behavioral risk factors and health indicators. This tool is intended to serve as a resource in the promotion of policy, system, and environmental changes.
County health educator http://countyhealthcalculator.org/	Tool for advocates and policy makers to understand and demonstrate how education and income affect health. (Virginia Commonwealth University Center on Human Needs)

12.2.2 Health Disparities and Health Equity

A definition of health disparity is available from the U.S. Department of Health and Human Services Strategic Plan, Healthy People 2020, and the National Partnership for Action to End Health Disparities (NPA). It states that "A health disparities is a particular type of health difference that is closely linked with social, economic, and/or environmental disadvantage. Health disparities adversely affect groups of people who have systematically experienced greater obstacles to health based on their racial and/or ethnic group; religion; socioeconomic status; gender; age; mental health; cognitive, sensory, or physical disability; sexual orientation or gender identity; geographic location; or other characteristics historically linked to discrimination or exclusion." (See http://minorityhealth.hhs.gov/npa/files/plans/hhs/hhs_plan_complete.pdf.)

Health Equity is defined as "an attainment of the highest level of health for all people. Achieving health equity requires valuing everyone equally with focused and ongoing societal efforts to address avoidable inequities, historical and contemporary injustices, and the elimination of health and healthcare disparities." (See http://minorityhealth.hhs.gov/npa/files/plans/hhs/hhs_plan_complete.pdf.).

Health varies intensely from community to community, with low-income and minority communities more likely to experience negative health outcomes. Understanding the social determinants of health is essential for developing strong public policies and activities that promote health equity and the elimination of health disparities. According to WHO, reducing inequities in health is important because health is a fundamental human right and its progressive realization will eliminate inequalities that result from differences in health status such as disease or disability. (See http://www.who.int/healthsystems/topics/equity/en/.)

In the United States, numerous efforts are ongoing to reduce disparities or eliminate inequities. These efforts emphasize the social determinants of health that considerably impact health. They include socioeconomic status, quality education, safe and healthy housing, access to affordable healthy and fresh foods, and access to and use of quality health services. Social determinants of health data can help clinicians and researchers identify the root causes affecting populations' health. Recognizing that moving from data to action can be challenging, the Centers for Disease Control and Prevention (CDC) has made available tools for putting social determinants of health into action. (See http://www.cdc.gov/socialdeterminants/tools/index.htm/.)

The National Partnership for Action to End Health Disparities (NPA) is the first national, multi-sector, community- and partnership-driven effort on behalf of health equity. This ongoing national effort is increasing the effectiveness of programs and mobilizing partners, leaders, and stakeholders toward eliminating health disparities. Government agencies are involved in the effort through the Federal Interagency Health Equity Team (FIHET). The FIHET provides leadership and guidance for national, regional, state, and local efforts that address health equity. There are ten Regional Health Equity Councils (RHECs) that engage HHS regional directors or administrators and are comprised of practitioners from various sectors (health care, education, transportation, technology, and business). The RHECs' primary role is to initiate action to implement the NPA's goals and therefore, advance the agenda to eliminate health disparities from the grassroots. (See http://minorityhealth.hhs.gov/npa/.)

Because big data is critical to the development of strategies to eliminate disparities and monitor progress, the FIHET's Data, Research and Evaluation Workgroup developed the first compilation of publically available federal datasets relevant to health disparities research and programs. The purpose of developing the Compendium of Federal Datasets Relevant to Disparities Research and Programs includes: (1) Creating a resource that facilitates identification and use of data relevant to research and programs aiming to end health disparities; (2) Providing easy access to information on publicly available data through a single, updated source of information; (3) Updating and building on ongoing federal data sharing efforts; (4) Stimulating research that connects information from different sources and supports work addressing health and social disparities in a meaningful way; (5) Encouraging data collection to address data gaps, facilitate the development and monitoring of quality improvement strategies, and the use of existing big data across and outside the Federal government; and (6) Informing programs and strategies, and alert healthcare and public health professionals as well as policymakers and the public on improvements and gap areas or on populations in need of attention. The inclusion of datasets from Federal agencies beyond HHS is a major feature of this product. The potential for adding new data to this resource, identifying new uses, promoting data linkage, and expanding research areas also serve the purpose of the overall HHS Health Data Initiative. The challenge still remains in eliminating barriers to making all information machine readable and accelerate translational impact, but work is ongoing. The Compendium is available at https://www.minorityhealth.hhs.gov/npa/templates/browse.aspx?lvl=1&lvlid=46.

Table 12.2 displays examples of health and healthcare disparities, some of which are associated with social determinants of health.

Table 12.2 Examples of Health and Health Care Disparities (National Partnership for Action to End Health Disparities; Bakken and Reame 2016)

Disparities category	Examples
General differences in or among population groups	Young people from racial and ethnic minority groups in the United States suffer disproportionately from a number of preventable diseases and health problems.
	American Indians and Alaska Natives have a lower life expectancy than other American perhaps due to economic adversity and poor social conditions.
	Older adults in the United States are not aware of the services recommended for their age group or may not know that these services are covered by Medicare.
	Life expectancy at birth among indigenous Australians is substantially lower (59.4 for males and 64.8 for females) than that of non-indigenous Australians (76.6 and 82.0, respectively).
	The prevalence of long-term disabilities among European men aged 80+ years is 58.8% among the lower educated versus 40.2% among the higher educated.
Specific differences in symptoms experiences	Higher rates of pain among multiracial groups with lung and colorectal cancer.
	Increased severity of pain in black patients with lung and colorectal cancer.
	Lower rates of 14 physical symptoms among Asians Americans.
	Decreased depression symptom severity among African Americans exposed to trauma.
	Risk for moderate lower urinary tract symptoms higher among Hispanic and Black men and lower in Asian men.
	Black women younger than 50 years more likely to report frequent and intense prodromal symptoms of myocardial infarction.
Specific differences in symptoms strategies	Lower rates of recognition of late life depression and treatment in Blacks.
Specific differences in symptoms outcomes	Time to reach functional decline in multiple sclerosis is shorter in Black Caribbeans compared to British Whites.

12.2.3 Using Big Data to Eliminate Disparities and Build Equity in Symptoms Management

Because health and healthcare disparities are still very wide among diverse populations, major opportunities exist within the health data initiative to improve data collection, make new data available to support studies, and improve or develop new quality improvement strategies, prevention or treatment interventions, all of which could result in significant translational impact.

Symptoms management is a priority area of nursing research and practice. It is also a major research focus for the National Institute on Nursing Research (NINR). NINR holds the Big Data in Symptoms Research Boot Camp annually, as part of the NINR Symptom Research Methodologies Series. This is a 1-week intensive research training course focusing on methodologies and strategies for incorporating novel methods into research proposals. (See http://www.ninr.nih.gov/training/trainingopportunitiesintramural/bootcamp#.VqhcJjHSnIU.)

Symptoms management relates to the understanding of the biological and behavioral aspects of symptoms such as pain and fatigue. New knowledge and strategies to improve the quality of life of people suffering from such symptoms are needed. Data including patient experiences with symptoms, symptom management strategies and outcomes are considered big data with complex consumers' self-reported information and *omic* data streams. Gender, race and ethnicity differences have been reported in health and health care including these specific areas of symptoms experiences, management strategies, and outcomes. For example, female carriers of 5-HTTLPR genotype showed a marginally significant protective effect for depressive symptoms and no significant influences of environment while males living in areas of relatively poor building conditions had higher depressive symptoms but no differences attributed to genotype (Uddin et al. 2011). Racial and ethnic differences in pharmacokinetics are another important area for investigation, as it is known that clearance rates affect treatment efficacy and risk for adverse events (Bakken and Reame 2016).

Numerous factors have stimulated the use of big data in research targeting the field of symptoms management (Collins and Varmus 2015; Bakken and Reame 2016). These include: (1) Increased availability of electronic health data with consumer-generated information. This information complements the traditional electronic sources of research data (e.g., surveys, clinical data warehouses, transaction data) and is generated from mobile devices, wearables, medical devices, and sensors, along with multiple sources of environmental and *omic* data; (2) Expanded research networks such as the Patient-Centered Outcomes Research Institute (PCORI) National Patient-Centered Clinical Research Network, which provides a variety of data sources through its Clinical Data Research Networks and Patient-Powered Research Networks. In addition, the National Center for Advancing Translational Sciences (NCATS) has launched a nationwide network—the Clinical Translational Science Award Accrual to Clinical Trials Network—composed of sites that share electronic health record data; (3) Precision medicine and biobanking-related initiatives; and (4) the NIH Big Data to Knowledge Initiative.

The use of big data and data science methods to advance the field of symptoms management is in its early development. Therefore, nurse scientists have a great opportunity to contribute to defining important questions to be addressed, applying relevant theories, contributing to the design of new data science methods, and considering the cultural, ethical, legal and social implications. One concern with the use of existing big data to address symptoms management relates to biases against underrepresented groups, such as depicted by differences in those represented in clinical trials and quantified self-data sources, which could limit the clear understanding of symptoms experiences and lack of cultural context. Racial and ethnic minorities, for example, are less like to participate in biobanks, limiting discoveries based on *omic* data sources (Brennan and Bakken 2015).

Nurses can have an important role in advancing data science by using big data to study symptoms management. As recommended by the Institute of Medicine (now the Health and Medicine Division of the National Academies of Sciences, Engineering and Medicine), it is essential to have the means to demonstrate the impact of nursing care (including nurse provider type) on patient outcomes (IOM 2010; Sensmeier 2015).

References

Agency for Healthcare Research and Quality. AHRQ issue brief: harnessing the power of data: how AHRQ is catalyzing transformation in health care. August 2015. http://www.ahrq.gov/research/findings/factsheets/informatic/databrief/index.html. Accessed 26 Jan 2016.

Aiken LH, Sloane DM, Bruyneel L, Van den Heede K, Griffiths P, Busse R, Diomidous M, Kinnunen J, Kózka M, Lesaffre E, McHugh MD, Moreno-Casbas MT, Rafferty AM, Schwendimann R, Scott PA, Tishelman C, van Achterberg T, Sermeus W. RN4CAST consortium. Nurse staffing and education and hospital mortality in nine European countries: a retrospective observational study. Lancet. 2014;383(9931):1824–30. doi:10.1016/S0140-6736(13)62631-8.

Bakken S, Reame N. The promise and potential perils of big data for advancing symptom management research in populations at risk for health disparities. Annu Rev Nurs Res. 2016; 34:247–60. doi: 10.1891/0739-6686.34.247

Bakken S, Jia H, Chen ES, Choi J, John RM, Lee NJ, Mendonca E, Roberts WD, Velez O, Currie LM. The effect of a mobile health decision support system on diagnosis and management of obesity, tobacco use, and depression in adults and children. J Nurse Pract. 2014;10(10):774–80. doi:10.1016/j.nurpra.2014.07.017.

Baro E, Degoul S, Beuscart R, Chazard E. Toward a literature-driven definition of big data in healthcare. Biomed Res Int. 2015;2015:639021. doi:10.1155/2015/639021.

Brennan N, Oelschlaeger A, Cox C, Tavenner M. Leveraging the big-data revolution: CMS is expanding capabilities to spur health system transformation. Health Aff. 2014;33(7):1195–201. doi:10.1377/hlthaff.2014.0130.

Brennan PE, Bakken S. Nursing needs big data and big data needs nursing. J Nurs Scholarsh. 2015;47(5):477–84. doi:10.1111/jnu.12159.

Clancy TR, Bowles KH, Gelinas L, Androwich I, Delaney C, Matney S, Sensmeier J, Warren J, Westra B. A call to action: engage in big data science. Nurs Outlook. 2014;62:64–5. doi:10.1016/j.outlook.2013.12.006.

Collins FS, Varmus H. A new initiative on precision medicine. N Engl J Med. 2015;372(9):793–5. doi:10.1056/NEJMp1500523.

Cortada IW, Gordon D, Lenihan B. The value of analytics in healthcare: from insights to outcomes. IBM Global Business Services, Life Sciences and Healthcare. Executive Report, January 2012.

Garcia AL. How big data can improve health care. Am Nurs Today. 2015 July;10(7):53–5.

Gartner. Gartner says solving "big data" challenge involves more than just managing volumes of data. STAMFORD, Conn. June 27, 2011. http://www.gartner.com/newsroom/id/1731916. Accessed 26 Jan 2016.

Holland CM, Foley KT, Asher AL. Can big data bridge the chasm? Issues, opportunities and strategies for the evolving value-based health care environment. Neurosurg Focus. 2015;39:1–2. doi: 10.3171/2015.9.FOCUS15497.

Institute of Medicine. The future of nursing: Leading changes, advancing health. Report recommendations. 2010. https://iom.nationalacademies.org/~/media/Files/Report%20Files/2010/The-Future-of-Nursing/Future%20of%20Nursing%202010%20Recommendations.pdf. Accessed 26 Jan 2016.

Kahn JP, Vayena E, Mastroianni AC. Opinion: learning as we go: lessons from the publication of Facebook's social-computing research. Proc Natl Acad Sci U S A. 2014;111(38):13677–9. doi:10.1073/pnas.1416405111.

Keenan, G. Big data in health care: an urgent mandate to CHANGE nursing EHRs!. Online J Nurs Inform (OJNI). 2014;18(1). http://ojni.org/issues/?p=3081.

Khoury MJ. Planning for the future of epidemiology in the era of big data and precision medicine. Am J Epidemiol. 2015;182(12):977–9. doi:10.1093/aje/kwv228.

Lyon D, McCain N, Elswick RK, Sturgill J, Ameringer S, Jallo N, Menzies V, Robins J, Starkweather A, Walter J, Grap MJ. Biobehavioral examination of fatigue across populations: report from a P30 Center of Excellence. Nurs Outlook. 2014;62(5):322–31. doi:10.1016/j.outlook.2014.06.008.

McKinsey Global Institute. Big data: the next frontier for innovation, competition, and productivity. Report, June 2011. Washington, D.C.

Marr B. Big data: the 5 Vs. 2015. http://www.slideshare.net/BernardMarr/140228-big-data-volume-velocity-vaiety-veracity-value. Accessed 26 Jan 2016.

Raghupathi W, Raghupathi V. Big data analytics in healthcare: promise and potential. Health Inf Sci Syst. 2014;2:3. doi:10.1186/2047-2501-2-3.

Redeker NS, Anderson R, Bakken S, Corwin E, Docherty S, Dorsey SG, Heitkemper M, McCloskey DJ, Moore S, Pullen C, Rapkin B, Schiffman R, Waldrop-Valverde D, Grady P. Advancing symptom science through use of common data elements. J Nurs Scholarsh. 2015;47(5):379–88. doi:10.1111/jnu.12155.

Roski J, Bo-Linn GW, Andrews TA. Creating value in health care through big data: opportunities and policy implications. Health Aff. 2014;33(7):1115–22. doi:10.1377/hlthaff.2014.0147.

Russell E, Johnson B, Larsen H, Novilla ML, van Olmen J, Swanson RC. Health systems in context: a systematic review of the integration of the social determinants of health within health systems frameworks. Rev Panam Salud Publica. 2013;34(6):461–7.

Sensmeier J. Big data and the future of nursing. Nurs Manag. 2015;46(4):22–7. doi:10.1097/01.NUMA.0000462365.53035.7d.

Shmueli G, Koppius OR. Predictive analytics in information systems research. MIS Q. 2011;35(3):553–72. doi:10.2139/ssrn.1606674.

Staggers N. Critical conversations about optimal design column. Online J Nurs Inform. 2013;1793. http://onji.orgissues/?p=2848.

Tan SSL, Gao G, Koch S. Big data and analytics in healthcare. Methods Inf Med. 2015;54(6):546–7. doi:10.3414/ME15-06-1001.

Uddin M, Koenen KC, Aiello AE, Wildman DE, de los Santos R, Galea S. Epigenetic and inflammatory marker profiles associated with depression in a community-based epidemiologic sample. Psychol Med. 2011;41(5):997–1007. doi:10.1017/S0033291710001674.

Zhu K, Lou Z, Zhou J, Ballester N, Kong N, Parikh P. Predicting 30-day hospital readmission with publicly available administrative database. A conditional logistic regression modeling approach. Methods Inf Med. 2015;54(6):560–7. doi:10.3414/ME14-02-0017.

Case Study 12.1: Clinical Practice Model (CPM) Framework Approach to Achieve Clinical Practice Interoperability and Big Data Comparative Analysis

Michelle Troseth, Donna Mayo, Robert Nieves, and Stephanie Lambrecht

Abstract This case study will share a common culture and professional practice framework to guide transformation work leveraging health information technology across multiple health settings in North America. An overview of the Clinical Practice Model (CPM) Framework's conceptual underpinnings will be shared followed by a detailed description of its Health Informatics Model with structured, mapped and tagged documentation content to support professional processes of care as well as national reference terminologies such as SNOMED-CT and LOINC. The implications of having real-time evidence-based content and clinical decision support at the point of care will be shared as well as the significance of clinical practice interoperability and big data comparative analysis for population health management.

Keywords Practice interoperability • Content interoperability • Within defined limits (WDL) • Evidence-based practice (EBP) • Clinical decision support (CDS) • Standardized clinical terminologies

12.1.1 Introduction

To revolutionize knowledge discovery, nurse leaders are recognizing the significance of standardizing clinical practice and big data is leading the way in sharable and comparable data to achieve better health outcomes (Delaney et al. 2015). The Clinical Practice Model (CPM) work began in 1983 by visionary nurse leader and healthcare futurist Bonnie Wesorick in Grand Rapids, MI. This case study shares the evolution of the CPM Framework and Models, which have provided a strong foundation for today's call to promote standardization, interoperability and contribution to the science of big data. The authors, all active participants in the University of Minnesota Big Data Science Nursing Knowledge Initiative, describe the attributes and value of implementing a culture and professional practice framework with a structured data model to accelerate big data capacity. The CPM Consortium has

M. Troseth, MSN, RN, DPNAP, FAAN (✉) • D. Mayo, MSN, RN • R. Nieves, BSN, RN, MBA, MPA, JD • S. Lambrecht, MS, RN-BC, MHA
Elsevier Clinical Solutions, USA
e-mail: m.troseth@elsevier.com

engaged over 400 organizations that have implemented the CPM Framework over time and have made significant contributions to the design, validation and expansion of all elements of the framework and models, including preparation for a data-driven, learning healthcare system. Lessons learned from the consortium will be shared including barriers to adoption of the standardized data as well as challenges in accessing the documented data for outcome and research purposes.

12.1.2 A Framework Approach

The CPM Framework (Fig. 12.1.1) has evolved over three decades and was designed to achieve the vision of sustainable culture and practice transformation and to address the challenges that all clinical settings encounter in achieving that vision. Benefits associated with using a framework driven approach are that it is (a) scalable, (b) data driven, and (c) can change as healthcare changes. It is also grounded in core beliefs, theories and principles and integrates the theoretical components of informatics with evidence-based practice within an overarching framework that is intentionally designed to support the patient, family, community, and caregivers to advance the culture and practice of care (Elsevier, CPM Resource Center 2011).

Each model is interrelated, action-oriented, outcome producing, replicable, evidence-based and technology-enabled to move healthcare organizations into a capacity-building mindset and into action (Wesorick and Doebbeling 2011; Christopherson et al. 2015). While each model plays a role in supporting sharable and comparable data, for the purpose of this case study, a detailed view of the Health Informatics Model will be provided followed by the Applied Evidence-Based Practice Model as both of these models enable the mapping to standardized clinical terminologies and big data science.

The Health Informatics Model, organized within a relational data base management system, is expressed in three different tiers: (1) Data Tier (physical layer) in which the content resides in a Structure Query Language (SQL) database where it is structured, mapped, and SQL-tagged to support the professional processes of care as well as national reference terminologies such as LOINC and SNOMED-CT; (2) Logic Tier (design layer), in which the rigorous content authorship process as described in the Applied Evidence-based Practice Model is abstracted to support principles of intentionally designed automation, provide context in the clinicians workflow and allow updating of content and change management as an ongoing process; and (3) Presentation Tier (end-user view at point of care) in which the standardized evidence-based content is expressed within any electronic health record (EHR) to support professional practice and documentation of services rendered. The Health Information Model includes documentation for six professional processes of care including capturing the patient story, planning care, providing care and education, evaluating progress toward goals, and providing professional exchange of patient information. The Health Information Model has been key to standardizing practice on the front end, creating practice and content interoperabil-

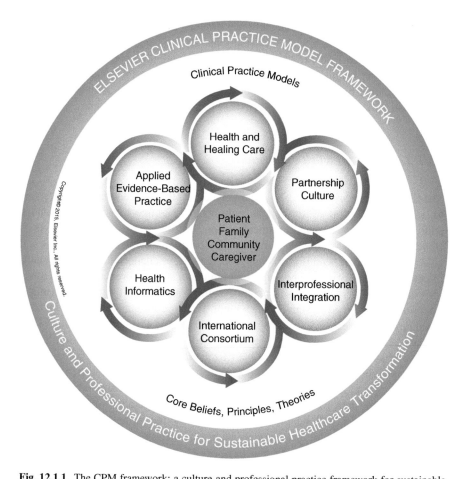

Fig. 12.1.1 The CPM framework: a culture and professional practice framework for sustainable healthcare transformation. Reprinted with permission

ity as well as expediting the mapping to common terminologies and standards on the back end. This approach is critical to exchanging data and creating big data sets generated by interprofessional teams who care for patients and families across the care continuum.

The Applied Evidence-Based Practice (EBP) Model is an action-oriented model that includes over 400 evidence-based clinical practice guidelines (CPGs) that are rigorously researched, graded and leveled to objectively identify the guideline elements that support the professional scopes of practice and processes of care across health professions. The CPGs provide the foundation for the plan of care and clinical documentation by leveraging structured notes, templates and flowsheets within the EHR, and consequently generate data supportive of big data science. The Applied EBP Model also supports content that provides standardized assessments and interventions, including research-based assessment scales (>60) that are standardized for multiple patient populations. The significance of having sustainable EBP environments to improve hospital performance metrics was recently identifiedby Melnyk et al. (2016). This calls for chief nurse executives to look at their current investment and integration of EBP into the EHR as one of the critical enablers of a sustainable EBP environment as well as big data research contributing to the body of evidence (knowledge discovery).

The Health Information Model and the Applied Evidence-Based Practice Model provide the capability to *intentionally design automation* via standardized clinical documentation and embedded clinical decision support (CDS) from the CPGs. This has enabled the ability to code standard assessment data from assessment scales and defined assessment parameters as well as the CPG content itself that is documented against in the EHR. The interrelated and action-oriented nature of these two models supports *practice interoperability* and *content interoperability*. Practice interoperability assures the same level of professional practice across all clinical settings and patient information exchange via a common culture and professional practice framework. Content interoperability assures consistency of content and associated coded-data that is exchanged accurately and effectively within and across various practice platforms.

12.1.2.1 Coding CPM Framework and Models to Standardized Clinical Terminologies

Always futuristic and with strong desire to advance professional practice, the first coding of CPM content to standardized terminology was in the late 1990s immediately following the first intentional design and implementation of CPM in an EHR. Coding was completed with standard nursing language and presented by Troseth and Hanson (1999) at the Second Conference on Nursing Diagnoses, Interventions and Outcomes (NANDA/NIC/NOC) Conference as an important concept worth exploring. Despite positive feedback from conference participants and standard experts, adoption proved to be difficult in the healthcare environment at the time due to factors such as lack of EHR vendor ability to organize and report, lack of healthcare organizations' ability to implement and report outcomes, and the beginning advancement of interdisciplinary approaches to care and documentation.

A decade later the HITECH Act of 2009 was enacted to tackle commercial, economic and technical barriers and to enhance information exchange. With its passage another window opened to realize the benefits of coding the CPM Framework documentation components to standardized clinical terminology. In addition, the passage of the Accountable Care Act (ACA) and the accepted terminologies of LOINC and SNOMED-CT provided the direction required to prioritize a coding roadmap that also aligned with the priorities of national nursing organizations, such as the American Nursing Association (ANA) and American Academy of Nursing (AAN).

12.1.2.2 Within Defined Limits (WDL) Assessments Mapped to Standardized Terminology

Patient assessments can vary nurse to nurse unless there are clearly defined assessment parameters. Assessing to specific parameters provides uniformity in *professional assessment standards*. This assures the standardization of a baseline assessment and decreases in variability of care. Within Defined Limits (WDL) provides a high level of assessment consistency between clinicians, and is legally sound because the definitions are predefined and clarify assessment expectations. It also improves usability and efficiency resulting in decreased documentation time. The CPM Consortium has been utilizing WDLs since the early 1980s and today it is used in over 400 healthcare settings. In 2013, the CPM Consortium underwent the latest review/revision via Collaborative Learning Communities (CLCs) from multiple sites and disciplines using current evidence and clinical expertise for each patient population. As part of the Nursing Knowledge Big Data Science Initiative, we advocated for, and have contributed to the Nursing Big Data Workgroup on *Encoding nursing assessments using LOINC and SNOMED CT*. An example of a CPM WDL for cardiac assessment coded to LOINC and SNOMED-CT (Fig. 12.1.2) shows the clinical impression (WDL parameters) and the associated codes.

12.1.2.3 CPM Clinical Practice Guidelines Mapped to Standardized Terminology

Each CPG supports scope of practice and reveals the value of nursing practice as part of the clinical documentation. The tool is designed to support the following professional processes of care: Care Planning, Assessment, and Intervention and Goal/Outcome Measurement. *Pressure Ulcer—Risk For* is provided as an example (Fig. 12.1.3) to show how the problem, outcomes and interventions can be coded to associated standardized terminologies using SNOMED-CT.

Question	LOINC Code	Answer	SNOMED ID
Cardiac WDL	80285-0	WDL S1 S2 Chest pain not present	449501000124109 26198000 111974005 161971004
Cardiac Rhythm	8884-9	Bradycardia Tachycardia	48867003 6285003
Heart Sounds	80276-9	S1 S2 Murmur	26198000 111974005 88610006

Fig. 12.1.2 Elsevier CPM assessment data: Within Defined Limits (WDL) for cardiac assessment with associated LOINC and SNOMED-CT codes. Cardiac WDL Definition: Regular rhythm, S1, S2; no reported chest pain. Used with permission

Problem	CPG (Problem)	SNOMED-CT
	Pressure Ulcer Rick (Using Braden scale)	At risk of pressure sore (285304000) Braden assessment scale (413139004)

Outcome	Outcome Row	SNOMED-CT
	Skin Integrity	Intact Skin (444884003)

Intervention	Intervention Row	SNOMED-CT
	Skin/Mucous Membrane Protection	Skin care management (228535001)
		Protection of patient (241718004)
		Mucous membrane structure (414751009)

Fig. 12.1.3 Elsevier CPM CPG data: "Pressure Ulcer—Risk For" with associated LOINC and SNOMED-CT codes for subset of data only. Used with permission

12.1.3   CPG Pressure Ulcer-Risk For- Example

It is important to note that the coded data elements are captured in the rows of clinical documentation behind the scenes from the clinician at the point of care as a result of the activation of the CPG on the Plan of Care. The CPM CPGs that address the priority chronic disease conditions (Heart Failure, Diabetes, etc.) across the continuum provide the foundation for big data comparative analysis for population health management along with other future coded assessment data such as Social and Behavioral Determinants of Health (SBDH) on the CPM Patient Profile.

12.1.4   The Challenges of Utilizing and Sharing Big Data

Many lessons have been learned over the past two decades of leveraging technology to advance professional practice and point-of-care transformation. There are three primary barriers to shift to a future state of adoption of standardized evidence-based clinical practice and big data analytics for knowledge discovery as a normative process:

1. **Customization of Clinical Content Compromises Data Harmonization**. One of the biggest challenges in providing standardized evidence-based clinical content is the tendency for organizations to review and change the data, frequently going back to "this is the way we have always charted this here". This pattern of variances in clinical documentation inhibits data collection despite having an initial common assessment tool and EHR vendor. The implication of customization for use of secondary data for research is described in Bowles et al. (2013) study across multi-hospital sites. As evidence-based clinical content sets grow and the relevance of "standardization of care" based on evidence is more under-

stood, organizations are realizing the value of "adopting vs. customizing". CPM has a process in place that welcomes suggested changes from CPM Consortium Sites along with their rationale based on the evidence. Most importantly, the CPM Transformation Services team is composed of experienced clinicians grounded in the CPM Framework and Models methodologies. They are able to provide the "why" behind the content design and help to educate and guide the transformational process of the intentionally designed automation. This helps to prevent the kinds of needless customization that impact practice interoperability and data harmonization.

2. **Slow Uptake of Latest Version of Clinical Content and Workflow Enhancements (Version Control).** Another barrier to success is that health-care organizations get behind in upgrading to the latest version of their CPM clinical content. Elsevier CPM is updated twice annually and there are strate-gies in place to efficiently deliver the latest updated content to healthcare orga-nizations. Sometimes organizations get behind due to competing priorities, more focus on the EHR implementation/software upgrades and roll-outs and/or lack of resources to assist with the content upgrade process. This is an area that has seen much improvement. More and more, organizations are realizing the value of expending the necessary resources to upgrade to the latest versions after experiencing the ramifications of getting behind on versions of clinical content (including the benefit of coding to standardized clinical terminologies).

3. **Lack of Data Extraction and Analytic Skills.** Last, not having more atten-tion, tools and skill sets to extract and analyze data at the local and enterprise levels has impacted the realized benefits of standardized evidence-based clinical content and coded terminologies. There needs to be more focus on the technical ability and skill sets to pull data out of the EHRs, such as report writing and data analysis, in order to conduct research on continuous data use and publish big data studies. This skill set will be key for nurses to develop and/or advocate for in order to make contributions to big data science.

12.1.5 Conclusion

The call and attention on big data science from nursing leadership could not be more timely and important. Collaboration on driving big data science is signifi-cant work beyond nursing as it will also set the pace for interprofessional educa-tion and practice priorities. We believe a framework approach is critical to transform the complex health system and that the knowledge from big data will further add to the evolution of the CPM Framework and Models as well as better health outcomes.

References

Bowles KH, Potashnik K, Ratcliffe SJ, Rosenberg M, Shih NW, Topaz M, Holmes JH, Naylor MD. Conducting research using the electronic health record across multi-hospital systems: semantic harmonization implications for administrators. JONA. 2013;43(6):355–60. doi:10.1097/NNA.0b013e3182942c3c.

Christopherson TA, Troseth MR, Lingerman EM. Informatics-enabled interprofessional education and collaborative practice: a framework-driven approach. JIEP. 2015;1(1):10–5. doi.org/10.1016/j.xjep.2015.03.002

Delaney C, et al. Nursing knowledge: big data science conference proceedings. 2015. http://www.nursing.umn.edu/prod/groups/nurs/@pub/@nurs/documents/content/nurs_content_5 04042.pdf. Accessed 13 Feb 2016.

Elsevier CPM Resource Center. The CPM framework™: culture and professional practice for sustainable healthcare transformation (Booklet). Grand Rapids, MI; 2011.

Melnyk BM, Gallagher-Ford L, Thomas BK, Troseth MR, Wyngarden K, Szalacha L. A study of chief nurse executives indicates low prioritization of evidence-based practice and shortcomings in hospital performance metrics across the United States. WVN. 2016;00:1–9. doi:10.1111/wvn.12133.

Troseth MR, Hanson DL. A concept worth exploring: integrated NANDA, NIC & NOC within a professional practice framework. Paper Presented at:the 2nd conference on nuring diagnoses, interventions and outcomes (NANDA/NIC/NOC). New Orleans, LA; April 1999.

Wesorick B, Doebbling B. Lessons from the field: the essential elements for point-of-care transformation. Med Care. 2011;49(Suppl):S49–58. doi:10.1097/MLR.0b013e3182239331.

Chapter 13
Big Data Impact on Transformation of Healthcare Systems

Gay L. Landstrom

Abstract Healthcare systems (HCS) have formed rapidly over the past decade. These systems have potential to capture, integrate, and analyze data in new and unique ways for use in improving patient care and understanding the environment of care. Data from multiple points on the continuum and from multiple healthcare organizations at the same level of care can be used to rapidly implement evidence-based practices, integrate the care of individual patients across the continuum, and improve the care of large populations of patients with common health issues. These big data sources can also aid in empowerment of patients and shared decision-making. In order to take advantage of these unique HCS opportunities, these systems need to address a variety of issues, including the necessity for leaders to have a vision of not only electronic system implementation, but also post-implementation vision of and invest in the integration and analysis of data.

Keywords Healthcare systems • Health systems • Big data • System integration

13.1 Introduction

The growth of healthcare systems (HCS), of various shapes, sizes, and legal structures, has been exponential over the last decade (Daly 2015). Hospitals have joined with other hospitals. Providers have joined their practices with hospitals, health systems, and even insurance companies. Long-term care providers have joined with home care companies, who in turn have joined with other providers along the continuum. Community hospitals of varying sizes and locations have joined with academic medical centers, finding ways to merge their missions for one strategic reason or another. These systems range from as small as two entities to hundreds of

G.L. Landstrom, Ph.D., R.N., N.E.A.-B.C.
Ascension St. John Providence, Warren, MI, USA
e-mail: gay.landstrom@ascension.org

© Springer International Publishing AG 2017
C.W. Delaney et al. (eds.), *Big Data-Enabled Nursing*, Health Informatics,
DOI 10.1007/978-3-319-53300-1_13

organizations joined to form HCS. Today, fewer and fewer providers of health care services are doing so in a solo manner. This trend is expected to continue through the near future.

These fledgling healthcare systems (HCS) are coming together for many reasons: financial security, purchasing power, leveraging of support systems across multiple entities, or even to avoid being left out of the ability to provide care in a region. Still others come together to form organizations that can sell care to payers, as ACOs or simply as organizations that have the parts of the care continuum needed to sell bundled services. The structures used to legally form these systems can vary as well, from complete asset mergers, to joint operating agreements, to simply strong affiliations or loose coalitions, and may include a combination of all these structures within a single system.

While these are all valuable reasons for forming HCS, the greatest benefits can be realized by the populations and individual people served by the systems. The potential to leverage the skill, knowledge, and expertise of healthcare providers and scale that ability across a broad population is profound in many cases. Technology and electronic systems further enable the leveraging of the data and knowledge in the systems that choose to invest in these structures. Some HCS have installed electronic health records (EHRs) across the members of their system, linking patient records with laboratory and radiology results, enabling clinicians in various locations to access the patient record simultaneously, and allowing the patient to utilize services of the health system across regions and states. Not only have HCS invested in EHRs, they have connected inpatient services with ambulatory clinic services, linking the providers and experts in the system to other healthcare organizations and individual patients through telehealth, and have begun to invest in remote medical surveillance of patients in their homes. Patients can input and retrieve data through portals, providing the opportunity for greater transparency and more meaningful patient empowerment in decision-making. All these investments are creating large caches of data, stored in data warehouses within organizations or in the cloud, waiting to be put to greater use in the care of patients and populations.

This chapter will explore how the formation of HCS and the joint leveraging of technology systems and the data they produce can bring added value to the care of the populations served by these systems, as well as adding to the global bodies of healthcare knowledge. The chapter will address big data based ways to improve patient care, some being realized today, with others dreamed of but not yet realized. Last, challenges will be discussed, requiring action from systems as well as national expert consensus bodies.

13.2 Limitations of the Past

In the time before the EHRs, health care had even more inefficiencies than we experience today. Some nurses remember when the acute care patient record was entirely on paper. The patient would leave the patient care unit to go to a physical therapy appointment or a diagnostic test and the paper chart would travel with the patient.

Any physician or nurse needing that chart would be forced to find it on whatever floor housed the department being visited.

As EHRs emerged, spurred on by the incentives of Meaningful Use dollars (CMS 2010), hospitals and ambulatory practices chose the vendors and systems that best matched their needs, workflow style, or budget. The result of these separate decision-making processes within individual organizations that would later choose to merge or collaborate was the selection of different electronic order entry and documentation systems. As the organizations came together to form HCS, the electronic picture often looked like a crazy quilt of mismatched products, products that had little ability to speak to one another.

As HCS grew larger and prioritized the need for accessing meaningful data in a common manner from every entity in the system, HCS began to invest in converting the crazy quilt of electronic systems to common platforms across the entire enterprise. While the investment in these common EHRs and other electronic systems was substantial, leaders held to the theory that common systems would improve care on a real-time basis, and would yield valuable and vast data for other purposes, such as quality improvement. As will be discussed later in this chapter, even with common systems and large amounts of data being captured, using that data to create information and knowledge can still be elusive.

13.3   How Healthcare Systems Come Together Electronically

HCS that seek integration of data typically focus on three core systems and a handful of other data capture and retrieval systems to meet more narrow needs. The core systems include: (1) financial reporting, payroll, and billing/revenue systems, (2) people management systems, and (3) electronic health records (EHRs), including provider order entry, results reporting, and documentation of care. The data capture and retrieval systems selected for integration are more varied, depending on the priorities of the HCS, and may include quality monitoring, discharge planning, policy and procedure libraries, platforms for deployment and documentation of education and development programs, risk management and claims systems, bed control, patient acuity and staffing systems, employee scheduling systems, and others.

As HCS form, it is common for early priorities to focus on those electronic systems that will facilitate the integration of financial data retrieval and ensure compliance with all required billing practices. Ensuring that new HCS are financially stable is a key priority during a time of great change, including change of ownership. Bringing all organizations and services within the HCS onto the same book of accounts allows visibility into the entire increasingly complex organization.

The management of the people assets of the organization is often, but not always, a priority for the growing HCS. Position control, salary and benefit administration, leave of absence management, and compliance with a host of laws and regulations, such as the Americans with Disabilities Act of 1990 (US Department of Justice Civil Rights Division 1990), are processes that must be handled with rigor, a task that becomes very difficult to handle manually as employee ranks swell through HCS formation.

EHRs have become an early priority for many HCS. Without common EHRs, retrieving data from disparate parts of different EHRs can be difficult. Adding to the difficulty in data retrieval is that data definitions can be different. Many health care organizations seeking to gain clinician adoption of the EHR build multiple ways for the individual to document patient data. While this sounds like a good plan from a change management standpoint, it often yields patient data being found in multiple locations in the EHR that is difficult to retrieve and is not comparable. The result is that data documented in the EHR can't be easily located for either individual real-time care or for retrospective retrieval, and data cannot support knowledge-informed practice.

While all of these electronic systems capture data that can used for generating new knowledge, improving care for the people served, or improving the work environment for healthcare providers and clinicians, more work needs to be done for HCS to turn the data into information. In health care, the clinicians using the EHRs often don't see the value in their daily work. In fact, clinicians often see the EHR as a source of additional and often frustrating work (Neff 2013). It will take effort to make this data valuable and useful, helping clinicians to solve problems and make patient care decisions within a particular healthcare context.

13.4 Big Data Emerging from Healthcare Systems

To better understand and discuss the kind of data being generated within HCS, it is helpful to look at big data in terms of volume, variety, velocity, and veracity (Gaffney and Huckabee 2014):

The *volume* of data being produced in HCS includes all of the data from the EHRs, physiologic monitoring data, medical devices, genomic data and more. At the 2015 Big Data in Nursing Conference (www.nursing.umn.edu), individual Chief Nursing Officers described their HCS generating petabytes (one million gigabytes) of data across the care continuum (Welton 2015). It is challenging to conceive of that volume of data being generated, and even more difficult to know how to access, integrate, and analyze the data in order to improve patient care.

The current *variety* of data types, languages, sources, and attributes creates challenges for gleaning knowledge (HIMSS 2015). Even with the efforts to standardize nursing language over the past several decades, the American Nurses Association alone recognizes 12 standard terminologies considered to support nursing practice (ANA 2014), to say nothing about additional languages used by other clinical disciplines. Standardization and normalization of data will be a prerequisite for any meaningful integration of the varied data and the creation of knowledge.

Data is being generated at such a *velocity* that data is accumulating in data warehouses while strategies for analysis are being developed. Without this analysis, the data adds no value to improving of patient care. Strategies must be developed for analyzing data in real-time to offer maximum value to individual patients, as well as broader populations.

Last, *veracity* must be an important consideration if we are ever to use the data to make patient care decisions. Accuracy and trustworthiness of the data are minimal requirements, demanding rigorous quality control and integrity of the data being collected, stored, retrieved and analyzed.

13.5 The Hope of Improving Health and Care Within Healthcare Systems Using Data

HCS have the potential to improve the health and care of those they serve through at least three different avenues: (1) utilizing electronic care support systems to enable rapid dissemination of evidenced-based care across all cross-continuum entities within the system, (2) integration of the individual patient story across the care continuum, and (3) integration of data from different sources to understand and manage populations of people (big data). Additional uses of HCS patient data are likely to be developed over time.

13.5.1 Rapid Dissemination of Evidence-Based Care

Changing clinical practice is one of the greatest challenges for healthcare leaders. Change within a single organization requires planning that addresses technology, processes, people and culture in order for change to be sustained over time (Schein 2004). HCS that want to rapidly change and integrate care across multiple entities of the same level (e.g., acute care hospitals) and across the continuum of care have the opportunity to use common EHRs as a platform for deploying clinical pathways or evidence-based guidelines of care, developing electronic alerts that can assist with risk identification and timely clinician decision-making, and analyzing outcomes data to measure impact on broad patient populations.

Trinity Health, based in Livonia, Michigan, is an example of an HCS that leveraged its EHR to provide evidence-based practices and decision support. This HCS had made a major investment in implementing a common EHR in its member acute care hospitals. In 2007, the organization began working with its clinicians to develop and deploy evidence-based electronic tools and decision support to improve care, achieving significant improvements in ventilator associated pneumonia (VAP), central line-associated blood stream infections (CLABSI), and catheter associated urinary tract infections (CAUTI). In 2010, analysis of system-wide patient data identified variation in practice for the care of sepsis patients. National data showed that severe sepsis and septic shock were preventable complications, adding an average $43,000 in unnecessary costs per patient, accounting for 40% of all ICU expenditures, and extending the patient length of stay by an average of 11 days (Zimmerman et al. 2004). Although mortality rates for sepsis patients within Trinity Health were below national averages, the organization launched a collaborative, interdisciplinary initiative across its 31-hospital system, and across ten states. The initiative would affect the care of patients across community-based acute care hospitals of varying sizes, from critical access hospitals through medium size teaching hospitals (Landstrom and Reynolds 2013), and would require partnership with the EHR vendor.

The Trinity Health sepsis collaborative was a large undertaking that involved several key initiatives and tactics, including:

- Identification of relevant literature and professional standards
- Identification of national sepsis work and any emerging guidelines
- Baseline assessments of the sepsis related practices within the Trinity Health hospitals
- Identification of accountable clinical executives at each hospital
- Formation of a sepsis team for the system-wide work, as well as individual implementation teams within each hospital
- Identification of a credible national expert from the Society of Critical Care Medicine's Surviving Sepsis campaign (Castellanos-Ortega et al. 2010) who could serve as a credible educator to launch the HCS initiative
- Analysis of sepsis patient data to establish a baseline metric

Trinity Health's sepsis work involved engaging the clinicians in each of the acute care hospitals through education, ongoing outcomes data, daily coaching by local champions and sepsis coordinators, and accountability mechanisms. Equally important was the work done with the EHR vendor to develop new real-time data analysis tools and alerts to assist clinicians with making patient care decisions.

The HCS worked with its EHR vendor to extract data from the client system and process it through an algorithm that compared actual patient data with set parameters. All emergency department and acute care patients' data were automatically screened against criteria for sepsis. Immediate notification of the provider and nurse occurred when a patient met the criteria for sepsis—two or more systemic inflammatory response syndrome (SIRS) criteria combined with a documented possible or actual source of infection. Providers were prompted by these alerts to consider the entire clinical picture and make a decision about diagnosis and treatment of sepsis when appropriate. Evidence-based plans of care were automatically pushed out to guide diagnostic testing and early goal directed therapy for rapid treatment of the sepsis. An additional electronic decision-making tool guided the provider to select the most effective antibiotic regimen based on the suspected source of infection. The EHR was modified to allow for real-time communication of the patient's status to all caregivers. Patient outcome data was monitored across the entire HCS and compared to the best national improvement rates within the Surviving Sepsis campaign.

Investment in accessing patient data and developing decision-making patient care tools yielded a strong return on investment for Trinity Health. Within 2 years, the HCS was saving $17 million in unnecessary costs of patient care and saving 400 lives annually (Landstrom and Reynolds 2013).

13.5.2   Integrating Individual Patient Care Data Across the Continuum

Discussions of the fragmented nature of health care in the United States have been going on for decades. This fragmentation affects the patient experience (Noest et al. 2014), patient safety (Goodman 2003), and certainly the ability to see patient-related information across the many sites of care across the continuum. In the recent past and in many current HCS, patients receiving treatment in an acute care hospital are discharged from that site of care, moved to their home or post-acute care site and provided follow-up care from a different part of the care continuum. Nurses in the acute care hospital do not have access to any information about the effectiveness of their care and interventions in most cases. Post-acute care providers may receive a faxed summary of the patient's care in the hospital setting, but often lack for any information about the unique experience or needs of the individual patient.

The development of HCS holds great promise to integrate patient information across all sites of care. The integration of EHR systems allows clinicians to have visibility of patient information across acute care, rehabilitation, post-acute care, skilled nursing facilities, home care, hospice, ambulatory practices, and all clinical disciplines.

HCS are often confronted by the decision to either adopt a fully integrated system or to select many best-in-class systems and connect those disparate systems through interfaces. Providers and individual disciplines often advocate for the best-in-class options. These specialty systems tend to be developed with deep knowledge of the goals and workflow of the individual service or discipline, appealing to individuals with the corresponding perspectives. However, connecting these best-in-class systems requires a number of one- and two-way electronic interfaces. These interfaces carry a cost upon initiation and every time one of the connected systems requires an upgrade. Costs include all the expert labor required to build and test these interfaces. Over time the costs can be substantial and must be taken into consideration when selecting the pathway to achieve the system integration required improving patient safety and quality of care for the entire population served by the HCS. Purchasing a fully integrated system means that the various parts of the EHR speak to one another without an interface. The downside of these integrated systems is that specialty pieces of the system sometimes fall short in the eyes of expert clinical users when compared with the corresponding best-in-class options. This important decision should be made considering all available information, with an eye to long-term costs and sustainability to the HCS, and the involvement of the anticipated clinician users of the system.

Closely related to the issue of integrating individual patient data across the continuum is the emerging opportunities seen in collecting "small data", sometimes referred to as the *quantified self* (Swan 2013). The quantified self refers to patients being involved in collected data related to their own biological, physical, behavioral, or environmental information. The sample size is one, but the depth of information on the one subject when all pieces of data are integrated has the potential to generate information previously inaccessible. This data could have great meaning for the science of behavior change and identifying new kinds of pattern recognition techniques. Data collection methods range from a variety of wearable electronic sensors, smart phones and biosensors to manual journals.

One additional trend related to the integration of data related to individual patients is the emergence of remote medical sensing. This technology is being tested by HCS, such as Kaiser Permanente and Dartmouth-Hitchcock (microsoft.com 2016), in order to generate real-time information about the health status of patients. This data is continuously monitored, with clinicians identifying concerning trends using evidence-based tools and connecting with the medical home or other health maintenance provider when intervention is warranted. This new development has the potential to identify a decline in the patient's physical status even before the patient becomes aware of any concerning symptoms. It also could help prevent the exacerbation of chronic illnesses and the physical and financial consequences of an illness episode. The enormous volume of data generated by remote medical sensing systems offers the potential for new insights into chronic illnesses and the interface of these illnesses with the daily lives of patients.

Big data can also play an important part in achieving full engagement of patients in decisions about their health care. *Shared decision-making* is a process that is patient-centered, fostering communication and information sharing, as well as empowering the patient to engage fully in making decisions regarding medical treatment options (Makoul and Clayman 2006). More recently, the concept of *coproduction of healthcare* has been introduced (Batalden et al. 2015) to describe the equality that should exist between the consumer of the health care (patient) and the professional (provider of the care). The model recognizes that patients own the execution of any health care plan. Civil discourse, co-planning, and co-execution are proposed to produce high-value health care and good health for all. Regardless of the model for engaging patients in decisions regarding their health and treatment plans, incorporating all available data in an integrated fashion will yield the best-informed decisions.

13.5.3   Integration to Manage Patient Populations

For the last few decades, researchers and healthcare leaders have been seeking information to help understand the health status of populations. Populations that tend to be high consumers of health care resources have been of particular concern. Risk assessment and resources allocation, disparities in care, and the burden of disease have been areas of focus, as have seeking to understand the impact of various interventions on lowering risk, lowering resource consumption, and disease exacerbation. This work has often been referred to as *population health*, defined as the health outcomes of a group of individuals, including the distribution of such outcomes within the group, but also including health outcomes, patterns of health determinants and the policies and interventions that link these two (Kindig and Stoddart 2003).

When HCS come together, tremendous potential is created for integrating and analyzing available patient data, gaining better understanding of the population that is intended to receive the benefits of the care provided by the HCS. Data from different sources across the continuum can help with pattern identification. For example, by aggregating data from ambulatory setting during stable controlled disease states and combining that with data from acute exacerbations of illness, catalyzing factors resulting in acute episodes can be identified and understanding gained about the negative impact of those acute disease exacerbations on overall health status decline. Various Emergency Department interventions during acute disease episodes can be studied relative to the long-term impacts on health. Resource allocations can be directed toward the best interventions from a continuum view rather than the limited view of a single site of care.

13.6 Challenges of Gleaning Information and Knowledge from the Data and Recommendations for Optimizing Data Within HCS

There is certainly the potential to use the immense data generated from patient care and other processes within HCS to better understand many dynamics of caring for people and the environments within which that care is provided. But issues need to be addressed in order for that data to become integrated, analyzed and useful for improvement. Data definitions, analytic capability and privacy issues are just a few of the barriers that require intentional work.

First, there is a strong consensus that not only do data standards need to be developed but more importantly must be implemented across healthcare institutions (HIMSS 2015; Welton 2015). We can no longer tolerate the waste found in using different languages to describe nursing care and patient outcomes. Standards for documenting nursing care are required in order to support data science research. These languages and documentation standards must be applied in all parts of the continuum, including acute care, long-term care, home care, ambulatory care, and all other sites of care, regardless of the EHR platform used. Nursing documentation needs to be transformed into structured discrete elements that still retain the ability to easily represent the patient's story. That is a combination of goals that will be challenging to achieve in the coming years.

Without strong and concurrent analytic infrastructure and processes, data will accumulate and have limited usefulness in providing new knowledge and improving patient care. HCS need to fully realize that the investment in implementing EHR systems alone is not enough. Investments must be made in the analytics infrastructure required to do real-time data integration, analysis, and knowledge generation.

Dartmouth-Hitchcock (DH) is one HCS that chose to make the investment in not only an EHR across the continuum, but also to create an Analytics Institute. Established in 2015, the DH Analytics Institute provides a centralized resource to support the clinical, academic, and research missions of the organization with data and analytics. The HCS implemented a new data warehouse as well as several analytical applications to access and analyze available data from clinical, financial, operational, patient experience, and other electronic systems in place across the continuum. In the near future, the HCS plans to develop a single portal where providers and leaders can view performance on any number of measures, allowing Analytics Institute resources to be focused on large data requests for quality improvement or research purposes.

One last issue needs to be included in this discussion of what must be addressed in order glean knowledge from the data generated by HCS. In all the excitement generated by the potential of using big data for improvement purposes, we still must address the issue of protecting the privacy of patient information. Connecting patient data across the continuum holds great promise for improvement of care for that patient, but only providers and caregivers with a legitimate reason should have access to all that information. Connecting the continuum and connecting many entities across the same level of care make the issue of security of the EHR more challenging. Well-conceived and tested processes for protecting patient information and maintaining appropriate levels of security in EHR access must be developed and monitored.

13.7 Conclusion

HCS possess the capability of doing what individual healthcare institutions have been unable to do in the past—connect data generated across entities and across the continuum in order to generate new knowledge and opportunities to understand processes of care and improve patient outcomes. There is the potential to improve and integrate care for individual patients, entire patient populations across healthcare provider settings, and to implement evidence-based practice in a rapid and impactful manner across the entire HCS. In order to realize those tremendous benefits, HCS leaders need to embrace a vision of the importance of gleaning information from the data that electronic systems produce and be prepared to make the financial investments necessary to extract, integrate, analyze and produce new knowledge from the data. This vision and commitment, combined with a willingness to accept data and documentation standards across all HCS, and commitment to protecting patient information despite the complexities involved, will help us begin to realize the potential power of healthcare big data.

References

American Nurses Association. ANA's position on the standardization and interoperability of health information technology to improve nursing knowledge representation and patient outcomes. Maryland, MD: Silver Spring; 2014. nursebooks.org
Batalden M, Batalden P, Margolis P, et al. Coproduction of healthcare services. BMJ Qual Saf. 2015;25(7):509–17. doi:10.1136/bmjqs-2015-004315.

Castellanos-Ortega A, Suberviola B, et al. Impact of the surviving sepsis campaign protocols on hospital length of stay and mortality in septic shock patients: results of a three-year follow-up quasi-experimental study. Crit Care Med. 2010;38:1036–43. doi:10.1097/SHK.0000000000000268.

Centers for Medicare and Medicaid Services. Medicare & Medicaid EHR Incentive Program: Meaningful Use Stage 1 Requirements Overview. 2010. https://www.cms.gov/Regulations-and-Guidance/Legislation/EHRIncentivePrograms/Downloads/MU_Stage1_ReqOverview.pdf. Accessed 1 Feb 2016.

Daly R. Healthcare deals surge in 2015. Healthcare Financial Management Association. 2015. http://hfma.org. Accessed 1 Feb 2016.

Gaffney B, Huckabee M. Part 1: what is big data? HIMSS data and analytics task force. 2014. http://himss.org/ResourcesLibrary/genResourcesFAQ.aspx?ItemNumber=30730. Accessed 1 Feb 2016.

Goodman GR. A fragmented patient safety concept: the structure and culture of safety management in healthcare. Hosp Top. 2003;81(2):2–29. doi:10.1080/00185860309598018.

HIMSS CNO-CNIO Vendor Roundtable. 2015. Guiding principles for big data in nursing. Big Data Principles HIMSS Nursing Informatics Community Workgroup:1–20.

Kindig D, Stoddart G. What is population health? Am J Public Health. 2003;93(3):380–3.

Landstrom GL, Reynolds MA. Translating evidence into clinical practice: sepsis. Paper presented at the International Nursing Administrative Research Conference, Baltimore, MD. Nov 14, 2013.

Makoul G, Clayman ML. An integrative model of shared decision making in medical encounters. Patient Education and Counseling. 2006;60(3):301–12.

Neff G. Why big data won't cure us. Big Data. 2013;1(3):117–23. doi:10.1089/big.2013.0029.

Noest S, Ludt S, Klingenberg A, et al. Involving patient in detecting quality gaps in a fragmented healthcare system: development of a questionnaire for Patients' Experiences Across Health Care Sectors (PEACS). Int J Qual Health Care. 2014;26(3):240–9. doi:10.1093/intqhc/mzu044.

Schein EH. Organizational culture and leadership. 3rd ed. San Francisco: Jossey-Bass; 2004.

Swan M. The quantified self: fundamental disruption in big data science and biological discovery. Big Data. 2013;1(2):85–99. doi:10.1089/big.2012.0002.

US Department of Justice Civil Rights Division. Americans with Disabilities Act. 1990. http://ada.gov. Accessed 31 Jan 2016.

Welton JM. Conference report: big data in nursing 2015. Voice of nursing leadership, Sep 2015:10–11.

Zimmerman JL, Vincent J-L, Levy MM, Dellinger RP, Carlet JM, Masur H, Gerlach H, Calandra T, Cohen J, Gea-Banacloche J, Keh D, Marshall JC, Parker MM, Ramsay G. Surviving sepsis campaign guidelines for management of severe sepsis and septic shock. Intensive Care Medicine. 2004;30(4):536–55.

Chapter 14
State of the Science in Big Data Analytics

C.F. Aliferis

Abstract Big data analysis is made feasible by the recent emergence and operational maturity and convergence of data capture, representation and discovery methods and technologies. This chapter describes key methods that allow tackling hard discovery (analysis and modeling) questions with large datasets. Particular emphasis is placed on answering predictive and causal questions, coping with very large dimensionalities, and producing models that generalize well outside the samples used for discovery. Within these areas the chapter emphasizes exemplary methods such as regularized and kernel-based methods, causal graphs and Markov Boundary induction that have strong theoretical as well as strong empirical performance. Other notable developments are also addressed, such as robust protocols for model selection and error estimation, analysis of unstructured data, analysis of multimodal data, network science approaches, deep learning, active learning, and other methods. The chapter concludes with a discussion of several open and challenging areas.

Keywords Big data • Big data analytics • Data mining • Machine learning • Predictive modeling • Causal modeling • Feature selection • Regularization • Overfitting Structured data analysis • Unstructured data analysis

14.1 Advances in Predictive Modeling and Feature Selection for Big Data

Predictive Modeling in recent Data Science literature refers collectively to classification and regression tasks. Classification is the assignment of pre-defined discrete category labels to entities based on their characteristics. For example, the assignment of disease labels to patients based on their clinical, lab and imaging tests.

C.F. Aliferis, M.D., Ph.D., F.A.C.M.I.
Institute for Health Informatics, University of Minnesota, Minneapolis, MN, USA
e-mail: califeri@umn.edu

© Springer International Publishing AG 2017
C.W. Delaney et al. (eds.), *Big Data-Enabled Nursing*, Health Informatics,
DOI 10.1007/978-3-319-53300-1_14

Regression is the continuous analogue of classification, in which a continuous value is assigned to each entity, for example the probability of death within 5 years from treatment of patients with stage I breast cancer[1] (Duda et al. 2012).

While both regression and classification have long and proven histories in statistics, applied mathematics, engineering, computer science, econometrics and other fields, in a Big Data Science context, analysts are routinely faced with the obstacle of *"curse of dimensionality"* (Aliferis et al. 2006). The curse of dimensionality is the set of problems caused by the combination of very large dimensionalities (i.e., number of variables) and relatively small sample sizes. Consider for example the task of creating a model for clinical diagnosis or prognostication using a discovery dataset with 10,000 variables and only 300 patients. In such circumstances conventional statistical and other analysis methods tend to collapse and analysis is either: (a) infeasible (for example with classical regression), (b) computationally very expensive (certain methods may not even terminate in any realistic time frame), (c) impractical (e.g., classical decision tree algorithms will run out of sample as each branch of the tree increases in depth, hence a good model will not be developed), (d) models will tend to be **"over fitted"**; that is to perform well on the discovery dataset but fail to generalize on new samples from the general population that was sampled to create the discovery dataset (Aliferis et al. 2006). Small sample/large dimensionality analysis scenarios are very common in practical Big Data mining situations and as the chapter will show this is a set of challenges that newer Big Data Analysis methods truly excel at.

In general, an analyst has a choice of how complex models she can explore for a given data modeling project. For example, in the regression problem, the analyst can choose among a straight line, a piecewise linear model, or a polynomial function of degree (e.g.,) equal to the number of training objects in the discovery dataset (Fig. 14.1).

Given a fixed sample size[2] for the discovery dataset there is a "sweet spot" in which the right model complexity is attained. Modeling with larger complexity will lead to overfitting and modeling with smaller than ideal complexity will lead to *under fitting* (i.e., suboptimal performance *both* in the discovery dataset *and* in the sampling population).

Modern Big Data analysis methods deploy powerful means for automatically balancing model complexity to the complexity of the process that generates the data (the "data generating process"). By doing so they manage to provide theoretical guarantees for good generalization, avoid over- and under-fitting, and run very fast. These operating characteristics are achieved even when the dimensionalities are very high (e.g., >10,000 variables) and samples size very small (i.e., as low as n = 50–200 instances in some application domains) (Dobbin and Simon 2005;

---

[1] Note that while "predictive modeling" from a generic linguistic perspective implies forecasting the future, recent use of the term in Data Science literature includes both prospective and retrospective classification and regression.

[2] The sample size is the second major element that determines model error according to statistical machine learning theory. For an introduction in the topic see (Aliferis et al. 2006) under "bias-variance" tradeoff.

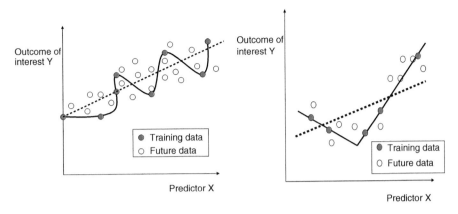

Fig. 14.1 The complexity of a model must be tied to the complexity of the data generating process. In the *upper part* the analyst chose an overly complex regression model (high-degree polynomial) that fits perfectly the discovery data (*blue points circles*) but fails to generalize to future samples (*white*). This is an example of overfitting. If the analyst chooses the straight dotted line, then the error in the discovery data increases but the model generalizes better to future data. In the *lower part*, the analyst chose the more complex of the two models depicted (piece-wise linear function). This model generalizes better than the simpler straight-line model. Had the analyst chosen the straight line model that would have been an example of under fitting (since there is a better model for this data)

Aliferis et al. 2009). In the examples provided in these references the use of microarray gene expression data to diagnose the type of cancer, differentiate from normal tissue and prognosticate clinical outcomes in patients with various types of cancer are all documented with thousands of gene expression variable inputs and sample sizes of <300.

We will briefly discuss Support Vector Machines (SVMs), one important class of such methods as exemplary Big Data Science analysis approaches to predictive modeling (Vapnik 2013; Statnikov et al. 2011, 2012). SVMs use two key principles (1) regularization and (2) kernel projection. **Regularization**: SVMs cast the classification or regression problems as a non-linear quadratic optimization problem where the solution to predictive modeling is formed as a "*data fit + parameter penalty*" mathematical objective function. Intuitively and as depicted in Fig. 14.2, each object used for training and subsequent model application is represented as a vector of measurements in a space of relevant dimensions (variables).

For example, a 22 year-old male without cancer can be represented via the vector [22, 1, 0] where the first dimension is age (22), the second gender (1 = female, 2 = male) and the third is disease status (0 = control, 1 = cancer). Classifying subjects into the Cancer or Control categories can now be formulated as a geometrical problem of finding a hyperplane (i.e., the generalization of a straight line from 2 dimensions to n dimensions) such that all the subjects above the hyperplane will have cancer and all subjects below the hyperplane will be normal controls. Translating this geometrical problem into linear algebra constraints is a straightforward algebraic exercise. Every variable has a weight and collectively these weights

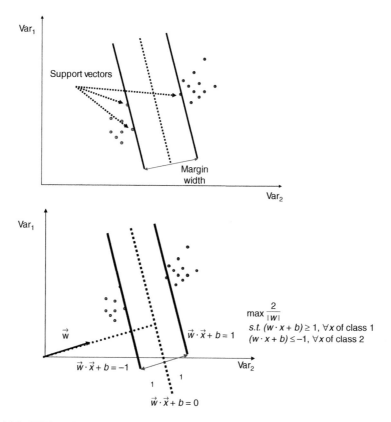

Fig. 14.2 SVMs and the classification problem as geometrical separation. In the *upper part* geometrical representation of a 2-class predictive modeling (classification problem) with 2 input dimensions (*Var1, Var2*). Each subject is represented by a dot (i.e., a 2-dimensional vector). Dark dots are patients and light ones controls. The line that separates patients from controls while maximizing the distance between classes is the solution to the SVM problem. The patients at the border of each class are the "support vectors". In the *lower part* we see the mathematical expression of the optimization problem of separating the two classes. Such problems are easily solved by modern software

determine the hyperplane (in other words the decision function that tells us whether a subject has cancer or is cancer-free). To ensure a good model, the SVM learning procedure requires that the hyperplane must be such that the number of misclassified subjects is minimized and this is the "data fit" part of the overall training function SVMs wish to minimize. In addition however, SVMs introduce a generalization-enforcing constraint, the so-called "regularizer" that is the total sum of squared weights of all variables that must also be minimized. Specifically, the sum of data fit and regularizer must be minimized subject to the locations and labels of training data fed in to the algorithm.[3]

---

[3] We omit for simplicity and brevity a variant of the above (the "soft margin" SVM formulation) which further allows for noisy data and mild non linearity in the data.

What does regularization achieve? One way to think of the effects of regularization is that by forcing weights to be as small as possible, all variables that *are not relevant or are superfluous* to the predictive modeling will have zero or near zero weights and are effectively "filtered" out of the model. Equivalently, the minimization of weights entails that the separation between classes is geometrically maximized and statistical machine learning theory shows that this often leads to more generalizable models. The formulation of data fit + parameter penalty is a very general one and there are many families of data mining methods, outside SVMs, that adopt it including statistics-inspired regression and classification methods (e.g., LARS and Lasso (Friedman et al. 2001)) and Bayesian methods (e.g., Sparse Bayesian Logistic Regression with Laplace priors (Genkin et al. 2007)). By simultaneously optimizing the model fit to the data and striving for smallest complexity of model possible for the task, excellent generalization and avoidance of overfitting are achieved.

14.1.1 Kernel-Based Transformation of the Data

SVMs and many other modern Big Data mining methods use this general method to deal with highly non-linear predictive modeling tasks. A non- linear task is one where there is not straight line (in 2 dimensions) or hyperplane (in >2 dimensions) that allows for clean separation among the data. Non-linear SVMs deal with this problem by mapping the original data into a new data space in which there is a hyperplane that separates the data. Once this hyperplane is found in the new space (the "feature space") the hyperplane is translated back into the original data space and we can use it for predictive modeling. Once translated into the original space the hyperplane becomes a non-linear surface (Fig. 14.3).

A rather remarkable property of SVMs and other methods that use kernels for modeling non-linear functions is that they can project huge data to vastly complicated spaces, find the solution there and then translate it back to an accurate model in the original data space in seconds (in most practical situations). Doing the same operation using traditional statistical modeling (e.g., via introducing in the model interaction effects and with model selection among alternatives) is both impractical and prone to overfitting (Harrell 2015). In addition, we mention that SVMs' complexity is not determined by the number of variables of the original data but by (a vastly smaller set of parameters <n), which further explains their ability to generalize well and avoid overfitting (Vapnik 2013). Statnikov et al. (2012) describes a variety of real-life applications of SVMs in the Health Sciences including (among other applications): cancer diagnosis from gene expression data, predicting lab results from mining the EHR, modeling the decision making of dermatologists for the diagnosis of melanomas and examining guideline compliance, and analyzing text (scientific articles) and citations to understand, prioritize, search and filter the literature.

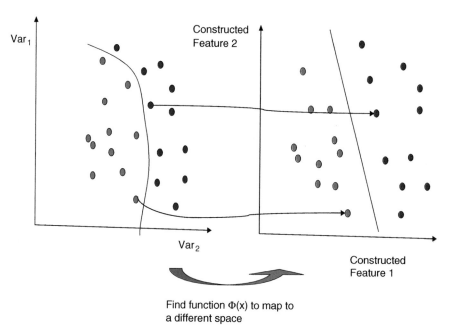

Fig. 14.3 Non-linearly separable classification by mapping to a different space. In the *left side* we see a non-linearly separable problem, that is, there is no straight line that accurately separates the two classes. The SVM (and other kernel techniques) use a mapping function that transforms the original variables (*Var1*, *Var2*) into constructed features, such that there exists a straight line that separates the data in the new space. Once the solution is found in the mapped space, it is reverse-transformed to the original variable space. It can be seen that the solution is a curved line. Because the mapping function is very expensive to compute, a special kernel function is used that allows solving the SVM optimization *without incurring the expense of calculating the full mapping*

14.1.2 Advances in Feature Selection

Variable selection for predictive modeling has received considerable attention during the last three decades in a variety of data science fields (Guyon and Elisseeff 2003; Kohavi and John 1997). Feature selection is a technique of choice to tame high dimensionalities in diverse big data applications. Intuitively, variable selection for prediction aims to select only a subset of variables for constructing a diagnostic or predictive model for a given classification or regression task. The reasons to perform variable selection include (a) improving the model predictivity and addressing the curse-of-dimensionality, (b) reducing the cost of observing, storing, and using the predictive variables, and finally, (c) gaining an understanding of the underlying process that generates the data. Often the feature selection problem for classification/prediction is defined as identifying the minimum-size subset of variables that exhibit the maximal predictive performance (Guyon and Elisseeff 2003). Variable selection methods can be broadly categorized into *wrappers* (i.e., heuristic search in the space of all possible variable subsets using a classifier of choice to assess each subset's predictive information), or *filters* (i.e., not using the classifier per se to select features, but instead applying statistical criteria to first select features and then build the classifier with the best features). In addition, as we saw with SVMs, some methods perform *embedded variable selection*, that is, they attempt to simultaneously maximize classification performance while minimizing the number of variables used. For example, *shrinkage regression* methods introduce a bias into the parameter estimation regression procedure that imposes a penalty on the size of the parameters. As explained earlier, the parameters that are close to zero are essentially filtered-out from the predictive model. A variety of embedded variable selection methods have been recently introduced. These methods are linked to a statement of the classification or regression problem as an optimization problem with specified error and regularization penalty functions. These techniques usually fall into a few broad classes: One class of methods uses the L^2-norm penalty (also known as ridge penalty), for example, the recursive feature elimination (RFE) method is based on the L^2-norm formulation of SVM classification problem (Guyon et al. 2002). Other methods are based on the L^1-norm penalty (also known as lasso penalty), for example, feature selection via solution of the L^1-norm formulation of SVM classification problem and penalized least squares with lasso penalty on the regression coefficients (Friedman et al. 2001). A third set of methods is based on convex combinations of the L^1- and L^2-norm penalties, for example, feature selection using the doubly SVM formulation (Wang et al. 2006) and penalized least squares with elastic net penalty (Zou and Hastie 2005). A fourth set uses the L^0-norm penalty, for example, feature selection via approximate solution of the L^0-norm formulation of SVM classification problem (Weston et al. 2003). The majority of feature selection methods in the literature and in practice are heuristic in nature in the sense that in most cases it is unknown what consists for optimal feature selection solution *independently of the class of models fitted*, and under which conditions an algorithm will output such an optimal solution.

Typical variable selection approaches also include forward, backward, forward-backward, local and stochastic search wrappers (Guyon and Elisseeff 2003; Kohavi and John 1997). The most common family of filter algorithms ranks the variables according to a score and then selects for inclusion the top k variables. The score of each variable is often the univariate (pairwise) association with the outcome variable T for different measures of associations such as the signal-to-noise ratio, the G^2 statistic and others. Information theoretic (estimated mutual information) scores and multivariate scores, such as the weights received by a Support Vector Machine, have also been suggested (Guyon and Elisseeff 2003; Guyon et al. 2002). Feature selection methods have found extensive application in recent years in the analysis of Big Data "omics" molecular datasets (e.g., gene expression, proteomics, methylation, miRNAs, microbiomics, etc), in the analysis of EHR data, the analysis of clinical text, scientific literature, social digital media and the WWW (indicative examples presented in Aliferis et al. 2003, 2010a; Aphinyanaphongs et al. 2005; Cooper et al. 1997; Fu et al. 2010; Guyon et al. 2002; Ray et al. 2014; Statnikov et al. 2012).

14.2   Advances in Causal Discovery with Big Data, Causal Feature Selection and Unified Predictive and Causal Analysis

Predictive modeling methods provide the ability to diagnose patients, prognosticate natural course of disease, predict response to treatments, stratify patients according to risk, forecast demand for health services at the population level, predict re-admissions, adverse events and a number of other factors with huge value to a "learning health system". They lack however a fundamental capability, that is to identify the *causal drivers of the above phenomena*. This capability along with the ability to *predict the effects of interventions to individual patients or healthcare systems* is provided by Causal Discovery mining methods (Pearl 2000; Spirtes et al. 2000; Aliferis et al. 2010a, b).

The topic of discovery of causal relationships by analysis of non-experimental (that is observational) data has been over the years a controversial one with the dominant attitude being that "correlation is not causation" and the implication that causality could not be ascertained reliably in the absence of randomized experiments. Indeed, some correlations are not causative while others are, and traditional statistical and predictive modeling methods cannot reliably differentiate among causal and non-causal relationships. However randomized experiments are in most settings expensive, unethical, impractical, or take too much time. Classical randomized experiments such as clinical trials or targeted biological experiments with model organisms are also too simplified (e.g., examining only two or a few variables at a time) and occasionally nonspecific (e.g., biological experiments silencing or inducing gene expression in many cases do not have absolute specificity for the target genes thus it is not always possible for experimenters to conduct experimental manipulations that uniquely target desired genes and proteins).

Usable and robust methods for differentiating between causative and non-causative (or confounded) correlations were not broadly known or available until the 1980s. Methods that are suitable, furthermore, for Big Data with thousands of variables or more where not available until 2002 (for local causal relationships) and 2006 (for full causal models). The core advances that have led to the very recent ability to discover causal relationships can be traced to the work of the Nobel Laureate Sir Clive Granger who in the 1960s devised a method for eliminating non causal time series correlations among two longitudinally observed variables X and Y in the presence of observed confounding time series Z (Granger 1969); the extensive work of Turing Award winner Judea Pearl and of Glymour, Spirtes and Scheines who in the 1980s presented fully-developed theoretical frameworks for mathematically representing causality in complex systems, providing several algorithms that could infer causal models and proving the correctness of these algorithms under broad assumptions (Pearl 2000; Spirtes et al. 2000). The prototypical PC algorithm was capable of learning correctly the causal model when no hidden variables are present and the IC* and FCI could the same even in the presence of hidden confounders. The work by Cooper and Herskovits (1992), Heckerman et al. (1995), and Chickering (2003) produced variants of the above methods that were based, instead of conditional independence tests and constraint propagation, on heuristic search over the space of causal models using Bayesian scoring functions to select the most likely causal models to have generated the data (notable algorithms are K2, GS and GES). The above algorithms would output complete and correct causal models (or equivalence classes of models that could not be differentiated without experiments from one another) and did not require temporal order information to orient causal direction in most practical cases. A problem that was remaining until 2000 was the lack of algorithms to scale to applicability to thousands of variables without sacrificing correctness of discovery. Specifically, at that time most correct causal discovery algorithms would run with approximately up to 100 variables thus rendering causal discovery with EHR data, WWW data, or omics data essentially impossible. Several algorithms or analysis protocols used at the time introduced a number of simplifications that would allow the analysis with high dimensionalities but at the expense of the correctness of the algorithm. An example was the Sparse Candidate Algorithm of Friedman et al. that was able to run in early gene expression platform data but introduced simplifying assumptions that resulted in many false positives and false positives in the output (Tsamardinos et al. 2006).

To address the challenge of scalability for big data, Aliferis and colleagues created in 2003 the first local algorithms (HITON-PC and MMPC) that could learn correctly the local causal relations around the response variable (i.e., direct causes and direct effects) (Aliferis et al. 2003; Tsamardinos et al. 2003). These methods also allowed for very small sample sizes and in 2006 they published the first algorithms that combined local with global learning and constrained and Bayesian techniques to produce the first accurate full causal model learning algorithm (MMHC) that could be run with thousands of variables (Tsamardinos et al. 2006). The original algorithm was tested with both health science and general Big Data (Tsamardinos et al. 2006) while extensions and modifications were tested with success against many pathway reverse engineering algorithms for discovery of Transcription Factor binding sites in multiple datasets where the ground truth was known (Narendra et al. 2011).

14.3 Unified Predictive-Causal Modeling and Causal Feature Selection

While predictive modeling and causal modeling are often viewed as two distinct types of modeling with different goals and challenges, these two types of modeling can be naturally unified when Causal Probabilistic Graphs are used as the modeling language. This type of model (also known as a "Causal Bayesian Network") has a number of desirable properties; by applying well established inference algorithms users of models can use the models predictively (Pearl 1988; Neapolitan 1990). Using causal probabilistic graphs for predictive modeling has the unique capability that any subset of variables in the model can be designated as query and any subset (of the remaining) variables can be designated as evidence (i.e., input variables). This is in stark contrast to standard predictive modeling algorithms that have one output variable and all of the remaining ones are inputs. In addition, by application of Pearl's "do calculous" the data scientist can ask queries of the type "what will be the effect of manipulating variable(s) X on the distribution of variable(s) Y?" (Pearl 2000).

A general caveat is that these types of calculations are expensive to the extent that *typically cannot be applied with large numbers of variables*. However a very practical state of the art approach to unify predictive and causal modeling in Big Data is via Markov Boundaries. The Markov Boundary of a response variable T (denoted MB(T)) is the minimal set of variables in the data such that all non Markov Boundary variables are independent of the response T once we know

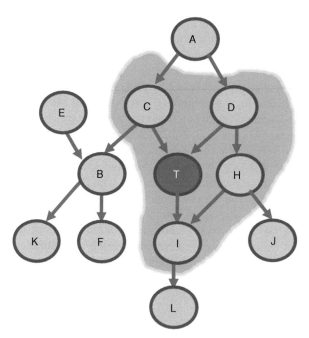

Fig. 14.4 The Markov Boundary is the set of variables that provides a principled and mathematically optimal way to reduce variable dimensionality, achieve optimal predictivity and discover direct causes and effects for a target/response variable of interest. The Markov Boundary of the response variable T is the set $\{C, D, I, H\}$. We can fit a model with these variables that is guaranteed to have maximum predictive information for T. All other variables can be discarded safely. It does not matter how many variables exist outside the Markov Boundary (seven in this example or infinite), they can all safely be ignored and still achieve optimal predictivity. Moreover the Markov Boundary has in the majority of distributions a local causal interpretation: it comprises the direct causes, direct effects and direct causes of direct effects of the response variable

MB(T).[4] (Pearl 1988). As shown by Tsamardinos and Aliferis (2003) the Markov Boundary of the response variable provides the optimal solution to the feature selection problem for the majority of distributions and practical analysis setups. These authors showed that discovery of the Markov Boundary allow not only optimal feature selection and predictive modeling *but also* local causal discovery. Fig. 14.4 shows the key points related to the Markov Boundary of the response variable.

The attractive properties of the Markov Boundary would be of little use if practical algorithms for its discovery did not exist. The first attempts to discover Markov Boundaries directly where by Koller and Sahami (1996) and Cooper and Aliferis in 1997 (Cooper et al. 1997). Both algorithms (K-S and K2 MB) where highly heuristic (thus approximate only) and non-scalable to large data. Margaritis and Thurn in 1999 introduced G-S, the first correct but still not scalable algorithm for direct discovery of Markov Boundaries (Margaritis and Thrun 1999). Tsamardinos and Aliferis introduced the first correct, sample efficient and very scalable algorithms (IAMB family, HITON, MMMB) and tested them on several diverse datasets (with up to 200,000 variables in initial experiments) achieving excellent predictivity and 1000-fold variable reduction (Aliferis et al. 2003; Tsamardinos et al. 2003). Subsequent comparisons with >100 algorithms and variants over >40 datasets fully validated the theoretical advantages of Markov Boundary induction (Aliferis et al. 2010a, b).

14.3.1   Synopsis of Other Important Big Data Mining Advances

The range and capabilities of modern Big Data Mining methods are so extensive that even a full volume would still not do them justice. We present here brief references to select few additional important classes of methods.

14.3.1.1   Advances in Integrated Analysis Protocols for Big Data

Analysis of Big Data is not complete just by having access to predictive modeling or causal modeling algorithms. A complete data mining protocol requires also a *model selection procedure* and an *error estimation procedure*. Model selection chooses the best analytic method for the task at hand and/or finds the best parameter values for the chosen method. Error estimators calculate the expected error of the best models produced by model selection, not in the training data (where it is easy to over fit and always produce perfectly fitting models) but in the general population where the discovery dataset was sampled from. A currently powerful and dominant protocol for big data is repeated nested n-fold cross validation (RNFCV). This protocol depicted in Fig. 14.5 has the following properties: is unbiased (i.e., error estimates are accurate

---

[4] The Markov Blanket of T is the set of variables in the data such that all non Markov Blanket variables are independent of the response T, once we know the Markov Blanket. Since the Markov Boundary is of interest in practice because it is the minimal Markov Blanket, it is common to see in the literature use of the term "Markov Blanket" when the more precise "Markov Boundary" is implied.

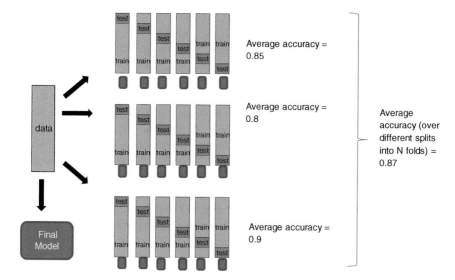

Fig. 14.5 The Repeated Nested N-Fold Cross Validation Procedure. This procedure performs model selection and error estimation. The discovery data is split in N pairs (in this example N = 6) such that each pair comprises a train and a test set and all tests sets when taken together equal the full data. Within each train set further random splits are performed (not shown) into a train-train and a validation subset. The validation subsets are used to check which parameters perform best in each one of the N folds as well as overall N folds. The best parameter combination is used to fit a model on the train set and check its error in the test set. The test set errors are averaged over the N folds. The whole process is repeated many times (typically 50 repeats is enough) and the final error estimate is the average over the many repeats. The final model selection is now obtained on all data using the same train-validate procedure once to select parameter values and the best parameters are applied on the full data to obtain the final model. This procedure is fully automated and the analyst has only to choose which methods the procedure will test and what parameter value ranges will be considered

asymptotically), can be fully automated in computer software, is relatively insensitive to variation of error estimates due to small sample, and can be used to simultaneously perform parameter and analysis optimization with error estimation.

Finally, the RNFCV can be applied with small and large sample sizes (although for very large n it is sufficient to use simplified versions such as the nested hold out method which is equivalent to non-repeated onefold nested cross validation). We note that when N = n (i.e., number of folds = sample size) this is called the *Leave One Out estimator*, a generally unbiased protocol that has higher variance than RNFCV. Another important approach is *Bootstrapping*, a general method for estimating properties of a sampling population from a random sample. Bootstrapping creates many samples from the discovery dataset sampling with replacement. Bootstrapping is not an unbiased estimator and requires knowledge of a data and learner distribution bias term needed to correct its error estimates, however. In most of the practical examples referenced previously various forms of RNFCV were applied. Specifically in Aliferis et al. (2009) the advantages of the protocol for prognosis of outcomes in various cancer patients is demonstrated. In Statnikov et al. (2011) detailed explanation of the RNFCV procedure is given whereas Statnikov et al. (2012) applied examples routinely use RNFCV as a model selection and error estimation procedure for clinical, text, and omics analysis tasks.

14.3.1.2 Advances in Big Data Methods for Unstructured Data

In this category we mention Natural Language Processing (NLP) methods that can process syntactically and semantically free text (Friedman et al. 1994, 2004; Friedman and Hripcsak 1999). Types of free text include scientific publications, care notes, WWW sites, online social media, and scientific articles among others. Whenever deep linguistic analysis is not needed but broad content determination is enough (for example categorizing articles according to disease or research categories, or determining patient phenotype from text) Machine Learning (ML) approaches are good alternatives to NLP. In many cases ML offers more accurate and easier to build models of the previous type. The aforementioned SVM predictive modelling method for example has excellent performance in many text analysis tasks. Another type of unstructured data has to do with citation networks, for example articles or WWW citations. Data mining tasks include predicting citations, understanding purpose of citations, distinguishing essential from non-essential references, etc. (Aphinyanaphongs et al. 2005; Fu and Aliferis 2010).

14.3.1.3 Polymodal and Multimodal Analysis Methods

Big Data includes not just clinical coded variables, but also text reports, imaging tests, high throughput molecular tests of various types (SNPs, gene expression, RNAseq/sRNA, metabolomics, chipSEQ, proteomics, DNA methylation, etc.), societal and environmental data including online digital media etc. The combination of all or some of these types of data requires specialized approaches for integrated analysis (Ray et al. 2014; Daemen et al. 2009; Gevaert et al. 2006). For example Ray et al. describe the combination of clinical data with gene expression, DNA methylation, proteomics, RNAseq, and other types for both diagnostic and prognostic modeling in cancer.

14.3.1.4 Clustering

The family of clustering techniques deals with grouping objects into meaningful categories (e.g., subjects into disease groups or treatment-response groups). Typically the produced clusters contain objects that are similar to one another but dissimilar to objects in other clusters (Duda et al. 2012). As has been shown mathematically there is no single measure of similarity that is suitable for all types of analyses nor is clustering the most powerful method for all types of analysis even when applicable. Thus this generally useful set of methods is known to be often abused or misused (Dupuy and Simon 2007).

14.3.1.5 Bayesian Methods

Bayesian methods offer alternatives to classical statistical methods both for parameter estimation and predictive modeling techniques. When coupled with causal semantics and appropriate discovery procedures (e.g., in the case of Causal Bayesian Networks) they can be unique tools for discovery of causal relationships as well. In the predictive modeling methods space we highlight the critical role that Bayesian Networks play (since they can model any distribution compactly and they can answer flexible queries) (Pearl 1988; Neapolitan 1990). In many cases simpler Bayesian classifiers can be more useful especially in smaller sample sizes or in distributions with special characteristics. Notable examples include Simple (aka "Naïve") Bayes models, TAN and BAN models (Cheng and Greiner 1999, 2001; Friedman et al. 1997).

14.3.1.6 Boosting

This family of methods operates by applying successive refinements on the decision function to eliminate residual errors. It can be also be thought as combining relatively simpler decision surfaces to create a complex decision surface. In large sample sizes boosted trees and other state of the art boosted methods exhibit superior performance and thus should be explored (Duda et al. 2012; Schapire 2003).

14.3.1.7 Decision Trees and Random Forests

Decision trees is an established modeling methodology that has many advantages including: ability to represent all distributions, easy to understand by humans (including non-experts), fast operation, and embedded feature selection (Duda et al. 2012). With the advent of Big Data classical Decision tree algorithms (e.g., ID3 and CART) were no longer applicable or as accurate due to curse of dimensionality issues. A newer version of Decision Trees, called Random Forests (RF) was introduced and is applicable to many Big Data predictive modeling settings (Breiman 2001). RFs build not one but many trees and take a vote (or average in the continuous case) over all trees to obtain a final model prediction. Each tree is built from a bootstrap sample sampled from the original discovery dataset. In addition, each tree at each branch construction step undergoes first a selection of a random subset of all available features. This aspect is counterintuitive but works well in practice. RFs do not over fit as the number of trees grows in the forest. RFs can over- or under-fit if the allowed size of each tree is bigger or smaller (respectively) than optimal for the data at hand. Finally, RFs have their own embedded error estimator (the "Out of Bag estimator"), which performs very well.

14.3.1.8   Genetic Algorithms (GAs)

Genetic Algorithms are heuristic search procedures in the space of models that the analyst wishes to consider. For example, the analyst may use GAs to find a good linear regression, a good SVM, a good Decision Tree or other model of choice. The search resembles the process of genetic evolution and can be shown to advance rapidly to better models (Mitchell 1997). On the other hand GAs are computationally very expensive and prone to get trapped to local optima (i.e., solutions that cannot be improved in the next reachable steps in any direction in the model search space, although a better solution does exist several steps away). GAs also are used when the analyst does not have a good insight about the process that generates the data, or about which method may perform well for the task at hand. When such insight exists it is typically better to use methods that have known properties that guarantee high performance for the desired analysis.

14.3.1.9   Artificial Neural Networks (ANNs) and Deep Learning

ANNs simulate biological neuronal systems and have a number of desirable properties: they can approximate any function, they can deal with very non-linear tasks and they have excellent empirical performance (Duda et al. 2012; Mitchell 1997). In recent years ANNs became less used since newer methods such as SVMs are easier to parameterize and faster to operate on most practical Big Data settings. Very recently *Deep Learning* approaches have created a resurgence of interest in ANNs. Deep Learning focuses on creating abstractions of original feature spaces that facilitate learning of complex functions and offering certain advantages such as robustness to image shifting (LeCun et al. 2015; Schmidhuber 2015).

14.3.1.10   Network Science

The field of network science offers a completely different approach to conventional predictive modeling and causal discovery methods. Network Science leverages the remarkable consistency in the properties of a broad array of systems that are adaptive and robust in nature (Albert et al. 2000; Holme et al. 2002; Newman et al. 2003; Barabasi and Oltvai 2004; Barabasi 2009; Barrenas et al. 2009). Systems that exhibit these measurable properties are called *Complex Adaptive Systems* (CAS) (Newman et al. 2003). The application of Network Science to problems of health and disease is called *Network Medicine* and its main idea follows: a disease represents a pathologic biological process that emerges, and is sustained over time, because it is embedded in a transformed biologic system that acquires adaptive properties. Accordingly, if such an adaptive system related to a given disease is identified, the capacity to determine its areas of vulnerability may reveal promising targets or new approaches for treatment. A typical network science analysis proceeds by building network representations of complex systems and then calculating a number of metrics on the network model. Such metrics include: Network Diameter, Characteristic Path Length, Shortest Path Distribution, Degree Distribution, and Clustering Coefficient. The specific structure and properties of the network model help the analyst identify drug intervention targets and other important system properties.

14.3.1.11 Active Learning

The field of Active Learning studies methods for the iterative collection of data and corresponding refinement of models until an accurate enough model is built or other termination criteria are met. Active Learning methods address both predictive modeling and causal discovery tasks (Tong and Koller 2002; Meganck et al. 2006).

14.3.1.12 Outlier Detection

Outlier (or novelty) detection methods seek to find extreme or otherwise atypical observations among data. "Super utilizer" patients is a prototypical example of an outlier that has great importance for healthcare. Numerous methods have been invented for outlier detection over the years in many fields including statistics, engineering, computer science, applied mathematics etc. and they are based on multivariate statistics, density estimation, "1-class" SVMs, clustering, and other approaches (Markou and Singh 2003).

14.3.1.13 Visualization

Visualization methods rely on the capability of the human visual apparatus to decode complex patterns when these patterns have been represented in convenient visual forms. Another use of visualization serves explanatory purposes; that is, presenting and explaining results that were obtained via computational means. Interactive data visualizations, where users are allowed to manipulate their views of the data to obtain more information, have been found to be rapid and efficacious in identifying early infection and rejection events in lung transplant patients (Pieczkiewicz et al. 2007). Data visualization can also be useful in displaying nursing care data, such as that coded with the Omaha System; and intraoperative anesthesia data, such as maintenance of blood pressure (Lee et al. 2015; Monsen et al. 2015). Evaluation techniques have been developed to gauge visualization effectiveness in clinical care (Pieczkiewicz and Finkelstein 2010). Significant challenges exist, however, in implementing visualization more widely in electronic health records, many of them resulting from the highly multivariable nature of case-oriented medical data (West et al. 2015).

14.4   Conclusions

14.4.1   Achievements, Open Problems, Challenges in Big Data Mining Methods

The progress of methods for big data mining has been nothing less than disruptive in recent decades, and practitioners have currently access to extraordinary methods. The present status of big data science in healthcare is one of convergence of advanced big data analysis methods with powerful computing and with explosive capability to capture and produce scientific and organizational data that merit analysis. This chapter provided an accessible introduction to recent advances and related practical methods for analysis of big data in nursing and the health sciences at large. The array of big data mining methods cover predictive, causal and other types of modeling with diverse types of big data. These capabilities have already started revolutionizing molecular biology and genomic medicine, as well as non-medical fields. They have proved the legitimacy of the novel big data mining,methods, yet healthcare and health sciences applications are still lagging behind. Part of the challenge relates to *educational gaps*, pointing to the need to ensure that researchers and healthcare leaders have sufficient familiarity with the core concepts and methods. *Organizational improvements* to coalesce data science programs and units to provide functional integrated collaborative science, service units and consulting services is also needed for more effective dissemination of available methods to the points of research or application need. The broad availability of *well-designed, validated and user friendly software* is an important bridge than needs crossing as well. Another set of challenges includes achieving the necessary improvements in practical analysis aspects such as *data design* (e.g., power-size analyses), and exploring unconventional data designs (e.g., methods that work directly with and are optimized for relational data, partially observed data, and data collected under a variety of non-random sampling regimes). Other worthy challenges for the future include developing quality control processes, and studying the comparative effectiveness of available algorithm, protocols, and tools to inform the creation of widely adopted best practices.

References

Albert R, Jeong H, Barabasi AL. Error and attack tolerance of complex networks. Nature. 2000;406:378–482.

Aliferis CF, Tsamardinos I, Statnikov A. HITON: a novel Markov blanket algorithm for optimal variable selection. In: AMIA 2003 annual symposium proceedings; 2003. p. 21–25.

Aliferis CF, Statnikov A, Tsamardinos I. Challenges in the analysis of mass-throughput data: a technical commentary from the statistical machine learning perspective. Cancer Inform. 2006;2.

Aliferis CF, Statnikov A, Tsamardinos I, Schildcrout JS, Shepherd BE, Harrell Jr FE. Factors influencing the statistical power of complex data analysis protocols for molecular signature development from microarray data. PLoS One. 2009;4(3):e4922. doi:10.1371/journal.pone.0004922.

Aliferis CF, Statnikov A, Tsamardinos I, Mani S, Koutsoukos XD. Local causal and Markov blanket induction for causal discovery and feature selection for classification. Part II: analysis and extensions. J Mach Learn Res. 2010a;11:235–84.

Aliferis CF, Statnikov A, Tsamardinos I, Mani S, Koutsoukos XD. Local causal and Markov blanket induction for causal discovery and feature selection for classification. Part I: algorithms and empirical evaluation. J Mach Learn Res. 2010b;11:171–234.

Aphinyanaphongs Y, Tsamardinos I, Statnikov A, Hardin D, Aliferis CF. Text categorization models for retrieval of high quality articles in internal medicine. J Am Med Inform Assoc. 2005;12(2):207–16.

Barabasi AL. Scale-free networks: a decade and beyond. Science. 2009;325:412–3. doi:10.1126/science.1173299.

Barabasi AL, Oltvai ZN. Network biology: understanding the cell's functional organization. Nat Rev Genet. 2004;5:101–13.

Barrenas F, Chavali S, Holme P, Mobini R, Benson M. Network properties of complex human disease genes identified through genome-wide association studies. PLoS One. 2009;4(11):e8090. doi:10.1371/journal.pone.0008090.

Breiman L. Random forests. Mach Learn. 2001;45(1):5–32. doi:10.1023/A:1010933404324.

Cheng J, Greiner R. Comparing Bayesian network classifiers. In: Proceedings of the 15th conference on uncertainty in artificial intelligence (UAI); 1999. p. 101–7.

Cheng J, Greiner R. Learning Bayesian belief network classifiers: algorithms and system. In: Proceedings of 14th biennial conference of the Canadian society for computational studies of intelligence; 2001.

Chickering DM. Optimal structure identification with greedy search. J Mach Learn Res. 2003;3(3):507–54.

Cooper GF, Herskovits E. A Bayesian method for the induction of probabilistic networks from data. Mach Learn. 1992;9(4):309–47.

Cooper GF, Aliferis CF, Ambrosino R, Aronis J, Buchanan BG, Caruana R, Fine MJ, Glymour C, Gordon G, Hanusa BH. An evaluation of machine-learning methods for predicting pneumonia mortality. Artif Intell Med. 1997;9(2):107–38.

Daemen A, et al. A kernel-based integration of genome-wide data for clinical decision support. Genome Med. 2009;1(4):39. doi:10.1186/gm39.

Dobbin K, Simon R. Sample size determination in microarray experiments for class comparison and prognostic classification. Biostatistics. 2005;6(1):27–38. doi:10.1093/biostatistics/kxh015.

Duda RO, Hart PE, Stork DG. Pattern classification. New York: John Wiley & Sons; 2012.

Dupuy A, Simon RM. Critical review of published microarray studies for cancer outcome and guidelines on statistical analysis and reporting. J Natl Cancer Inst. 2007;99(2):147–57. doi:10.1093/jnci/djk018.

Friedman C, Hripcsak G. Natural language processing and its future in medicine. Acad Med. 1999;74(8):890–5.

Friedman C, Alderson PO, Austin JH, Cimino JJ, Johnson SB. A general natural-language text processor for clinical radiology. J Am Med Inform Assoc. 1994;1(2):161.

Friedman N, Geiger D, Goldszmidt M. Bayesian network classifiers. Mach Learn. 1997;29(2):131–63.

Friedman J, Trevor H, Tibshirani R. The elements of statistical learning, vol. 1. Berlin: Springer; 2001.

Friedman C, Shagina L, Lussier Y, Hripcsak G. Automated encoding of clinical documents based on natural language processing. J Am Med Inform Assoc. 2004;11(5):392–402. doi:10.1197/jamia.M1552.

Fu LD, Aliferis CF. Using content-based and bibliometric features for machine learning models to predict citation counts in the biomedical literature. Scientometrics. 2010;85(1):257–70. doi:10.1007/s11192-010-0160-5.

Genkin A, Lewis DD, Madigan D. Large-scale Bayesian logistic regression for text categorization. Technometrics. 2007;49(3):291–304.

Gevaert O, De Smet F, Timmerman D, Moreau Y, De Moor B. Predicting the prognosis of breast cancer by integrating clinical and microarray data with Bayesian networks. Bioinformatics. 2006;22:e184–90. doi:10.1093/bioinformatics/btl230.

Granger CW. Investigating causal relations by econometric models and cross-spectral methods. Econometrica. 1969;37(3):424–38.

Guyon I, Elisseeff A. An introduction to variable and feature selection. J Mach Learn Res. 2003;3:1157–82.

Guyon I, Weston J, Barnhill S, Vapnik V. Gene selection for cancer classification using support vector machines. Mach Learn. 2002;46(1–3):389–422.

Harrell F. Regression modeling strategies: with applications to linear models, logistic and ordinal regression, and survival analysis. New York: Springer; 2015.

Heckerman D, Geiger D, Chickering DM. Learning Bayesian networks: the combination of knowledge and statistical data. Mach Learn. 1995;20(3):197–243.

Holme P, Kim BJ, Yoon CN, Han SK. Attack vulnerability of complex networks. Phys Rev E. 2002;65:056109.

Kohavi R, John GH. Wrappers for feature subset selection. Artif Intell. 1997;97(1):273–324.

Koller D, Sahami M. Toward optimal feature selection. In: Proceedings of the international conference on machine learning; 1996.

LeCun Y, Bengio Y, Hinton G. Deep learning. Nature. 2015;521(7553):436–44. doi:10.1038/nature14539.

Lee S, Kim E, Monsen KA. Public health nurse perceptions of Omaha System data visualization. Int J Med Inform. 2015;84(10):826–34. doi:10.1016/j.ijmedinf.2015.06.010.

Margaritis D, Thrun S. Bayesian network induction via local neighborhoods. Adv Neural Inf Process Syst. 1999;12:505–11.

Markou M, Singh S. Novelty detection: a review—part 1: statistical approaches. Sig Process. 2003;83(12):2481–97.

Meganck S, Leray P, Manderick B. Learning causal bayesian networks from observations and experiments: A decision theoretic approach. MDAI, 2006;3885:58–69.

Mitchell TM. Machine learning, vol. 45. Burr Ridge, IL: McGraw Hill; 1997. p. 995.

Monsen KA, Peterson JJ, Mathiason MA, Kim E, Lee S, Chi CL, Pieczkiewicz DS. Data visualization techniques to showcase nursing care quality. Comput Inform Nurs. 2015;33(10):417–26. doi:10.1097/CIN.0000000000000190.

Narendra V, Lytkin N, Aliferis C, Statnikov A. A comprehensive assessment of methods for de-novo reverse-engineering of genome-scale regulatory networks. Genomics. 2011;97(1):7–18. doi:10.1016/j.ygeno.2010.10.003.

Neapolitan RE. Probabilistic reasoning in expert systems: theory and algorithms. New York: Wiley; 1990.

Newman MEJ, Barabasi AL, Watts DJ. The structure and dynamics of networks. Princeton, NJ: Princeton University Press; 2003.

Pearl J. Probabilistic reasoning in intelligent systems: networks of plausible inference. San Mateo, CA: Morgan Kaufmann Publishers; 1988.

Pearl J. Causality: models, reasoning, and inference. Cambridge, UK: Cambridge University Press; 2000.

Pieczkiewicz DS, Finkelstein SM. Evaluating the decision accuracy and speed of clinical data visualizations. J Am Med Inform Assoc. 2010;17(2):178–81.

Pieczkiewicz DS, Finkelstein SM, Hertz MI. Design and evaluation of a web-based interactive visualization system for lung transplant home monitoring data. In: AMIA annual symposium proceedings; 2007. p. 598–602.

Ray B, Henaff M, Ma S, Efstathiadis E, Peskin ER, Picone M, Poli T, Aliferis CF, Statnikov A. Information content and analysis methods for multi-modal high-throughput biomedical data. Sci Rep. 2014;4. doi:10.1038/srep04411.

Schapire RE. The boosting approach to machine learning: an overview. In: Nonlinear estimation and classification. New York: Springer; 2003. p. 149–71.

Schmidhuber J. Deep learning in neural networks: an overview. Neural Netw. 2015;61:85–117.

Spirtes P, Glymour CN, Scheines R. Causation, prediction, and search, vol. 2. Cambridge, MA: MIT Press; 2000.

Statnikov A, Aliferis CF, Hardin DP, Guyon I. A gentle introduction to support vector machines. In: Biomedicine: theory and methods, vol. 1. Singapore: World Scientific; 2011.

Statnikov A, Aliferis CF, Hardin DP, Guyon I. A gentle introduction to support vector machines. In: Biomedicine: case studies and benchmarks, vol. 2. World Scientific; 2012.

Tong S, Koller D. Support vector machine active learning with applications to text classification. J Mach Learn Res. 2002;2:45–66.

Tsamardinos I, Aliferis CF. Towards principled feature selection: relevancy, filters and wrappers. In: Proceedings of the ninth international workshop on artificial intelligence and statistics (AI & Stats); 2003.

Tsamardinos I, Aliferis CF, Statnikov A. Time and sample efficient discovery of Markov blankets and direct causal relations. In: Proceedings of the ninth international conference on knowledge discovery and data mining (KDD); 2003. p. 673–8.

Tsamardinos I, Brown LE, Aliferis CF. The max-min hill-climbing Bayesian network structure learning algorithm. Mach Learn. 2006;65(1):31–78.

Vapnik V. The nature of statistical learning theory. New York: Springer Science & Business Media; 2013.

Wang L, Zhu J, Zou H. The doubly regularized support vector machine. Stat Sin. 2006; 16:589–615.

West VL, Borland D, Hammond WE. Innovative information visualization of electronic health record data: a systematic review. J Am Med Inform Assoc. 2015;22(2):330–9. doi:10.1136/amiajnl-2014-002955.

Weston J, Elisseeff A, Scholkopf B, Tipping M. Use of the zero-norm with linear models and kernel methods. J Mach Learn Res. 2003;3(7):1439–61.

Zou H, Hastie T. Regularization and variable selection via the elastic net. J R Stat Soc Series B (Stat Methodol). 2005;67(2):301–20.

Part IV
Looking at Today and the Near Future

Charlotte A. Weaver

In Part IV, we present the work of early innovators in the application of big data analytics within the healthcare domain that demonstrates its potential, usefulness and power to discover new insights and answer challenging questions that have eluded clinicians, managers and researchers heretofore. Additionally, new web and cloud platform technologies have made possible this aggregation of massive databases from disparate sources not seen previously (see Murphy and Barry; and Harper and McNair). These data sources include bringing data together from multiple venues/provider organizations (post-acute care, ambulatory clinics, Emergency Departments, and acute care); claims data from insurance and government payers, and from numerous large healthcare provider organizations using different electronic health record systems.

Included in Part IV are three nurse-led initiatives from different large health systems. These systems include: the Veterans Administration; Kaiser Permanente; and Cerner Corporation in partnership with insurers, government payers, national laboratory centers, and multi-facility, client organizations. Weir and colleagues present their work on the question of delirium risk and detection using the Veterans Health Administration's massive VINCI database. In addition to reporting on different methodologies used, their focus includes detailing the full team that does the statistical analysis and has the expertise needed for doing these types of analyses as an important part of the story that needs to be understood by those seeking to be ready to do big data analytics within their own settings.

Soriano, O'Brien and Chow from Kaiser Permanente (KP) also focus on delirium in their description of a nurse-defined research initiative for early identification and more effective treatment of delirium patients across their healthcare system. This multi-year initiative started by querying the KP database of over 5 million lives and 20 years of data. The KP story offers up an excellent example for other nurse executives within large healthcare organizations on how to use data analytics to address complex care issues.

Harper and McNair offer a thorough discussion on the value of using analytics methods to merge data from multiple sources to answer questions that heretofore have not been available to nursing. They explore partnerships between EHR vendors and their healthcare organization clients in this journey, as well as the nurse informatician's roles in ensuring the success of the partnership endeavor. Four case studies are presented that illustrate the breadth and creativity of specific healthcare organization's initial forays into big data analytics: Smith and Snavely tackle reducing alarm fatigue in a community-based health system; Giard and Sutton capture actual keyboard times to provide a machine-generated measure on nurse documentation times; Jenkins describes an innovative methodology for deriving patient-specific, direct-nursing cost metrics; and, Bober and Harper describe Advocate Health Care's undertaking to develop a tool that scores a patient's readmission risk.

Murphy and Barry offer an applied look at how analytics behind the scene will soon be the new norm for point of care. In an exploration of the combined powers of web-based technologies and analytics housed in new mobile applications, they list the ways mobile apps will enable precision nursing at point of care and support team care across the care continuum.

Part IV chapters and case studies illustrate how the early pioneers are building cross- discipline teams and industry partnerships to bring the resources, "know how", technology and data sources together to make their "big data analytics" possible. These are valuable exemplars and stories. These current day initiatives speak to opportunities that nurse leaders, healthcare systems, payers and research centers can reference as they start their own analytics journey in pursuit of optimum quality outcomes, minimizing waste and error, and promoting safe and enjoyable workplace environments.

Chapter 15
Big Data Analytics Using the VA's 'VINCI' Database to Look at Delirium

Charlene Weir, Joanne LaFluer, Bryan Gibson, and Qing Zeng

Abstract Acute Mental Status Change (AMSC) or delirium is a fluctuating disturbance of cognition and/or consciousness with reduced ability to focus, sustain, or
shift attention and is associated with poor patient outcomes. Managing these patients
is a central nursing concern because patients with AMSC involve significant nursing
resources and increased monitoring. Big Data can be used to develop cognitive support interventions at all levels, ranging from individual nurses to the team to
institutional-level stewardship. However, developing and designing this support
requires transformation of big data resources to match the information needs at

Funding acknowledgement: Veterans Health Administration Health Services Research &
Development: # CRE 12-321.

C. Weir, Ph.D. (✉)
VA IDEAS Center of Innovation, Salt Lake City, UT, USA

Department of Biomedical Informatics, University of Utah School of Medicine,
Salt Lake City, UT, USA
e-mail: charlene.weir@utah.edu

J. LaFluer, Ph.D.
VA IDEAS Center of Innovation, Salt Lake City, UT, USA

School of Pharmacy, University of Utah, Salt Lake City, UT, USA

B. Gibson, D.P.T., Ph.D.
Department of Biomedical Informatics, University of Utah School of Medicine,
Salt Lake City, UT, USA

School of Pharmacy, University of Utah, Salt Lake City, UT, USA

George Washington University, Washington, DC, USA

Q. Zeng, Ph.D.
VA IDEAS Center of Innovation, Salt Lake City, UT, USA

George Washington University, Washington, DC, USA

287

these different levels of analysis. This chapter illustrates how the Veterans Health Administration's, large data repositories are being used to address the goals of improved classification of patients with delirium: by combining text data with ICD9 codes (International Classification of Diseases); creating algorithms that can successfully identify patients at risk; by examining patterns of real patients; building useful alerts and supplying the metrics needed to create and maintain a stewardship program.

Keywords Decision support • Nursing • Delirium • Big data analytics • Nursing informatics • Analytic methods

15.1   Introduction

Researchers excited by the concept of Big Data are often confused as to its defining characteristics. Is Big Data research defined by the <u>size and complexity of the data</u> (millions of rows and thousands of sources)? Or is it the <u>kind of question</u> being asked (e.g. population-based analytics)? Or is Big Data distinguishable by the <u>methods used</u> to establish scientific validity (e.g. machine learning). In this chapter we will make the case that Big Data research is all three. We will summarize work conducted in the Veterans Health Administration (VHA) using its large databases with a diverse set of methods and a variety of questions focusing on a specific clinical condition, delirium. We have three sub-studies that together comprise an overall research program that uses Big Data to: (1) identify patients at risk for delirium; (2) improve classification of this often poorly measured clinical condition; and (3), design a stewardship program to help monitor incidence and quality.

15.1.1   The Problem with Delirium

Delirium or Acute Mental Status Changes (AMSC) is defined as a fluctuating disturbance of cognition (memory, language, orientation) and/or consciousness with reduced ability to focus, sustain, or shift attention (APA 2013). Delirium may impact 14–56% of all hospitalized elderly patients (Inouye 1998, 2006), and is associated generally with poor patient outcomes, including: increased length of stay, discharge to a nursing home, increased likelihood of falls and accidents and increased likelihood of dementia.

Several factors make delirium a difficult clinical problem. The first is the significant burden that confused patients have on nursing staff because they require more monitoring and increased staffing (Carr 2013). The second issue is that delirium often goes undetected, and even when detected is not often documented accurately. Rates of delirium may be as high as 32–67% on general medical units, 65% in emergency units (Elie et al. 2000) and 86% in nursing homes (Inouye et al. 1999; Agarwal et al. 2010). However, across 22 studies, Steis and Fick found that nurses' recognition of delirium symptoms ranged from only 26–83% (Steis and Fick 2008). Additionally, documentation of signs of delirium often remain unaddressed. In prior work by the authors of this chapter, ICD9 codes only agreed with chart review in 30% of the cases where clinicians noted in text that the patient was confused or delirious (Spuhl et al. 2014).

Another difficult aspect of delirium is that effective prevention and treatment requires sharing of information across the healthcare team. Achieving a common awareness is very difficult in an age where information overload from electronic medical records is ubiquitous. Clinicians often do not read each other's notes and the electronic medical record requires substantial time to sift and sort through (Carnes et al. 2003; Holroyd-Leduc et al. 2010; Khan et al. 2012). However, in studies showing improved outcomes from delirium programs, improvements appear to be associated with a multidisciplinary approach, including enhanced communication and shared awareness of goals across the healthcare team (Inouye et al. 1999; Chen et al. 2011; O'Hanlon et al. 2014; Pitkala et al. 2006).

In summary, delirium is a concept that is vaguely referenced, inconsistently documented and incompletely communicated. We hypothesize that developing cognitive support systems for users of the electronic health record (EHR) can ameliorate these problems. In order to develop such cognitive support systems, we are using statistical tools that uncover patterns identified by analyzing thousands of records and integrate text and structured data and generate algorithms based on these analyses, to improve the classification of mental status, specify risk factors in the population of interest and to improve the shared display of information.

15.1.2 Big Data Can Help

Big Data approaches can help in the care of delirium in three unique and important ways. First, by using knowledge extracted from analysis of thousands of narrative records, we can create algorithms that classify patients with acute mental status changes more accurately than ICD9 codes alone. Most of the information about mental status is embedded in descriptive narrative notes and by creating rules for identification and discrimination of mental status concepts, we can improve identification of patients who have had episodes of AMSC. In other words, we improve the diagnostic accuracy in the medical record. Creating those rules requires thousands of real patient records to identify complex patterns.

Secondly, once we have improved our classification, we can then identify patients who are at risk in order to change provider behavior. The algorithms needed to create alerts that assess risk require accurate classification. These algorithms also require an alert design that has a high positive-predictive value, meaning that they are correct most of the time and are not providing too many false positives, thereby avoiding alert fatigue (Shojania et al. 2010). Delirium researchers have demonstrated a significant preventative impact on the incidence of delirium due to interventions initiated by nurses as well as other care team members as well (Inouye et al. 1999). Thus, identifying an effective alert targeted at nurses has the most potential of providing the benefit to improving care.

Finally, Big Data analytics can help with developing and maintaining stewardship programs. Stewardship refers to institutional level programs that use analytics to continuously monitor the appropriateness of care, patient outcomes and expenditures of resources in order to maximize value. In the case of acute mental status changes, the data required to monitor the cost/benefit of ACMS patients' care includes: the numbers of patients identified at risk; the number of at-risk patients where preventive actions were taken; the resources used to increase monitoring; and, relevant patient outcomes—such as falls, length of stay and inappropriate medication use. Any proposed stewardship program must use Big Data to be effective because of the need to analysis complicated multivariate relationships. In other words, a complicated analysis of resource management related to mental status changes (staffing, sitters and room assignment), requires testing in a Big Data environment.

15.1.3   VHA Data Resources

The Veteran Health Administration is the United States' largest, integrated healthcare system, serving 8.3 million Veterans annually in 152 medical centers and 1400 community-based outpatient clinics and community living centers (http://www.va.gov/health/findcare.asp).

Veterans Health Administration's VINCI (Veterans Informatics and Computing Infrastructure) is a research-led, initiative supported by the VA (Veterans Administration) Health Services and Research Development (HSR&D) office and supported by the VA Office of Informatics and Technology. It provides researchers with an integrated national database and analytic tools organized into a secure workspace. VINCI holds records for more than 20 million unique veterans, including more than 2 billion clinical documents from providers, 1.6 billion diagnostic codes (ICD9), data on 1.7 billion pharmacy prescriptions, and more than 6 billion lab tests (both orders and results). Vinci also includes over 50 different commercial, open source, government, and custom software programs for software development, collaboration, data processing and data analysis. The available applications in VINCI include MatLab, Nvivo, R, SAS, Stata, SPSS, and R (https://www.hsrd.research.va.gov/for_researchers/vinci/).

The Corporate Data Warehouse (CDW) program is an additional source for institutional-level data. The CDW provides site-specific clinical data infrastructure for ongoing clinical care for hospitals and clinics, as well as the four Regional Data Warehouses (RDW1-4). The goal of the CDW is to improve business management decision-making, clinical decision-making, quality and patient safety through monitoring and tracking, as well as providing the information needed to lead population-based stewardship.

15.1.4 Case Study 1: Identifying Patients at Risk for Delirium

The first step in designing cognitive support for nurses is to create models that identify patients at risk for delirium that are useful and relevant. Creating alerts that identify patients at risk must have sufficiently high, positive-predictive value (PPV) that clinicians will not ignore them (i.e. if every patient triggers the alert then it is a non-informative nuisance). The approach we used was to build risk models on the VHA's VINCI data that predict a nurse-relevant outcome, specifically, patients with delirium who required a sitter. Sitters are expensive in an inpatient setting and utilize substantial nursing resources (Carr 2013). Creating a model highly predictive of a sitter requirement would likely enhance the PPV of an alert and maximize informativeness.

Designing algorithms to support alert design requires careful consideration of how these alerts would be implemented. Besides creating alerts that have high predictive value, we also needed to have alerts fire on a daily basis, because mental status can change rapidly. Hence, we aimed to build predictive models for *each* hospital day in order to integrate these alerts with daily staffing decisions. Nursing staff levels vary daily and staffing models would require a daily estimate of patients at risk of acute mental status changes as it is one of the main drivers of acuity formulas for nursing staffing requirements.

15.1.4.1 Methods

We developed a predictive model for patients likely to have delirium and to require a sitter using a national cohort of Veterans ages 40+ who were hospitalized in Veterans Health Administration hospitals from October 2011 through September 2013. The model was developed by randomly sampling 50,000 patients from the full patient cohort during that time period and fitting a Cox Proportional Hazards Regression model in each sample using backward selection. Candidate risk factors that were considered included characteristics with an empirically demonstrated association with inpatient delirium found in published studies (Inouye 1998, 2006; McCusker et al. 2002; Siddiqi et al. 2006). Among that list we identified those variables available in the CDW data resources (both text and structured data) that were associated with inpatient delirium. Multiple analyses were conducted, including development on test and validation cohorts.

15.1.4.2 Results

We found that out of 700,180 Veterans' hospitalizations, 9806 (1.4%) had an order for sitters or restraints. Final risk factors are summarized in Table 15.1 for Day 2. Accuracy (C statistic) for the risk rule tool was maximized at a threshold of 33

Table 15.1 Final predictive variables for daily sitter use

Variable	Category with highest index weight
Age	80+
Benzodiazepines	Exposure
Bicarbonate	Below normal
Carbon dioxide	Above normal
Corticosteroids	None
Creatinine	Above normal
Lactate	Above normal
Neutrophils	Above normal
Oxygen order	Yes
Sedative hypnotics	Yes
WBC	Above normal

(C = 73.1%). At that threshold, specificity was 81.2% and sensitivity was 65.1%, but positive predictive value (PPV) was low (7.35%). PPV was increased to above 35–40% at higher cut-points where specificity was maximized (i.e., scores ≥175). At a score of 175, 0.4% of all patients would test positive for delirium risk.

Notice that the list does not include some well-known predictors, such as dementia or prior delirium as these washed out as predictors on a daily basis. Current work is being conducted to validate the model on a gold standard mental status assessment of patients in the hospital to extend the clinical usefulness of a risk predictor.

15.1.4.3 Conclusion

Using the big data resources of VINCI supported the design and testing of a nurse-specific risk predictor calibrated daily and based on nursing resource use. Although testing this alert in a real setting has not yet been done, the fact that the model is based on data that is readily available in the VA and linked to the electronic medical record greatly enhances the translation of big data analytics to daily clinical care decisions.

15.1.5 Case Study 2: Improving Classification Using Natural Language Processing

The second phase of our overall study is to examine the incidence of references in clinical documentation as text data on acute mental status changes. Big data and natural language processing (NLP) techniques are of particular significance to the study of mental health problems using EHR documentation. The description in text of mental health symptoms and treatment responses have a much higher variation than the description of physical symptoms and treatment responses and are typically

<u>only</u> documented using free text notes. The current accuracy for structured data extraction of delirium incidence is about 30%. We aimed to improve accuracy by integrating text information with structured data. Most mental health symptoms are self-reported by patients and thus often documented using patient language. For instance, depression as a symptom (in contrast to a diagnosis) may be described as "feeling down, feeling blue, depressed, low spirit, heavy hearted, in a funk, despair, etc.". To capture this variation and to accurately classify patients with delirium, we needed to apply NLP techniques to large datasets of electronic progress notes.

15.1.5.1   Method

To identify descriptions relevant to delirium from the clinical notes, the simplest method is to perform a keyword search using terms such as "delirium" or "delirious." One can expect this method to have high precision (the patients retrieved have a high likelihood of having delirium) but low recall (many patients with delirium are missed) because delirium is often described by various complex symptoms, behavioral descriptions and informal language that varies across provider roles and patient situations (e.g. "the patient is agitated and is a poor historian"). Despite the limitations of this method, we used it as our "first pass" approach.

To improve classification, we experimented with *topic modeling* (Blei 2012). Topic modeling is a machine learning approach, meaning a form of artificial intelligence where the computer learns the relationships in text without being explicitly programmed. Often the computer is given a few examples and then engages in an iterative process of finding understandable patterns. The Topic Modeling method is particularly useful for clinical conditions that have vague referents in text and for which it is not possible to write strict rules. The central component of the thematic structure is referred to as "*topics*". The basic assumption of topic modeling is that documents are mixtures of topics and a topic is represented by a sequence of words that are semantically close to each other. Blei has an easy to understand description of topic modeling with vivid explanations that help the lay person understand this methodology and how to apply it in clinical documentation research (Blei 2012).

Two types of topic modeling methods were used in our experiments. The first is an unsupervised machine learning method called *Latent Dirichlet Allocation (LDA)* (Blei et al. 2003). It is one of the most popular topic modeling methods and has been studied extensively in biomedical research studies (Arnold et al. 2016; Huang et al. 2015). The second type modifies LDA and is developed by Dr. Qing Zang's group in the VA that leverages the International Classification of Diseases (ICD) codes associated with clinical notes.

Although ICD codes are known to have low sensitivity, we can only use the ICD9 codes to "jumpstart" the topic modeling approach. The modified LDA method can be considered an example of "weakly supervised" learning. Supervision refers to a method of starting the process of text identification with an accepted referent. This method has two variants, both of which were tested in this study. The first ICD method, which we call ICD-1 method, generates topics directly based on the ICD codes assigned to the documents. The second method, which we call ICD-2 method, refines the basic ICD topics using a *topic coherence* measure.

To create a reference standard two nurse experts annotated the delirium-related text in 300 documents. In addition, the second expert rated the documents based on the presence/absence of delirium (1- certainly not there to 7- certainly present) and the importance of delirium to the overall purpose of the document.

We trained these three different topic modeling methods using a Pittsburgh EHR Dataset containing 100,866 clinical documents (TREC 2011). Each document consisted of three parts: admission ICD code, discharge ICD codes (multiple) and the text. That means that we programmed the computer to use a starter set of terms and then go through the text in multiple iterations finding a stable pattern set of associated terms (clusters of related terms).

15.1.5.2   Results

We evaluated the three topic modeling methods together with the keyword search method in two ways: (1) document rank correlation and (2) sentence relevance determination. The selection of these evaluation tasks reflects the needs of our clinical use case. To provide point-of-care, decision support, we needed to identify both the documents and the specific text segments relevant to delirium. The keyword search query we used throughout the evaluation process was a single word "delirium."

As shown in Table 15.2, the unsupervised and weakly supervised topic modeling methods were able to identify delirium with similar accuracy as a human expert.

As expected, the keyword search had the best precision but the worst recall. The ICD-1 performed worst on document rank correlation, precision and F-score. This may be due to the fact that the original assignment of the ICD codes corresponding to delirium in the clinical notes had very poor accuracy. Since ICD-1 generates topics directly from the ICD code assignments, its topic quality on delirium topics was also affected. ICD-2 performed the best in document rank correlation, recall and F-score. The results are illustrated in Table 15.3.

Table 15.2 Examples of topics relevant to delirium

Method	Topic terms
LDA	Delirium isperdal agitation agitated mg confused delirious risperdal iv risperdal time confusion restraints dose mental status questions risperidone multifactorial dementia unable
ICD-1	Alcohol withdrawal delirium thiamine tremens withdrawal dts alcoholic alcohol intoxication burst fracture hepatic_encephalopathy free_flap chronic_alcoholism mandibulectomy dt sas chronic_alcohol_abuse alcohol_level benzodiazepines alcoholism nonoperative
ICD-2	Delirium haldolisperelirious confused agitated confusion multifactorial isperdal restraints answers questions, antipsychotics

Table 15.3 Comparison of sentence relevance determination between machine and human

	Document rank correlation	Sentence relevance determination		
		Precision	Recall	F-score
LDA	0.268	0.455	0.481	0.468
ICD-1	0.144	0.383	0.470	0.422
ICD-2	**0.446**	0.757	**0.612**	**0.677**
Keyword search	0.388	**0.984**	0.285	0.442
Second expert		0.600	0.554	0.576

15.1.5.3 Conclusion

Building an accurate and generalizable method to capture data from text requires large numbers of documents and methods tailored to a specific domain. In this example, we demonstrated that even though we started with structured data that is known to be imprecise, we were able to create a model using Big Data that supports improved classification of patients with ACMS. This work requires a team of people who have the necessary skills—clinicians, informaticists, computer scientists, and data science experts. The clinicians must add their clinical knowledge to the work in order to interpret NLP and machine learning programs, the computer programmer systems expert is required to access and organize the large data files and the data science statistician is needed in order to run the machine learning programs.

Currently, we are taking the next step and validating this model on a small subset of manual mental status assessments by clinical experts (a referent standard). The results will further refine the algorithm. The final goal will be a model that can identify text noting mental status changes in real clinical time in order to share that information in a display for all members of the team. The result (according to our hypothesis) is improvement in shared situation awareness of the patient's status across roles without requiring extensive searching in the medical chart.

15.1.6 Case Study 3: Building a Stewardship Program

The necessary data infrastructure and analytics required to institute a stewardship program that activates the twin goals of quality of care and resource use for patients with AMSC is largely missing in most institutions. Transforming big data into usable "online" information at the institutional level for active monitoring requires careful calibration to match data displays, data access and data interpretation to the stewardship tasks.

15.1.6.1 Methods

We assessed stewardship readiness as part of a national VA phone interview to Nurse Managers of Medical/Surgical wards on the topic of ACMS. Fifty-eight Nurse Managers from 58 VA hospitals selected to represent geographic distribution nationally, hospital size and academic affiliation were given a structured interview on the topic of their data practices for mental health information.

15.1.6.2   Results

In general, we found that mental status assessment was limited to orientation for 93% of respondents. Formal tools, such as the Richmond Agitation and Sedation Scale (RASS) and the CAM (Confusion Assessment Method) were rarely used (5%) and only 10% of the 58 VA hospitals participating in the survey reported policies for prevention. The methods for assessing mental status and the timing at which the assessments occurred were viewed as being a matter of nurses' individual clinical judgment. Since nurses are often the major recorders of mental and emotional status of patients, this variability limits the final usefulness of big data analytics. Tracking ACMS at the ward or institutional level were rare (5%) and none of the 58 VA hospitals participating reported a stewardship program that analyzed the incidence of risk and prevalence, the cost of interventions (e.g. sitters) with the rates of outcomes (e.g. falls and length of stay), and interventions.

15.1.6.3   Conclusion

Although there is a general consensus regarding the importance of assessing patients at risk, monitoring the incidence of delirium and tracking preventive activities for ACMS, we found that few VA hospitals have this program, and we suspect this reflects the current state across most healthcare systems A central thesis of this chapter is that the necessary data has not been made available in a form that is usable and tractable at the institutional level. Big data analytics will have this potential, but the key is to match the analytic products to the clinical information requirements.

15.2   Overall Discussion

These three studies demonstrate the complexity of using Big Data to transform care at the bedside. Key issues for making this transformation include: (1) the quality of the data itself; (2) matching data use and algorithm development to the clinical context in which it will be used (individual nurse level versus institutional stewardship level) and (3) integrating the patient story into Big Data analysis.

15.2.1   Quality of Data

The poor quality of structured data sources for mental status attributes is not unique (Cappiello et al. 2004). High quality data is foundational for valid outcomes in big data analyses and many researchers working in this domain have had to establish procedures and valid methods for dealing with inaccurate and missing data (Ca and Zhu 2015).

The sources of bad quality data are numerous, but in the case of mental status, a significant cause may be the failure to precisely document patient's mental and emotional status clearly in order to effectively communicate and classify patients. And while much of the methods detail presented here are technical in nature, the take away for nurse leaders, clinicians and nurse informaticists is that the data is no better than the documentation practices. In today's clinical settings, we tend not to document assessments specific interventions in the EHR. Nurses need better cognitive support tools and integrated standard protocols to evoke best practice interventions and associated documentations.

15.2.2 Matching Data Analytics to the Question and Producing Actionable Information

The overall goal of this work is to develop and implement interventions to provide improved cognitive support and stewardship analytics to nurses in the area of mental status, and particularly delirium. The first specific question nurses and doctors have with the presence of confusion or agitation is how likely is the patient is to have ACMS—is this a change from baseline? The second question relates to whether the delirium is caused by a modifiable source. Big data analysis could help in making the more fine-grained analysis regarding the distribution of modifiable causes across different populations.

Identifying patterns in big data related to mental risk and prevalence could have direct implication for nurses given the current lack of specificity in language and documentation, especially in terms of the daily planning and organization of nursing staffing. Confusion, wandering, and behavioral outbreaks present significant risk for the patient and increased workload for nurses. Interviewees in this study often expressed concern over how the patient's mental status impacted staffing and assignment decisions. Research indicates that when appropriate levels of staffing of registered nurses on inpatient units is achieved, patients are safer and have better outcomes (Aiken et al. 2002a). Nurses need precise and measurable information about mental status in order to inform decisions at the shift and institution level. The tension between costs and staffing requirements may lead to heavier workloads on nurses raising the issue of patient safety (Aiken et al. 2000, 2002b). Comparative decisions require accurate and consistent measurement of a targeted patient group over time.

15.2.3 Integrating the Patient's Story

Finally, natural language processing and other machine learning techniques have the potential to improve the integration of the patient's "story" (e.g. their context, emotional and social characteristics) with physiological data. Accurate understanding of the patient's mental status requires more contextual information—how much more depends on the question and is not yet known. However, exploring questions such as the relationship between social, emotional and changes in mental status will not only improve classification, but bring more robust methods for understanding risk and evaluating interventions.

15.2.4   Overall Conclusion

This chapter describes how leveraging big data resources could have a transformative affect on nursing practice for the care of patients with AMSC in three important ways: classification, alerts and stewardship programs. First, researchers could use big data resources to develop and test algorithms that improve identification of patients with AMSC or delirium by integrating structured data with narrative text. Because most references to a patient's mental status are recorded in text, using topic modeling with natural language processing based loosely on ICD9 codes greatly enhances classification. Secondly, once classification has been improved, efforts to develop an actionable alert risk for AMSC (e.g. AMSC due to a modifiable cause) could have sufficient positive predictive value to be useful to nurses. Finally, once data models have been created based on real patient conditions, online and continuous monitors can be developed to inform program evaluation, resource use and quality of care. The three cases (identifying risk, classifying correctly and building a stewardship program) presented here, illustrate the difficulty and complexity of developing big data analytic models that can be useful in clinical practice. But they also show the promise of using these new analytic tools and evolving their application in EHR environments to provide needed cognitive and decision support to our front line nurses with the benefits of safer and more effective care, improved patient outcomes and reduced costs to healthcare systems.

References

Agarwal V, O'Neill PJ, Cotton BA, Pun BT, Haney S, Thompson J, et al. Prevalence and risk factors for development of delirium in burn intensive care unit patients. JAMIA. 2010;31(5):706–15.

Aiken LH, Clarke SP, Sloane DM. Hospital restructuring: does it adversely affect care and outcomes? J Nurs Adm. 2000;30(10):457–65.

Aiken LH, Clarke SP, Sloane DM, Sochalski J, Silber JH. Hospital nurse staffing and patient mortality, nurse burnout, and job dissatisfaction. JAMA. 2002a;288(16):1987–93.

Aiken LH, Clarke SP, Sloane DM. Hospital staffing, organization, and quality of care: cross-national findings. Int J Qual Health Care. 2002b;14(1):5–14.

APA. Diagnostic and statistical manual of mental disorders. Arlington, VA: American Psychiatric Association Publishing; 2013.

Arnold C, Oh A, Chen S, Speier W. Evaluating topic model interpretability from a primary care physician perspective. Comput Methods Prog Biomed. 2016;124:67–75.

Blei DM, Ng Y, Jordan MI. Latent Dirichlet Allocation. J Mach Learning Res. 2003;3:993–1022.

Blei D. Probabilistic topic models. Commun ACM. 2012;55(4):77–84.

Ca L, Zhu Y. The challenges of data quality and data quality assessment in the big data era. Data Sci J. 2015;14(2):1–10.

Cappiello C, Francalanci C, Pernici B. Data quality assessment from user's perspective. New York; 2004.

Carnes M, Howell T, Rosenberg M, Francis J, Hildebrand C, Knuppel J. Physicians vary in approaches to the clinical management of delirium. J Am Geriatr Soc. 2003;51(2):234–9.

Carr F. The role of sitters in delirium: an update. Can Geriatr J. 2013;16(1):22–36.

Chen CC-H, Lin M-T, Tien Y-W, Yen C-J, Huang G-H, Inouye SK. Modified hospital elder life program: effects on abdominal surgery patients. J Am Coll Surg. 2011;213(2):245–52.

Elie M, Rousseau F, Cole M, Primeau F, McCusker J, Bellavance F. Prevalence and detection of delirium in elderly emergency department patients. CMAJ. 2000;163(8):977–81.

Holroyd-Leduc J, Abelseth G, Khandwala F, Silvius J, Hogan D, Schmaltz H, et al. A pragmatic study exploring the prevention of delirium among hospitalized older hip fracture patients: applying evidence to routine clinical practice using clinical decision support. Implement Sci. 2010;5(1):81.

Huang Z, Dong W, Duan H. A probabilistic topic model for clinical risk stratification from electronic health records. J Biomed Inform. 2015;58:28–36.

Inouye S. Delirium in hospitalized older patients: recognition and risk factors. J Geriatr Psychiatry Neurol. 1998;11:118–25. [PubMed]

Inouye S. Delirium in older patients. N Engl J Med. 2006;354:s1157–65.

Inouye SK, Bogardus ST Jr, Charpentier PA, Leo-Summers L, Acampora D, Holford TR, et al. A multicomponent intervention to prevent delirium in hospitalized older patients. N Engl J Med. 1999;340(9):669–76.

Khan BA, Zawahiri M, Campbell NL, Fox GC, Weinstein EJ, Nazir A, et al. Delirium in hospitalized patients: implications of current evidence on clinical practice and future avenues for research—a systematic evidence review. J Hosp Med. 2012;7(7):580–9.

McCusker J, Cole M, Abrahamowicz M, Primeau F, Belzile E. Delirium predicts 12-month mortality. Arch Intern Med. 2002;162(4):457–63.

O'Hanlon S, O'Regan N, MacLullich A, Cullen W, Dunne C, Exton C, et al. Improving delirium care through early intervention: from bench to bedside to boardroom. J Neurol Neurosurg Psychiatry. 2014;85(2):207–13.

Pitkala KH, Laurila JV, Strandberg TE, Tilvis RS. Multicomponent geriatric intervention for elderly inpatients with delirium: a randomized, controlled trial. J Gerontol A Biol Sci Med Sci. 2006;61(2):176–81.

Shojania K, Jennings A, Mayhe WA, Ramsay C, Eccles M, Grimshaw J. Effect of point-of-care computer reminders on physician behaviour: a systematic review. CMAJ. 2010;182:E216–E25.

Siddiqi N, House A, Holmes J. Occurrence and outcome of delirium in medical in-patients: a systematic literature review. Age Ageing. 2006;35(4):350–64.

Spuhl J, Doing-Harris K, Nelson S, Estrada N, Del Fiol G, Weir C. Concordance of Electronic Health Record (EHR) data describing delirium at a VA Hospital. AIMA 2014 Symposium; 2014.

Steis M, Fick D. Are nurses recognizing delirium? A systematic review. J Geronol Nurs. 2008;34(9):40–8.

TREC. http://www-nlpir.nist.gov/projects/trecmed/2011/.

Chapter 16
Leveraging the Power of Interprofessional EHR Data to Prevent Delirium: The Kaiser Permanente Story

Rayne Soriano, Marilyn Chow, and Ann O'Brien

Abstract With the emergence of big data in healthcare, there are multiple opportunities to improve quality, safety, and affordability through leveraging robust amounts of patient data and the application of evidence-based practice and research. Current literature on Electronic Health Record (EHR) systems has focused on the data entry burden for clinicians; however, little is known about how this data is made actionable for interprofessional teams to leverage the information and work together to prevent adverse patient outcomes and intervene at the point of care. This chapter describes how pharmacists, physicians, nurses, data analysts and informatics teams at Kaiser Permanente worked together to overcome the challenges of data silos and leverage real-time knowledge from the EHR to prevent delirium in the hospital. This collaboration resulted in the design and use of a delirium risk score and care protocols by the care team. Our story includes the journey in using multiple sources of data in the EHR and leveraging each discipline's expertise to become a coordinated, knowledge-worker team.

Keywords Nursing informatics • Interdisciplinary team care • Information sharing Big data • Analytics • Electronic health record • Delirium avoidance in acute care Quality and safety improvements

16.1 Introducing the Delirium Picture

Gladys, a 75-year-old woman was admitted to the hospital after falling at home and with symptoms of hypotension and shortness of breath. She was admitted to the ICU for 3 weeks for pneumonia and respiratory failure. She was intubated for 10 days and was placed on a Propofol and Ativan drip. Prior to her hospitalization, Gladys lived alone and was active in her community. She also walked about 5 miles a day. Today she has been transferred to your medical surgical unit. When you approach

R. Soriano, Ph.D., R.N. (✉) • M. Chow, R.N., Ph.D. • A. O'Brien, R.N., M.S.N.
Kaiser Permanente, Oakland, CA, USA
e-mail: Rayne.Q.Soriano@kp.org

© Springer International Publishing AG 2017
C.W. Delaney et al. (eds.), *Big Data-Enabled Nursing*, Health Informatics,
DOI 10.1007/978-3-319-53300-1_16

Gladys, she does not move much. When you talk to her she is somewhat confused. Her eyes are squeezed shut and she does not open them when you speak to her. When you begin your evaluation and are about to reposition her in bed, she groans. She also seems to be calling on God between groans, but she is not speaking clearly. When you try to move her, she clenches her fists. After leaving a message to the attending physician regarding Gladys' condition, the physician orders more Ativan for agitation. Pharmacy verifies the medication and you give Gladys her medication. Unfortunately, as you were attending to other patients, Gladys tries to get up from bed to use the bathroom and falls. Because of this event and her slowly deteriorating condition, Gladys' hospital stay is further prolonged and eventually she is discharged to a home care facility. Despite members of Gladys' care team doing their part to intervene based on her symptoms, Gladys' delirium went unrecognized leading to a poor outcome in her care.

16.2 Introduction

In Gladys' case and many other patients admitted in the hospital, all members of the care team are using robust EHR systems across the continuum of care—including inpatient, ambulatory, home, and virtual settings, yet we still find opportunities in the industry to improve care coordination to deliver effective, safe, and timely care. How might we optimize team-based care in all settings to be more patient centric through wise clinical decision support and improved data sharing?

The Health Information Technology for Economic and Clinical Health (HITECH) and Meaningful Use incentive program has stimulated the growth of EHR adoption over the past few years (Centers for Medicare and Medicaid Services 2013). The U.S. Department of Health and Human Services (2013) defined the EHR as real-time, patient-centered records that make information available instantly, bringing together everything about a patient's health in one place, however, the literature suggests that EHRs as a source of information may be underutilized, with reports of structural and system challenges (Agno and Guo 2013; Ash and Bates 2005; Berner and Moss 2005; DesRoches et al. 2008; Grabenbauer et al. 2011; Kossman 2006; Kossman and Scheidenhelm 2008; Moody et al. 2004; Rogers et al. 2013). Additionally, the literature points out that process workarounds have led to increased complexity and challenges for providers in accessing information due to the persistence of paper documentation and non-integrated systems (Ash and Bates 2005; Keenan et al. 2013). Beyond the implementation of EHR systems, the Meaningful Use initiative was developed as an incentive program to assure that EHRs are used according to standards that achieve quality, safety, and efficiency measures (Centers for Medicare and Medicaid Services 2013). Unfortunately, the focus has been on the implementation of EHRs and each clinical domain has built tools that support their respective workflows, regulatory requirements and data needs. When data is not linked to real-time knowledge and the context of how data fit together for patient-centric care, the potential value of EHR systems is often not experienced by front-line clinicians.

As organizations strive to meet the Triple Aim with the utilization of EHRs and health information technology (Berwick et al. 2008) there is an urgent need to move beyond the current siloed approach of entering and reviewing data that is common in the industry to one of integrating relevant and contextual information and dedicating resources towards inter-professional collaboration. This chapter describes an interdisciplinary approach to leverage clinical and informatics professionals in utilizing EHR data to design a tool and a workflow to identify and prevent delirium at Kaiser Permanente.

16.3 The Impact of Delirium

Delirium is sudden severe confusion due to rapid changes in brain function that occur with physical or mental illness (American Delirium Society 2015). Delirium is most often caused by physical or mental illness, and is usually temporary and reversible. Many disorders cause delirium. Often, the conditions are ones that do not allow the brain to get oxygen or other substances (Smith and Seirafi 2013). Unrecognized delirium may lead to negative outcomes, such as increased morbidity, longer hospital stays, and increased health care costs.

According to the American Delirium Society: Among hospitalized patients who survived their delirium episode, the rates of persistent delirium at discharge are 45% at one month, 33% at three months 26% and at six months 21%.

- Compared to hospitalized patients with no delirium (after adjusting for age, gender, race, and comorbidity), delirious patients have:

  - Higher mortality rates at one month (14% vs. 5%), at six months (22% vs. 11%), and 23 months (38% vs. 28%)
  - Longer hospital lengths of stay (21 vs. 9 days)
  - A higher probability of receiving care in long-term care setting at discharge (47% vs. 18%), six months (43% vs. 8%), and at 15 months (33% vs. 11%)
  - A higher probability of developing dementia at 48 months (63% vs. 8%)

- Seven million hospitalized Americans suffer from delirium each year or approximately 20% of inpatients have delirium at any one time (Ryan et al. 2013).
- Delirium is misdiagnosed, detected late or missed in over 50% of cases across healthcare settings (Kean and Ryan 2008).

In summary, delirium is a relatively frequent, challenging and debilitating problem in hospitalized patients resulting in significant health, social and financial cost to patients, families and health care organizations and this is why Kaiser Permanente targeted the falls and mortality in this high-risk population.

In 2013, Kaiser Permanente convened an interdisciplinary team to improve the accuracy of identification of inpatient delirium and develop a comprehensive model for prevention. The analytics team from Kaiser's Utility of Care Data Analysis (UCDA) team collaborated with clinical informaticians and data scientists to extract and analyze the EHR records associated with delirium to identify factors placing patients at highest risk.

As part of this analysis, the team developed control charts which trended falls outcomes, (since our evidence showed a correlation between delirium and falls in the hospital), restraint use, and length of stay. They also examined the time between fall events for patients and performed a matched comparison between patients with a positive Confusion Assessment Method (CAM) result (indicating the presence of delirium) and a comparable cohort from the prior year. They then used propensity scores to determine the prior year comparison group matching on the CAM positive profile including age, history of delirium, dementia, and delirium occurrence during their stay, surgery during their stay, discharge status, DRG, and DRG weight.

Based on their analysis, those who were at highest risk for developing delirium included:

• Patients who are 75 years and older
• Patients who are 65 years and older and had surgery in the current admission
• Patients with a history of delirium
• Patients with a diagnosis of dementia.

Along with these risk factors, medications can also contribute to delirium (deliriogenic), especially in classes such as opioids, benzodiazepines, and antihistamines (Clegg and Young 2011).

16.4   Discovering the Delirium Story Through Multiple Sources of Information

Studies on the information needs of clinicians have highlighted the various sources and types of information needed to provide care across the continuum. The literature on what information clinicians need for their care delivery reflects that they require many types of data from varying sources in order to effectively care for patients. Studies of nurses, for example, revealed that a range of information sources and activities was appraised and used during practice (Bonner and Lloyd 2011); information access was influenced by the specific situation, the surroundings, and the personal preferences of the nurse (Carter-Templeton 2013); and the process of information seeking for routine and non-routine decisions was different. Nurses making routine decisions relied on information from their experience and an assessment of the patient, whereas participants making non-routine decisions experienced more uncertainty about their decisions (O'Leary and Mhaolrunaigh 2012). Beyond medication information, pharmacists had positive attitudes regarding evidence-based practice (Weng et al. 2013) and were found to need more information about the patient and situation, such as laboratory results and other clinical data (Trinacty et al. 2014). Pharmacists also reported needing pharmacogenomic information during their practice (Romagnoli et al. 2016).

16.5   Accessing Data in the EHR

Clinicians such as physicians use various methods to access information, depending on the source and the need, and on whether routine or non-routine. Physicians optimize their information-seeking process by accessing resources they believe maximize their information gain and aided in their medical reasoning and decision-making process (Kannampallil et al. 2013). Residents predominantly used a patient-based information-seeking strategy in which all relevant information was aggregated for one patient at a time. In contrast, nurse practitioners and physician assistants used a source-based, information-seeking strategy in which similar (or equivalent) information was aggregated for multiple patients at a time (Kannampallil et al. 2014). Pharmacists also used software such as a listserv to find solutions for complex problems and as a source for mentorship (Trinacty et al. 2014).

Challenges to accessing information. Despite having multiple sources of information, clinicians industry-wide are challenged in accessing clinical information from various systems. These challenges include:

- A lack of time (Argyri et al. 2014; Clarke et al. 2013; Cook et al. 2013; Gilmour et al. 2012; Marshall et al. 2011);
- Inaccessible information (Koch et al. 2012; Marshall et al. 2011);
- Unfamiliarity with computers (Argyri et al. 2014);
- Lack of skills for searching information (Argyri et al. 2014; Clarke et al. 2013);
- Effort of entering data (Michel-Verkerke 2012);
- Not enough computer terminals to meet the needs of all members of the interdisciplinary team (Cook et al. 2013; Gilmour et al. 2012);
- The volume of available information; and
- Not knowing which resource to search (Cook et al. 2013).

In light of the benefits and features of EHRs for improving access to information to improve quality, there is an emerging body of literature from hospital settings which reveals that EHR use is problematic for caregivers in large federal and academic health systems, as well as stand-alone hospitals. Structural challenges with accessing information in the EHR include: problems filtering information (Berner and Moss 2005; Grabenbauer et al. 2011); lack of context-sensitive, decision support and issues with the usefulness of the data (Berner and Moss 2005); lack of usable functionality (DesRoches et al. 2008; Rogers et al. 2013); lack of interoperability (Ash and Bates 2005); lack of technical support (Agno and Guo 2013); slow system response, need for multiple screens, overuse of checkboxes and "copy and paste" documentation (Kossman 2006; Kossman and Scheidenhelm 2008); and, various system interface challenges (Moody et al. 2004).

16.6   The KP Discovery Journey

During the exploratory phase of the Kaiser Permanente Delirium initiative, it became apparent that multiple disciplines were involved in the prevention and management of delirium in the hospital, including pharmacists, nurses, and physicians. Along with the numerous professional roles involved in the patient's care, each had their own documentation types, structure and location of data in the EHR.

While physicians entered orders, coded diagnoses and documented clinical notes, pharmacists entered medication interventions, and nurses documented delirium risks scores, assessments, and care plan interventions. Despite the large volumes of delirium data being entered by various care givers, the EHR did not support data transparency in a way that promoted interdiscipline, clinician communication or care coordination. Due to the lack of real-time, contextual information at the point of care, it was challenging to identify the patients at highest risk for developing delirium or all the care team members involved in their care. In Gladys' case, although the nurse entered the positive CAM score in the EHR, the physician was not aware of the patient's positive CAM profile due to a lack of decision support and data transparency between care providers in the EHR, resulting in an MD order for Ativan despite its deliriogenic properties. The pharmacist also verified the medication without ever realizing that the nurse had documented a positive CAM score, indicating the presence of delirium. Although the nurses caring for Gladys entered their assessments, concerns, and interventions in the EHR, there were no prompts to update the delirium assessment nor were they presented any decision support regarding deliriogenic medications. Although each discipline could document and see their part in delirium prevention and management, there was not a shared place that presented one view of the patient's story and how all team members contributed to the shared goal.

In the case of accessing delirium information, caregivers were limited to their own views of the EHR. Physicians accessed patient histories and other notes and were relegated to waiting for a phone call from nurses or pharmacists when a patient began exhibiting signs of delirium. Meanwhile, as nurses were entering information in the EHR, they had no way of knowing at any given time which patient(s) were at the highest risk for developing delirium. Nurses do not routinely have access to other clinicians' data when developing the patient's plan of care. Pharmacists shared a similar frustration in that they only saw screens with medication orders, but had no way of knowing that nurses rated a patient at a high risk of developing delirium. These challenges are not uncommon, and reflect a current state that can be found across the health system industry regardless of size or EHR vendor used (O'Brien et al. 2015).

16.7 Transforming Care with Actionable Information

Benefits motivating the implementation of EHRs include improvements in quality and safety through the transparency of clinical information, more efficient workflows, improvements in clinical communication, and advances in clinical documentation leading to aggregate clinical data and decision-support tools for clinicians (Institute of Medicine 2000, 2003, 2012; Thakkar and Davis 2006; Kutney-Lee and Kelly 2011; Chaudhry et al. 2006; Amer 2013; LaBranche 2011).

The data and information from EHRs are foundational elements for monitoring and improving quality at the point of care. The Institute of Medicine (IOM) (2003) highlighted that more immediate access to computer-based, clinical information, such as laboratory and radiology results, can reduce redundancy and improve quality. The IOM (2003) report further explains that the availability of complete, patient health information at the point of care delivery, together with clinical decision support systems such as those for medication order entry, can prevent many errors and adverse events from occurring.

In 2003, the IOM was tasked to develop criteria for core functionalities that EHRs must possess in order to achieve quality improvement. Among all of the functionalities identified, data and information were called out as foundational elements. The IOM (2003) report also emphasized that the health information and data captured by an EHR system must evolve over time as new knowledge becomes available—both clinical knowledge and knowledge regarding the information needs of different users. In 2012, the IOM reaffirmed the importance of Health Information Technology (HIT) and EHRs by highlighting that, "clinicians expect health IT to support delivery of high-quality care in several ways, including storing comprehensive health data, providing clinical-decision support, facilitating communication, and reducing medical errors. Safely functioning health IT should provide easy entry and retrieval of data, have simple and intuitive displays, and allow data to be easily transferred among health professionals (pp. 3–4)." In order to derive actionable information from the EHR and design solutions to enhance the prioritization of care for patients at highest risk for delirium and improve the transparency of delirium risk information between caregivers, operational and quality leaders at Kaiser Permanente engaged informatics experts as part of this journey.

16.8 An Interdisciplinary Approach to Delirium Prevention

Against the backdrop of different disciplines involved in preventing delirium and having multiple sources of delirium information, our delirium project included pharmacists, physicians, nurses, data analysts, and informatics experts. This interdisciplinary team collaborated to face the challenges of leveraging EHR data to identify patients at highest risk for delirium and to present a holistic patient view to the entire care team. As a starting point, the team learned about each other's workflows, documentation elements, and views in the EHR. This was an important revelation as to the differences of information sources between each discipline.

Developing the interdisciplinary delirium risk score. Based on the risk factors identified for delirium, the project team highlighted elements that were already present in the EHR, such as the patient's age and medical history. They also acknowledged that surgical status, delirium risk scores, neurologic assessments, and medications were all data elements that were entered in the EHR throughout the course of a patient's hospitalization. Since all of the information, including the risk factors, Confusion Assessment Method (CAM) score (Inouye 2003), nursing assessment, and medications are documented in the EHR, the informatics team developed an interdisciplinary, delirium risk score. This risk score combined multiple sources of information to provide all caregivers more rapid and contextual information. All disciplines had access to the same information in the same format to immediately identify the patients at the highest risk for delirium and to intervene accordingly. By leveraging the power of integrated data from multiple care givers, discreet yet disparate data was transformed into actionable information embodied in the interdisciplinary delirium risk score. The next step was to make the interdisciplinary assessment operational so that the care team could immediately see all hospitalized patients prioritized based on the presence of the following risk factors and this information was made visible to all care team members:

1. Highest Risk—Positive CAM assessment AND deliriogenic medication(s) given
2. High Risk—Risk factors present and deliriogenic medication(s) given OR Positive CAM assessment and no deliriogenic medication(s) given
3. Moderate Risk—Risk factors present and no deliriogenic medication(s) given OR no risk factors present and deliriogenic medication(s) given
4. Low Risk—Negative CAM assessment and No risk factors present OR no deliriogenic medication(s) given.

Enhancing clinical decision support. Using the interdisciplinary delirium risk score as a foundation, clinical decision support tools were developed to improve the ability to prevent delirium. For nurses delivering care, the informatics team developed automated alerts to trigger a delirium care plan so that the most appropriate clinical interventions could be applied based on the assessments and risk scores entered in the EHR. For pharmacists, the multidisciplinary delirium risk score was used as a foundation to build a prioritization dashboard, giving them insights into which patients had deliriogenic medications as part of their hospital orders along with the nursing risk assessments for those patients. For physicians, the multidisciplinary delirium risk score served as a tool to identify patients who were at highest risk of delirium, which allowed them to make changes to medication orders as needed, or if no alternative existed, work with the geriatric nurse specialists and nursing staff to implement non-pharmacologic interventions to prevent delirium.

In order to make the information from the interdisciplinary delirium risk score actionable, the following new workflows were developed and implemented for each discipline. Upon seeing a high-risk delirium indicator for any patient in the EHR dashboard:

- The pharmacist will:

  - Review the medications and notify the attending physician and geriatric clinical nurse specialist (**if available**) via phone call or EHR in-basket message
  - Present possible medication alternatives to the attending physician
  - Document their intervention in the EHR

- The attending physician will:

  - Review the medication(s) with pharmacy
  - Change the deliriogenic medication in the EHR as appropriate
  - If no alternative medications possible, work with the geriatric clinical nurse specialist (**if available**) and primary nurse to provide non-pharmacologic delirium interventions

- The primary nurse will:

  - Validate the CAM score for the patient
  - Apply the appropriate care plan interventions for the patient
  - Perform pharmacologic and non-pharmacologic interventions in collaboration with pharmacist and attending physician
  - Continue to document the CAM per policy and update the plan of care based on the patient's condition

16.9 Measuring Success of the Interdisciplinary Delirium Risk Score

In developing the interdisciplinary delirium risk score, questions from various disciplines arose including workflow implications and effects on current patient workload. Physicians asked about the expectations that medications would be modified and raised concerns that this was not always possible. Pharmacists raised concerns about a potential lack of resources to handle the volume of patients identified as having a high risk of developing delirium. Nurses raised concerns about their bandwidth to coordinate between physicians and pharmacists for all high-risk delirium patients since they were already so busy caring for other patients in their assignment. In order to answer these operational questions of the delirium risk score, the team of clinical leaders, data analysts and clinical informaticists developed a measurement strategy, including process metrics for each discipline. Baseline information was gathered in order to set the stage of the current state of delirium care prior to introducing the delirium risk score. These descriptive data included: the number of patients in the hospital who had a coded diagnosis of delirium; rate of deliriogenic medications prescribed; the quality of documentation of the nursing assessments; and, CAM risk scores. Success measures were then developed to help evaluate the usability and value of the delirium risk score. Now that the tool was developed and metrics were identified, the delirium team engaged all of their hospital regions to plan pilots and measure the effectiveness of the delirium risk score.

Before starting the pilots, next steps include developing an interprofessional training curriculum to educate teams on how to utilize the delirium risk score tools and establish process measures that examine the usefulness of the information and the usability of the tools within the workflows of pharmacists, nurses, and physicians. With the ability to see delirium risk scores in real time, it is important to evaluate effects on clinical efficiency and communications between team members in order to truly find the value in using these tools. Once the tools become hardwired into the workflows of clinicians, potential outcome measures could also include length of stay, incidence of delirium, complications related to delirium, and readmission rates (Gleason et al. 2015).

16.10   Summary

Kaiser Permanente and the industry are on a journey to improve our quality, safety and affordability through the prevention of inpatient delirium regardless of clinical setting. Although the first generation of EHRs have focused on implementation and mandated documentation requirements, the quest for highly reliable and evidence-based personalized care can only be achieved with improved care coordination enabled by intelligent EHR systems.

EHRs are intended to support the safe delivery of evidence-based and personalized care. The challenges of seeking clinical information in silos and the first generation of EHRs were reviewed. With the emergence of predictive analytics and early warning scores using EHR data, the promise of best care at lower cost is within our reach. With increasing attention being given to care coordination industry-wide, the Kaiser Permanente delirium pilot provides an exemplar of best practices including interprofessional data sharing, data visualization, data integration into a risk tool, smart and timely clinical decision support and a culture of team-based care.

References

Agno CF, Guo KL. Electronic health systems: challenges faced by hospital-based providers. The health care manager. Health Care Manag. 2013;32(3):246–52.

Amer KS. Quality and safety for transformational nursing: core competencies. Upper Saddle River, NJ: Pearson; 2013.

American Delirium Society. https://www.americandeliriumsociety.org/about-delirium/healthcare-professionals (2015). Accessed 27 Feb 2016.

Argyri P, Kostagiolas P, Diomidous M. A survey on information seeking behaviour of nurses at a private hospital in Greece. Studies Hlth Tech Inform. 2014;202:127–30.

Ash JS, Bates DW. Factors and forces affecting EHR system adoption: Report of a 2004 ACMI discussion. JAMIA. 2005;12(1):8–12.

Berner ES, Moss J. Informatics challenges for the impending patient information explosion. JAMIA. 2005;12(6):614–7.

Berwick DM, Nolan TW, Whittington J. The triple aim: care, health, and cost. Health Aff (Millwood). 2008;27(3):759–69.

Bonner A, Lloyd A. What information counts at the moment of practice? Information practices of renal nurses. J Adv Nurs. 2011;67(6):1213–21.

Carter-Templeton H. Nurses' information appraisal within the clinical setting. CIN. 2013;31(4):167–75. quiz 76-7

Centers for Medicare and Medicaid Services. Meaningful use Baltimore, MD: Centers for medicare and medicaid services. 2013. http://www.cms.gov/Regulations-and-Guidance/Legislation/EHRIncentivePrograms/Meaningful_Use.html. Accessed 27 Feb 2016.

Chaudhry B, Wang J, Wu S, Maglione M, Mojica W, Roth E, et al. Systematic review: impact of health information technology on quality, efficiency, and costs of medical care. Ann Int Med. 2006;144(10):742–52.

Clarke MA, Belden JL, Koopman RJ, Steege LM, Moore JL, Canfield SM, et al. Information needs and information-seeking behaviour analysis of primary care physicians and nurses: a literature review. Hlt Inform Libr J. 2013;30(3):178–90.

Clegg A, Young JB. Which medications to avoid in people at risk of delirium: a systematic review. Age Ageing. 2011;40(1):23–9.

Cook DA, Sorensen KJ, Wilkinson JM, Berger RA. Barriers and decisions when answering clinical questions at the point of care: a grounded theory study. JAMA. 2013;173(21):1962–9.

DesRoches C, Donelan K, Buerhaus P, Zhonghe L. Registered nurses' use of electronic health records: findings from a national survey. Medscape J Med. 2008;10(7):164–72.

Gilmour JA, Huntington A, Broadbent R, Strong A, Hawkins M. Nurses' use of online health information in medical wards. J Adv Nurs. 2012;68(6):1349–58.

Gleason LJ, Schmitt EM, Kosar CM, Tabloski P, Saczynski JS, Robinson T, et al. Effect of delirium and other major complications on outcomes after elective surgery in older adults. JAMA. 2015;150(12):1134–40.

Grabenbauer L, Skinner A, Windle J. Electronic Health Record Adoption—Maybe it's not about the money: physician super-users, Electronic Health Records and patient care. App Clin Inform. 2011;2(4):460–71.

Inouye SK. The Confusion Assessment Method (CAM): training manual and coding guide. New Haven: Yale University School of Medicine; 2003.

Institute of Medicine. To err is human: building a safer health system. Washington, DC: National Academies Press; 2000.

Institute of Medicine. Key capabilities of an electronic health record system: letter report. Washington, DC: National Academies Press; 2003.

Institute of Medicine. Health IT and patient safety: building safer systems for better care. Washington, DC: National Academies Press; 2012.

Kannampallil TG, Franklin A, Mishra R, Almoosa KF, Cohen T, Patel VL. Understanding the nature of information seeking behavior in critical care: implications for the design of health information technology. Artificial intelligence in medicine. 2013;57(1):21–9

Kannampallil TG, Jones LK, Patel VL, Buchman TG, Franklin A. Comparing the information seeking strategies of residents, nurse practitioners, and physician assistants in critical care settings. JAMIA. 2014;21(e2):e249–56.

Kean J, Ryan K. Delirium detection in clinical practice and research: critique of current tools and suggestions for future development. J Psychosom Res. 2008;65(3):255–9.

Keenan G, Yakel E, Dunn Lopez K, Tschannen D, Ford YB. Challenges to nurses' efforts of retrieving, documenting, and communicating patient care information. JAMIA. 2013;20(2):245–51.

Koch SH, Weir C, Haar M, Staggers N, Agutter J, Gorges M, et al. Intensive care unit nurses' information needs and recommendations for integrated displays to improve nurses' situation awareness. JAMIA. 2012;19(4):583–90.

Kossman SP. Perceptions of impact of electronic health records on nurses' work. Studies Hlt Tech Inform. 2006;122:337–41.

Kossman SP, Scheidenhelm SL. Nurses' perceptions of the impact of electronic health records on work and patient outcomes. Comput Inform Nurs. 2008;26(2):69–77.

Kutney-Lee A, Kelly D. The effect of hospital electronic health record adoption on nurse-assessed quality of care and patient safety. J Nurs Adm. 2011;41(11):466–72.

LaBranche B. Rapid clinical information drives patient safety. Nurs Mgt. 2011;42(12):29–30.

Marshall AP, West SH, Aitken LM. Preferred information sources for clinical decision making: critical care nurses' perceptions of information accessibility and usefulness. Worldviews Evid Based Nurs. 2011;8(4):224–35.

Michel-Verkerke MB. Information quality of a Nursing Information System depends on the nurses: a combined quantitative and qualitative evaluation. IJMI. 2012;81(10):662–73.

Moody LE, Slocumb E, Berg B, Jackson D. Electronic health records documentation in nursing: nurses' perceptions, attitudes, and preferences. CIN. 2004;22(6):337–44.

O'Brien A, Weaver CA, Settergren T, et al. EHR documentation: the hype and the hope for improving nursing satisfaction and quality outcomes. Nurs Admin Q. 2015;39(4):333–9.

O'Leary DF, Mhaolrunaigh SN. Information-seeking behaviour of nurses: where is information sought and what processes are followed? J Adv Nurs. 2012;68(2):379–90.

Rogers ML, Sockolow PS, Bowles KH, Hand KE, George J. Use of a human factors approach to uncover informatics needs of nurses in documentation of care. IJMI. 2013;82(11):1068–74.

Romagnoli KM, Boyce RD, Empey PE, Adams S, Hochheiser H. Bringing clinical pharmacogenomics information to pharmacists: a qualitative study of information needs and resource requirements. International journal of medical informatics. 2016;86:54–61.

Ryan DJ, O'Regan NA, Caoimh RO, Clare J, O'Connor M, Leonard M, et al. Delirium in an adult acute hospital population: predictors, prevalence and detection. BMJ. 2013;3(1):1–8.

Smith JP, Seirafi J. Delirium and dementia. In: Marx JA, Hockberger RS, Walls RM, et al., editors. Rosen's emergency medicine: concepts and clinical practice. Philadelphia, PA: Elsevier; 2013. p. 1398–408.

Thakkar M, Davis DC. Risks, barriers, and benefits of EHR systems: a comparative study based on size of hospital. Perspectives in health information management. AHIMA. 2006;3:1–19.

Trinacty M, Farrell B, Schindel TJ, Sunstrum L, Dolovich L, Kennie N, et al. Learning and networking: utilization of a primary care listserv by pharmacists. Can J Hosp Pharm. 2014;67(5):343–52.

U.S. Department of Health and Human Services. Benefits of EHRs. 2013. http://www.healthit.gov/providers-professionals/learn-ehr-basics. Accessed 27 Feb 2016.

Weng YH, Kuo KN, Yang CY, Lo HL, Chen C, Chiu YW. Implementation of evidence-based practice across medical, nursing, pharmacological and allied healthcare professionals: a questionnaire survey in nationwide hospital settings. Implem Sci. 2013;8(112):1–10.

Chapter 17
Mobilizing the Nursing Workforce with Data and Analytics at the Point of Care

Judy Murphy and Amberly Barry

Abstract Mobile apps for nurses are beginning to be seen as critical components of providing effective and efficient patient-centered care and assisting in the transformation to value-based care delivery. Mobile apps work to augment the data and processes in the electronic health record (EHR) and assist in care coordination across the continuum; providing anytime, anywhere access to patient information and the plan of care in all venues—acute, long term, community and home. Mobile enables a new way to care for patients: not digitizing an existing process, but designing an end-to-end experience that allows nurses to have access to data and analytics to make the right decisions and execute tasks at the point of care. New mobile capabilities allow nurses to rise above the fragmentation that exists today and focus on the patient, using the best practice and the latest evidence. This chapter describes how mobile not only changes the way nurses work, but the way they think and communicate with other clinicians and with patients.

Keywords Mobile • Mobile app • Mobile health • Mobile healthcare • mHealth eHealth • Connected health • Handheld computer • Mobile device • Smartphone Tablet • Point of care • Nursing workflow • Workflow support • Communication Data • Information • Analytics • Nursing informatics

17.1 Introduction

This chapter describes mobile technology's and mobile apps' impact on the way nurses do their work and the way health is maintained and health care is delivered. By exploring the topic from differing perspectives, it provides the reader with insights into this rapidly evolving area. In particular, it considers the importance of mobile health (mHealth) to the profession of nursing and its impact on clinical

J. Murphy, R.N., F.A.C.M.I., F.A.A.N., F.H.I.M.S.S. (✉) • A. Barry, R.N., P.H.N.
IBM Global Healthcare, Armonk, NY, USA
e-mail: murphyja@us.ibm.com

© Springer International Publishing AG 2017
C.W. Delaney et al. (eds.), *Big Data-Enabled Nursing*, Health Informatics,
DOI 10.1007/978-3-319-53300-1_17

practice, care coordination, and patient-centered care. It also describes how nurses can contribute to the development of mHealth systems, ensuring that they are effective and efficient tools for both nurses and patients. Topics discussed in this chapter include:

- Background
- Mobile Infrastructure
- Mobile Impact on Nurses' Roles and Processes
- Apps for Nurses in all Care Venues
- Apps for Patients
- The Value of Mobile with the Power of Analytics
- Summary

This chapter will explore what being in the "digital age" means to nursing. The convergence of the telecommunications and computer industries on mobile devices has caused dramatic changes in how we communicate, process and use information in all aspects of our lives. Now social networks, sensor technologies, wearable devices and the "Internet of Things" (IoT) are expanding, evolving and being incorporated for use in healthcare. These continuous advances in mobile technologies, the emergence of the patient-controlled personal health record, and the ability to put analytics in the hands of patients allow them to be active participants in the management of their health and a full partner in their care. This creates a new ecosystem for health and health care—one that nursing can support and flourish in and one that mobile technology will enable.

17.2 Background

mHealth refers to mobile health technology and the domain is one which is dynamic and evolving. mHealth is defined by the National Institutes of Health (NIH) as "the diverse application of wireless and mobile technologies designed to improve health research, health care services and health outcomes" (NIH 2014). HIMSS defines mHealth as "the generation, aggregation, and dissemination of health information via mobile and wireless devices and the sharing of that information between patients and providers" (HIMSS 2015). The World Health Organization (WHO) states mHealth is the "use of mobile and wireless technologies to support the achievement of health objectives" (WHO et al. 2015). The mHealth Working Group of the mHealth Alliance takes this a step further by saying "mHealth is innovative, appropriate, integrated, evidence-based, scalable, interoperable and sustainable. mHealth supports improved health services and contributes to improved health services and access to better health outcomes, particularly for underserved populations in developing countries" (mHealth Working Group 2016). Each of these definitions reflects a commonality of purpose and intent of the technology and the associated apps for data access and empowerment of the end user.

Since the year 2000, the promise of reduced healthcare costs and improved patient outcomes associated with mHealth has inspired many to build their app development business models around activities such as remote patient monitoring, mobile alerts and clinical reminders. These early movers entered the market too early as its conditions did not support scalability of the existing solutions. With the launch of the Apple App Store in 2008, the mHealth market was able to enter into the early commercialization phase. Ultimately, the Apple App Store allowed mHealth solution providers to reach out to a mass market and grow their size and income.

Apple and its fast growing number of copycats transfigured the entire customer and developer experience, making mobile applications easy to download and use, as well as easier to develop and distribute. This ease of development applies to mHealth solutions also. Both major platforms, Apple and Android, are by far the leading mobile operating systems for mHealth apps (and all apps) today. Over the last few years Android has seen tremendous growth in the number of apps listed in the Health & Fitness and Medical sections in Google Play. In 2015, the healthcare research firm, IMS Institute for Healthcare Informatics (IMS Institute 2015), counted over 165,000 apps listed in the mHealth sections of Apple and Android. However, the market is still somewhat fragmented, with 40% of health apps having fewer than 5000 downloads, and just 36 apps accounting for nearly half of all health app downloads.

Not only are consumers taking advantage of smartphones to manage and improve their own health with wellness and fitness apps, they also are using them for medical conditions and chronic disease tracking and management as well. A significant number of mHealth applications are also designed for use by health care professionals, including apps for medical reference, continuing education, remote monitoring, decision-making, analytics, and result review. Figure 17.1 shows the distribution of mHealth app categories from a 2014 study.

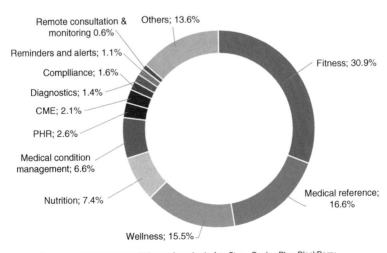

Source: research2guidance, 808 apps form Apple App Store, Goolge Play, BlackBerry App World and Windows Phone Store (March 2014)

Fig. 17.1 mHealth app category share (research2guidance 2014)

17.3 Mobile Infrastructure

17.3.1 Mobile Device and App History

In 1973, the first mobile phone call was made by Martin Cooper of Motorola to Joel Engel of Bell Labs. The device used weighed 2.5 pounds and measured 9 × 5 × 1.7 inches. It took two more decades before a mobile phone could include both telephone and app features, and IBM Simon, the first commercial product to combine phone and app features, was launched in 1994 (Business Insider 2015). Today, over 90% of Americans own a mobile phone (Pew 2014), 65% of Americans own a smartphone (Smith 2015), and there are 1.5M apps in the Google Play Store and 1.5M apps in the Apple Store (Statistica 2015). There have been numerous operating systems since 1994—some have persisted and some not. Early systems included EPOC, Palm and Blackberry. Today, the dominant players are Android (Google), iOS (Apple), Windows Mobile and Fire (Amazon).

Ubiquitous and reliable wifi is also an important enabler for widespread use of mobile apps. Standards for wireless communication were first published in 1997, but it wasn't until 1999 when Apple introduced wi-fi as an option on its new iBook computers that it actually took off. Other laptop computer-makers quickly followed suit, and wifi caught on broadly with consumers in 2001. For the next 15 years, wifi technology advances promoted more reliable connections and increased data transfer speeds

Smartphones and mobile apps are now widely used for navigating numerous important life activities, from calendaring and email to researching a health condition and accessing educational resources, following breaking news. They are use to participate in social media such as Twitter or Facebook. Smartphones help users navigate the world around them, using GPS for turn-by-turn driving directions to assistance with public transit. This is especially true for young adult users, who have deeply embedded mobile devices into the daily context of their lives.

17.3.2 History of Mobile in Healthcare

Healthcare has a long and circuitous history with mobile. Ever since wifi first became available, nurses and other clinicians have been chasing mobile devices and apps that fit their very mobile workflow. In the early days, mobility manifested itself in "Computers on Wheels" essentially taking laptops with extended batteries, putting them on carts, and connecting them to the network via wifi. This allowed nurses to bring the computers into the patient rooms for data access and data entry at the point of care. But there were several problems. The carts were heavy and not comfortable to push or pull, particularly on carpeted floors or over thresholds to enter rooms. They also had to be plugged in to recharge each day. Wifi was new, and the connection periodically dropped, causing the loss of any data entered and requiring logging in again. Finally, the applications were not user friendly and not conducive to using in front of the patient. So their use was met with mixed clinical success and carts were often left in hallways or at the nurses' station.

Healthcare also saw the use of special function mobile devices for tasks such as point of care, blood/urine testing and specimen collection, or communication devices like pagers and hands free phones. But these applications were never broadly deployed due to cost and singularity of use. Expecting providers to carry multiple devices for individual specific functions was not realistic and did not scale across entire organizations.

Today, with the wide availability of new mobile devices such as tablets and smartphones, with hardened wifi infrastructures and with the proliferation of new user-friendly mobile apps, the use of mobility in healthcare is changing for both patients and nurses. These smaller devices are easy to carry around, have good battery life, and less chance of interfering with the nurse-patient interaction. In the next section there are descriptions of some examples of the mobile apps being seen in healthcare today.

17.4   Mobile Impact on Nurses' Roles and Processes

Mobile apps have become critical components of providing effective and efficient patient-centered care and assisting in the transformation to value-based, delivery of care. The implications for use vary far and wide depending upon focus of the app, the clinical role in which it is used, the problem that is being addressed and the workflow process it is supporting. In some cases these apps allow clinicians and patients to rise above the tethered, ineffective structure of current environments and workflows, and forego digitalizing an ineffective existing process. The new experience enabled by mobile technologies not only allows access to data and analytics at points of care never before enabled, but decision making and task execution, changes outcomes with a new level of satisfaction for both the clinician and the patient.

Implications for mobile apps based on role, venue and problem needing to be addressed has not only driven the development of specific mobile apps but has enabled a 10,000 foot view of overall healthcare needs to allow for broad transformational apps that can, using big data analytics, link several workflow streams, care providers and populations. Some of these apps have been in response to healthcare reform and the regulations that drive toward better transitions of care, or a means to report on quality and patient satisfaction. Other apps focus on solving a specific problem within a clinical specialty, disease process or patient teaching need.

Over the past few years, we have seen proliferation of new apps for nursing and patient uses at an impressive rate as described in the previous section, but we are just beginning to scratch the surface in usage. According to a survey of 500 healthcare professionals, a market research firm found that although only 16% currently use mobile apps with their patients, 48% indicated that they will introduce mobile apps to their practice in the next five years. The survey further reflects that 81% of healthcare professionals believe that health apps will increase their knowledge of patients' conditions and 45% said they think the apps will increase the efficiency of patient treatment (Research Now Group Inc. 2015). These statistics not only demonstrate a probable increase in usage, but the belief in the value of apps in both quality of care and efficiency.

As this trend continues, mobile apps will be able to deliver the results of big data analytics to the point of care in order for clinicians to look at patient care through a different lens and with an increased focus on specificity and personalization—both driving development toward improved patient satisfaction. However, this trend is also being balanced in design and purpose with the need to link to a broader view to solve issues with gaps in care, population health insights and 360-degree views of patients. The potential for the apps and the balance between specificity and broader purpose is best illustrated when evaluating the mobile apps as they relate to potential impact on nursing processes and specific roles.

17.5   Apps for Nurses: Education

Considering the gamut of nursing apps, we have seen development in the areas of education and practice. For example, in the primary education for nurses, mobile apps enable quick access to educational resources such as medical definitions, disease and procedure information, nursing content, medication references, 3D anatomy images and language translation. They also support virtual education used during clinical practicums to support and enhance learning, and provide access to services such as licensure exam study guides (NCLEX Mastery 2016).

One app example that is changing the way nursing students "visualize" anatomy in a more adaptive and explorative way is Heart Pro III (3D4Medical 2016), an app that allows the user to view animations, as well as zoom, rotate or cut parts of the heart in graphic 3D. This flexible viewability with stunning and clear 3D imagery is ideal for students and educators, and even enables drag-and-drop quiz formats for further study. The added educational value with the ability to visualize animation of heart disease and function for endocarditis, instead of just reading or viewing stagnant imagery, is remarkable. As a side benefit, the enhanced understanding that animation can provide in a mobile app is also very useful for the nursing student or nurse when educating patients.

Another example of a mobile educational tool used with nursing students is Eko Core (Eko Devices Inc. 2016), which combines a digital attachment to the traditional analog stethoscope with a mobile app for auscultation instruction. This device enables transitioning between analog and digital modes and wirelessly connects the stethoscope to a mobile device. This Bluetooth enabled, wireless tool allows heart sounds to be visualized in the app for enhanced auscultation teaching. With features to record, save, and share, the patient's heart sounds can be kept for future reference, allowing students to access a library of annotated heart sounds so that they can hone auscultation skills anytime, anywhere. Beyond the teaching aspect, the tool can also help to coordinate care with specialists and integrate results into the patient's electronic health record (EHR).

These apps are only a small sample of the broad move to mobile options for nursing education, as it follows the e-learning movement exhibited in all forms of education today. One study conducted with nursing students evaluated the effectiveness of mobile-based discussion versus computer-based discussion on self-directed learning readiness, academic motivation, learner-interface interaction, and flow state. This randomized controlled trial recognized that students reacted positively to mobile technology and the potential for portability and immediacy. As has been identified by other clinicians' experiences with technology, it was also reported that students tended to reflect negatively on their learning if they viewed the technology as time-consuming or contributing to delays in response time. The conclusion from this study was that effective teaching using mobile-based discussion and e-learning tools must consider portability, immediacy, and interactivity (Lee 2015).

17.6 Apps for Nurses: Practice

In nursing practice, mobile apps targeting specific roles and nursing specialties offer a unique opportunity to simplify a level of complexity that has, in the past, been mastered only through a lengthy learning curve based on experience, time and personal mentoring. Now, the support of mobile resources and apps can expedite the learning curve for the practicing nurse. Considering the educational apps mentioned above and adding role and specialty based apps, nurses practicing today have many more opportunities to digest, evaluate and bring into practice more valuable and targeted information. Apps can also support ongoing, continuing education for nurses in an era of rapidly expanding healthcare research and ever-increasing healthcare knowledge. When considering the care venue or nursing specialty, apps for the practice of nursing have proven valuable in all venues; primary care, acute care, home care and in care coordination across venues. These will be expanded upon in the following sections.

17.6.1 Primary Care

There are many mobile apps that are being increasingly used by nurses in primary care. With the healthcare delivery and payment reform movement, primary care services are in high demand and new primary care models are developing. One model that has gotten a great deal of attention and implementation is Patient-Centered Medical Home (AHRQ PCMH Resource Center 2016), where the focus is on patient-centric care delivered by interprofessional teams. In this model, care is not physician or office-centric, therefore it is important to share clinical information and care/treatment plans between team members and across time and physical sites. This patient-centric system often requires data from disparate sources to be combined, analyzed and presented to the team member in a specific care venue for a specific patient that brings the full historical and social determinants of health into the view for consideration and action. In this way, mobile technology and analytics are a perfectly matched pair.

One example of a mobile app that supports care coordination in the PCMH model is Flow Health (Flow Health Inc. 2015), which enables real-time patient updates to be pushed to mobile devices. Admissions and discharges from the emergency department or hospital are available to the PCMH care teams, facilitating the closing of potential gaps in care before readmissions or complications arise. PCMH care teams can securely message within each timeline post to coordinate care and yet messages automatically expire to maintain private off-the-record conversations.

Another example of the use of mobile in primary care involves the logging and tracking of patients' polypharmacy, which traditionally has been a time-consuming and inaccurate process involving routine re-creation of med lists and medication reconciliation. One group of mobile apps that seek to help with this issue are apps for medication identification, and one example is MedSnapID (MedSnap 2016). This app uses a smartphone camera to generate definitive medication histories that not only identify the medications by image, but also recognize serious drug interactions early in the patient interview process. The app can integrate with the patient's electronic health record to speed the medication history-taking and documentation process. Not only do these apps support a more efficient and accurate medication reconciliation process for primary care, but they can also have additional benefits for fraud and abuse identification. When combined with MedSnap Verify Services (MedSnap Verify 2016), the app can help providers and drug manufacturers identify counterfeit medications and "fingerprint" them to track down their source. Suspected medication photos are loaded into a database which then analyzes the pills' size, color, shape, and other attributes to verify their authenticity. The National Association of Boards of Pharmacy (NABP) reports that reducing fraud not only lowers the cost of care, but that the patient safety aspects of these point-of-care apps can help improve the quality of care, while enhancing nursing satisfaction and efficiency (NABP 2015).

For primary specialties such as obstetrics (OB), mobile apps have been developed that simulate the pregnancy wheel for estimating delivery date (EDD). These apps are more efficient and potentially more accurate due to the number of data inputs, and they are also simply easier to use than their paper-based counterparts. One example is the mobile app from the American Congress of Obstetrics and Gynecology, The ACOG App (ACOG 2016). This app provides valuable resources and up-to-date information from the leading experts in women's health and OB care. It also has an EDD Calculator, and uses data from last menstrual period and first accurate ultrasound to determine estimated due date.

Another OB-specific app, Prevent Group B Strep (CDC 2016), targets the prevention and management of group B strep infection in mothers and newborns. Patient characteristics are entered and the app enables access to patient-specific guidelines for clinicians at the point of care. The app was developed in collaboration with and is endorsed by the American College of Obstetricians and Gynecologists

(ACOG), the American Academy of Pediatrics (AAP), the American College of Nurse-Midwives (ACNM), and the American Academy of Family Physicians (AAFP)—and is being distributed by the Centers for Disease Control and Prevention (CDC). The app also supports intrapartum prophylaxis information by displaying appropriate antibiotic regimens.

17.6.2 Acute Care

As primary care mobile apps support efficiency and quality for the volumes of patients seen every day in clinics, acute care nursing apps support the often hectic workflow of the nurses when managing high volumes of information pushed through electronic health records (EHRs) in hospitals. Many of the large EHR vendors have yielded to their clients' demands for mobility enablement in their platforms, offering iOS and Android enabled access to information in the EHR through apps they have created for the clinicians. Initially, the functionality offered included data retrieval views for patient lists and result review. This limited scope has expanded to include searching and secure messaging. Today, the EHR-tethered apps allow writing back to the EHR database and include functions such as barcode med administration, e-prescribing, order entry, and documentation entry including the capture of clinical photos.

There are also independent mobile apps being used in the acute care setting. The Hospital RN app (Apple+IBM 2016) frees nurses from having to carry multiple, single-function devices by consolidating communications and alarm management on a single device. Pages, patient requests, critical labs, safety alerts, device and bed alarms are summarized and prioritized for immediate disposition and are pushed through notifications in the app. The app also streamlines the nurses' workflow by consolidating interventions from across all assigned patients and allowing dynamic re-sorting of the interventions based on time (e.g., all due at 9 a.m.) or on priority (e.g., all stats), thereby improving coordination and saving nursing time.

A companion app, the Hospital Lead app (Apple+IBM 2016) summarizes and displays the information that charge nurses need regarding all the patients and all the staff on a nursing unit. An intuitive dashboard combines multiple hospital databases into a single view, and shows key information such as occupied/vacant beds, patients in the Emergency Department awaiting admission, staff assignments, patients past the average length of stay for their diagnosis, patients scheduled to be discharged today that do not yet have a discharge order, and patients with barriers to discharge identified. Push notifications deliver escalated issues from staff so they can be assisted with or assigned to others, and a prioritized task list for the charge nurse can be maintained.

For specialty areas of the acute care setting such as the ICU, mobile apps that simplify complicated medical formulas, scores, scales and classifications have evolved. MedCalX (MedCalX™ 2015) is a medical calculator available on mobile platforms that has continued to develop over 15 years. Carefully crafted for mobile devices, with the goal to keep it simple for the clinician, yet support workflow and decision making in critical moments, MedCalX contains an exhaustive list of calcu-lations. Bringing a multitude of information together such as CAM-ICU, APACHE II Score, Glasgow Scoring System, CURB-65 Score and SOFA Score, to name a few, this app makes calculations quick and easy, and provides formulas and refer-ences to substantiate its answers. Users can also add notes and flag certain calcula-tors as favorites to enhance the user experience.

17.6.3   Home Care

Although both primary and acute care nursing have distinct needs for mobility, nowhere have mobile apps been more anticipated and more helpful than in home care. With no consistent physical venue, tethered documentation tools and carrying paper-based forms and reference materials have been issues for these inherently mobile nurses for decades. Nurses that travel to their patients and may also need remote access for monitoring when they cannot be in the home have seen many new apps that make their work more efficient, cost effective and supportive to better outcomes.

Home care EHRs have existed and been used in the home on traditional laptops for over a decade, but now, as was described with the acute care EHRs, these home care EHRs have also evolved to be web-based platforms that allow nurses internet connectivity to the EHR and to document care, access clinical data, share data with care team and revise care/treatment plans using internet connected mobile devices (Weaver and Teenier 2014). Documentation is faster and more simplified with the user interfaces found in mobile, while still maintaining the protocols and regulatory documentation checks to guide practice for home care. Nurses can trend progress at the push of a button and can even take photos of wound progression and attach them to the patient's record. If there is no connectivity in a patient's home, records can be downloaded to mobile devices prior to the visit and uploaded once connections are reestablished. With mobile devices that are less bulky than laptops, face time with patients is maximized and can enhance patient satisfaction.

Besides scheduling and time reporting enabled on hand held devices, GPS tech-nology, such as CellTrak (CellTrak Technologies, Inc. 2016), provide driving direc-tions and track mileage. When a caregiver arrives for a visit, the mobile app performs electronic visit verification, presents the care/treatment plan, making complete point-of-care documentation easy, and enables real-time communication with the office and other care team members via alerts and secure messaging. Similar functionality can be seen in the HomeRN app (Apple+IBM 2016) and DeVero's Home Health and Hospice system (DeVero 2016), integrating some of the regular features of the smartphone like GPS, time tracking, emailing and texting seamlessly into the app.

It is not the purpose of this chapter to discuss telehealth and remote monitoring in the home, but it is important to acknowledge that these are adjunctive tools used by those providing care in the home to communicate with and monitor patients when they cannot be present. This is a very large and growing area—today there are many fitness and health-related monitoring devices being connected to health-related apps and used for remote monitoring and tracking. But as the Internet of Things (IoT) continues to evolve and connect "smart sensors" that capture or monitor data and automatically trigger certain events, we can expect to see an explosion in this space. Forbes has estimated a $117 billion market for IoT in healthcare by 2020 (McCue 2015).

17.6.4 Care Coordination

Beyond the venues of primary, acute and home care, the overall care coordination of the patient between venues demonstrates one of the biggest challenges in healthcare today, the almost inevitable gaps in care and information as the patient transitions between care environments. Several methodologies, technologies and team-based care models focus on how to close these gaps, one of these being the tremendous growth in population health management. The Institute for Healthcare Improvement (IHI) defines population health as the health outcomes of a group of individuals, including the distribution of such outcomes within the group. These groups are often geographic populations such as nations or communities, but can also be other groups such as employees, ethnic groups, disabled persons, prisoners, or any other defined group (Lewis 2014). Healthcare reform and the Triple Aim (Better Health, Better Care, Lower Cost) have brought a level of focus on effective mobile apps to support the management of both individuals and populations using analytics and engagement. By first using data analytics to understand the population you are addressing, then using data analytics to get a 360 view of the individual you are caring for, you can determine the appropriate engagement strategies to provide the most impact based on the segment of population being addressed—healthy well vs. at risk vs. chronic disease vs. acute management—and based on the individual's preferences. Finally, you can evaluate your strategies in order to find patterns and opportunities for quality improvement.

When usability is expanded across the many varied population healthcare team roles, from health coach to care coordinator to community based health facilitators, the mobile needs for data analysis and care coordination still require disease specific drill down capabilities, while maintaining a broad view of the patient within the population. This is when big data analytics intersect care coordination workflows to bring insight into action for better population outcomes. After a comprehensive process of acquiring, aggregating, and standardizing clinical and operational data, analytics and risk models can be applied to that data, then insights for understanding patient and population outcomes become achievable. Utilization patterns emerge demonstrating geographic and socioeconomic patterns related to emergency and acute care access. Stratifying the data for patient disease registries and enabling work lists with mobile devices at the point of care allow care coordinators to engage patients with effective outreach opportunities. Proactively managing patient populations with automated, multi-modal communication campaigns that reach patients in an effective, persistent manner brings mobility to the population care team members as well to the patients, enhancing both efficiency and effectiveness of the valuable insights. Mobility is an essential component for the patient engagement required for population health management.

One example representing the explosion in disease specific mobile apps engaging patients with insightful robust data focuses on the treatment of asthma. With over 100 mobile apps to consider, digital health technology is rapidly expanding for asthma management in both pediatrics and adults. Asthma management apps may include references and education, symptom diaries, medication reminders and early warning software for asthma triggers. There are apps evolving that aspire to a level of customization to the patient's asthma action plan that will impact more tailored management, taking the population of asthma to a patient-centric view. The Propeller (Propeller Health 2016) mobile app uses a sensor to keep track of medication use, with a record of the time and place. The sensor is a small device that attaches to the top of an existing inhaler, and when synced with an individual's smartphone, event logs are automatically generated rather than being dependent on an individual entering log data. Used with both rescue and controller medications for tracking symptoms and adherence respectively, the objective data is collected, and the feedback provides patient and provider insight for personalized management and education on ways to improve asthma control. Asthma specific apps provide visual output that can be discussed with a healthcare provider either during a clinic visit or sent remotely when a change in the asthma action plan is indicated.

Another population health mobile app that focuses on population by age factors, iGeriatrics (The American Geriatrics Society 2016) combines the American Geriatrics Society's free clinical information offerings into one easy-to-use application. Aimed at healthcare providers and covering a wide range of topics such as Beers criteria, Geriatrics Cultural Navigator, GeriPsych Consultant, Prevention of Falls Guidelines and Management of Atrial Fibrillation, these apps for the older adult population apply age-specific, cross-cultural assistance for care providers. With over 100 medication tables, treatment algorithms, and commonly used medical calculators, medication safety for the aging adult is addressed as well. The on-going updates for information contained in these apps are supported with a subscription in order to continually stay current on the aging population and available services. Tailored specifically to the geriatric population, this app is perfect for the adult gerontology acute care nurse practitioner. Combining disease registries with age-associated mobile apps such as iGeriatrics enables a new level of population health management specificity.

17.7 Apps for Patients

As described in the previous section, patient coordination of care improved by mobile access is only one impact that mobility has had on healthcare. Consumerism and the focus on effective usability have driven the patient mobile app options to a whole new and far-reaching level. The wearable fitness devices not only allow for accountability for patients around activity and nutritional goals, but also provide clinicians with key lifestyle information and insights near real-time to make a larger impact than what was previously available. These mobile apps for patients and the general healthy population have seen dramatic growth. Wearable startups are revolutionizing healthcare and not just fitness. In fact, Forbes estimates that by the end of 2016, the entire wearable technology market will gross almost $2 billion in revenues (Stoakes 2015).

Some patient-focused apps are disease specific and are focused on chronic disease management. The American Society of Nephrology tested a mobile app for chronic kidney disease patients to help them determine if over-the-counter or prescription medications are safe to take based on their specific condition. When considering a population with a particular disease, the effectiveness of the app is not only related to the data analysis capabilities and decision-making process, but the ease of use of the device. With only a five percent error rate, this particular mobile app proved promising to the research group. As demonstrated in this study at the intersection of clinician and patient focused apps, the effectiveness outside of the original app purpose, such as driving down cost or managing disease, will be based on usability of the app (Diamantidis et al. 2015).

Another patient-focused app example is a wound care app that allows diabetic patients to analyze pictures of their wound to learn what type it is as well as the proper way to dress and care for the affected area (Advanced Tissue 2014). The three-minute analysis completed using this app becomes an efficient way to teach as well as communicate specifics about the wound to care providers between visits or when access to care is limited. This is just one of many mobile options that promote self-care. Consumerism and advocacy have also brought mobile apps for selecting insurance plans, evaluating nursing homes, finding and evaluating physicians, as well as connecting to support groups for health related issues.

17.7.1 Patient Portals

It is interesting to note that patients who own digital health devices are five times more likely to regularly use a patient portal to access personal health information than non-device owners, according to research from Parks Associates, which indicates that device ownership can be a driver for online engagement (Parks Associates 2015). Many organizations such as Parks Associates are doing research on factors impacting the use of portals in an attempt to understand the disappointing use of patient portals to date. As Meaningful Use has driven EHR adoption, use of the associated EHR-tethered patient portals has seen a disappointingly low uptake by patients. With inconsistent outcomes and widely varied implementation efforts, portals have a new focus on access, usability and reward through the use of mobile devices. If mobile technologies can enable a higher use, innovative device deployment may be the next focus for healthcare payers and providers to ensure communication, bill payment, compliance and increased quality of care. LifeMap Solutions and the Icahn School of Medicine of Mount Sinai initially developed an app for asthma patients using their iPhones for a research study. Not only was the Asthma Health App found to be a good research tool, but researchers also determined that the app helped to improve the education and awareness of asthma patients, as well as the best ways to treat asthma (Traxler 2015).

Considering the many opportunities described that are enabled by mobile apps and the potential for providing more effective and efficient patient-centered care, it is not difficult to apply value to mobility. When mobile apps align to nurses' roles and responsibilities in primary care, acute care, population health management or home-based care, they quickly become critical components in the transformation to value-based delivery of care. Reimagining the tethered, ineffective structure of current EHR environments, mobility is allowing access to insightful data and analytics at points of care, enhancing decision making, enabling efficient task execution, and proving the value in outcome realization and cost effectiveness.

17.8   The Value of Mobile with the Power of Analytics

The mobile app examples in the previous sections demonstrate many different types of benefits for nurses and patients based on the specific focus of the app: patient safety and best practice enablement, clinical decision support, convenience, improved accuracy, enhanced productivity, consolidated communication and alarm management, streamlined workflow, efficient/anytime/anywhere data access and entry, and data integration and device management. But there are some broader value drivers for mobile as well.

To take advantage of the new opportunities mobile technologies enable, healthcare organizations are focused in four areas. First, they are investing in mHealth initiatives—from telemedicine, remote monitoring, mobile apps and innovative medical devices—to extend services beyond traditional settings, which increase access, convenience and efficiencies. Second, they are developing new engagement and health strategies delivered through secure mobile technologies that place the individual at the center of their own health and wellness efforts to drive better health outcomes. Third, they are creating a streamlined, high performance experience and a better decision-making process for their clinicians. Finally, they are gaining insight through data analytics and cognitive systems to provide more personalized and proactive interventions for individuals, with improved outcomes for populations.

17.8.1   Extend Healthcare Services

New mobile technologies and mHealth initiatives are extending reach and capacity of the healthcare industry to provide services in remote and underserved areas. Mobile networks are empowering individuals, requiring healthcare organizations to provide multi-channel approaches, varied settings and convenient locations to define new relationships, coordinate and improve efficiency of service. Connectivity across device and technology are enabling the industry in unprecedented ways to engage individuals in their own health and wellness, whether aging at home or across the care continuum.

17.8.2 Patient Engagement

Whether motivated by financial incentives, easier access to healthcare services or changing cultural norms, individuals today are beginning to engage more directly in managing their own health and well-being. Access, activation and engagement are increasing as people are becoming more connected than ever with an array of providers through easy-to-use connected mobile devices. These empowered individuals are demanding value from their providers, and are demonstrating willingness to partner with them to optimize their health outcomes.

Mobile technologies are enabling new systems of engagement and creating real transformation. Today's mobile technologies facilitate reducing disparities in access to care, and compel individuals to become advocates for their own health. There is a dramatic shift by payers and providers to focus more on consumers by redefining relationships to engage and empower individuals in their own health decisions, better coordinate resources and extend the reach of their services across their communities.

17.8.3 Decision Support

Accurate, comprehensive data from various venues of care, technology platforms and care providers, when aggregated and enabled with insight for care decisions, allow a more comprehensive and contextual awareness for care. This is exemplified in a seminal study showing that nurses who used a mobile application with evidence-based decision support were significantly more likely to diagnose depression, weight issues and tobacco use than nurses not using such tools (Bakken et al. 2014). With new modalities, technology and sources of integration, nurses are confronted with more information every day related to patient data and decisions for care. It is not surprising that most find it difficult to find the right information, in the right context, at the right time in order to take the right action. Mobile apps with the underlying power of big data analytics can help provide focus to determine and support the most effective and efficient actions.

17.8.4 Insight through Analytics

Healthcare is an enormous industry, prime for disruption and new forms of insight through analytics. So when using mobile apps, there is another very important consideration beyond the specific use they are being employed for. This consideration focuses on the data collection and aggregation for analytics. Mobile has an unprecedented ability to collect data in healthcare—from what is entered to what is incorporated through device integrated and IoT.

These new and unprecedented amounts of data—across populations and at the level of the individual—are creating opportunity for deeper insight, personalization and earlier interventions. Leaders are leveraging mobile to increase speed and responsiveness to meet the needs of the individual, care providers and care givers. Information when it is needed, where it is needed and how it is needed, with personalized, private and secure interactions, is thus earning the trust and confidence of the patients served.

17.9 Summary

Mobile devices and apps are already invaluable tools for nurses, other health professionals, and patients. As their features and uses expand, they are expected to become even more valuable and widely incorporated into every aspect of clinical practice. Future mobile apps are projected to include even larger databases that can be mined for new insights, as well as clinical decision support that will aid in decision-making and provide guidance to the nurse in order to improve patient outcomes and support the Triple Aim.

References

3D4Medical. http://applications.3d4medical.com/heart_pro (2016). Accessed 20 Mar 2016.
ACOG. http://www.acog.org/About-ACOG/News-Room/ACOG-and-GBS-App (2016). Accessed 20 Mar 2016.
Advanced Tissue. Mobile apps for wound care and management. 2014. http://www.advancedtissue.com/mobile-apps-wound-care-management/. Accessed 28 Mar 2016.
AHRQ PCMH Resource Center. https://pcmh.ahrq.gov/ (2016). Accessed 20 Mar 2016.
American Geriatrics Society. http://www.americangeriatrics.org/publications/shop_publications/smartphone_products/ (2016). Accessed 20 Mar 2016.
Apple+IBM. http://www.ibm.com/mobilefirst/us/en/mobilefirst-for-ios/industries/healthcare/home-rn/; http://www.apple.com/business/mobile-enterprise-apps/healthcare.html (2016). Accessed 20 March 2016.
Bakken S, et al. The effect of a mobile health decision support system on diagnosis and management of obesity, tobacco use, and depression in adults and children. JNP. 2014;0(10):774–80.
Business Insider. http://www.businessinsider.com/worlds-first-smartphone-simon-launched-before-iphone-2015-6 (2015). Accessed 20 Mar 2016.
CDC. http://www.cdc.gov/groupbstrep/guidelines/prevention-app.html (2016). Accessed 20 Mar 2016.
CellTrak Technologies, Inc. http://www.celltrak.com/ (2016). Accessed 20 Mar 2016.
DeVero Home Health System. http://www.devero.com/ (2016). Accessed 28 March 2016.
Diamantidis CJ, et al. Remote usability testing and satisfaction with a mobile health medication inquiry system in CKD. Clin J Am Soc Nephrol. 2015;10(8):1364–70.

Eko Devices Inc. https://ekodevices.com/whitepapers/Eko_for_Education_Whitepaper.pdf (2016). Accessed 20 Mar 2016.

Flow Health Inc. https://www.flowhealth.com/providers (2015). Accessed 20 March 2016.

mHealth Working Group. https://www.mhealthworkinggroup.org/about (2016). Accessed 20 Mar 2016.

HIMSS. mHealth. http://www.himss.org/mhealth (2015). Accessed 20 March 2016.

IMS Institute Report. Patient adoption of mHealth: Use, evidence and remaining barriers to mainstream acceptance, page 1. 2015. http://www.imshealth.com/en/thought-leadership/ims-institute/reports/patient-adoption-of-mhealth. Accessed 20 Mar 2016.

Lee M. Effects of mobile phone-based app learning compared to computer-based web learning on nursing students: pilot randomized controlled trial. Healthc Inform Res. 2015;21(2):125–33.

Lewis N. IHI Leadership Blog. 2014. http://www.ihi.org/communities/blogs/_layouts/ihi/community/blog/itemview.aspx?List=81ca4a47-4ccd-4e9e-89d9-14d88ec59e8d&ID=50. Accessed 20 Mar 2016.

McCue MJ. $117 billion market for internet of things in healthcare By 2020. Forbes/Tech. 2015.

MedCalX. http://medcalx.ch/ (2015). Accessed 20 Mar 2016.

MedSnap. https://medsnap.com/medsnap-id/ (2016). Accessed 20 Mar 2016.

MedSnap Verify. https://medsnap.com/verify/ (2016). Accessed 20 Mar 2016.

NABP. Technology abounds for combating counterfeit meds, abuse. 2015. http://www.nabp.net/news/technology-abounds-for-combating-counterfeit-meds-abuse. Accessed 20 Mar 2016.

NCLEX Mastery. 2016. http://www.nclexmastery.com/. Accessed 20 March 2016.

NIH. Introduction to mHealth online training course and mHealth at the NIH. 2014. http://mhealthinsight.com/2014/07/09/the-national-institutes-of-health-launches-a-mhealth-online-training-course/. Accessed 20 Mar 2016.

Parks Associates. http://www.parksassociates.com/blog/article/chs-2015-pr13 (2015). Accessed 20 Mar 2016.

Pew. Mobile technology fact sheet. 2014. http://www.pewinternet.org/fact-sheets/mobile-technology-fact-sheet/. Accessed 20 Mar 2016.

Propeller Health. https://www.propellerhealth.com/ (2016). Accessed 20 Mar 2016.

Research Now Group Inc. http://www.researchnow.com/en-gb/PressAndEvents/News/2015/april/are-mobile-medical-apps-good-for-our-health-infographic.aspx?cookies=disabled (2015). Accessed 20 Mar 2016.

Research2guidance report. The state of the art of mHealth app publishing. In: mHealth App Developer Economics. 2014, page 12. http://research2guidance.com/r2g/research2guidance-mHealth-App-Developer-Economics-2014.pdf. Accessed 20 Mar 2016.

Smith A. U.S. smartphone use in 2015. Pew Research Center. 2015. http://www.pewinternet.org/2015/04/01/us-smartphone-use-in-2015/. Accessed 20 Mar 2016.

Statistica. http://www.statista.com/statistics/276623/number-of-apps-available-in-leading-app-stores/ (2015). Accessed 20 Mar 2016.

Stoakes U. Startups making a name for themselves in the $1 billion wearables market. Forbes/Entrepreneurs. 2015.

Traxler. http://lungdiseasenews.com/2015/10/02/asthma-health-app-iphone-offers-valuable-features-insights-patients-doctors-researchers/ (2015). Accessed 20 Mar 2016.

Weaver C, Teenier P. Rapid EHR development and implementation using web and cloud-based architecture in a large home health and hospice organization. In: Nursing Informatics 2014. Proceedings of the 12th International Congress on Nursing Informatics in Taipei, Taiwan. Amsterdam, IOS Press. 2014. p. 380–387.

WHO, UN Foundation, Johns Hopkins University Global mHealth Initiative. mHealth MAPS toolkit. 2015. http://www.who.int/life-course/publications/mhealth-toolkit/en/. Accessed 20 Mar 2016.

Chapter 18
The Power of Disparate Data Sources for Answering Thorny Questions in Healthcare: Four Case Studies

Ellen M. Harper and Douglas McNair

Abstract The practice of using discrete, quantitative data from multiple data sources and evidence to support clinical decisions began centuries ago when Florence Nightingale invented polar-area diagrams to show that many army soldiers' deaths could be traced to unsanitary clinical practices, and therefore, were preventable. Today, the massive volume of healthcare data that is generated from our healthcare systems requires sophisticated information technologies for storing, aggregation and analyses. Additionally the complexities of data generated from unconnected, disparate systems present challenges because new data science and big data analytics methods must be used to unlock their potential for providing answers. The nurse's unique challenge is to make sense of all the data coming from disparate sources and derive useful actionable information. This chapter and its accompanying four case studies address how the everyday use of information generated through big data analytics can inform practice for frontline nurses: for effectiveness and quality improvement; for changing care delivery that requires sharing information within a care team and across settings; and, for academic researchers for the generation of new knowledge and science.

Keywords Nursing informatics • Knowledge generation • Big data • Big data analytics • Disparate data sources • EHR • Healthcare analytics • Quality and cost improvements • Applied analytics for nursing

E.M. Harper, DNP, RN-BC, M.B.A., FAAN (✉)
School of Nursing, University of Minnesota, Minneapolis, MN, USA
e-mail: ellenharper123@gmail.com

D. McNair, M.D., Ph.D.
Cerner Math, Kansas City, MO, USA

© Springer International Publishing AG 2017
C.W. Delaney et al. (eds.), *Big Data-Enabled Nursing*, Health Informatics,
DOI 10.1007/978-3-319-53300-1_18

18.1 Introduction

The practice of using discrete, quantitative data from multiple data sources and evidence to support clinical decisions began centuries ago when Florence Nightingale invented polar-area diagrams to show that many army soldiers' deaths could be traced to unsanitary clinical practices, and therefore, were preventable. Nightingale used the diagrams to convince policy-makers to implement reforms that eventually reduced the number of deaths as captured in "Causes of Mortality," 1855. Nightingale was able to personally collect, sift through, and analyze the mortality data because the volume of information was manageable. Today, the massive volume of healthcare data that is generated from our healthcare systems requires sophisticated information technologies for storing, aggregation and analyses. Additionally the complexities of data generated from unconnected, disparate systems, sometimes from different parts of the globe, present challenges. Systems must be designed to do real-time, data processing in order to deliver knowledge in a timely way. Getting the right information to the right person at the right time so that informed decisions can be made is how care is impacted. However, before nurses in clinical practice, leadership and research can embrace the use of big data science and analytics, they must be knowledgeable about how to do so. The key element is having nurses involved in health information policy so that nursing data is included in clinical data repositories (CDRs) for analytics and research. Thus "getting the word out" widely throughout nursing domains will be vital to bridge the gap between what new technologies have made possible and the current state of nursing practice.

Every nurse is experiencing seismic changes across the healthcare landscape, fueled by a powerful confluence of economic, demographic and regulatory forces. At the same time we are in the midst of an acute transformation with the adoption and use of health information technology (health IT). As healthcare organizations have grown, the data associated with them has increased quickly. Most health systems have data in multiple systems, formatted in multiple ways, and spread out so much that it is hard to find answers. The nurse's unique challenge is to make sense of all the data coming from disparate sources and derive useful, actionable information. Big data is often described in terms of the "3Vs"—velocity, volume, and variety, as illustrated in Fig. 18.1 with descriptive examples for each domain.

The volume of data generated in today's healthcare environment will be ever increasing due to use of mobile devices and wearables extending care delivery into the community and home, as well as mandates for population health and consumer engagement in their health. Thus, the ability to store, aggregate, and combine data, and then to use the results to perform deep analyses have become more important to health systems and their operational leaders. Nursing is a key stakeholder in this shift and can be greatly empowered by adopting these new tools. Big data provides

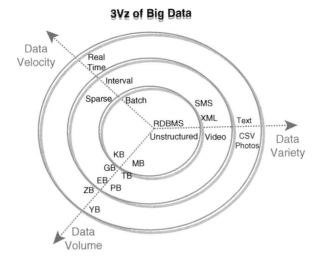

Fig. 18.1   Source: http://blog.sqlauthority.com

an opportunity to find new discernment into existing data as well as guidelines to capture and analyze your future data. Secondary uses or re-use of electronic health record (EHR) data can provide answers for:

- Enhancing individuals' healthcare experiences;
- Expanding knowledge about diseases and treatments;
- Strengthening the healthcare systems' effectiveness and efficiency;
- Supporting public health and security goals; and
- Aiding businesses in meeting customers' needs (Safran et al. 2007).

The Triple Aim, developed by the Institute for Healthcare Improvement (IHI), is a framework to simultaneously pursue three dimensions toward optimizing a health system's performance: improving the patient experience of care (including quality and satisfaction); improving the health of populations; and, reducing the per capita cost of health care (IHI Triple Aim 2015). In addition to the goals of the Triple Aim, healthcare systems are taking on risk for managing patient populations' health and resource utilization. This risk sharing has generated the need for big data, healthcare business intelligence (BI), and analytics to inform healthcare system performance. More nurses are embracing the concept of evidence-based practice, a system in which treatment decisions for individual patients are made based on scientific evidence. In many cases, aggregating data sets into big data algorithms is the best source of evidence, and nurses are prompting other care team disciplines and stakeholders to take action as well. Nursing science and nursing practice has much to gain from data science initiatives (Brennan and Bakken 2015).

18.2 Nursing Informatics as a Valuable Resource and Analytics Team Member

The field of nursing informatics (NI) has much to offer in bridging the worlds of care delivery and data analytics. The application of NI knowledge is essential to enable capturing health and care data in a structured way to accomplish the vision of accurate, reliable, and clinically meaningful measurement across systems and settings of care. The American Nurses Association (ANA) recognized NI as a specialty that "*integrates nursing science with information management and analytical sciences to identify, define, manage, and communicate data, information, knowledge, and wisdom in nursing practice*" (ANA 2015). The Health Information and Management Systems Society (HIMSS) has also been a strong supporter in encouraging healthcare organizations to utilize nurse informaticists (HIMSS 2015; Sensmeier 2015). ANA and HIMSS recognize the value of nurse informaticists in the build, design and implementation of an EHR for accurate concept representation, optimal use of clinical decision support in workflow and coding data to support evidence-based practice, research, and education. Through this work, nursing informaticists support physicians, nurses, consumers, patients, the interprofessional healthcare team, and other stakeholders in their decision-making. Nurse informaticists work across clinical and technical teams to ensure that standard terminologies and coding work is done. It is only through the use of information structures, information processes, and information technology (ANA 2015) that all healthcare data is made available for analysis. In today's world, every nurse leader needs an NI partner as an integral member of their leadership team. The same is true of EHR vendors. The adage—data does not automatically lead to knowledge—highlights the need for EHR vendors to partner with NI nurses for optimum use of data and analytics. In the case studies that follow, NI experts from health systems sought Cerner NI experts to partner with them to solve real problems. Each case study illustrates how essential this partnership is to the successful design, development, and adoption of patient-centered information technology (IT). Without input from nurses, IT projects can become mainly about the technology itself. With NI nurses' input, these projects promote better patient care, increase the clinician end users' sense of fulfillment, and yield higher return on investment for the organization (Brock and Wilcox 2015). NI staff engagement has played an important role in strengthening the research-to-practice bond, providing clinicians and researchers the opportunity to use valid and reliable practice data for further research (Lang 2008; Lundeen et al. 2009).

Big data and nursing informatics. In the 'big data' era, bringing volumes of disparate data together, ontologies and data normalization are fundamental to reliable use of "as-collected" EHR information in observational research (Redeker et al. 2015; Sensmeier 2015; Soares 2013). Healthcare data included in CDRs are increasingly used for quality reporting, business analytics and research. However, data generated by nurses and interprofessional clinical practice are seldom standardized in EHRs nor included in CDRs. The fragmentation of patient data has launched efforts to effectively exchange patient health information between various stakeholders, between health systems, across state lines and the enterprise (Kasthurniathine et al. 2015). With the increasing emphasis on interprofessional practice and care coordination across settings, CDRs need to include data from all members of the care team. By harmonizing data to support research there can be a better understanding of each discipline's impact on patient safety, effectiveness, and efficiency.

Generally speaking, most healthcare organizations' data comes from clinical, financial and operational applications. On their own, each type of data has a specific use. Put together, organizations can start to answer larger, thorny questions such as workflow optimization, staffing needs, efficiency and care quality. This chapter includes four case studies authored by nurses in practice and industry. These case studies demonstrate the power of disparate data sources and analytics methods applied so as to answer complex queries significant to health care (Case study 2015).

18.3   The Knowledge Framework and NI

Given the importance of big data and analytics amidst the healthcare industry today, areas of rapid expansion and future developments in NI fall into this knowledge framework: (a) knowledge acquisition; (b) knowledge processing; (c) knowledge dissemination and improvement; and (d) knowledge generation.

Knowledge acquisition. New technologies and mobile device applications are rapidly influencing how clinical information is being captured and are beginning to have significant impact on NI and nursing research (Henly et al. 2015; Higgins and Green 2008; Gerrish and Lacey 2013; Kaushal et al. 2014). New technologies impacting this trend include: monitoring equipment and body-area sensor networks (Momber et al. 2015; Rezaie and Ghassemian 2015); mobile health ('mHealth'); electronic health ('eHealth') services mediated by consumer-facing SMART (Substitutable Medical Applications & Reusable Technologies) on FHIR (Fast Healthcare Interoperability Resources) apps for soliciting self-reported data from smartphones and tablets; and wearables (small, unobtrusive devices that facilitate high-frequency monitoring and telemetry from the home). These technologies and applications are all starting to contribute to the corpus of information available for care and for research (Hackl et al. 2015). Automated monitoring and patient and caregiver self-reported data can reduce the burden of manual documentation for nursing personnel, albeit accompanied by incremental demands on nurses' cognition, attention, and interpretive effort. Mobile technology is increasingly used for assessing, prompting, educating, and engaging individuals who have chronic conditions (Farrar 2015; Georgsson and Staggers 2015; Hopia et al. 2015). However, the uptake of such technology may be modest, and long-term adherence to its use may be problematic. To date, these aspects have received little attention in the nursing informatics literature. Notably, the Thirteenth International Congress for Nursing Informatics (http://ni2016. org/) in Geneva in 2016 had as its theme "eHealth for all". The American Nursing Informatics Association meeting in San Francisco in 2016 (https://www.ania.org/ ania-2016-conference) likewise emphasized population health and technology-mediated eHealth preventive services.

No one can predict the future, but it is interesting to project where technology and innovation may take nurses and the nursing profession. Increasingly, genomics-based precision health and personalized health interventions (Cullis 2015) entail evaluating the safety, effectiveness, and cost-effectiveness of "long tail" interventions (Anderson 2008). Today we live in a world of ubiquitous information and communication technology, in which retailers have virtually infinite shelf space and consumers can search through innumerable options. Thus popular products become disproportionately profitable for suppliers, and customers become even likelier to converge in their tastes and buying habits. Yet in treatment options, patients receiving a given intervention or treatment regimen accumulate quite slowly because so many variations are offered. Or in contrast, a clinical trial is designed for traditional "one size fits most" treatments. Studies for evaluating new personalized interventions or plans of care are more akin to study designs for rare diseases (Gagne et al. 2014). The small patient populations can diminish business interest in the advancement of medications and therapies. Yet even for those rare illnesses or disorders where financial support is ample and companies are engaged, operational methodology and data restrictions control the ability to generate evidence on patient outcomes. In this future setting, extended clinical data (ECD) are needed to meet the operational and research needs of health systems and nursing leadership (Polit and Beck 2016). By ECD, we mean structured or semi-structured data documented in flowsheets or other online expressions that represent interprofessional patient findings, patient assessments, personalized goals, personalized interventions, and outcomes. Examples of these data include physiological concepts (e.g., pain, pressure ulcers, tissue perfusion, bowel elimination), as well as psychological (e.g., cognitive function, coping, spiritual distress), and functional categories (e.g., activities of daily living, self-care, nutrition) along with health related problems (e.g., medication management, fall risk, omics-based risks). The term 'omics' refers to genomics, transcriptomics, proteomics, epigenomics, exposomics, microbiomics, and metabolomics (Neufe et al. 2015). Presently, HL7 (2015) and other standard development organizations have not yet published standards addressing various kinds of ECD that are generated in telehealth and personalized/precision health (Phillips and Merrill 2015).

Knowledge dissemination and improvement. Effectively communicating data, information, knowledge, and wisdom in nursing practice (ANA 2015) was mentioned earlier in the chapter; however, it is not a specific focus of the accompanying four Case Studies. Our shared perspective is that communication and dissemination methods ought to receive greater attention in the future, as part of advancing NI and advancing improvements in care. Social media-based case management and virtual nursing are online endeavors and fundamentally depend on good workgroup applications. Today's web-based software tools (e.g., https://basecamp.com/; https://trello.com; https://www.oracle.com/applications/primavera/index.html; http://www.projectlibre.org/projectlibre-documentation; http://metacommunications.com/; http://www.sciforma.com/product/project_portfolio_management_overview) make collaborative authoring, critiquing, and improvement of content for professional nursing practical and available on a virtual, geography-independent basis.

The Accountable Care movement and electronic clinical quality measures (eCQM) for nursing require yet more kinds of new, online tools to facilitate fair and responsive stewardship of nursing outcomes measures. Using the ISO 9241-11 usability standard, the System Usability Scale (SUS) and similar instruments for assessing satisfaction and measuring user characteristics will likely grow in adoption and facilitate progressive interdisciplinary, collaborative practice (Pilon et al. 2015).

Knowledge generation. Big Data, Hadoop and No-SQL databases represent a completely different software ecosystem that allows for massively parallel, vertical computing within a framework for high-performance and agile processing of information at massive scale. These database infrastructures have been specifically adapted to the heavy demands of big data. Graph-theoretic databases often used to map genome data, temporal databases with built-in support for handling data involving time, and unstructured and semi-structured data are all becoming more prevalent in NI (Brennan and Bakken 2015). With data sets that are interoperable and progressively larger in scope, there is an increasing need for nursing nomenclatures and ontology that are mapped to the adopted standards of Systematized Nomenclature of Medicine (IHTSDO 2015) and Logical Observation Identifiers Names and Codes (LOINC 2015), to power improved decision support and knowledge discovery in the learning health system. In this connection, privacy-protected de-identified clinical data can be seen as a public good, in much the same way as clean air and water and other resources (McNair 2015). Privacy issues must be addressed in a robust and publicly transparent manner, and the technology to do this is reasonably well understood (Dwork and Roth 2014; El Emam and Arbuckle 2013). At the same time we are in the midst of an acute transformation with the adoption and use of health information technology (health IT) which is generating increasing amounts of patient care data available in computable form. Yet in the current state, we find ourselves unable to objectively provide the data that nursing care does make a difference in safety, quality and patient and nursing staff outcomes. However, the SNOMED and LOINC standards represent the tools that will allow us to codify nursing data and open it up for analyses and big data analytics. Thus, there is a quiet confidence that nurses' knowledge and use of technology will continue to improve as a collective, not only for our practice but also in the communities we serve. There are great opportunities for nurses to use the EHRs to support nursing's workflows, enhance critical thinking and turn data into information and knowledge that can impact care significantly.

In a caring-centered, knowledge-enabled future, nurses have an increasingly key role in health services design and research. Systems biology and personalized omics applications are being added to nursing, so that nursing as a high-touch, caring-oriented profession also has an increasing knowledge-worker component to its essence (De Chesnay and Anderson 2015; Unerti et al. 2015). Changing demographics, underserved populations, disparities and vulnerabilities, and Pay-for-Value initiatives are all affecting health and care services. In the United States, health care reform and Obamacare (Patient Protection and Affordable Care Act 2010) bring increased attention to preventive services (Vogt et al. 2004; Waitman et al. 2011). The success of these new science developments in the context of regulatory and legislative health system reform mandates rely heavily on nursing knowledge, on nursing management, on adequate nurse staffing and efficient workflow, and apt nursing practice and policies. Increasingly, some types of research may lead not only to changes in nursing practice but also to discovery of new nursing concepts and evolution of nursing theory, as illustrated by the contributions provided in the four Case Studies that follow this chapter. Nurse researchers perform analyses to improve health, identify and characterize factors that contribute to good and bad outcomes, and determine the comparative safety and effectiveness of various interventions. These researchers also discover new relationships between the process and content of care services and the financial and clinical outcomes they yield (Albert 2015; Conley et al. 2015).

To accurately address complex patterns of clinical findings, process variables and outcomes, nurse researchers increasingly need to go beyond conventional empirical means of identifying the independent variables and the dependent outcomes variables. Often, input and output patterns of prevention and care services delivery defy the insight and intuition of any one human being or standard biostatistical measures. Frequently, this is because the relevant patterns are quite "dilute" or rare. A nurse may never have seen a particular pattern in her/his entire practice life. Such uncommon patterns simply do not occur often enough to be fully understood or regarded as significant. Even several dozen investigational centers' or institutions' case volume may not be sufficient to detect the patterns. However, several hundred centers' data—such as those accumulated in big data HIPAA-compliant de-identified repositories—are sufficient to detect and analyze these low frequency patterns with adequate statistical power.

Other situations involve patterns that occur frequently but throughout multiple "venues" and episodes of care, such that no one person's scope of practice encompasses all the elements of a sequence or trajectory. And yet other times, the patterns have items missing from instance to instance and from case to case. Modern research tools, including 'machine learning' algorithms (Lantz 2013; Murphy 2012) will enable detection and analysis of patterns like this. In the context enabled by such tools, nursing informatics is increasingly becoming a kind of citizen/crowd science. The four case studies by this chapter's contributors are emblematic of these aspects of nursing research. For instance, workflow and operations research studies (McNair 2015) are becoming more common now that some EHR systems incorporate 'instrumented' code in the software so that nurse engagement with the system as well as with the patient is automatically measured as a byproduct of using the system.

There are a number of current national and regional efforts to share data between CDRs for the purpose of supporting research activities including identification of cohorts for clinical studies and comparative effectiveness of research from multiple sites. These include:

- Commonwell Health Alliance (http://www.commonwellalliance.org/)
- i2B2 (https://www.i2b2.org/about/; Warren et al. 2012)
- ViPAR network (Carter et al. 2015)
- eMERGE network (https://emerge.mc.vanderbilt.edu/)
- Health Care Systems Research Network (formerly HMO Research Network (HMORN), (http://www.hcsrn.org/en/)
- Clinical Data Research Networks (CDRNs) making up Patient-Centered Outcomes Research Network (PCORnet), (http://www.pcornet.org/; Fleurence et al. 2014)
- MiniSentinel (http://www.mini-sentinel.org/; Behman et al. 2011)
- Observational Medical Outcomes Partnership (OMOP) (http://omop.org/ ; Stang et al. 2010)
- Observational Health Data Sciences and Informatics (OHDSI) (http://www.ohdsi.org/).

All of the above listed networks have generally focused on a restricted set of data. That includes patient demographics such as age, gender, race, marital status, and geographic location; medical diagnoses; medical or surgical procedures; encounters; and laboratory orders and results, as well as medications. However, these domains represent only a relatively small fraction of the data about the patients and the care provided to them that is captured in the EHR. To date, those networks' datasets have not included nursing interventions, flowsheet items, nursing documentation, nursing outcomes and diagnoses, stage or severity or acuity of illness, patient goals, patient education, patient or family satisfaction, operational workflow data of nurses or other clinician users of EHR systems that mediate delivery of care and prevention services, provider prescribing patterns, staffing patterns, data on local air pollutants and other diurnal environmental exposures, or other data that are now captured in some EHRs.

Researchers have traditionally used evidence from randomized controlled trials (RCTs) to determine the efficacy of a treatment or intervention under idealized conditions (Terry 2011; Von Ballmoos et al. 2015). By contrast, all of the Case Studies are examples of observational research study designs. Observational designs are often used to measure the effectiveness of an intervention in "real world" scenarios. Modifications of existing designs, including both randomized and observational ones, are used for comparative effectiveness research in an attempt to give an unbiased estimate of whether one treatment is more effective or safer than another for a particular population. A systematic analysis of study design features, risk of bias, parameter interpretation, and effect size for all types of randomized and non-experimental observational studies is needed to identify specific differences in design types and potential biases. Anglemyer et al. (2014) report results from 15 systematic reviews citing 4406 studies and found similar results reported by randomized controlled trials (RCT). They concluded that, on average, there is little evidence of differences in conclusions between well-designed observational studies and RCT studies. While any well-designed study can produce high-quality evidence, there are multiple challenges with prevention intervention RCTs or observational studies (Golfam et al. 2015). New hybrid designs that carry benefits of both randomized and observational methods may be the best way forward (Frakt 2015). Even regulatory agencies that have in the past preferred RCTs are now searching for alternative approaches, such as observational studies (Hershman and Wright 2012).

Knowledge processing. With health reform driving more proactive and preventive ambulatory care and curtailing reactive and acute care, it is not surprising that care venue boundaries of nursing services are increasing. Oyri and colleagues have referred to this expanding boundaries trend as a future state characterized by 'ubiquitous nursing' (Oyri et al. 2007). In that connection, professional nursing practice and tools are evolving via telehealth, mHealth and eHealth nursing as noted above. Software-based tools are beginning to appear that promote greater nursing labor efficiency (McNair 2015), reduced fragmentation and increased continuity of care and more nurse time spent in meaningful contact with the patient, either at the bedside in acute and residential settings or virtually in mHealth or eHealth settings. Team communications will likely change as a result of this evolving work mix (Booth et al. 2015). Research focused on team communication may reveal undesired consequences of workflow changes and inter-professional communication patterns and enable those to be properly addressed.

Increasingly, NI must include risk of bias assessments in the evaluation of evidence according to the Grading of Recommendations Assessment, Development, and Evaluation system (GRADE) or similar system (Brunwtti et al. 2013; Guyatt et al. 2011). Despite improvements in methodology to reduce and manage bias, there can be issues with over-optimistic reporting of preliminary results; sample sizes that are too small to statistically power the reported decisions and actions; well-intentioned but misapplied machine-learning methods that neglect the time-honored statistical methods and technology evaluation policies; questionable modelling decisions; and problematic validation methodologies (Grove and Cipher 2016). Additionally, there can be substantial issues (a) in the statistical analysis feature selection process, via LASSO or gradient boosting or other methods, whereby nursing researchers can select features before modelling begins, and (b) in the validation of the models on the same data used for model development and training. Thus, with the availability of big data there is increasing impetus for improved methods for properly processing clinical information from disparate interoperable sources. Big data analytic methods are the domain of the new data scientists and are key for managing the analytic process to ascertain reliable, new knowledge and actionable decisions from the data (Flach 2012). 'Bigger' offers new opportunities to ask, and successfully answer, important questions, but bigger is not always 'better'. Thoughtful, careful critiquing goes hand-in-hand with creative, successful innovation. In light of all of these factors, a most exciting era for frontline nurses, operational leaders, NI and nursing research lies straight ahead.

18.4   Conclusion

This chapter offers a positive view of the future with intensive, continuing development of technology and improving global economies, such as would make possible ongoing attention to prevention and bettering the public health. Healthcare is in a period of significant transformational activity through the accelerated adoption of healthcare technologies including big data and 'omics; new reimbursement systems that emphasize quality measurement, shared savings and care coordination; and more extensive use of mobile and wearable technologies by patients, providers, family member caregivers, and others. In particular, the change-resistant nature of the healthcare system continues to create barriers to direct consumer involvement and engagement necessary for transformational change. In this connection, the historic patient advocacy role of the nurse means that NI has strategic importance in achieving engagement and transformation.

As nurses we recognize that nursing is both an art and a science, and that health IT is a tool to be leveraged to enhance the critical thinking skills that nurses bring to the delivery of care. Unlike paper documentation, data is no longer regarded as static or stale, whose usefulness is finished once the primary purpose (documentation of care, support for charging, etc.) for which it was collected was achieved. Secondary use of data has become a raw material used to create innovation (Harper 2013). The opportunity to capitalize on the vast amount of health and care data that are captured, stored and processed is now a reality and will be demonstrated through the four case studies that follow. These case studies demonstrate the power of partnership when the expert data analytic and NI resources from an EHR vendor are made available to its clients to address initiatives most pressing to that unique organization. The case studies also serve as exemplars of best practice on vendor/healthcare systems partnership. Additionally, extraordinary ideas, collaboration, and learning occur when nurses in academic, practice and informatics are brought together to share what they have done that works, and, equally as important, what does not work in creating the learning health environment. These case studies demonstrate the use of data to answer questions and enlighten areas that were previously invisible.

References

Albert NM, editor. Building and sustaining a hospital-based nursing research program. New York: Springer; 2015.

American Nurses Association. Nursing informatics: scope and standards. Silver Spring, MD. 2015. nursebooks.org.

Anderson C. The long tail: why the future of business is selling more for less. New York: Hyperion/Hachette; 2008.

Anglemyer A, Horvath HT, Bero L. Healthcare outcomes assessed with observational study designs compared with those assessed in randomized trials. Cochrane Database Syst Rev. 2014. http://www.cochranelibrary.com/cochrane-database-of-systematic-reviews/table-of-contents-cdsr.html. Accessed 20 Jan 2016.

Behman RE, Benner JS, Brown JS, McClellan M, Woodcock J, Plaatt R. Developing the Sentinel System—a national resource for evidence development. N Engl J Med. 2011;364(6):498–9. doi:10.1056/NEJMp1014427.

Booth RG, Andrusyszyn MA, Iwasiw C, Donelle L, Compeau D. Actor. Network theory as a sociotechnical lens to explore the relationship of nurses and technology in practice: methodological considerations for nursing research. Nurs Inquiry. 2015. doi:10.1111/nin.12118.

Brennan PF, Bakken S. Nursing needs big data and big data needs nursing. J Nurs Sch. 2015;47(5):477–84. doi:10.1111/jnu.12159.

Brock VP, Wilcox D. Incorporating technology as a tool for improving quality of care. Am Nurse Today. 2015;10(9):11–2.

Brunwtti M, Shemitt I, Pregno S, Oxman AD, Jaeschke R. GRADE guidelines: 10. Considering resource use and rating the quality of economic evidence. J Clin Epi. 2013;66(2):140–50.

Carter KW, Francis RW, Carter KW, Francis RW, Bresnahan M, Gissler M, Grønborg TK, Gross R, Gunnes N, Hammond G, Hornig M. ViPAR: a software platform for the virtual pooling and analysis of research data. Int'l J Epi 2015;8;dyv193. Accessed 19 Jan 2016.

Case studies. http://writing.colostate.edu (2015). Accessed 21 Nov 2015.

Cullis P. The personalized medicine revolution: how diagnosing and treating disease are about to change forever. Vancouver, BC: Greystone Books Ltd; 2015.

Conley YP, Heitkemper M, McCarthy D, Anderson CM, Corwin EJ, Daack-Hirsch S, Taylor JY. Educating future nursing scientists: recommendations for integrating omics content in PhD programs. Nurs Outlook. 2015;63(4):417–27.

De Chesnay M, Anderson B. Caring for the vulnerable: perspectives in nursing theory, practice and research. Boston: Jones & Bartlett; 2015.

Diagram of the causes of mortality in the army in the east. https://ihm.nlm.nih.gov (1855). Accessed 21 Nov 2015.

Dwork C, Roth A. Algorithmic foundations of differential privacy. New York: NOW Publishers; 2014.

El Emam K, Arbuckle L. Anonymizing health data: case studies and methods to get you started. 1st ed. Sebastopol, CA: O'Reilly Media; 2013.

Farrar FC. Transforming home health nursing with telehealth technology. Nurs Clin N Am. 2015;50(2):269–81.

Flach P. Machine learning: the art and science of algorithms that make sense of data. Boston: Cambridge University Press; 2012.

Fleurence RL, Curtis LH, Califf RM, Platt R, Selby JV, Brown JS. Launching PCORnet, a national patient-centered clinical research network. JAMIA. 2014;21(4):578–82.

Frakt AB. An observational study goes where randomized clinical trials have not. JAMA. 2015;313(11):1091–2.

Gagne JJ, Thompson L, O'Keefe K, Kesselheim AS. Innovative research methods for studying treatments for rare diseases: methodological review. Br Med J. 2014;349:g6802–12.

Georgsson M, Staggers N. Quantifying usability: An evaluation of a diabetes mHealth system on effectiveness, efficiency, and satisfaction metrics with associated user characteristics. JAMIA. 2015;23(1):5–11.

Gerrish K, Lacey A. The research process in nursing. New York: John Wiley & Sons; 2013.

Golfam M, Beall R, Brehaut J, Saeed S, Relton C, Ashbury FD, Little J. Comparing alternative design options for chronic disease prevention interventions. Eur J Clin Invest. 2015;45(1):87–99.

Grove SK, Cipher DJ. Statistics for nursing research: a workbook for evidence-based practice. Atlanta: Elsevier Health Sciences; 2016.

Guyatt GH, Oxman AD, Vist G, Kunz R, Brozek J, Alonso-Coello P. GRADE guidelines: 4. Rating the quality of evidence—study limitations (risk of bias). J Clin Epidemiol. 2011;64(4): 407–15.

Hackl WO, Rauchegger F, Ammenwerth E. A nursing intelligence system to support secondary use of nursing routine data. Appl Clin Infor. 2015;6(2):418–28.

Harper EM. The economic value of health care data. Nurs Admin Q. 2013;37(2):105–8.

Health Information and Management Systems Society. Guiding principles for big data in nursing: using big data to improve the quality of care and outcomes. http://www.himss.org/big10 (2015). Accessed 1 Dec 2015.

Henly SJ, McCarthy DO, Wyman JF, Stone PW, Redeker NS, McCarthy AM, Alt-White AC, Dunbar-Jacob J, Titler MG, Moore SM, Heitkemper MM. Integrating emerging areas of nursing science into PhD programs. Nurs Outlook. 2015;63(4):408–16.

Hershman DL, Wright JD. Comparative effectiveness research in oncology methodology: observational data. J Clin Onc. 2012;30(34):4215–22.

Higgins JP, Green S. Cochrane handbook for systematic reviews of interventions. Chichester: Wiley-Blackwell; 2008. http://www.cochranelibrary.com/cochrane-database-of-systematic-reviews/table-of-contents-cdsr.html. Accessed 28 Dec 2016

HL7. Introduction to HL7 standards. 2015. http://www.hl7.org/implement/standards/index.cfm?ref=nav. Accessed 15 Oct 2015.

Hopia H, Punna M, Laitinen T, Latvala E. A patient as a self-manager of their personal data on health and disease with new technology: Challenges for nursing education. Nurs Ed Today. 2015;35(12):e1–3. doi:10.1016/j.nedt.2015.08.017.

Institute for Healthcare Improvement. IHI triple aim. http://www.ihi.org/engage/initiatives/tripleaim/Pages/default.aspx (2015). Accessed 20 Nov 2015.

Kasthurniathine SN, Mamlin B, Kumara H, Grieve G, Biondich P. Enabling better interoperability for healthcare: Lessons in developing a standards-based application programming interface for electronic medical record systems. J Med Sys. 2015;39(11):1–8.

Kaushal R, Hripcsak G, Ascheim DD, Bloom T, Campion TR, Caplan AL, Tobin JN. Changing the research landscape: the New York City clinical data research network. JAMIA. 2014;21(4):287–90. doi:10.1136/amiajnl-2014-002764.

Lang NM. The promise of simultaneous transformation of practice and research with the use of clinical information systems. Nurs Outlook. 2008;56:232–6.

Lantz B. Machine learning with R. Birmingham, UK: Packt Publishing; 2013.

Location Observation Identifiers Names and Codes (LOINC). https://loinc.org/ (2015). Accessed 15 Jan 2016.

Lundeen S, Harper E, Kerfoot K. Translating nursing knowledge into practice: an uncommon partnership. Nurs Outlook. 2009;57:173–5.

McNair D. Enhancing nursing staffing forecasting with safety stock over lead time modeling. Nurs Admin Q. 2015;39(4):291–6.

Member C, Legako K, Gilchrist A. Identifying medical wearables and sensor technologies that deliver data on clinical endpoints. J Clin Pharmaco. 2015. doi: 10.1111/bcp.12818.

Murphy KP. Machine Learning: A Probabilistic Perspective (1st ed.). Cambridge: MIT Press; 2012.

Neufe PD, Bamman MM, Muoio DM, Bouchard C, Cooper DM, Goodpaster BH, Hepple RT. Understanding the cellular and molecular mechanisms of physical activity-induced health benefits. Cell Met. 2015;22(1):4–11.

Oyri K, Newbold S, Park H, Honey M, Coenen A, Ensio A, Jesus E. Technology developments applied to healthcare/nursing. Stud Hlth Tech and Inform. 2007;128:21–37.

Phillips A, Merrill J. Innovative use of the integrative review to evaluate evidence of technology transformation in healthcare. J Biomed Inf. 2015;58:114–21. doi:10.1016/j.jbi.2015.09.014.

Pilon BA, Ketel C, Davidson HA, Gentry CK, Crutcher TD, Scott AW, Rosenbloom ST. Evidence-guided integration of interprofessional collaborative practice into nurse-managed health centers. J Prof Nurs. 2015;21(4):340–50.

Polit D, Beck C. Nursing research: generating and assessing evidence for nursing practice. Boston: Wolters-Kluwer; 2016.

Redeker NS, Anderson R, Bakken S, Corwin E, Docherty S, Dorsey SG, Grady P. Advancing symptom science through use of common data elements. J Nurs Sch. 2015;47(5):379–88. doi:10.1111/jnu.12155.

Rezaie H, Ghassemian M. Implementation study of wearable sensors for activity recognition systems. Hlthcare Tech Lett. 2015;2(4):95–100. doi:10.1049/htl.2015.0017.

Safran C, Bloomrosen M, Hammond WE, Labkoff S, Markel-Fox S, Tang PC, Detmier DE. Toward a national framework for the secondary use of health data: an American Medical Informatics Association white paper. JAMIA. 2007;14(1):1–9.

Sensmeier J. Big data and the future of nursing knowledge. Nurs Mgt. 2015;46(4):22–7.

International Health Terminology Standards Development Organization (IHTSDO). SNOMED CT. http://www.ihtsdo.org/snomed-ct (2015). Accessed 15 Jan 2016.

Soares S. Big data governance: an emerging imperative. Boise, ID: McPress; 2013.

Stang PE, Ryan PB, Racoosin JA, Overhage JM, Hartzema AG, Reich C, Welebob E, Scarnecchia T, Woodcock J. Advancing the science for active surveillance: rationale and design for the observational medical outcomes partnership. Ann Int Med. 2010;153(9):600–6.

Terry AJ. Clinical research for the doctor of nursing practice. Jones & Bartlett Learning: Sudbury, MA; 2011.

Unerti KM, Schaefbauer CL, Campbell TR, Sentero C, Siek KA, Bakken S, Vieinot TC. Integrating community-based participatory research and informatics approaches to improve the engagement and health of underserved populations. JAMIA. 2015;23(1):60–73.

Patient protection and affordable care act. 2009. http://www.hhs.gov/healthcare/about-the-law/read-the-law/index.html. Accessed 8 Nov 2015.

Vogt TM, Elston-Lafata J, Tolsma DD, Greene SM. The role of research in integrated health care systems: The HMO research network. Permanente. 2004;J10(4):10–7.

Von Ballmoos MC, Ware JH, Haring B. Clinical Research Quo Vadis? Trends in reporting of clinical trials and observational study designs over two decades. J Clin Med Res. 2015;7(6):428–34.

Waitman LR, Warren JJ, Manos EL, Connolly DW. Expressing observations from electronic medical record flowsheets in an i2b2 based clinical data repository to support research and quality improvement. AMIA Annu Symp Proc. 2011:1454–63.

Warren JJ, Manos EL, Connolly DW, Waitman LR. Ambient findability: developing a flowsheet ontology for i2B2. In: NI 2012: Proceedings of the 11th International Congress on Nursing Informatics, Montreal. American Medical Informatics Association, Washington DC. 2012. http://www.ncbi.nlm.nih.gov/pmc/articles/PMC3799091/. Accessed 10 Dec 2015.

Case Study 18.1: Alarm Management: From Confusion to Information

Kevin Smith and Vicki Snavely

Keywords Alarm safety • Nursing informatics • Alert fatigue • Patient safety

18.1.1 Introduction

For the last several years, alarm fatigue hazards have been on the radar of safety institutions such as the Association for the Advancement of Medical Instrumentation, Emergency Care Research Institute and The Joint Commission. Indeed, The Joint Commission made alarm safety a top priority by including the issue as a National Patient Safety Goal in 2014 and 2015 (TJC 2014 patient safety goal 2014, TJC Patient safety goal 2016–2015). In this context, NCH Healthcare System (NCH), one of the most progressive systems in Naples, FL, worked with its information technology (IT) provider, Cerner Corporation, to create and implement a process improvement initiative designed to mitigate alarm fatigue.

In September 2013, NCH launched our alarm management journey by sending secondary alerts from primary devices (such as cardiac monitors) to the nurse's communication device (Apple iPhones®). Reporting of the number and types of alerts data was reviewed shortly after go live to some surprise. Besides the sheer volume of alerts, many were nuisance alerts (alerts that required no intervention). In response to a potential alert fatigue issue, NCH formed an interdisciplinary Alarm Safety Committee to provide structure and governance to the decisions and policies that ultimately made up the NCH Alarm Management policy. Throughout this data-driven process, NCH minimized alarm fatigue and made the nursing units a quieter, safer care environment. This case study describes the work and results of those projects.

18.1.2 Testing New Technology

The project started by moving from a centralized telemetry monitoring system to a decentralized model using the secondary alerting capability of our EHR provider's *CareAware AlertLink*™. The solution uses a communication platform, *CareAware Connect*™ to route the real-time alerts from devices automatically to the patient's assigned nurse's NCH issued iPhone. As with any new and emerging technology, NCH was happy to be selected as a development partner by our EHR provider.

K. Smith, B.S.N., R.N., CNML, CVRN-BC (✉)
Naples Healthcare System, Naples, FL, USA
e-mail: kevin.smith@nchmd.org

V. Snavely, R.N.
Cerner Corporation, Kansas City, MO, USA

Together we learned and collaborated on the communication of secondary alerting within an overall communication strategy and testing the capabilities of a data communication platform (iBus™) and warehouse.

A key goal was to meet the requirements set forth by The Joint Commission (2016, 2014), and the National Patient Safety Goals. Our first step was to create an Alarm Safety Committee with representatives and nurses from NCH and from our EHR partner. The committee objectives were:

- Establish alarm management as an organizational priority
- Identify and prioritize the most important alarms to manage based on our alarm data history, current best practices, and industry standards
- Create policies/procedures to ensure alarm safety
- Educate the staff to the new alarm management workflow

The interdisciplinary committee was headed by the system Chief Nursing Officer, and included Directors of Nursing, Biomed, Respiratory Therapy, Safety/Quality, Nursing Education and an internal Joint Commission representative.

Once all the device alarms were centralized through the single communication platform, NCH was surprised at the volume and types of alarms that were being generated. Without centralizing these into one place it was very hard to create the alarm picture. Our first priority was reducing the number of alarms safely. As a Magnet organization, we started with a literature review on best evidence-based practice. Based on literature recommendations, NCH selected four prevalent types of lethal dysrhythmia alarms: asystole, ventricular tachycardia, ventricular fibrillation and pause. Even with limiting to only these lethal alarm types being sent to the primary device, the numbers were still considered too high. NCH decided to postpone the secondary alerting initiative in order to better understand and manage the entire alarm ecosystem. The committee felt there was more to be learned from the data.

18.1.3 Data-Driven Monitor Management

The Alarm Management Shift Report was developed to analyze NCH's alarm ecosystem. The report displayed the frequency, number, and type of alarms for each patient's bed by the hour. This data was instrumental in highlighting and identifying the nonactionable alarms being generated. Because the devices were all set to the manufacturer setting and not customized to the patients, the committee determined that a process change was needed to reduce the alarm noise.

In June 2014, staff received 30–45 min training sessions on the telemetry monitors. Using the evidence-based foundation and clinical scenarios, nurses were shown how to adjust the generic monitor parameters to patient specific alarm limits. The teams relied heavily on a data driven process as they continued to analyze the alarm history data and impact on practice. Through multiple iterations of reviewing the alarm history data, including the types of alarms that fired, alarm load (which beds and types of patients had the most alarms), and prevalence of misfires, a new process and policy for alarm management was developed. For example, the team focused on tachycardia alarms, which were high-volume and are often generated as artifact (not a true alarm). A process change was made. The electrodes were changed every night on all patients and replaced with lead hygiene (a specific skin preparation of washing and drying of the skin with a towel) to ensure good connectivity ("AACN alarm management" 2013). The process change was able to significantly reduce nuisance alarms (Fig. 18.1.1).

Our alarm history data confirmed that more than 80 percent of the alarms came from less than 20 percent of the patients. This was in line with other industry research ("AACN alarm management" 2013). Thus, the need to identify the patients with the most alarms became a priority. Working closely with our EHR partner we were able to develop a "heat map" view, which gives a graphical representation of which patients were generating the most alarms. The larger the box, the more alarms had been generated from that patient's room. The patient in room 4119 had the highest number of alarm fire (Fig. 18.1.2).

While the heat map view provided a good overview, it did not deliver the granularity that NCH needed to be able to manage their alarms in "near-real-time" fashion. Important data on which alarm fired, the time, and how often it fired was needed to understand what had generated all of the alarm noise. It was from this need and continued data analysis that the Alarm Management Shift Report was created. The report ultimately helped NCH analyze the frequency, number, and types of alarms generated by every patient for each hour. By organizing the data into information that was meaningful and actionable, nurses were able to make decisions to better manage the alarm load.

The teams spent several months on data analysis for process improvement validation and every patient at any time of the day. Any primary cardiac monitor device that fired an alarm and was on the NCH IT network and connected to our EHR partner's communication platform displayed on the report. This milestone was significant in moving the utilization of retrospective, unit-level data to a near real-time report (Fig. 18.1.3).

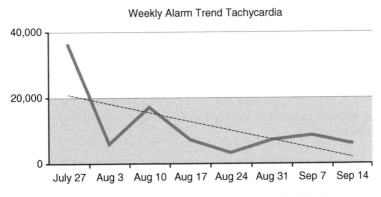

Fig. 18.1.1 Shows the decrease in tachycardia alarms with improved lead hygiene

Fig. 18.1.2 The "heat map" view of the alarm activity

In April 2015, our EHR partner set up a HIPAA-compliant email distribution system that sent the reports to the unit charge nurses three times a day. The charge nurses, in turn, used the report to identify patients whose cardiac monitor setting or lead hygiene could be adjusted to reduce the alarm load for their shift. The Alarm Management Shift Report gave NCH new insights into alarm activity and helped clinicians identify and address practices that led to a dramatic reduction in non-actionable, nuisance alarms. The ability to see the alarm activity as it was generated for each patient provided the staff with the information needed to make changes.

18.1.4 Results

NCH measured the impact of the Alarm Management process on a 45-bed telemetry unit. Over a four-month period, the pilot telemetry unit lowered its total number of alarms from 255,912 in January 2015 to 79,486 in April 2015 (Fig. 18.1.4).

This decrease in nuisance alarms was the result of practice change that included the following:

- Full assessment of the current state of alarm management for NCH
- Gap analysis between evidence-based best practices and current state
- Implementation of policies for patient specific alarm limits

Shift Reports - Initial Alarms

Date Range: All to next 24 hours.
Last Run:3/6/2015 7:30:05 PM

Hours 18–23 are dated 03/05/2015; hours 00–18 are dated 03/06/2015.

Room	Event Type	18	19	20	21	22	23	00	01	02	03	04	05	06	07	08	09	10	11	12	13	14	15	16	17	18	Sum	%
4146	HEART_RATE_HIGH	26	20	46	2	-	-	-	-	6	-	3	4	4	-	13	10	-	-	-	1	6	8	-	-	3	30	182	40,81
	NO_TELEM	-	-	-	-	3	-	-	-	-	-	-	-	-	-	-	-	-	-	-	-	-	-	3	-	19	6	31	6,95
	PVC_HI	-	-	-	-	-	-	-	-	-	-	-	2	3	4	12	14	-	3	-	-	6	-	1	-	1	5	51	11,43
	SILENCE	-	-	1	-	-	-	-	-	-	-	-	-	-	-	-	1	-	1	1	1	1	1	1	1	-	2	11	2,47
	TACHY	25	-	43	-	-	-	-	-	2	-	2	5	5	-	16	13	-	-	-	-	6	9	2	-	2	27	157	35,2
	UNSILENCE	1	-	-	-	-	-	-	-	2	-	2	-	-	-	3	1	-	-	-	-	-	-	1	-	1	1	12	2,69
	V-TACH	1	-	-	-	-	-	-	-	-	-	-	-	-	-	-	-	-	-	-	-	-	-	-	-	-	-	1	0,22
	VT_HIGH	-	-	-	-	-	-	-	-	-	-	-	-	-	-	-	-	-	-	-	-	-	1	-	-	-	-	1	0,22
4147	ACCELERATED_VENT	-	1	-	-	-	-	-	-	-	-	-	-	-	-	-	-	-	-	-	-	-	1	-	-	-	-	2	0,04
	HEART_RATE_HIGH	-	-	55	170	103	57	7	23	20	33	29	31	29	32	55	13	33	117	135	149	73	13	2	-	10	19	1.208	21,85
	NO_TELEM	-	-	1	-	-	-	-	-	-	-	-	3	-	-	-	-	-	1	-	-	-	-	-	-	13	9	27	0,49
	PVC_HI	3	3	72	236	137	96	34	57	67	76	76	79	105	59	96	51	72	231	253	242	140	57	35	-	47	47	2.371	42,88
	SILENCE	-	-	-	6	1	-	-	-	-	-	12	-	4	1	5	1	1	1	1	1	2	1	3	-	1	4	45	0,81
	TACHY	-	-	59	180	120	70	15	47	50	50	81	77	67	42	71	27	64	187	192	198	105	19	4	-	23	34	1.782	32,23
	UNSILENCE	-	-	-	6	1	-	-	-	-	-	12	-	3	1	5	1	1	2	1	1	2	1	3	-	1	4	45	0,81
	V-TACH	-	-	-	2	-	-	-	-	-	-	-	-	-	-	-	-	-	-	-	1	-	-	-	-	-	-	3	0,05
	VT_HIGH	-	-	4	2	5	-	-	-	-	1	-	1	-	1	1	4	-	3	2	6	5	4	2	-	1	4	46	0,83

Fig. 18.1.3 Alarm Management Shift Report

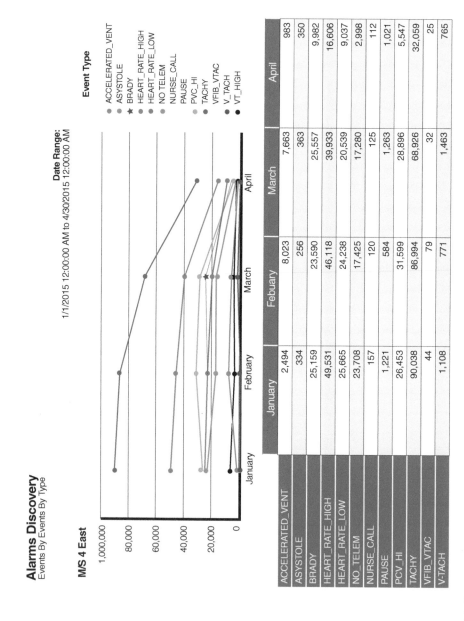

Alarms Discovery
Events By Events By Type

MS 4 East

Date Range:
1/1/2015 12:00:00 AM to 4/30/2015 12:00:00 AM

Event Type

● ACCELERATED_VENT
● ASYSTOLE
● BRADY
● HEART_RATE_HIGH
● HEART_RATE_LOW
● NO TELEM
● NURSE_CALL
● PAUSE
● PVC_HI
● TACHY
● VFIB_VTAC
● V_TACH
● VT_HIGH

	January	Febuary	March	April
ACCELERATED_VENT	2,494	8,023	7,663	983
ASYSTOLE	334	256	363	350
BRADY	25,159	23,590	25,557	9,982
HEART_RATE_HIGH	49,531	46,118	39,933	16,606
HEART_RATE_LOW	25,665	24,238	20,539	9,037
NO_TELEM	23,708	17,425	17,280	2,998
NURSE_CALL	157	120	125	112
PAUSE	1,221	584	1,263	1,021
PCV_HI	26,453	31,599	28,896	5,547
TACHY	90,038	86,994	68,926	32,059
VFIB_VTAC	44	79	32	25
V-TACH	1,108	771	1,463	765

Fig. 18.1.4 Shows the decline in alarms accomplished with the Alarm Management Shift Report and other interventions

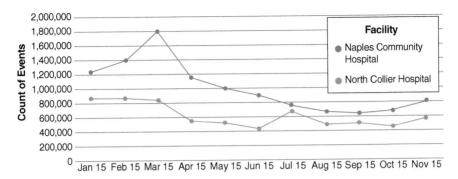

Fig. 18.1.5 Overall alarm reductions

- Implementation of policies for lead hygiene
- Data-driven alarm safety committee led by key stakeholders
- Education for staff on improved monitor management
- Introduction of alarm management data as part of shift hand-off

Figure 18.1.5 demonstrates the overall reduction in alarm events from January 2015 to November 2015. Over the 10-month period, NCH showed a decrease in overall system alarm volume by 63% from combined hospital totals of 2,065,927 in January to 1,315,637 in November. A lesson learned was ongoing data analysis is important to sustainability. The slight increase in October and November correlated back to an intake of new nurse residents. NCH identified a plan to reeducate each new nurse resident.

18.1.5 Conclusion

NCH was able to embed our alarm management best practices through analytics, broader stakeholder engagement, process improvement, and technology tuning into the nurses' daily activities. Our alarm management team has come a long way, with the dramatic reduction of alarms. We have changed the whole culture of alarm safety at NCH. Now, when an alarm sounds, the staff knows it means something.

References

AACN practice alert, alarm management. 2013. http://www.aacn.org. Accessed 21 Nov 2015.
The Joint Commission. 2014 national patient safety goals NPSG.06.01.01. Improve the safety of clinical alarm systems. 2014. http://www.jointcommission.org. Accessed 22 Nov 2015.
The Joint Commission. 2016 national patient safety goals NPSG.06.01.01. Improve the safety of clinical alarm systems. 2015. http://www.jointcommission.org. Accessed 22 Nov 2015.

Case Study 18.2: Nursing Time in the Electronic Health Record: Perceptions Versus Reality

April Giard and Darinda Sutton

Keywords Nursing informatics • Electronic time capture • Nurse documentation time • Nurse documentation time by task

18.2.1 Introduction

Eastern Maine Health System (EMHS), an integrated health delivery system serving the state of Maine, is committed to improve the health and well-being of patients by providing high quality, cost effective services. As our healthcare organization embarks on the second decade of EHR usage, optimization of our EHR systems is paramount to increase clinical adoption and enhance workflow efficiencies. EMHS nursing leadership engaged with our EHR partner, Cerner Corporation, on this groundbreaking case study to answer the question, "How are nurses spending their time using an EHR?"

The work of nurses has been studied by many in an effort to determine what percent of time, as well as what categories of nursing activities and interventions, are allocated over the course of a shift or period of time. Most of the studies published have relied on time and motion studies using direct observations of nursing workflow, surveys for nurse perceptions of how they spend their time, or self-reported timings. Hendrich et al. (2008) found documentation accounted for the greatest amount of time (35.3%), while medication administration process was second at 17.2% of a nurse's time within a 10-hour shift. Sutton (2004) published similar results with documentation at 20% of the nurse's shift, medication administration at 17% and direct patient care was 15%. Common denominators of these studies were nurses used paper medical records and manual documentation.

The potential for increased documentation time during and after an EHR implementation, in an already labor-intensive and demanding environment, generated new studies (Korst et al. 2003; Strauss 2013). Yet, some studies showed essentially no change in the amount of time nurses spent documenting with the implementation of the EHR (Hakes and Whittington 2008; Ward et al. 2011; Korst et al. 2003). Yee et al. (2012) concluded that nurses spend 19% of their time completing documentation, regardless of electronic charting usage. Kuziemsky (2015) discussed the nurses' perception of their care delivery and what workflow changed with the implementation of the EHR to 'nursing the computer' versus providing nursing care to the patient. Kossman (2008) described that nurses' perceived their workflow had

A. Giard, M.S.N., NP-BC, NEA-BC (✉)
Eastern Maine Health System, Brewer, ME, USA
e-mail: agiard@emhs.org

D. Sutton, M.S.N., RN-BC
Cerner Corporation, Kansas City, MO, USA

changed to a focus on the computer rather than on the patient's needs. Strauss (2013) explored the patient's experience of the nurse–patient relationship when nurses use an electronic health record to communicate with them.

This case study describes the electronic method of capturing the amount of active time (in minutes) logged into the EHR system and stratified by categories of nursing activities (i.e., documentation, chart review, medication administration) of the bedside clinician. The automated collection of data was made possible by embedded Response Time Measurement System (RTMS) timers in the EHR software. The timers capture the start and stop timings of keyboard and mouse activity allowing the system to calculate time for each nurse. This high-performing and cost-effective big data processing methodology collects data on where in the EHR the nurse "clicked" and how much time was spent for each category of work. Our goal was to use the information to highlight areas where we could streamline workflow and drive efficiencies, while also enhancing the overall usability and experience of the patient and nurse. The implications of the study are discussed in relationship to EMHS nurses' perception that they spend a significant part of their shift using the EHR to carry out and document patient care.

18.2.2 Methods

With the advancement of data collection tools, EMHS was able to collect the transactions report for the amount of "active" time that was spent by each nurse using the EHR and organized into the category of work. The initial validation of the RTMS timers was conducted by Cerner prior to external validation across a number of health systems. Cerner added the Citrix Smart Auditor software tool to the computers used by nurses to essentially record their EHR activity over an entire shift. The Smart Auditor recordings were viewed by Cerner staff, who used a stopwatch to record the time and category of work observed. The stopwatch timings were compared to the timer data for each patient chart that was opened for the time period. Cerner's internal validation accounted for approximately 700 Smart Auditor observed hours, across 70 clients and a total of 6395 shifts were analyzed. External validation of the timers was conducted across three health systems, one of which was EMHS. All three organizations mirrored the internal validation process on a smaller scale, over approximately 21 months.

An analysis of the workflow for RNs delivering direct patient care was conducted to identify the major categories of work, regardless of venue of care. Seventeen categories of work were identified.

1. Alerts
2. Nursing Organizer/Activity lists
3. Charging
4. Chart Review
5. Discharge
6. Documentation
7. Histories
8. Chart printing
9. Medication Administration

10. Medication Reconciliation
11. Messaging/electronic Communication
12. Orders
13. Patient Discovery
14. Patient Education
15. Problems and Diagnosis
16. Registration & Scheduling
17. Sign Review

Each of the 17 categories of work was isolated to the EHR section, view and functionality used by the nurse. The Cerner engineers placed RTMS electronic timers into the software that automatically captured active time for each category of work. Active time in the EHR was defined as keyboard strokes, mouse clicks and mouse miles. Mouse miles can further be defined as measure of the distance the mouse has travelled between two anchored points. 1 mouse mile = 1 pixel. 1700 or more mouse miles per minute will count as active time. For example, when the nurse clicked onto the flowsheet and started to document, a software timer started. When the last documentation detail was signed the timer stopped and a total time was calculated as "documentation" category of work. The micro data from each nurse was aggregated according to the category of work and summarized. All data from nurses was then rolled up by their EHR security position (e.g. Oncology RN, OR RN).

Early on we knew we needed to account for the "passive" time nurses would be in the EHR but doing other important patient care activities, planning for care and simply taking the time for critical thinking. The following activity ratios were used to determine active vs. passive time.

Time	If <45 s between RTMS timers, capture time as active
	If >45 s between RTMS timers, then check clicks, keystrokes, mouse miles
Clicks	Clicks can be right, left or other clicks of the mouse.
	3 or more clicks per minute will count as active time
Keystrokes	Keystrokes are any key action on the keyboard.
	15 or more keystrokes per minute will count as active time
Mouse miles	Mouse miles measure the distance the mouse has travelled between two anchored points. 1 mouse mile = 1 pixel.
	1700 or more mouse miles per minute will count as active time

18.2.3 Results

The case study looked at the results of the active time nurses spent on the specific activities included in each work category. The study included 2029 EMHS nurses over a two-month period of time. The active time was calculated for each nurse per day, which included some variability due to difference between shift lengths. The totals from each nurse were then rolled up by their EHR security position. Figure 18.2.1 reveals the percentage of active time for all nursing positions for each work category. Similar to other studies, the greatest amount of time, 45%, was in the documentation work category versus the other activities such as chart review, 22%, medication administration, 10%, etc.

The ability to view the active time by security position is shown in Fig. 18.2.2. The Registered Nurse position had an average active time in the EHR of 1 h and 34 min per day. The Registered Nurse position represented the nurses working on the units considered general medical/surgical. In comparison, an Oncology RN spent 1 h and 35 min, an OR RN spent 1 h and 12 min, a behavioral health nurse (Acadia Nursing) spent 1 h and 19 min and the nursing student spent 1 h and 3 min. The new results of the study did not correlate with the EMHS nurses' perceptions that they spend a significant part of their shift using the EHR to carry out and document patient care.

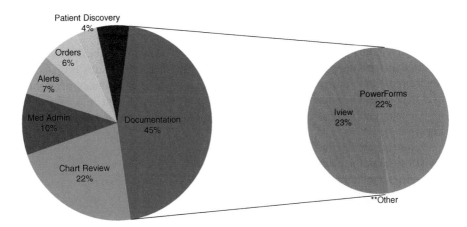

Fig. 18.2.1 Active time in EHR—all nurses

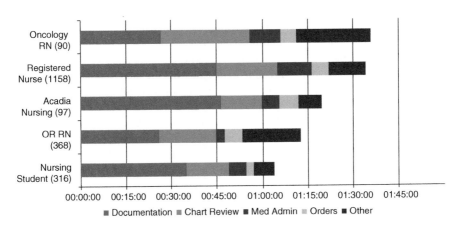

Fig. 18.2.2 Active time in EHR by nursing position—per day

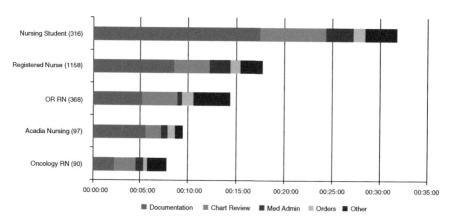

Fig. 18.2.3 Average time per patient

Understanding that acuity and patient assignments would vary, the data was also reported as the average time each nurse spent per patient. The results revealed the Registered Nurse position spent, on average, 17 active EHR minutes per patient over their shift. The nursing student spent the most time per patient at 31 min, while the Oncology RN spent on average, the least at 7 min and 46 s per patient (see Fig. 18.2.3).

Before presenting the final results to the nursing leadership at each hospital in the EMHS system, the nursing leaders were asked how much time they perceived that staff nurses spend in the EHR. Most responses ranged from 4 to 8 h a day. When the results were presented, there was dismay and disbelief from the nursing leaders. Many found it difficult to comprehend as they had such strong perceptions of how much time nurses spend in the EHR. It was important that many of the nurses that helped validate the tool where in attendance. They were able to validate that the data was indeed correct and how valuable it was for them to have been included in the process of validation. A key communication to nursing leaders was that we have not begun the work to study the total time spent in the EHR, nor its relationship to quality of the documentation or patient outcomes. These are areas for future study.

Once the nursing leaders were able to accept what the data findings meant, they started to examine why their perceptions were so different than what the data showed. Computers are now a ubiquitous tool for the clinicians providing direct patient care. They are in the patient's room, in the hallway, or at the nurse's station and always close at hand, which may account for the perception that they were using them (active time) a great deal more than they were. Other nurses realized that even though the computer was in the room with the patient, they would frequently be doing beside care and only using the computer for the brief moments needed to document. Other realizations came to mangers that saw nurses at the nursing station with the computer open and made the assumption the nurses were in the EHR actively charting, versus using that time to plan care, find information in the chart, do critical thinking and prioritize activities, etc. Nurses started to think about the computer differently and even though they are frequently referring to it and it is physically with them, the presence of the computer does not mean they are actively using it.

18.2.4 Conclusion

New health information technologies (HIT) have the potential to create a better work environment for nurses by improving the efficiency, effectiveness, and quality of care. As HIT continues to evolve, the amount of discrete patient-level data that can be extracted from the EHR has increased. Unlike the limitations of the time and motion studies with small sample size and work-sampling methods, our case study was able to automate the retrieval of a large sample size of discrete data without the time, effort and cost of conducting direct time studies.

This case study represents a significant contribution to nursing science with the ability to discretely describe the active engagement of nurses using the computer to deliver care. It is clear there is opportunity for more work to be done, both in the perception of active and passive time in the EHR and how efficient the EHR workflows are designed. The valuable information from the case study helped EMHS understand that there is still work to do to incorporate the tool (computer) into the workflow so it is not perceived as intrusive or taking up time but instead providing direct patient care. The case study gave EMHS more questions to be answered. How are nurses spending their time if it is not the EHR taking their time? What is preventing them from being able to spend more time at the bedside? EMHS plans to focus on initiatives and research studies impacting the nursing practice by utilizing these methods and tools moving forward.

References

Hakes B, Whittington J. Assessing the impact of an electronic medical record on nurse documentation time. CIN. 2008;26(4):234–41.

Hendrich A, Chow M, Skierczynski B, Zhenqiang L. A 36-hospital time and motion study: how do medical-surgical nurses spend their time? Permanente J. 2008;12(3):25–34.

Korst LM, Eusebio-Angeja AC, Chamorro T, Aydin CE, Gregory KD. Nursing documentation time during implementation of an electronic medical record. In: Anderson J, Aydin E, editors. Evaluating the organizational impact of healthcare information systems. New York: Springer; 2005. p. 304–14.

Kossman SP, Scheidenhelm SL. Nurses' perceptions of the impact of electronic health records on work and patient outcomes. CIN. 2008;26(2):69–77.

Kuziemsky CE. A model of tradeoffs for understanding health information technology implementation. In: Botin L, Bertelsen P, Nøhr C, editors. Techno-anthropology in health informatics: methodologies for improving human-technology relations. Amsterdam: ISO Press eBooks; 2015. p. 116–28. doi: 10.3233/978-1-61499-560-9-116. Accessed 3 Nov 2015

Strauss B. The patient perception of the nurse-patient relationship when nurses utilize an electronic health record within a hospital setting. CIN. 2013;31(12):596–604.

Sutton DE. Cases in redesign, II UPMC Health System: Transforming care through clinical documentation. In: Ball M, Weaver C, Keil J, editors. Healthcare information management systems. 3rd ed. London: Springer; 2004. p. 119–27.

Ward MM, Vartak S, Schwichtenberg T, Wakefield DS. Nurses' perceptions of how clinical information system implementation affects workflow and patient care. CIN. 2011;29(9):502–11.

Yee T, Needleman J, Pearson M, Parkerton P, Parkerton M, Wolstein J. The influence of integrated electronic medical records and computerized nursing notes on nurses' time spent in documentation. CIN. 2012;30(6):287–92.

Case Study 18.3: Identifying Direct Nursing Cost Per Patient Episode in Acute Care—Merging Data from Multiple Sources

Peggy Jenkins

Keywords Nursing informatics • Big data analytics • Analytic methods • Direct nursing costs

18.3.1 Introduction and Background

This case study illustrates an analytic approach using data from multiple EHR systems to derive direct nursing cost metrics at the patient level. Additionally, it presents an innovative method to measure direct nursing cost using existing data sources that link individual patient to specific nurse. The age of big data brings opportunity for nursing leaders to delineate structure, innovate and transform our work in ways unattainable previously. Using patient-level data creates new opportunities to ask and answer questions that were never available before because the data was missing or the systems could not be merged. Included in the discussion is a comparison of episode-of-care costs across units and hospitals providing benchmarking data to better understand the optimum use of nursing resources to affect quality patient outcomes. Linking cost data to quality data may provide better science on the value of nursing. The methodology allows exploration of staffing questions such as "What is the direct nursing cost for patients with DRG 190 on unit A compared to unit B?" plus a plethora of additional questions.

18.3.2 Definition of Direct Nursing Cost per Acute Care Episode

A review of nursing cost literature dating back several decades (Wilson et al. 1988; Witzel et al. 1996) can be confusing and daunting because numerous definitions of nursing cost and methodologies to derive cost are found (Chiang 2009; Pappas 2007; Welton et al. 2009). The definition of direct nursing cost used in this case study is the amount of assigned time per shift a nurse spends in direct care or care-related activities with a patient multiplied by the actual nursing wage. Direct nursing cost per patient shift was aggregated to (1) direct nursing cost per day and (2) acute care episode, defined as admission-to-discharge on the study unit (Jenkins and Welton 2014).

P. Jenkins, Ph.D., R.N.
College of Nursing, University of Colorado, Boulder, CO, USA
e-mail: Peggy.jenkins@ucdenver.edu

18.3.3 Data Sources and Data Management Plan

Data were extracted from three databases: (1) Medical Management System, (2) Human Resources, and (3) Clairvia Care Value Management system (Clairvia). Information technology staff from the hospital mined data from the Medical Management and Human Resource systems. Cerner staff extracted data from Clairvia after obtaining all approvals from hospital personnel. The Clairvia software suite is made up of six modules that are used in over 1200 healthcare organizations. One of the modules is the demand-driven patient assignment module that organizes data at the patient and nurse level. Every nurse assigned to a patient during a hospital stay can be recorded. Variables collected include patient, room, acuity level, nurse assigned, hours assigned by nurse, diagnosis, nurse skill level, and census. Information about nurses was collected in the human resources database. Variables such as salary, benefits, education level, certification, and years of service were included in the database. Patient characteristic data was collected from the Medical Management System and included patient age, gender, DRG, and complication code. The nurse researcher received data in seven Excel files. A unique patient and nurse identifier was used to merge the files in STATA version 12 software. The researcher used the Long model for data management (Long 2009) consisting of four steps: (1) planning, (2) organization, (3) computing, (4) documentation.

18.3.4 Architecture for File Merger

The architecture used for file merger is illustrated in Fig. 18.3.1.

Microeconometric methodology (Cameron and Trivedi 2010) informed data cleansing, merger, and analysis. Data cleansing required:

- Patients who were not on the study unit from admission to discharge were dropped from the study. In future studies, patient data from all units a patient occupies during an acute care episode could be used if the patient assignment software is available on every unit in the hospital.
- Patients with missing intensity data were dropped because multiple variables are not imputed in microeconometric methodology.

The final merged file contained 44,771 shift observations and included data for 3111 patients and 150 nurses.

18.3.5 Construction of Outcome Variable

The nursing cost per acute care episode (ncace2) outcome variable was constructed using STATA and a four-step process. First, the main merge file was used in which a variable, nursing cost per shift (ncsh), was built multiplying

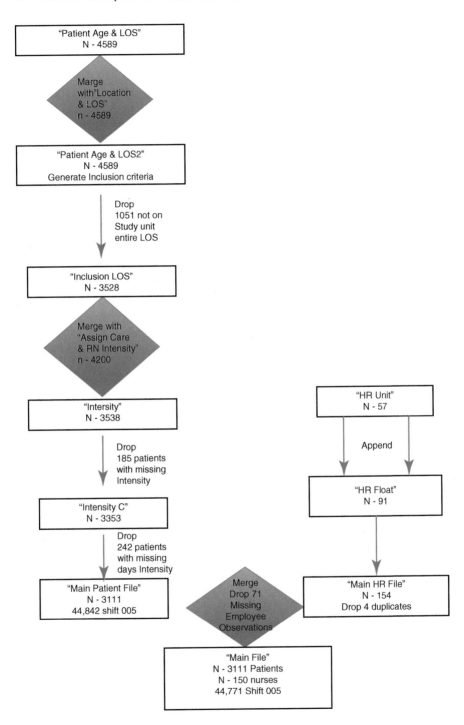

Fig. 18.3.1 File merge architecture

nursing intensity by nursing wage. Second, the new variable, ncace, was generated using the sum command for nursing hours per shift sorted by patient encounter number. The command used was "by enc_no: generate ncace = sum (ncsh)". For each patient, nursing cost was summed each shift, so the last shift record equaled the total cost of the acute care episode. Third, the nursing cost per acute care episode was generated for each patient by pulling the last observation by patient encounter number. STATA code used was "by enc_no: generate ncace2=ncace if _n = _N. Fourth, the values for nursing cost per episode of care were kept and other values were dropped. This resulted in 3111 values, which matched the master file patient encounter number.

18.3.6 Data Analysis

Data was analyzed using STATA version 12 software. Descriptive and inferential statistics were used to answer four research questions and the analysis has been published previously (Jenkins and Welton 2014). STATA do-files provided a tool to systematically record commands for each step of the data management and analysis process. Research must be reproducible and, given the complexity of data for this project, a method to automate data replication was necessary. STATA do-files were used to automate the research process and provide documentation for replication. First, individual do-files were created for each step of data management. Long's (2009) work on data workflow analysis guided the process used for creating do-files. Files were named and numbered so they were run in chronological order. Files were carefully debugged individually. Do-files for data management and data analysis were completed separately. Second, a master do-file was created to allow an easy method to reproduce the research. The master do-file was carefully created and debugged to assure data was cleansed, organized, labeled, summarized, and tabulated with integrity.

18.3.7 Key Findings

Detailed study results have been published previously (Jenkins and Welton 2014). A key finding was significant variability in nursing intensity (0.36–13 h) and cost per patient day ($132–$1455) for similar patients. Higher cost nurses were not assigned sicker patients ($F(3, 3029 = 87.09)$, $p < .001$), $R2 = 0.124$. Mean nursing direct cost per day was $96.48 (SD $55.73). The direct cost per day is low in this study because patients transferring to or from other units during the acute care episode were excluded. In future research, direct nursing cost per patient measured from admission to discharge across units could be studied if patient assignment software was used on all nursing units. Direct nursing cost for the entire acute care episode could then be measured and benchmarked.

The study provides a method to measure direct nursing cost by patient for an acute care episode and to explore relationships of nursing and patient characteristics and direct nursing cost. Patient-level data opens up the possibility to describe, explore, and test a broad range of research questions that cannot be studied using unit- or hospital-level data. The data in this study suggests the patient assignment may not be based on experience or educational level of the nurse. The data does support that nursing years' experience has a positive relationship with direct nursing cost.

18.3.8 Discussion

Is there a way to use data linking *patient-to-nurse* to assure all patients are receiving an optimal amount of nursing intensity each shift? The granularity of the data in this study illuminates disparities in the amount of nursing time delivered per patient. For instance, one RN worked a double shift and provided 23.07 intensity hours to a combination of 13 patients. Because the RN provided 5 h of care during one shift to one patient, some of the other patients only received 0.5 h of care. Is 0.5 h of RN direct care optimal nursing intensity? What is optimal nursing intensity for patients with certain diagnoses? What is the best use of higher-cost nurses with more experience? Linking patient outcomes to direct nursing cost is an area for future nursing research. The data used in this study provides a rich source for nurse scientists and administrators to experiment with to move toward clearer understanding of the best use of nurses to influence positive patient outcomes.

References

Cameron AC, Trivedi PK. Microeconometrics using stata. College Station, TX: Stata Press; 2010.

Chiang B. Estimating nursing costs—a methodological review. Int J Nurs Stud. 2009;46(5):716–22.

Long JS. The workflow of data analysis using Stata. College Station, TX: Stata Press; 2009.

Jenkins P, Welton J. Measuring direct nursing cost per patient in the acute care setting. J Nurs Admin. 2014;44(5):257–62.

Pappas SH. Describing costs related to nursing. J Nurs Admin. 2007;37(1):32–40.

Welton JM, Zone-Smith L, Bandyopadhyay D. Estimating nursing intensity and direct cost using the nurse-patient assignment. J Nurs Admin. 2009;39(6):276–84.

Wilson L, Prescott PA, Aleksandrowicz L. Nursing: a major hospital cost component. Hlth Serv Res. 1988;22(6):773.

Witzel PA, Schultz GL, Ryan SA. A cost estimation model for measuring professional practice. Nurs Econ. 1996;14(5):286–314.

Case Study 18.4: Building a Learning Health System—
Readmission Prevention

Marlene A. Bober and Ellen M. Harper

Keywords Nursing informatics • Big data • Analytics • Readmission risk profile
• Readmission prevention

18.4.1 Introduction

Advocate Health Care (Advocate) is a large health system located in the Chicago metropolitan area. It is the largest health system in Illinois and one of the largest healthcare
providers in the Midwest, operating more than 12 hospitals, 250 sites for care delivery,
and one of the largest medical groups, home health, and hospice companies in the area.
Our case study is particularly important considering the reimbursement guidelines
enacted by the Center for Medicare & Medicaid Services (CMS) that are holding hospitals accountable for unplanned, preventable hospital readmissions. Increasingly CMS
(2014) uses readmission rates as a metric to gauge quality of care, and hospitals with
high rates of readmissions are subject to financial penalties under these guidelines.

   In 2012, Advocate Health Care joined forces with our EHR partner, Cerner
Corporation, to form the Advocate Cerner Collaboration (ACC). Uniting their healthcare
technology and data management capabilities with Advocate's population risk and clinical integration expertise, the ACC focused on creating data-driven predictive models and
advanced analytic tools to enhance patient care throughout the continuum of care. Due
to the higher than the national average hospital readmission rates in the Chicago metropolitan area and the lack of a prediction model using reliable data that can be easily
obtained, they selected the Readmission Risk Model solution as their first project.

   The future widespread adoption of EHRs will make increasing amounts of
clinical information available in computable form. Secured and trusted use of this
clinical data, beyond its original purpose of supporting the health care of individual patients, can speed the progression of knowledge from the laboratory
bench to the patient's bedside and provide a cornerstone for healthcare reform
(Friedman et al. 2010). The Readmission Prevention solution is designed to generate and apply the best evidence for each patient and provider; to drive the process of discovery as a natural outgrowth of patient care; and to ensure innovation,
quality, safety, and value in health care. The solution is a suitable example of a
learning healthcare system.

M.A. Bober, R.N., M.S. (✉)
Acute Enterprise Care Management, Advocate Health Care, Downers Grove, IL, USA
e-mail: marlene.bober@advocatehealth.com

E.M. Harper, DNP, RN-BC, M.B.A., F.A.A.N.
School of Nursing, University of Minnesota, Minneapolis, MN, USA

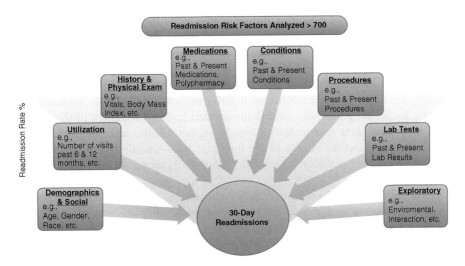

Fig. 18.4.1 ACC readmission risk prediction model

18.4.2 Methods

The Readmission Prevention Solution has three key aspects. First it uses the ACC published 30-Day Readmission Risk Model (Choudhry et al. 2013). The model uses clinical and administrative data to generate a risk score when the patient is admitted that predicts the likelihood of being readmitted to the hospital within 30 days of discharge. Second, using clinical decision support, it scores and strati-fies patients into high, moderate, and low risk for 30-day readmission. And third, it formalizes the Interdisciplinary Plan of Care (IPOC) clinical workflow includ-ing interventions and documentation of readmission prevention education for dis-charge planning. The solution closes the loop between people, process, and technology, the three elements for a successful organizational transformation of practice.

The primary dependent variable for the prediction model is hospital readmis-sions within 30 days from the initial discharge. Independent variables were seg-mented into eight primary categories: Demographics and Social Characteristics, Hospital Utilization, History & Physical Examination (H&P), Medications, Laboratory Tests, Conditions and Procedures (using International Classification of Diseases, Ninth Revision, Clinical Modification codes ICD-9 CM), and an Exploratory Group (Fig. 18.4.1).

The readmission risk score is calculated within 2 h of admission and updated every 2 h to reflect the most current patient data from the electronic health record (EHR). Variation occurs as patient health information is updated in the

EHR. A change in the score will either indicate an improvement or worsening of the patient's likelihood of readmission. The greatest score variability was seen when new lab results were posted and past/current medications are updated. If a patient was admitted and only administrative data was available the risk score will still calculate, but there is an expected increase (and therefore, variability) in the score once additional health data is updated. Lastly, patients who had a longer length of stay also demonstrated higher variability of their score because more data (particularly procedures, lab results, and medications) are posted throughout the course of the inpatient stay. A readmission risk score of 10% or greater is considered high risk, 6.9–9.9% is moderate risk, and less than 6.9% is low risk.

The patient's risk score automatically populates and refreshes throughout the day where it displays to nurses in the nursing inpatient summary as a near real time indicator. If the score is high risk, care managers initiate the Readmission Prevention Interdisciplinary Plan of Care (IPOC) which includes interventions and education designed to decrease the risk such as:

- Customized care plan that incorporates the patient's discharge needs
- Ensure that education provided is at the appropriate health literacy level
- Incorporate the Teach Back method while providing education
- Ensuring medications are accurately reconciled and education provided is understood
- Ensuring that the patient has a follow up visit scheduled
- Provide the patient (as appropriate) with condition-specific (Acute Myocardial Infarction, Heart Failure, Chronic Obstructive Pulmonary Disease and Pneumonia) education options specifically geared toward readmission prevention.

Although the primary nurse is responsible to ensure discharge arrangements for follow-up visits are completed and documented, an interdisciplinary team approach is required to achieve a successful discharge transition.

18.4.3 Results

As shown in Fig. 18.4.2, a decrease in the overall readmission rates has been observed across all risk categories and within the majority of payers since implementation in June 2013. The increased readmission rates observed in October, November, and December 2014 were attributed/related to high census at many facilities due to the 2014 flu season.

18.4.4 Discussion

The findings revealed opportunities for future work. Simple, effective education interventions, both general and condition-specific, coupled with added communication between clinicians, proved to be very effective. Up to a 20 percent reduction in

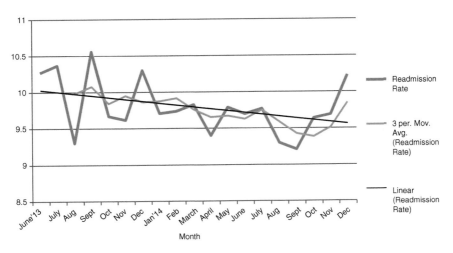

Fig. 18.4.2 Readmission rates by month: June 2013—December 2014

readmission rates was observed in the high-risk population when patients received the education intervention, compared to those who did not. The Readmission Prevention IPOC is mandatory for patients with high-risk scores and strongly recommended for other in patients that may be deemed high risk by the nurse's clinical judgment. This process change increased the number of high- and moderate-risk patients receiving condition-specific education. As the effectiveness of the intervention was analyzed, we found increased adoption is not only important for quality improvements and patient outcomes, but also to improve measurement of treatment effect. It is important to note that strong interdisciplinary collaboration with nursing colleagues was necessary to ensure that the readmission prevention interventions were successfully delivered to patients.

The readmission prevention education for patients with specific conditions reached statistical significance during the second evaluation period (Fig. 18.4.3). Among high-risk patients with heart failure, those who received heart failure readmission prevention education prior to discharge experienced a 21% decrease in readmission.

As a learning health system, there have been several rounds of enhancements to the readmission prevention workflow. After completion of the first evaluation, RN care managers at all hospitals had additional training on workflow changes. Clinical teams responded favorably with an increase in adoption; however, desired levels of adoption were not reached during the second round of evaluation. Following the upgrade, use of the Readmission Prevention IPOC (primary metric used to measure adoption of the solution) for high-risk patients increased from 30% to 40% and all hospitals from 45% to 75%. The clinical user's feedback has been an important aspect of the learning. With requested changes by the clinical teams, indicators demonstrate improved user experience, driving continued use of the solution as well as increased adoption.

The Readmission Prevention Solution enables consistent and standardized interventions to be delivered throughout the Advocate System. Robust rates of adoption indicate the solution has been well integrated into the workflow and is

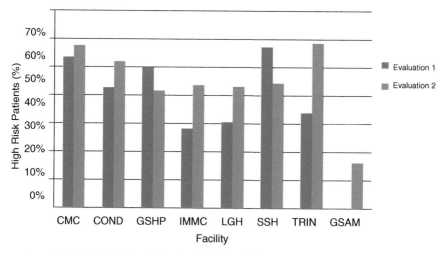

Time Period of Evaluation 1 = Jun – Dec 2013; Evaluation 2 = Jan – Dec 2014

Fig. 18.4.3 Percent of high-risk patients that received readmission prevention education prior to discharge

well suited for future efforts. Additional interventions specific to transitional care visits and the post-acute care network are being considered. System improvements, positive outcomes, and increased adoption suggest that the Readmission Prevention Solution has made a positive impact. The structure and process are now in place to truly enable continuous process improvement as future opportunities are identified.

18.4.5 Conclusion

The Readmission Prevention Solution is fully implemented at eight Advocate facilities and continues to show value through standardizing readmission prevention workflow, driving effective education interventions, and helping to make the process of preventing readmissions predictable. Results from the case study indicate that condition specific (heart failure) types of readmissions are being prevented by consistent and standardized readmission prevention education. Now, as trends of increased acuity of the patient population shifts, targeted secondary interventions must be identified to mitigate readmissions for additional high-risk conditions. Once identified, these interventions can be included to enhance the solution. In addition to the successes observed in process and workflow, the solution upgrade provided an example of the positive impact that collaboration can bring. RN care managers provided feedback that was incorporated into the upgraded version resulting in increased adoption of the solution. Additionally, the education interventions significantly benefited the condition-specific patient subgroups. As Advocate continues to reduce readmission rates, the Readmission Prevention solution will serve as an effective platform for interdisciplinary collaboration toward readmission prevention and discharge planning. The ACC continues our significant focus and effort toward meaningful use of health information technology to improve health care. This case study has helped establish a foundation of systems thinking around coordination of care and population health.

References

Centers for Medicare & Medicaid Services. Readmissions reduction program. 2014. https://www.cms.gov/medicare/medicare-fee-for-service-payment/acuteinpatientpps/readmissions-reduction-program.html. Accessed 21 Nov 2015.

Choudhry SA, Li J, Davis D, Erdmann C, Sikka R, Sutariya B. A public-private partnership develops and externally validates a 30-day hospital readmission risk prediction model. Online J Public Health Inform. 2013;5(2):219.

Friedman CP, Wong AK, Blumenthal D. Achieving a nationwide learning health system. Sci Trans Med. 2010;2(57):57cm29.

Part V
A Call for Readiness

Roy L. Simpson

In this final part, the focus is forward-looking to future trends in healthcare delivery, education, research, and the vendor marketplace through the lens of new emerging technologies and big data's impact. Nursing education, credentialing, curriculum changes in nursing programs needed at undergraduate, doctoral and graduate levels to prepare nurse leaders to participate in and to use big data are addressed by McCauley and Delaney, Berkowitz, and Eckardt and Henly. Specific informatics curriculum changes needed for nurses are detailed by Warren, and certification initiatives for health professionals undertaken by the American Medical Informatics Association are described in the case study by Gadd and Delaney. In a delightful look at healthcare as the last holdout of resistance to the trends for industries to transition from local to global, Sermeus chronicles the emerging technologies, social media communities, and analytic tools that he projects will pull big data and big data science in nursing and healthcare into a more global sphere as well. In a concluding overview, Warren, Clancy, Weaver and Delaney offer a future look at the implications and opportunities created by the entry of big data technologies and methodologies into healthcare. These changes come with challenges for nursing education curriculum at all levels and for nurse researchers to rapidly adopt and embrace new networked data resources and analytic methods, team approach and partnerships with data scientists and industry. In healthcare delivery, technology merges with healthcare reform and reimbursement structural changes to fundamentally shift where healthcare happens and the power dynamic between patient and providers, and makes team care an imperative. Steep and rapid change is happening within the U.S. healthcare system, and it holds the promise that healthcare providers and patients alike will realize individual, family and population health and enjoy the newly invented system much more than the past.

Chapter 19
What Big Data and Data Science Mean for Schools of Nursing and Academia

Linda A. McCauley and Connie W. Delaney

Abstract The Age of Data is upon us, promising sweeping changes in all areas of business, including healthcare. The Shaw quote reminds us that data (facts) are fundamental to change. The Fourth Paradigm: Data-Intensive Scientific Discovery (Hey et al. The fourth paradigm: data-intensive scientific discovery, 2009) builds on the importance of data. Our science is powered by advanced computing capabilities and team work. The age of data-intensive discovery encompasses the transition from hypothesis-driven to data-driven science. The convergence of statistics, computer science and physical and life sciences is a reality today. What does it mean for nursing to become more data intensive? What will it mean for the practice of nursing and how do academic institutions adjust to prepare the nurses of the future to function well in a data-rich world? The potential is great to harness the massive stores of data on biological systems as well as social determinants of health; patients could change the way that care is provided. However our history tells us it may not be an easy transition. Nursing will need a large number of data-savvy professionals who can lead the profession forward; academic nursing must change to meet this challenge. In this chapter, the critical need for workforce training will be examined and the role that schools of nursing and academia will play to prepare the work-

L.A. McCauley, RN, Ph.D., FAAN, FAAOHN (✉)
Nell Hodgson Woodruff School of Nursing, Emory University, Atlanta, GA, USA
e-mail: lmccaul@emory.edu

C.W. Delaney, Ph.D., RN, FAAN, FACMI
School of Nursing, University of Minnesota, Minneapolis, MN, USA

© Springer International Publishing AG 2017
C.W. Delaney et al. (eds.), *Big Data-Enabled Nursing*, Health Informatics,
DOI 10.1007/978-3-319-53300-1_19

force of the future will be described. Competencies across different types of nursing education programs will be reviewed and exemplars of innovative curricular change will be described.

Keywords Academic nursing • Nursing workforce • Competencies

> *"Progress is impossible without change, and those who cannot change their minds cannot change anything."*
>
> George Bernard Shaw

19.1 Why is Big Data Important for Academic Nursing?

While technology is advancing fast in clinical settings and in the biological sciences, nurses often lack the technology and analytic skills needed to adapt to this rapid change. The role of nursing academia has always been to prepare nurses for practice, not only upon graduation, but also in preparation for where the profession and healthcare are going. However, it is doubtful that most nursing students graduating today realize the data richness of healthcare and that they are entering one of the most data-intensive industries, comparable in many respects to the defense and national security areas. Much of the discussion within schools of nursing is still focused on the impact of the electronic health record (EHR) on the nurse-patient interaction (Korst et al. 2003; Moreland et al. 2012; Rantz et al. 2011; Stevenson et al. 2010), with nurses still expressing the opinion that the technology interferes with and does not enhance care. However, evidence shows that the electronic health record enhances care quality.

The lag in the integration of technology, informatics, and the EHRs into schools of nursing' curricula has been noted by Fetter (2009) and Jones and Donelle (2011). Many nurses view informatics and data analytics as a highly specialized area of graduate study. While more than 6000 nurses are considered part of the **Healthcare Information and Management Systems Society (HIMSS** 2014), there were only approximately 1039 *ANCC* board-certified *informatics* nurses in the U.S. as of 2013, according to the 2013 Role Delineation Study: Nursing Informatics—National Survey Results. This is an inadequate number to advance the practice of almost 3 million practicing nurses in 2012.

Most of the curricular change in schools of nursing has focused on helping prepare pre-licensure students to work with EHRs and other highly sophisticated technologies used in clinical settings. While this emphasis is appropriate and certainly part of the technology and informatics essentials for new graduates, who are the nurses who will have the advanced skills to analyze health data, identify patterns, and generate new methods for collecting, analyzing, and interpreting data? How can academic programs prepare nurses with the technological, informatics and data analytic knowledge and skills required for knowledge-based care across a variety of health care settings?

Faculty are a central component to address preparation of a data saavy workforce. While faculty in some schools have strong data analytical skills, they tend to be highly engaged in research programs and consequently have minimal contact with students across the curriculum. Moreover, there are minimal curricular pathways to introduce and integrate information technology and informatics within and across the academic programs. Even if EHR teaching/learning platforms such as Cerner's Academic Education Solution software, Elsevier's SimChart or Lippincott's DocuCare program are purchased for student use, there may not be specific curricular tracking to integrate informatics competencies using these teaching platforms.

Major professional nursing associations have published policy statements supporting competency in nursing informatics (AACN 2006, 2008, 2011; ANA 2014; NLN 2008). The competency statements in these resources are broad and allow schools great flexibility in sequencing content across their curricula. Consequently an increasing number of nursing programs offer classes and certifications in the area, often as part of the basic core curriculum with little consistency across programs.

19.2 Undergraduate Education

Following the American Association of Colleges of Nursing (AACN) identification of "information management and application of patient care technology" as a core competency of undergraduate nursing education, academic institutions have quickly adjusted their curricula (AACN 2008). At the undergraduate level the majority of the curricular initiatives in informatics have focused on preparing nurses to use information management tools in practice and to access and input information in the electronic databases for healthcare. Miller et al. (2014) have described the preparedness of novice nurses to effectively use EHRs in acute care settings. While 90% of new nurses report receiving EHR training in their workplace, few novice nurses (20%) report coursework on EHR use during nursing school (Miller et al. 2014). Academic institutions are increasingly beginning to recognize the need for meaningful and quality learning opportunities related to the use of the EHR, but initiatives are often hampered given the lack of informatics expertise among the nursing faculty. Moreover, while software packages support the integration of simulated assignments for assessments and care plans into the curriculum, faculty and other curricular design resources are needed to develop structured curricular components.

Some examples have been published on strategies to expose undergraduate students to the power of big data analytics as a tool to assess the outcomes, costs, or quality of patient care. Curricular challenges related to teaching novice nurses to harness the large stores of data on patient care, determine effective interventions, and identify needed improvements do exist. Structured learning exercises are needed to teach students how to manage and transform data from large databases to support care decisions and to improve practice based on evidence. Student knowledge of meta-analysis, Cochran reviews and the more comprehensive big data approach to using data to improve patient care delivery are needed (Geurden et al. 2012).

19.3 Master's Education

Traditionally, informatics has been a specialty area of graduate nursing education. The AACN includes the use of patient-care technologies to deliver and enhance care and the use of communication technologies to integrate and coordinate care as a competency for all master's graduates (AACN 2011). Master's-prepared nurses are expected to be able to integrate organizational science and informatics to make changes in the care environment to improve health outcomes. Curricular content to support these competencies might include system science and integration, information science, technology assessment, and quality processes and improvement science. The AACN Essentials document describes five broad areas of informatics and healthcare technologies including:

- Use of patient care and other technologies to deliver and enhance care;
- Communication technologies to integrate and coordinate care;
- Data management to analyze and improve outcomes of care;
- Health information management for evidence-based care and health education;
- Facilitation and use of electronic health records to improve patient care.

While the knowledge and skills in each of these broad areas are essential for all master's- prepared graduates, the extent and focus of each will vary depending upon the nurse's role, setting, and practice focus. Particularly relevant to big data is that all master's-prepared nurses are expected to be prepared to gather, document, and analyze outcome data that serves as a foundation for decision making and the implementation of interventions or strategies to improve care outcomes. The master's-prepared nurse uses statistical and epidemiological principles to synthesize the data, information, and knowledge to evaluate and achieve optimal health outcomes. Many of the challenges of integrating informatics and big data content in the undergraduate curriculum also exist for the specialty areas of graduate education. Many advanced practice nurses serving as faculty and/or preceptors for master's students have very limited education in technology and informatics and view the EHR as a barrier to efficient care in varied clinical settings. While a great potential exists to develop a focus on data analytics in master's education to prepare graduates who will be managing the care of panels of patients and/or practicing in states with limited scope of practice, limited preparation of the faculty is a barrier.

19.4 Nursing Informatics Graduate Specialty

The American Nurses Association (ANA) (2015) defines the nursing informatics specialty as a specialty that "integrates nursing science with multiple information and analytical sciences to identify, define, manage and communicate data, information, knowledge and wisdom in nursing practice"(ANA 2014). The ANA Nursing Informatics: Scope and Standards of Practice, 2nd Edition (2014) describes 16 nursing informatics standards, which provide a framework for evaluating practice outcomes and goals, with a set of specific competencies accompanying each standard that serves as essential evidence of compliance with that standard. Nursing informatics supports consumers, patients, and providers in their decision-making in all roles and settings. This support is accomplished through the use of information structures, information processes, and information technology. While informatics nurses primarily worked in hospital settings in the past, today they are increasingly employed in diverse inpatient and community settings as well as industry and corporate environments. In all these settings, the informaticians' work is often conducted in an inter-professional environment with strong interaction with information systems personnel.

Accredited graduate-level educational programs for nursing informatics were first offered in 1989. Some universities have elected to embrace an interdisciplinary approach by offering graduate studies in health informatics or the specialty of biomedical informatics. There is a growing number of graduate-level programs in nursing informatics in the U.S. The top ten programs are listed by U.S. News & World Report (http://grad-schools.usnews.rankingsandreviews.com/best-graduate-schools/top-nursing-schools/nursing-informatics-rankings). AACN describes graduate programs in nursing informatics among those that prepare graduates for indirect patient care roles with a focus on aggregates, systems, or organizations (AACN 2011). Graduates of nursing informatics programs are prepared to sit for the basic certification in nursing informatics (ANCC 2013).

Demands for informaticians to have knowledge of data science methodologies for database management and analytics are acute (Hey et al. 2009). To augment basic certification, as in nursing informatics, the American Medical Informatics Association (AMIA) has launched an initiative to advance the preparation of informaticians and speak to the related implications for curriculum changes. Through advanced health informatics certification these informaticians with validated advanced expertise in the specialty will support big data/data science across the health science disciplines and the healthcare systems. In November 2015, the AMIA Board of Directors endorsed recommendations that, among other purposes, defined the scope of certification and recommended the name Advanced Health Informatics Certification (AHIC) to identify the certificate.

There is no one accepted list of competencies for nursing informatics, which reflects the rapid advancements in the area, and the movement of some knowledge/skills once thought as specialized and now needed for all practicing nurses or for other health disciplines as well. Informatics faculty need to stay abreast of the competencies for nursing informatics and be able to integrate content across programs, for specialist education and for all levels of nursing education. All nurses need informatics to support their area of practice. While nursing informaticians often assist nurses to understand and adapt to changes in information and technology, there is untapped potential for the role that nurse informaticians can play in assisting all nurses in learning the skills needed to extract data from systems and to improve their ability to do data visualization to communicate more effectively with patients, healthcare teams and other stakeholders.

19.5 Doctorate in Nursing Practice (DNP)

Strong data management skills are an essential component of the DNP degree (Sylvia and Terhaar 2014) in order for these nurses to transform practice at the system level and improve outcomes. Schools of nursing expect DNP graduates to have the analytical stills to gather, manage, and analyze large datasets with information on clinical, operational, or financial performance. The AACN has also integrated this competency in their DNP Essentials document (AACN 2006), describing the important role that DNP graduates play in using information systems/technology and patient care technology to improve and transform healthcare. Learning outcomes related to information systems/technology prepare the DNP graduate to apply budget and productivity tools, practice information systems and decision supports, and web-based learning or intervention tools to support and improve patient care. Specifically, the DNP program prepares the graduate to:

1. Design, select, use, and evaluate programs that evaluate and monitor outcomes of care, care systems, and quality improvement including consumer use of healthcare information systems.
2. Analyze and communicate critical elements necessary to the selection, use and evaluation of healthcare information systems and patient care technology.
3. Demonstrate the conceptual ability and technical skills to develop and execute an evaluation plan involving data extraction from practice information systems and databases.
4. Provide leadership in the evaluation and resolution of ethical and legal issues within healthcare systems relating to the use of information, information technology, communication networks, and patient care technology.
5. Evaluate consumer health information sources for accuracy, timeliness, and appropriateness.

19.6 PhD Education

PhD programs prepare nurses at the highest level of nursing science to conduct research to advance the science of nursing, lead team science, foster translation of the science to contribute to healthcare improvements via the development of new knowledge and scholarly products that provide the foundation for the advancement of nursing science (AACN, July 2014). While the areas of inquiry that encompass nursing science are broad, in recent years the role of big data and nursing data science has gained increased attention. Many biomedical big data initiatives are housed at the National Institutes of Health (NIH) in the Office of Data Science. Big data refers to the complexity, challenges, and new opportunities presented by the combined analysis of data. In biomedical research, these data sources include the diverse, complex, disorganized, massive, and multimodal data being generated by researchers, hospitals, and mobile devices around the world (NIH, accessed 22 May 2016). The National Institute for Nursing Research (NINR, accessed Jan 2016) was the first NIH institute to recognize the contributions that nursing can make in the area of big data by offering a weeklong "bootcamp" open to graduate students, faculty and clinicians. The bootcamp was designed to increase knowledge in big data methodologies for use in research and to discuss strategies for incorporating novel methods into research proposals. Content areas included human genomics, precision medicine, healthcare analytics including patient outcomes, empowering research with nursing and health informatics, symptoms science and big data, research use of clinical data, electronic data methods, ethics, privacy and security, data integration, genome sequencing and outcomes of big data.

19.7   Challenges Ahead

The major barrier facing the integration of big data and data science in schools of nursing is the current lack of faculty expertise in informatics and data analytics. Evidence shows that many nursing faculty lack the expertise necessary to teach informatics content at even the baccalaureate level (NLN 2015). While the early years of nursing and health informatics looked at informatics as only a specialty area held by a few nurses, there is clear evolution and need for all nurses to be competent in this area. This need requires that schools adopt several strategies to increase this content in the curricula, and expand the knowledge and skill of the faculty.

Numerous professional development opportunities exist for faculty through professional organizations and flexible learning alternatives such as the AMIA 10 x 10 program. Additionally, the National Nursing Informatics Deep Dive Program, funded by the Gordon and Betty Moore Foundation and provided by the University of Minnesota School of Nursing in collaboration with AACN, was a unique initiative to increase the number of nursing school faculty prepared to teach nursing informatics in their program curricula. Between 2012 and 2015 a total of 3139 faculty, health system nurse educators and clinical nurses participated in the various activities including On-site Conferences/Workshops, Webinars and webinar downloads, Nursing Informatics Courses for credit, Nursing Informatics Deep Dive Program webpage downloads, and certificate courses. A website to support nursing faculty teaching informatics has been developed at the University of Minnesota School of Nursing. The website contains slides from all of the events presented in the last two years using National Nursing Informatics Deep Dive

Program materials, as well as a crosswalk that aligns the *AACN Essentials for Informatics and Patient Care Technology, The QSEN Competencies for Informatics* and *The TIGER Competencies for Practicing Nurses*. The site also contains numerous sample assignments, links to informatics standards and professional websites and instructional videos on a variety of subjects, such as the electronic health record, standardized nursing languages, workflow, consumer informatics, telehealth and other key emerging areas. Webinars and WebExs developed under this grant are also available on the website and are linked to the national AACN website and the QSEN national website located at Case Western Reserve School of Nursing.

Another major barrier to integration of big data and healthcare technology into the curricula is the lack of meaningful practice opportunity for nursing faculty. Traditionally faculty practice was primarily in the realm of advanced practice and the need for advanced practice nursing faculty to maintain their specialty certification, to see patients and teach students simultaneously. Many schools have adopted a broader definition of faculty practice. The Johns Hopkins University School of Nursing has a broadened definition that defines faculty practice as the provision of direct and indirect nursing care with the goal of integrating the four missions (practice, research, teaching and service) of the School of Nursing. This broader definition provides expanded opportunities for faculty to develop, implement and expand programs of scholarship, translate research into practice and to disseminate new practice knowledge. Collaborative relations are forged to integrate the practice and academic sides of nursing and interprofessional interaction. This broadened definition allows more faculty to serve as role models to students and other health care professionals and to expand nursing practice into unique settings and partnerships. In addition to maintaining clinical expertise and advanced practice certification, these practice opportunities enhance education by providing a clinically relevant curriculum, including continual improvement of informatics, technology and big data concepts that reflect practice and how rapidly it is occurring.

An interdisciplinary approach is also essential. Partnerships with other disciplines can provide opportunities to integrate coursework into the curricula. Data science offers an excellent opportunity for interprofessional education. The learning can be enhanced by engaging students from multiple disciplines together to examine questions that can be asked from large databases. Biostatisticians who have traditionally worked with nursing scientists can be a strong resource to assist nursing faculty in introducing and supporting students in exercises with large data sets. Partnerships with computer scientists and engineering have proliferated.

Within doctoral programs, the analysis of large data sets is gaining a new respectability. Historically, in many schools of nursing, PhD students were encouraged to collect their own primary data for the experience of designing a study, obtaining IRB approval and collecting data and managing data, even if the sample was not generalizable and was too small to have meaningful impact. Knowledge discovery, including data mining and visualization, were viewed as a non-productive fishing expedition, not belonging in the realm of science. While cutting-edge research schools have embraced data science, it does mean that schools must invest in having strong analytical expertise to support students who want to work with large datasets, state- of-the-art analytical tools, and access to large data (Olsen et al. 2007; Skiba 2011).

One outstanding exemplar of this is OptumLabs, an open, collaborative research and innovation center founded in 2013 by a partnership between Optum and The Mayo Clinic. The OptumLabs research collaborative is made up of over 20 partners, including the University of Minnesota School of Nursing, whose research teams conduct secondary data analysis using the OptumLabs Data Warehouse (OLDW). Its core linked-data assets include de-identified claims data for privately insured and Medicare Advantage enrollees, de-identified electronic medical record (EHR) data from a nationwide network of provider groups, and a de-identified database of consumer lifestyle information. The database contains longitudinal health information on more than 150 million enrollees, representing a diverse mixture of ages, ethnicities and geographical regions across the United States.

19.8   Curriculum Opportunities

Can big data science be effectively integrated across educational levels? To increase understanding of how to effectively teach nursing students at all levels to appreciate and use data analytics, exemplars are needed of curricular strategies that are appropriate at different levels of the curriculum. For example in addition to the traditional focus on individual patients and entering accurate data into the EHR, students need academic exercises in which they reflect on the aggregate data that is being entered and the extent that the data can be used to measure the effectiveness of nursing care for groups of patients or for the health system. As nurses work with new technology that collects real-time physiological data from patients, students need academic exercises that allow them to see the real-time data aggregated across time and for multiple patients. Students need to reflect to what extent this aggregated real-time data can be linked to other sources of data to reveal patterns of health outcomes in groups of patients whose care is being managed in population health. Schools of nursing should evaluate to what extent their students are taught to evaluate whether the data in the EHR is nursing sensitive and being collected appropriately. There is opportunity to teach both undergraduate and graduate students the power of clinical data and related analytics to expand personalized care and population health.

The AACN Essentials for Information Management and Patient Care Technologies for undergraduate students (AACN 2008) focus on teaching students to demonstrate skill in using technology, information systems and communications devices that support safe nursing practice. Schools of nursing assure that students are familiar with the technology and can document interventions related to achieving nurse-sensitive outcomes. However, it is less clear where they gain expertise in understanding output from EHRs and how to work with spreadsheets and healthcare databases. Pre-licensure students are taught evidence-based practice and through research courses learn to search research publication databases for relevant studies. Course work in introductory research and statistics give students beginning skills in evaluating research studies and the evidence they provide regarding safe and effective care. Less emphasis is placed on the use of clinical data and data representing social determinants of health to demonstrate ways to improve patient care outcomes and create safe care environments.

The emphasis on population health and population health management offers excellent opportunities to integrate data science into nursing curricula. Students need opportunities to see how cost-benefit and effectiveness studies are conducted with large data sets. Creative methods to introduce quality metrics and dashboards assessing data trends over time are needed. As students rotate across different clinical settings there should be discussion of different EHRs, the information available in the patient report, the use of standardized nursing terminology, and the extent to which patient information is digitalized. Schools of nursing should integrate concepts of outcome measurement, practice-level improvements, surveillance, population health, decision support, and research in all levels of the curriculum with appropriate and leveled competencies.

Genetics offers an example of big data and how curriculum enhancement and innovation might take place. The ANA recognizes that advances in genetics and genomics are being translated into nursing practice every day including screening, diagnosis, treatment, pharmacogenomics and lifestyle choices (ANA 2009). The ANA emphasizes that an understanding of the ethical challenges encountered by those receiving genetic and genomic-based health care is an essential competency for all nurses. However before nurses can appreciate the ethical issues related to the use of genomic data, they have to have a basic understanding of the big data being generated by genomic and metabolomics techniques. These new areas of big data provide immense potential for identifying the relationships between a disease and its genetic, environmental and/or health-based risk factors. The emerging data science offers not only increased understanding of the mechanisms underlying the development and progression of disease but also how conditions are related to different risk factors such as lifestyle and environmental exposures. While researchers will untangle the complex relationship between genes, exposures and metabolic pathways, nurses will be expected to integrate and apply this knowledge in their practice. Without an understanding of the big data approach that generates the knowledge, nurses will be ineffective in translating the science to the patients they care for, and in applying ethical reasoning.

There are also multiple opportunities to integrate data visualization into nursing curricula. All students need to understand and interpret the visual displays of data in their practice (Skiba 2014). It is quite possible that the faculty who have skills in data entry and working with technology in the clinical area are not the same faculty with the knowledge or skill in data visualization. Students need opportunities to see the output of data analytics, recognize the major patterns emerging from the data, and translate those data to a more simplistic format of graphs, figures, or diagrams that are understandable to the patients or groups with whom they are communicating. Skiba (2014) provides a beginning list of online sources of examples of data visualization. These can become a starting point for integrating this skill into all levels of nursing curricula.

The ANA offers a number of learning tools to help nurses in this area and references the G2C2 website for case studies of the application of genomic knowledge in clinical settings (http://g-2-c-2.org/). The ANA report of a consensus panel outlining genetic/genomic nursing competencies could serve as a blueprint for integrating a similar document describing the competencies for nurses on big data. While this would overlap to come extent with the genomics competencies, it would provide more detail on other aspects of big data such as metabolomics and the exposome, and focus on the integration of such data within the EHR (American Nurses Association 2016). The genomic information in EHRs is key in implementing a vision of individualized or personalized medicine by enabling the discovery of subsets of patients who have unusual or definable clinical trajectories that may not be typical for others with their diagnosis (Roden and Denny 2011).

19.9   Conclusion

Clearly awareness of the import of big data and data science has proliferated in nursing. Major nursing organizations have established competency expectations and early certification and accreditation efforts to support the evolution of informatics competency and big data science. Addressing faculty preparation and the expansion of informaticians and data science experts is critical to the profession's ability to engage in knowledge discovery in the new world and advance the nursing lens and impact on the quality, safety, and experience of health care.

References

American Association of Colleges of Nursing (AACN). The essentials of baccalaureate education for professional nursing practice. 2008. http://www.aacn.nche.edu/education-resources/BaccEssentials08.pdf

American Association of Colleges of Nursing (AACN). The essentials of doctoral education for advanced nursing practice. 2006. http://www.aacn.nche.edu/dnp/Essentials.pdf.

American Association of Colleges of Nursing (AACN). The essentials of master's education in nursing. 2011. http://www.aacn.nche.edu/education-resources/MastersEssentials11.pdf.

American Association of College of Nursing (AACN). The research-focused doctoral program in nursing pathways to excellence. 2014. http://www.aacn.nche.edu/education-resources/PhDPosition.pdf.

AMIA. https://www.amia.org/education/10x10-courses. Accessed 22 May 2016.

AMIA. 2016. https://www.amia.org/. Accessed 4 June 2016.

American Nurses Association (ANA). Consensus Panel on Genetic/Genomic Nursing Competencies. Essentials of genetic and genomic nursing: competencies, curricula guidelines, and outcome indicators; 2009.

American Nurses Association (ANA). Nursing informatics: scope and standards of practice. 2nd ed. American Nurses Association; 2015.

American Nurses Credentialing Center (ANCC). www.nursecredentialing.org/.../NurseSpecialties/Informatics/.../Informatics-2013RDS (2013). Accessed 25 May 2016.

American Nurses Association (ANA). Nursing informatics: scope and standards of practice. 2nd ed. American Nurses Association; 2014.

American Nurses Association (ANA). Genetics/genomics nursing: scope and standards of practice. 2nd ed. Washington, DC: ANA; 2016.

Fetter MS. Collaborating to optimize the nursing students' agency information technology use. Comput Inform Nurs. 2009;27(6):354–64.

G2C2. http://g-2-c-2.org/. Accessed 20 May 2016.

Geurden BJ, et al. How relevant are the Cochrane Database of Systematic Reviews to nursing? Int J Nurs Pract. 2012;18(6):519–26. doi:10.1111/ijn.1.2012.

Health Information and Management Systems Society (HIMSS). 2014 nursing informatics workforce survey.2014.http://s3.amazonaws.com/rdcms-himss/files/production/public/FileDownloads/2014-Nursing-Informatics-Workforce-Survey-Executive-Summary.pdf

Hey T, Tansley S, Tolle K. The fourth paradigm: data-intensive scientific discovery. 2009. http://research.microsoft.com/en-us/collaboration/fourthparadigm/.

Jones S, Donelle L. Assessment of electronic health record usability with undergraduate nursing students. Int J Nurs Sch. 2011;8(1):1–18. doi:10.2202/1548-923X.2123.

Korst LM, et al. Nursing documentation time during implementation of an electronic medical record. J Nurs Admin. 2003;33(1):24–30.

Miller L et al. 2014. Novice nurse preparedness to effectively use electronic health records in acute care settings: critical informatics knowledge and skill gaps. Online J Nurs Inform. 2014;18(2). http://www.himss.org/novice-nurse-preparedness-effectively-use-electronic-health-records-acute-care-settingscritical

Moreland PJ, et al. Nursing satisfaction with implementation of electronic medication administration record. Computers, Informatics, Nursing. 2012;30(2):97–103. doi:10.1097/NCN.0b013e318224b54e.

National League for Nursing (NLN). A vision for the changing faculty role: Preparing students for the technological world of health care. 2015. https://www.nln.org/docs/default-source/about/nln-vision-series-(position-statements)/a-vision-for-the-changing-faculty-role-preparing-students-for-the-technological-world-of-health-care.pdf?sfvrsn=0.

National League for Nursing (NLN). Preparing the next generation of nurses to practice in a technology-rich environment: An informatics agenda. 2008. http://www.nln.org/docs/default-source/professional-development-programs/preparing-the-next-generation-of-nurses.pdf?sfvrsn=6.

NIH. https://datascience.nih.gov/bd2k/about/what. Accessed 22 May 2016.

NINR. https://www.ninr.nih.gov/training/trainingopportunitiesintramural/bootcamp#.VqTzIk1gn-c. Accessed Jan 2016.

Olsen L, Aisner D, McGinnis JM. The learning healthcare system: workshop summary (IOM Roundtable on Evidence-Based Medicine). Washington, DC: National Academies Press; 2007. doi:10.17226/11903.

Rantz MJ1, Alexander G, Galambos C, Flesner MK, Vogelsmeier A, Hicks L, Scott-Cawiezell J, Zwygart-Stauffacher M, Greenwald L.The use of bedside electronic medical record to improve quality of care in nursing facilities: a qualitative analysis. Comput Inform Nurs. 2011;29(3):149–56. doi:10.1097/NCN.0b013e3181f9db79.

Roden DM, Denny JC. Integrating electronic health record genotype and phenotype datasets to transform patient care. Clin Pharm Therap. 2016; doi:10.1002/cpt.321.

Skiba D. Informatics and the learning healthcare system. Nurs Educ Perspect. 2011;32(5):334–6. doi:10.5480/1536-5026-32.5.334.

Skiba DJ, Barton AJ, Knapfel S, Moore G, Trinkley K.Infusing informatics into interprofessional education: the iTEAM (Interprofessional Technology Enhanced Advanced practice Model) project. Stud Health Technol Inform. 2014;201:55–62. PMID: 24943525.

Stevenson J, et al. Nurses' experience of using electronic patient records in everyday practice in acute/inpatient ward settings: A literature review. Health Inform J. 2010;16(1):63–72. doi:10.1177/1460458209345901.

Sylvia M, Terhaar M. An approach to clinical data management for the doctor of nursing practice curriculum. J Prof Nurs. 2014;30(1):56–62. doi:10.1016/j.profnurs.2013.04.002.

US News & World Report Rankings. http://grad-schools.usnews.rankingsandreviews.com/best-graduate-schools/top-nursing-schools/nursing-informatics-rankings (2016). Accessed 24 May 2016.

Case Study 19.1: Informatics Certification and What's New with Big Data

Cynthia Gadd and Connie White Delaney

Abstract Demands for informaticians to have knowledge of data science methodologies for database management and analytics are acute. This case study addresses the American Medical Informatics Association (AMIA) initiative to advance the preparation of informaticians and speaks to the related implications for curriculum changes. Through advanced health informatics certification these informaticians with validated advanced expertise in the specialty will support big data/data science across the health science disciplines and the healthcare systems.

Keywords Certification • Interprofessional informatics • Accreditation • Health informatics curriculum

19.1.1 Introduction

Many forces in the healthcare industry coupled with the evolution of technologies are accelerating the promise of big data and the potential of data science to transform knowledge discovery and quality, efficiency, person-centric care, and population health. Three of the most prominent forces include:

- Rapidly rising costs and payment reform that are forcing payers and health-care providers to shift from a fee-for-service approach to a values-based, health-focused system that prioritizes patient outcomes (including rewarding providers for targeted treatments that actually work).
- Clinicians who are continuing to move towards evidence-based medicine and health, reviewing data from a multitude of sources, including genomics, to provide support for diagnostic and treatment decisions, as well as outcomes and quality measurement.

C. Gadd, Ph.D., MBA, MS, FACMI
Department of Biomedical Informatics, Vanderbilt University,
Nashville, TN, USA
e-mail: cindy.gadd@vanderbilt.edu

C.W. Delaney, Ph.D., RN, FAAN, FACMI
School of Nursing, University of Minnesota,
Minneapolis, MN, USA

- The surge in adoption of EHRs, in the wake of HITECH and ACA, that has achieved rates of implementation equaling 76% of hospitals (Office of the National Coordinator for Health Information Technology, 2016a) and 83% of physician practices (Office of the National Coordinator for Health Information Technology, 2016b), as reported by *HealthIT.gov* for the most recent years available.

These forces and others demand data scientists to inform the wise use of big data to support clinicians and consumers to make better decisions—from personalized treatments to preventive care. Key healthcare applications of data science include segmenting populations to target action, genomic medicine, predictive analytics and preventive measures, patient monitoring, home devices, mobile technologies, self-motivated care, disease modeling, and enhanced EHRs. (http://www.mastersindata-science.org/industry/health-care/ Retrieved 29 Jan 2016). "Data scientists are among the most in-demand employees in the healthcare industry," said Thomas Burroughs, Ph.D., professor and executive director of Saint Louis University Center for Health Outcomes Research (SLUCOR) "Although the healthcare system is creating an unprecedented amount of digital data today, leaders are struggling to turn 'big data' into usable information to improve patient care quality, patient care experience, and health system costs and efficiency. Data scientists are pivotal to transforming health, quality, and the healthcare industry." (http://www.slu.edu/news-data-science-graduate-program. Retrieved 29 Jan 2016). The development of data science expertise among the informatics professionals within the many *health informatics* disciplines (e.g. dentistry, medicine, nursing, pharmacy, public health) is and will continue to be central to the transformation of the healthcare system and care of people.

19.1.2 AMIA's Path Toward Establishing Advanced Health Informatics Certification

Parallel to this timeframe of health care transformation, health informatics has undergone another transformation in which opportunities for formal education beyond the master's and doctoral degrees and professional recognition, such as certification, are on a meteoric rise that is commensurate with workforce demand. *Certification* is the process by which individuals demonstrate that they have competency in a field and that they are proficient in performance of a specific position, professional role, and/or task. The process of establishing certification within a field creates a gold standard and a clear set of expectations for the knowledge and skills that individuals should possess to be considered competent. In the field of health informatics, certification is not currently a requirement to perform certain roles and/or tasks. Rather certification is a signal of competency and experience. In time, organizations seeking to hire qualified individuals may give preference to or require certification.

AMIA has been working on the issue of certification for individuals who practice clinical and health informatics since 2005 (Shortliffe et al. 2015). AMIA's efforts led to the medical subspecialty of clinical informatics, the American Board of Medical Specialties certification, and the AMIA Clinical Informatics Board Review Program. (Detmer and Shortliffe 2014; https://www.amia.org/clinical-informatics-board-review-course (retrieved 29 Jan 2016)).

In 2011, the AMIA Academic Forum, the membership unit within AMIA dedicated to serving the needs of post-baccalaureate biomedical and health informatics training programs, created a Task Force on Advanced Interprofessional Informatics Certification (AIIC). The Task Force issued a consensus statement in February 2012 that established three basic principles:

- A pathway to certification for individuals not eligible for the clinical subspecialty certification is critical.
- Such a pathway should focus on the core informatics content that is relevant to all professions.
- Interprofessional informatics certification should be at the graduate level, based on the same core content used for the subspecialty certification (Gardner et al. 2009), have the same rigor as the subspecialty certification process, and convey the same level of assurance of competency as the subspecialty certification.

As part of its work, the Task Force conducted an environmental scan of the health informatics certification landscape in Sept–Oct 2011, which identified several professions for which there were no existing certifications: MD nonboard-certified and trained in biomedical informatics (e.g., with MS, PhD, or nondegree fellowship); non-MD PhD in biomedical informatics; dental informatics; medical librarianship (although there is a credentialing process); and public health informatics (although informatics is a crosscutting competency in public health certification). The professions that have existing certifications include: nursing informatics (ANCC/ANA) (although at the basic level); information science/computer science (CPHIMS/HIMSS); health information management (CCHIIM/AHIMA); and pharmacy informatics (ASHP) through accredited residency programs. The eligibility criteria for certification exams varied but graduate degrees plus experience were not unusual.

The Task Force hypothesized that the commonalities of informatics competencies across specializations and professions are quite large relative to their differences and therefore recommended that the AIIC exam be discipline neutral. Individual specializations would be free to develop specialized designations (e.g., a certification or credential) that meet the needs of their constituencies. The AIIC and these discipline-specific designations should enhance rather than compete with each other.

In December 2014, AMIA's Board of Directors convened a multi-disciplinary work group to build on the effort of the AMIA Academic Forum AIIC Task Force and recommend the core content and eligibility pathways for AIIC. This AIIC Work Group includes representatives from clinical informatics research, dentistry, nutrition, nursing, osteopathy, pharmacy, public health, and radiology. To inform their deliberations, work group members solicited input from stakeholder groups within these professions and related AMIA Working Group Chairs (e.g., the Nursing Informatics WG) and over 66 individuals responded from the various disciplines. Simultaneously, AMIA staff analyzed options for establishing a trusted, professionally neutral home for developing and administering the AIIC examination. This neutral organization would establish the final core content and eligibility pathways for AIIC.

AMIA's commitment to health informatics certification was reinforced by the following organization goals:

- Strengthen the profession of health informatics by creating a way for advanced practitioners to demonstrate their expertise.
- Establish advanced certification in health informatics that meets the needs of individuals from diverse health professions and is equivalent in rigor to the clinical informatics medical subspecialty.
- Create a clear pathway for professionals seeking advanced certification, including those who need more education and training.
- Dedicate significant resources toward the realization of the certification program launch.

In November 2015, the AMIA Board of Directors endorsed recommendations of the AIIC Work Group that:

- Defined the scope of certification and recommended the name *Advanced Health Informatics Certification (AHIC)* to identify the certificate.
- Recommended approaches for developing health informatics Core Content.
- Proposed rigorous quantitative and qualitative eligibility requirements intended to span diverse health informatics professions.

Additionally, AMIA staff identified strategies to create a certifying entity and laid the groundwork for implementing AHIC.

19.1.3 Advanced Health Informatics Certification (AHIC)

The purpose of the newly renamed Advanced Health Informatics Certification (AHIC) is to establish a clear set of expectations for the knowledge and skills that must be mastered to be proficient in a field and provide a recognized mechanism for individuals to demonstrate their proficiency and experience to potential employers. The focus of the certification is on professionals who work to improve the health of individuals and populations by applying informatics knowledge and skills to the operational aspects of information and knowledge problems that directly affect the practice of healthcare, public health, and personal wellbeing.

In 2016, the most immediate actions to insure the establishment of the AHIC were to:

- Initiate a process for finalizing Core Content that would serve as the basis for the AHIC examination;

- Publish proposed Eligibility Requirements for potential applicants of the AHIC; and
- Create a certifying entity and develop awareness and a clear value proposition for AHIC among health informatics professionals.

The certifying entity will make final decisions on Core Content, Eligibility Requirements, and other aspects of advanced health informatics certification.

Implications of AHIC for changes in health informatics education are most obvious in the need to develop Core Content to guide the creation of future AHIC examinations and in the relationship between certification of individuals and accreditation of the programs in which individuals are educated and trained. In December 2014, the Work Group initiated a high-level review of the clinical informatics subspecialty (CIS) Core Content (Gardner et al. 2009), which needs to be updated to reflect current practices, models, and technologies that were not as firmly established in 2008 when the document was developed. Prominent among these areas in need of updating is data science and its specific healthcare applications, such as predictive analytics and disease modeling. Fortunately, there are many supportive initiatives afoot, including the NIH Big Data to Knowledge (BD2K) funded programs in data science education and workforce development. (https://datascience.nih.gov/bd2k/funded-programs/enhancing-training (retrieved 29 Jan 2016). Moreover, other educational resources, such as expanding the ONC open HIT curriculum, are key. (https://projectreporter.nih.gov/project_info_description. cfm?aid=8935847&icde=0 (retrieved 29 Jan 2016).

Complementary attention to data science will be necessary during the development of a competency framework for assessing programs seeking to become accredited. While the Core Content for a certification exam and the competency framework for assessing health informatics master's programs have distinct purposes, programs will need to consider both documents for successful accreditation and successful certification pass-rates for their graduates.

Acknowledgements

The AMIA Board of Directors, under the guidance of Gil Kuperman, Blackford Middleton, and Thomas Payne, AMIA Presidents Don Detmer, Ted Shortliffe, Kevin Fickenscher, and Douglas Fridsma, and the Academic Forum provided pivotal leadership to AMIA's efforts to establish advanced certification for clinical and health informatics professionals. The Academic Forum Task Force on Advanced Interprofessional Informatics Certification, the Academic Forum Roles and Functions Task Force, and the Advanced Health Informatics Certification Work Group provided critical input to the AMIA Board on how to best structure AHIC. Members of the AMIA Academic Forum Task Force on Advanced Interprofessional Informatics Certification (2011–2012) and the Advanced Interprofessional Informatics Certification Work Group (2014–present) have provided critical input to the AMIA Board on how to best structure AHIC.

References

American Medical Informatics Association. 1988. https://www.amia.org/clinical-informatics-board-review-course. Accessed 29 Jan 2016.

Detmer DE, Shortliffe EH. Clinical informatics: prospects for a new medical subspecialty. JAMA. 2014;311(20):2067–8.

Gardner RM, Overhage JM, Steen EB, et al. Core content for the subspecialty of clinical informatics. J Am Med Inform Assoc. 2009;16:153–7.

Gardner RM, et al. https://datascience.nih.gov/bd2k/funded-programs/enhancing-training. Accessed 29 Jan 2016.

http://www.mastersindatascience.org/industry/health-care/. Accessed 29 Jan 2016.

https://projectreporter.nih.gov/project_info_description.cfm?aid=8935847&icde=0 (2014). Accessed 29 Jan 2016.

Office of the National Coordinator for Health Information Technology. Non-federal acute care hospital health IT adoption, Health IT Dashboard. 2016a. http://dashboard.healthit.gov/dashboards/hospital-health-it-adoption.php. Accessed Feb 2016.

Office of the National Coordinator for Health Information Technology. Office-based physician health IT adoption. Health IT Dashboard. 2016b. http://dashboard.healthit.gov/dashboards/physician-health-it-adoption.php. Accessed Feb 2016.

Saint Louis University. http://www.slu.edu/news-data-science-graduate-program (2015). Accessed 29 Jan 2016.

Shortliffe EH, Detmer DE, Munger BS. Clinical informatics: emergence of a new discipline. In: Finnell JT, Dixon BE, editors. Clinical informatics study guide: text and review. New York: Springer; 2016. p. 3–21.

Case Study 19.2: Accreditation of Graduate Health Informatics Programs

Judith J. Warren

Abstract The role and value of accreditation of academic programs are discussed along with their relationship to professional organizations, federal regulations and practice. For a successful accreditation process to work, professional organizations must define their domains and identify standards and competencies that identify quality performance in their disciplines. Accreditation organizations must define objective standards and evidence guidelines that identify a program that provides quality education. Employers and government must insist on quality education by hiring or funding students whose programs demonstrate the achievement of accreditation. Finally, the impact of big data and data science on nursing informatics education is presented.

Keywords Accreditation • Education • Professional organizations • Competencies

19.2.1 Introduction

Accreditation is part of the strategy that develops diverse, flexible, robust and respected higher education. The accreditation process assures and improves the quality of higher education. Organizations designed to conduct accrediting evaluations collaborate with experts to create a set of standards thus insuring the relevance of the accreditation. When a program successfully completes an accreditation review, the program is able to advertise their ability to provide appropriate instruction, student support, resources, and other services to meet the educational goals of the students (CHEA 2015). For students, accreditation is very important. The educational programs must be accredited for students to qualify for Title IV financial aid (US Department of Education 2016).

Accreditation is a quality process for programs in higher education. Accreditation assures that teaching, student achievement, curricula, academic support and other criteria meet certain levels of excellence and quality. Standards, used for this evaluation, are developed by the accrediting organizations, whose governing boards are composed of experts in the discipline being evaluated. The standards are developed and then submitted to the public for input and review. Programs submit self-studies

J.J. Warren, Ph.D., RN, FAAN, FACMI
University of Kansas School of Nursing, Kansas City, KS, USA
e-mail: jjwarren@live.com

based on these standards to the accrediting organization for evaluation. This submission is then followed with an onsite visit by site visitors who submit their evidence to the accrediting organization. Program accreditation is important for students as it provides value related to not only judging quality, but also obtaining employment, receiving student aid and transferring credits. For nursing there are two official accreditation organizations: Commission on Collegiate Nursing Education (CCNE) (CCNE 2015) and Accreditation Commission for Education in Nursing (ACEN) (ACEN 2016). For informatics there is one official accreditation organization: Commission on Accreditation for Health Informatics and Information Management Education (CAHIIM) (CAHIIM 2015). CCNE accredits baccalaureate, masters and professional doctorate degrees in nursing. ACEN accredits diploma, associate, baccalaureate, masters and professional doctorate degrees in nursing. CAHIIM accredits masters' degrees in health informatics and degrees in health information management. All three organizations are accredited by the Council for Higher Education Accreditations (CHEA) (CHEA 2016). CCNE is also accredited by the United States Department of Education. While CCNE and ACEN do not accredit nursing informatics specialties within the general nursing graduate programs, they do have requirements for informatics education for all students. For this reason, CAHIIM, which accredits graduate health informatics programs, is important to recognize quality informatics specialty programs in nursing.

CHEA is the primary national voice for accreditation and quality assurance to the U.S. Congress and U.S. Department of Education, as well as to the general public, opinion leaders, students and families. CHEA provides leadership for identifying and articulating emerging issues in accreditation and quality assurance. They also provide a national forum to address issues of mutual interest and concern in accreditation. CCNE, ACEN, and CAHIIM actively participate in this forum. CHEA is the only nongovernmental higher education organization in the United States that undertakes this responsibility for the public (CHEA 2015b). Encouragement is being given to accreditors to change their standards on curriculum from content-focus to outcome-focus. This new view requires programs to insure that their curricula support student achievement of the competencies specified by the professional organizations (CHEA 2016). For nursing informatics, these professional organizations are American Association of Colleges of Nursing (AACN) (AACN 2015), National league for Nursing (NLN) (NLN 2016) and American Medical Informatics Association (AMIA) (AMIA 2015).

Nursing competencies for use in curriculum development and accreditation evaluation are developed by AACN. These are published as "Essentials" and are freely available on their web site (AACN 2016). The "Essentials" are developed through a process of committees and consensus conferences. They represent the best thoughts about competencies for nursing practice for undergraduate and graduate students (both masters and professional doctoral degrees). Participation of administrators and faculty is essential to the process of developing these competencies.

Nursing interpretative statements for curriculum development and accreditation evaluation are developed by NLN. Standards for Accreditation have quality indicators and interpretive guidelines that guide faculty in creating their programs. The quality indicators are available on their web site (NLN 2016). These standards have been developed through group consensus of the membership.

While there are no official competencies for informatics professionals, AMIA has recently initiated a process for their development (AMIA 2016). AMIA has also become a member of CAHIIM to support accreditation of programs for this domain (AMIA 2015). AACN does not develop competencies for the specialty of nursing informatics, nor does NLN. In the same way as nurse midwives and nurse anesthetists, nurse informaticians are looking to have more than CCNE or ACEN accreditation. The addition of CAHIIM accreditation insures a strong informatics program. Nursing informatics programs are considering accreditation for their graduate programs in informatics. Accreditation discussions for nursing practice doctorates in nursing informatics have begun, as this is not yet a CAHIIM service (personal communication with the University of Minnesota School of Nursing),

CCNE, ACEN and CAHIIM also belong to and participate in the Association of Specialized and Professional Accreditors (ASPA). This organization is dedicated to enhancing quality in higher education through specialized and professional accreditation. To guide the accreditors, ASPA has developed a Code of Good Practice. This code describes best practices for member organizations for establishing relationships with their programs and institutions. CCNE, ACEN, and CAHIIM endorse and adhere these principles (ASPA 2015).

As accreditors, CCNE, ACEN, and CAHIIM track the evolution of new knowledge and professional requirements. The knowledge and ability to work with big data using data science is a new expertise desired by employers and the public. This trend has exploded due to new hardware and software capacities in handling very large amounts (petabytes) of stored and real-time data flows. The large amounts of data in electronic health records and genomic databases provide data and information to health care providers who are able to develop knowledge from these databases. New medical devices can export their data directly to electronic health records or other databases used in health care. Patients now use devices to record fitness activities, glucose monitoring, vital signs, food consumption, therapy logs, and a variety of other data. The Internet and social media have created online communities providing information and social support for patients and families. The use of the information developed in these communities is now accessible to understand the impact of disease on daily living. Analysis and use of this data is becoming critical in today's world in identifying best practices and guiding safe, effective patient care. While specific competencies have not been developed in this field, many informatics programs are incorporating this content into their curricula. The accreditors are monitoring this trend to determine impact on standards development. Professional organizations are monitoring this new field to determine if their competencies should change or if a new competing discipline is emerging. See Fig. 19.2.1 for a depiction of the relationships between organizations.

Fig. 19.2.1 Relationships between organizations for accreditation activities

19.2.2 Accreditation Standards

CCNE, ACEN, and CAHIIM collaborate through CHEA and ASPA to develop criteria for well-defined accreditation standards that will identify how well an educational program is performing. This collaboration establishes a national education strategy that insures quality education. While each accreditor may organize their standards differently, the basic categories of what is evaluated are the same. Standards are developed for the following: the sponsoring educational institution, governance of the program, program mission and goals, program evaluation and improvement, faculty, curriculum, program director, resources, students, and compliance with fair practices. To assist programs applying for accreditation, materials and resources are available on each web site (ACEN 2016; AMIA 2015; CCNE 2015).

As faculty are responsible for curriculum, they are most interested in the standard concerning curriculum. This standard is based on a relationship with a professional organization. The professional organization represents the views and practice of the professional. The professional organization is responsible for developing the competencies required to practice. A standard on curriculum will state that the curriculum must build on the professional competencies, thus insuring student outcome-based performance. Faculty can participate in the evolution of this standard by being active in the professional organization's competency work and by volunteering to be site visitors for the accreditor. This synergy between the profession and the accreditor insures the quality for which program accreditation was created.

Administrators are responsible for creating a program that is responsive to society's demand for well-educated graduates. As knowledge, innovation, and technology evolve, the need for adding new content becomes a challenge: how to add this new material in an already full curriculum. This dilemma is well described in a classic text, The Saber-Tooth Curriculum by Abner Peddiwell (1939). This is a fable about a faculty who taught the fighting of saber-tooth tigers long after the tigers were extinct, because tradition demanded the content. Today's educators use program evaluation, an accreditation standard, to determine how well their program is meeting societal needs and how the curriculum should evolve. Program improvement, another accreditation standard, focuses on monitoring trends and professional practices to create needed improvements in the curriculum. The excitement about big data and data science is not only being monitored by academic programs but also by professional organizations and accrediting organizations. ACEN, AACN and AMIA have workshops and presentations at their conferences and provide forums for professionals seeking jobs in these new fields of big data and data science. As they explore these new fields, discussions concerning competencies are beginning. As the competencies are developed and added to the official view, then CCNE, ACEN, and CAHIIM, who are also monitoring these new fields, will begin to revise accreditation guidance by pointing to these new competencies in the curriculum standards. Thus, the relationship between practice, education, and accreditation is enriched.

Another standard concerning faculty specifies the credentials needed by them to insure that a quality education is provided. Formal education achievement, licensure, and certification are the common components of credentialing. Since big data and data science are new and academic offerings and programs in these areas are just now emerging, a further method of evaluation of faculty competence in these fields must be added—evidence of lifelong learning. Faculty development programs need to be in place to support learning in regard to big data and data science. These programs need to supplement faculty reading and discussions. Evidence of interacting with experts in these fields, whether they are at conferences or online modalities, needs to be documented. Faculty development is probably the greatest challenge for administration. Learning is a labor-intensive process requiring time and resources. Faculty time and desire for mastery of new content and skills need to be encouraged and rewarded. Including learning experiences to the curriculum for big data and data science will require an investment in faculty.

Having sufficient resources is the next standard to be considered in the accreditation standards. What is the evidence for resources for teaching and learning in big data and data science? In the Big Data Primer (See Chap. 3) we learned that the "Vs—volume, velocity, and variety" defined big data. So to teach big data, a program needs to have data sources that meet these criteria for student learning. Not only are very large data sets needed but also the hardware (appropriate data architecture, servers, input devices, and output device) and software (database management systems for big data, programming languages, and data analytics) to support learning to engage in big data analysis need to be available. As previously mentioned in this book, new partnerships with industry or the government may need to be forged to provide access to these resources. Data science requires faculty who are knowledgeable in this domain: expertise in statistics or epidemiology is not sufficient. As with big data, resources for data science include software able to assist the student with data visualization, data wrangling, analytics, decision support, and business intelligence. These resources are in addition to resources normally evaluated in accreditation activities. No longer do we need only libraries and skill laboratories (as in the days of old), we need access to the above resources and partnerships. We need high-speed access to the Internet for online education from academia and industry. These requirements require more funding than schools of nursing are used to having. Plus, this funding is about data and analytics, not funding to support learning about patient care as in the past. Strategy will be required to create the value proposition for this budget at a time when all other programs at a university are being challenged with big data and data science needs in their disciplines. The future of nursing will be in learning to use big data to improve patient care and outcomes.

19.2.3 Recommendations for Future Accreditation Requirements

Since big data and data science are new fields in nursing and informatics, determining the quality of programs offering instruction is in its infancy. There is a developmental process that needs to occur to put accreditation standards in place. First the question of whether big data and data science is new or a fad must be answered. Companies are partnering with academia to meet their needs for data scientists (Bengfort 2016). Over the last few years Gartner has moved big data from its place in the peak of inflated expectations to the middle of the trough of disillusionment on the HypeCycle for Technology curve, thus indicating beginning adoption (Gartner 2016). They list big data and data analytics as two of the five major trends to watch. Evidence of healthcare organizations using big data approaches is reflected in the new positions for data scientists. The National Institutes of Health has a new center for Big Data to Knowledge (BD2K) and has awarded grant money in this domain (NIH 2015). AACN, AMIA, and NLN have papers and workshops at their conferences. Faculty are beginning to develop skills and expertise. The next hurdle is to modify existing competencies to include this knowledge and skill. This process usually takes one to two years. Then AACN, NLN, and AMIA will notify CCNE, ACEN, and CAHIIM of the new competencies. Once CCNE, ACEN, and CAHIIM receive the new competencies, programs will be given another one to two years to come into compliance.

The following recommendations are made concerning accreditation, big data and data science:

Competency Work

- Develop competencies specific to big data and data science
- Determine which competencies are for all students and which are for nursing data scientists
- Link big data, as big data requires a team approach, to interprofessional education

Implementation Work

- Provide for faculty development and experience
- Develop partnerships that share resources for big data and data science work
- Reward faculty leadership in developing these programs
- Identify forms of faculty scholarship in big data and data science

Accreditation Work

- Identify evidence for determining faculty competence
- Produce guidance for accreditation application preparation
- Train site visitors to evaluate evidence of quality education in big data and data science programs

19.2.4 Conclusion

As big data and data science are gathering more impact in health care and education, expect to see more employment opportunities in these areas. Students will want to receive quality education for employment. Employers will want to know that potential employees have achieved a quality education. Professional organizations will develop competencies. Accreditors will include big data and data science within their scopes of work. All of this will occur at a rapid rate. Begin now to get ready for the future.

References

Accreditation Commission for Education in Nursing. http://www.acenursing.org. Accessed 1 Feb 2016.
American Association of Colleges of Nursing. http://www.aacn.nche.edu. Accessed 28 Dec 2015.
American Association of Colleges of Nursing. *The Essentials Series.* http://www.aacn.nche.edu/education-resources/essential-series. Accessed 5 Jan 2016.
American Medical Informatics Association. http://www.amia.org. Accessed 28 Dec 2015.
American Medical Informatics Association. Health informatics accreditation committee. https://www.amia.org/health-informatics-accreditation-committee. Accessed 5 Jan 2016.
American Medical Informatics Association. AMIA Joins CAHIIM. https://www.amia.org/news-and-publications/press-release/amia-joins-cahiim-leadinformatics-program-accreditation. Accessed 28 Dec 2015.
Association of Specialized and Professional Accreditors. Code of good practice. http://www.aspa-usa.org/code-of-good-practice. Accessed 28 Dec 2015.
Bengfort J. Big Data 101: How higher ed is teaching data science. 2013. http://www.edtechmagazine.com/higher/article/2013/10/big-data-101-how-higher-ed-teaching-data-science Accessed 11 Jan 2016.
Commission on Accreditation for Health Informatics and Information Management Education. http://www.cahiim.org. Accessed 28 Dec 2015.
Commission on Collegiate Nursing Education. http://www.aacn.nche.edu/ccne-accreditation. Accessed 28 Dec 2015.
Council for Higher Education Accreditation. Value of accreditation. http://www.chea.org/pdf/Value%20of%20US%20Accreditation%2006.29.2010_buttons.pdf. Accessed 28 Dec 2015.
Council for Higher Education Accreditation. http://www.chea.org/pdf/chea-at-a-glance_2015.pdf. Accessed 28 Dec 2015.
Council for Higher Education Accreditation. http://www.chea.org/accreditation_toolkit/accreditation_toolkit.pdf. Accessed 15 Jan 2016.
Council for Higher Educations (2015). Recognized accrediting organizations. http://www.chea.org/pdf/CHEA_USDE_AllAccred.pdf#search="recognition". Accessed 1 Feb 2016.

Gartner. Gartner's Hype Cycles for 2015: Five megatrends shift the computing landscape. https://www.gartner.com/doc/3111522?ref=SiteSearch&sthkw=hype%20cycle%20big%20data&fnl=search&srcId=1-3478922254. Accessed 5 Jan 2016.

National Institutes of Health. Big data to knowledge. https://datascience.nih.gov/bd2k. Accessed 28 Dec 2015.

National League for Nursing Accreditation Commission. http://www.nln.org/accreditation-services/proposed-standards-for-accreditation. Accessed 1 Feb 2016.

Peddiwell A. The saber-tooth curriculum. New York: McGraw-Hill; 1939.

US Department of Education, Office of Student Aid. https://studentaid.ed.gov/sa/home. Accessed 4 Feb 2016.

Chapter 20
Quality Outcomes and Credentialing: Implication for Informatics and Big Data Science

Bobbie Berkowitz

Abstract This chapter explores the links among methods to measure expert nursing practice, patient outcomes, and data systems that enable an examination of factors that lead to patient outcomes. Certification and the nursing credential are discussed as examples of methods to designate expertise in practice. However, the literature shows that the ability to link the credential to patient outcomes is yet to be refined. The Institute of Medicine's Standing Committee on Credentialing Research in Nursing provides the background for research on credentialing and outcomes along with a discussion on the work of the National Quality Forum in developing measures for performance. Measurement sets for quality performance are experiencing considerable precision but fall short as a way to measure the work of nursing and its direct influence on quality. Recommendations are suggested for continued work on pathways to evidence and refining measures to capture the influence on credentialing in quality.

Keywords Credentialing • Certification • Quality • Patient outcomes • Measurement

20.1 Introduction

The role of data, particularly data related to patient outcomes and provider performance, has taken on a critical role in education, practice, and research within nursing. Access to data is essential to setting educational priorities, preparing expert clinicians, framing research questions, and pondering the future of health care

B. Berkowitz, PhD, RN, NEA-BC, FAAN
Columbia University School of Nursing, New York, NY, USA
e-mail: bb2509@columbia.edu

© Springer International Publishing AG 2017
C.W. Delaney et al. (eds.), *Big Data-Enabled Nursing*, Health Informatics,
DOI 10.1007/978-3-319-53300-1_20

knowing that the complexity of our health system is changing rapidly along with the demand for expert professional nurses. Data helps answer questions about what our priorities should be within education, practice and research and how best to assure expertise in these areas. This has become particularly important in seeking ways to measure quality performance as the move to value-based payment for health care is well underway. Brennan and Bakken (2015) remind us that data, particularly data from large data sets, has shown value in illuminating new ways of examining and understanding the patient experience and hold great promise for discoveries about nursing practice. However, in what order do we manage the readiness for quality performance management? Do we create the tools, data sets, and expertise to manage and measure nursing performance and patient outcomes or do we begin in the academic environment with the expertise of our clinician educators and our scientists? The reality is that all of these challenges require simultaneous management, application, and innovation. The opportunities envisioned by the nurse scientist, clinician, educator, and more recently the nurse informaticist for the future may be limited only by lack of access to the tools to put the knowledge to work. This chapter explores several of many issues related to this ability to imagine the future and then create the required expertise. In particular we examine high-quality performance by professional registered nurses and how to measure it. The first issue relates to current thinking about the role of certifying nursing knowledge and whether certification is sufficient to assure high-quality patient outcomes. The second issue is whether the link between quality patient outcomes and a specific credential can be extracted from data that establish evidence linking this relationship. From what we know about linking certification and quality, it seems evident that we need to continue to develop the process, data, and interoperability of systems to create a reliable signal that quality nursing practice exists within the healthcare system and as new evidence enters the clinical environment the health system will "learn" new ways of assessing and assuring quality. First we review quality performance as examined through a certification process. We then examine the link between credentialing and patient outcomes. We finally examine the processes that measure quality and look at the data that may open a window on quality performance.

20.2   High-Quality Performance

We know that many factors within the practice environment impact high-quality nursing performance and patient outcomes including the role that the certification process may play. Kizer (2010) examined factors that characterized a "world-class medical facility", referring in part to a method for assessing plans for new health-care facilities. He cited as an example the process utilized to plan for the Walter Reed Army Medical Center and National Naval Medical Center in 2005. This process considered the physical environment in particular and a design that builds quality into its processes and practices of care including diagnostic capability, treatment quality, technology, expertise of staff, and organizational culture. These physical and technological aspects of a care environment may be particularly critical for nurses who spend a great deal of their time managing the physical environment and its associated technology and appropriate treatments. Kizer also noted that while compliance with accreditation and certification standards was important, it was not sufficient. Cook and West (2013) discussed outcomes research in medical education and urged caution in making a direct link between what a physician does in treating a patient and the patient outcome. They stated that so many other factors are at play, including patient preferences for and compliance with treatment and variation in how a disease may vary across demographics.

Credentialing individual registered nurses includes the license to practice nursing achieved after completing educational requirements and passing the National Council Licensure Examination (NCLEX) for registered nurses and specialty credentialing that recognizes specific areas of expertise, specialization, and advanced practice. Specialty certification generally requires additional experience, expertise, and education along with requirements for periodic re-certification to maintain the credential.

Continuing nursing education has been a longstanding method for maintaining competence for evidence-based practice. Continuing education is a standard component of the renewal of a credential. The American Nurses Credentialing Center (ANCC) is one of the leading professional nursing organizations devoted to creating and applying methods for recognizing and promoting excellence in nursing practice. Two primary methods used are individual credentialing and organizational credentialing. The central focus of individual credentialing is a certification process for 25 specialty programs including advanced practice. ANCC defines continuing nursing education as "those learning activities intended to build upon educational and experiential bases of the professional registered nurse for the enhancement of practice, education, administration, research, or theory development to the end of improving the health of the public and RN's pursuit of their professional career goals" ANCC (2011, p 10).

The Magnet Recognition Program® is ANCC's model for organizational credentialing. It was introduced initially in the 1980s and strengthened in the early 1990s to create a roadmap to health system commitment to the quality of their nursing workforce. A great deal of research has been done and disseminated on the impact and outcomes of Magnet® designated health systems on outcomes for nurses and patients. An extensive review of these outcomes has been published by Drenkard L. (2010), Kelly et al. (2011), and McHugh et al. (2012). These studies reported positive outcomes for creating enhanced work environments for nurses in terms of recruitment and retention (higher retention, lower burnout), higher proportion of BSN prepared and specialty credentialed nurses, and higher nurse staffing. McHugh et al. (2012) reported that Magnet® designated hospitals show lower mortality rates than non-Magnet® designated hospitals and they theorized that this difference is due in part to better work environments for nurses and that these work environments influence nursing performance and patient outcomes. Lundmark et al. (2012) conducted a comprehensive review of the literature on nurse credentialing including organizational credentialing (Magnet® designation) with similar findings: lower RN turnover, higher nurse job satisfaction, lower nurse burnout, higher perceived unit effectiveness, higher nurse-perceived quality of care, higher nurse-ranked safety of care, lower nurse-reported adverse events, and lower rate of sharps injuries.

20.3   Credentialing and Patient Outcomes

The link between patient outcomes, nursing practice and credentialing was explored in depth by the Institute of Medicine Standing Committee on Credentialing Research in Nursing sponsored by the American Nurses Credentialing Center (ANCC). As reported by Drenkard (2013), the focus of the Standing Committee would include an examination of emerging priorities for nurse credentialing research, relevant research methodologies and measures, the impact of individual and organizational credentialing on outcomes, and strategies for advancing credentialing research. The work of the Standing Committee was reported in "Future Direction of Credentialing Research in Nursing: Workshop Summary" Institute of Medicine (2015). The results of this workshop confirmed that it is still not clear that the additional education and testing typical of credentialing requirements has a direct impact on quality outcomes. The review of the findings from this report is useful for thinking about the value of designating a credential as evidence of competent expert practice. If we begin with the premise that there is intrinsic value in a process that designates competence than we must find valid methods to link practice with specific competencies. Perhaps more important is the ability to link evidence of competence and expertise to patient outcomes. As described earlier, within the profession of nursing a system to test for and designate competent practice has been created through a network of organizations that develop, test, and certify clinicians in basic to advanced practice. The certifying bodies grant a credential as evidence of attaining specific standards or acquiring certain levels of expertise. Whether a specific credential for a practicing nurse holds value beyond the satisfaction of passing the certifying exam can be debated. However, many health systems have attached value to members of the nursing staff who hold a credential. Some have provided rewards such as formal recognition, higher salaries and promotions. The health system itself may draw benefits from having a nursing workforce that holds credentials (i.e., Magnet® designation). And yet, we continue to ask the question: why is the credential valued and rewarded? The IOM Standing Committee was particularly interested in whether there was evidence that the certification process with its steps to a credential including educational and practice requirements and testing could predict quality outcomes for patients. Although it may be important that a nurse who succeeds in obtaining a credential gains from the knowledge, skills, and confidence required to achieve and maintain the credential, the IOM Standing Committee was interested in whether research linked the credential to outcomes for patients and health systems.

The IOM Standing Committee commissioned three scholarly papers to guide discussion on frameworks for nurse credentialing research, nurse credentialing research designs, and data harmonization for credentialing research. These papers addressed what is currently known about how credentialing research could be conceptually framed (Needleman et al. 2014), how we might go about designing research that could seek links among certification and evidenced-based outcomes (McHugh et al. 2014) and why data harmonization is critical for credentialing research (Hughes et al. 2014). Together these papers addressed the key question: does the credential lead to evidence-based practice and what designs and data are appropriate?

Measuring quality. Needleman et al. (2014) created a diagram with processes to link credentialing with outcomes. The pathways included nursing competence and performance, the organization of work, and the organizational structure (culture, leadership, and climate). Of course, much of this diagram depends on a data set that enables extraction of these variables. The outcomes in the diagram (patient, nurse, organization, population) are drawn in part from the literature on the Triple Aim (Berwick et al. 2008). These aims are a function of the patient experience, cost, and the improvement of population health and have greatly influenced the development of measures that promote the payment of healthcare services by the Center for Medicaid and Medicare Services (CMS). The National Quality Forum (NQF) is an organization whose mission is to improve America's healthcare system through the utilization and application of evidence-based measurement National Quality Forum (2015a). NQF also recommends measures for public reporting and provides recommendations related to quality improvement priorities, electronic measurement and tools for performance measurement. They review and endorse measures used by a number of Federal, state, and private organizations. NQF has a specific focus on performance measures, practices for providing healthcare, frameworks for expectations in meeting standards, and reporting guidelines on how to select and report performance data (Measuring Performance, National Quality Forum, 2015b). For example, NQF provides guidance to CMS on measurement selection to be used for public reporting and value-based purchasing. NQF's structure is designed to evaluate and endorse measures intended to promote quality outcomes throughout health systems, the same systems where nurses influence patient outcomes. NQF-endorsed measures operate in multiple settings, often at the system level (hospitals, post-acute care, ambulatory care). Therefore, the data that are measured may be at the system level such as cost data, practice level data tracking unit based infection rates, or at the provider level measuring patient experience such as pain management.

While the sets of measures within the NQF (625 currently endorsed measures) are vast and comprehensive for provider and system performance, their application specific to a nurse's performance and resulting patient outcomes is difficult to single out. Some of this can be explained by the fact that measures specific to the patient are often the result of teams of nurses and other providers who care for a patient. Measures associated with performance, for example, are generally related to healthcare processes and structure. Linking these processes and structures to an individual provider such as a nurse is more complex.

Access to and use of data. As noted by McHugh et al. (2014) research to study the impact of practice is dependent on data, particularly if we want to measure the relationship of a variable such as credentialing to enhanced patient outcomes. As was noted previously, a credentialed nurse may practice within a team of nurses, not all credentialed. How do we parse out the difference the credential makes? Data related to credentialing that could be linked to specific practices and teams should be included in data sets at the unit level. McHugh et al. (2014) discussed that the data on nurse specific measures must be available across hospitals in order to examine the influence of variables, including credentialing, on patient outcomes. The authors made the case that data and analytics across systems, including electronic health records, are generally not standardized from system to system. The authors provided a number of examples of the difficulty working with healthcare data and noted that much of the data collected in healthcare environments is challenged with unique definitions, content, timeframe, clinical focus, granularity, and the ability to differentiate among provider types. You can imagine how difficult it is to compare variables across systems when the measures and data are not standardized. If we want to study the effect of credentialing on patient outcomes, we need to link operational, economic, and patient outcome data to understand the full extent of nursing influence. The measures also need to be standardized and included in data repositories that are accessible to researchers who have expertise in informatics and the ability to utilize large data sets.

The literature is quite clear that understanding the elements of the health system including organizational and provider performance and the outcomes for patients is highly dependent on robust data. Health information technology is essential in today's complex care environments. NQF (Health IT) characterizes the focus of health technology to include:

- Harmonization criteria to assure consistency across quality measures included in the electronic health record;
- Ability for the electronic health record to support the measurement of concepts such as care coordination and patient safety—two critical elements for patient outcomes;
- Standardization of clinical concept so that outcomes can be monitored and communicated.

NQF is exploring a number of other related concepts critical to health information technology, but the nature of current measurements sets helps us understand why it is difficult to measure and collect data about the performance of a nurse, credentialed or not, and link that with how an individual patient responds.

Englebright et al. (2014) explored the difficulty of measuring and collecting data on nursing performance. The authors were interested in assuring that nursing care was documented in the electronic health record (EHR) in a way that would capture the basics of current practice. They conducted a study to define and select practices that represented "basic nursing care" in a 170-bed community hospital. A group of expert nurses determined which activities were part of all nursing care regardless of the patient's diagnosis through an examination of nursing theory and nursing practice regulation for basic care. In part, their premise was that nursing contribution to patient outcomes and system quality had not been well defined in a way that could be accurately reflected in the EHR.

There is so much more for us to learn and imagine through the harnessing of the metrics of big data in creating the environment and essential tools and competencies to provide quality patient outcomes. In this equation for creating quality, the registered nurse is an essential element. Is it possible to utilize the power of data science including the EHR along with ways of visualizing data to explain how a nurse with certain characteristics adds value? Brennan and Bakken (2015) call for new ways of structuring and using data that enable more precision for answering critical question about nursing phenomenon. Perhaps the use of big data will allow the processes for building competence to come to light, including our educational endeavors, lifelong education, and methods for linking competence to outcomes.

20.4 Conclusion

The conversation will continue as to whether a certification process assures a nurse's competence to achieve the highest quality care as evidenced by patient outcomes. The process has been complicated in part by limitations in access to appropriate research methods and to an informatics platform with the necessary data sets. The ability to measure the contribution of nursing practice to the achievement of quality patient outcomes is predicated on many factors including high-performing academic institutions for nursing education, the development of the knowledge base for practice, and the translation of evidence into nursing practice. It appears that a rigorous process of credentialing expertise is a pathway to assuring quality but we need to create the best possible data science methods to measure that expertise.

Professional nursing with its complement of clinicians, scientists, educators, leaders, and policy experts might consider responsibility for a least three commitments going forward:

1. Continue to seek ways to establish the pathways and corresponding evidence for outcomes driven by nursing practice.
2. Assure precision in reporting high-quality outcomes by exploring and refining measures that characterize actions nurses take while providing evidence-based care.
3. Utilize data science methods to drive our understanding of nursing's contribution to quality including standardizing measures and tools across the multiple environments where nursing care occurs.

References

American Nurses Credentialing Center. ANCC primary accreditation application manual for providers and approvers. Silver Spring, MD. 2011. Accessed 2013.
Berwick D, Nolan T, Whittington J. The triple aim: care, health, cost. Health Aff. 2008;27(3):759–69. doi:10.1377/hlthaff.27.3.759.

Brennan P, Bakken S. Nursing needs big data and big data needs nursing. J Nurs Sch. 2015;47(5):477–84.

Cook D, West C. Reconsidering the focus on "outcomes research" in medical education: a cautionary note. Acad Med. 2013;88(2):162–7. doi:10.1097/ACM.0b013e31827c3d78.

Drenkard K. The business case for Magnet®. JONA. 2010;40(6):1–9. doi:10.1097/NNA. 0b013e3181df0fd6.

Drenkard K. Nurse credentialing research: a huge step forward. JONA. 2013;43(1):4–5. doi:10.1097/NNA.0b013e31827903c3.

Englebright J, Aldrich K, Taylor C. Defining and incorporating basic nursing care actions into the electronic health record. J Nurs Sch. 2014;4(1):50–7.

Hughes R, Beene M, Dykes P. The significance of data harmonization for credentialing research. Discussion Paper. Washington, DC: Institute of Medicine; 2014.

Institute of Medicine. Future directions of credentialing research in nursing: workshop summary. Washington, DC: National Academies Press; 2015.

Kelly L, McHugh M, Aiken L. Nurse outcomes in Magnet® and non-Magnet hospitals. JONA. 2011;41(10):428–33. doi:10.1097/NNA.0b013e31822eddbc.

Kizer K. What is a world-class medical facility? Am J Med Qual. 2010;25(2):154–6. doi:10.1177/1062860609357233.

Lundmark V, Hickey J, Haller KL, Hughes R, Johantgen M, Koithan M, Newhouse R, Unruh L. A national agenda for credentialing research in nursing. American Nurses Credentialing Center Credentialing Research Report. Silver Spring, MD; 2012.

McHugh M, Kelly L, Smith H, Wu E, Vanak J, Aiken L. Lower mortality in Magnet® hospitals. Med Care. 2012;51(5):382–8. doi:10.1097/MLR.0b013e3182726cc5.

McHugh M, Hawkins R, Mazmanian P, Smith H, Spetz J. Challenges and opportunities in nurse credentialing research design. Discussion Paper. Washington, DC: Institute of Medicine; 2014

National Quality Forum. http://www.qualityforum.org/ (2015a). Accessed Jan 2016.

National Quality Forum. Measuring Performance. 2015b. http://qualityforum.org/Measuring_ Performance/Measuring Performance.aspx. Accessed Jan 2016.

Needleman J, Dittus R, Pittman J, Spetz J, Newhouse R. Nurse credentialing research frameworks and perspectives for assessing a research agenda. Discussion Paper. Washington, DC: Institute of Medicine; 2014.

Chapter 21
Big Data Science and Doctoral Education in Nursing

Patricia Eckardt and Susan J. Henly

Abstract Knowledge and skill development ensuring that nursing scientists are optimally prepared to conduct big data research and that advanced practice nurses are prepared to use findings from big data research to optimize the health outcomes for individuals, families, and communities is a priority need in doctoral education. The purpose of this chapter is to identify critical issues involved in incorporating "big data" science into research- and practice-focused doctoral programs in nursing. The need for preparation of nurses in big data science, data science methods, and translation of findings from big data nursing research into practice will be summarized. Common core knowledge and skills and competencies specific to preparation of advanced practice nurses and nursing scientists will be emphasized. Resources for incorporating big data into doctoral curricula will be identified.

Keywords Doctoral education in nursing • Nurse informaticians • Big data • Nursing informatics • Data science • Practice-focused nursing doctorate • Research-focused nursing doctorate

21.1 Introduction

Big data is central to recent advances in diverse scientific disciplines. For example, the genomic era in biology is grounded in big data. Nursing science across scales from molecules to society and nursing care from prevention interventions to critical care are

P. Eckardt, Ph.D., R.N. (✉)
Heilbrunn Family Center for Research Nursing, Rockefeller University, New York, NY, USA
e-mail: peckardt@rockefeller.edu

S.J. Henly, Ph.D., R.N., F.A.A.N.
University of Minnesota School of Nursing, Minneapolis, MN, USA

© Springer International Publishing AG 2017
C.W. Delaney et al. (eds.), *Big Data-Enabled Nursing*, Health Informatics,
DOI 10.1007/978-3-319-53300-1_21

increasingly informed by or based on big data approaches. In the health sciences, big data comprises images, waveforms, text (nursing notes) and numerical information (flowsheets) from electronic health records (EHRs), laboratory and omics records, telehealth monitoring systems, interactive surveys, and data from research studies; big data is from multiple sources and possess many forms. Existence of big data offers tremendous opportunity to understand and treat complex health problems from population to individual level. However, the size, dimensionality and heterogeneity of big data are not easily stored or analyzed with traditional software or analytical methods. When contrasted with traditional small or large data sets, the defining attribute of big data is lack of straightforward structure for conceptualization of properties and inherent available information to be wrangled into created knowledge by traditional methods. The opportunities and difficulties of big data are creating exciting new challenges for doctoral nursing education.

21.2 About Big Data and Nursing

21.2.1 Ubiquity of Big Data

Health-related big data is creating massive between-subject as well as equally massive within-subject data (Azmak et al. 2015; Edelman et al. 2013; Henly et al. 2015a; Westra et al. 2015a). Big data drives healthcare policy and clinical decision-making (Cios and Nguyen 2016; Coffron and Opelka 2015). These data have been increasingly collected, aggregated, and integrated over the past thirty-five years, with an unprecedented increase in momentum in the US following the 2009 American Reinvestment Recovery Act (ARRA) Health Information Technology for Economic and Clinical Health Act (HITECH) mandate for meaningful use of data 2009 (Beaty and Quirk 2015). Precision medicine—health care tailored to the individual—is emerging as the defining illustration of the omnipresence and influence of big data in nursing and health care (Jameson and Longo 2015). Precision medicine initiatives, also known as personalized medicine or therapies and initiatives (Chinese Lung Cancer Collaborative Group 2015; Hong et al. 2012; Lei et al. 2012; Wu et al. 2013) in the US, China, and elsewhere will integrate data from medical records, real-time sensory devices and the Internet of Things, genomic sequences, samples of flora that live in and on patients' bodies, designed studies, and survey data about lifestyles, attitudes, knowledge, and health beliefs and practices to personalize health care. In the US, the Big Data to Knowledge (BD2K) initiative is designed to support advancement of tools to support broad use of biomedical big data, disseminate methods and software, and enhance training in the use of big data (Bourne et al. 2015).

21.2.2 Definitions

Even with the ubiquity of big data, a universally accepted definition of big data does not exist; it is said to possess volume (great size), velocity (speed; quick change), variety (many forms), veracity (trustworthiness vs. messiness), and value (potential for information, knowledge, or wisdom); (e.g., Higdon et al. 2013). A principal facet of definitions is that data is of a size and dimensionality incapable of being arranged and understood using traditional means (Cios and Nguyen 2016; Magee et al. 2006; Sanders et al. 2012). Data science overlaps with computer science, statistics, data mining and predictive analytics, and machine learning for the systematic use of big data (Jordan and Mitchell 2015); it includes both philosophical and analytical approaches to working with big data (Brennan and Bakken 2015).

21.2.3 Nursing Interface with Big Data

Nurses are involved in the use of big data as generators of big data, as scientists involved in knowledge discovery, and as clinicians using findings to help patients maintain and improve their health while containing costs. For example, complex systems science has been used to improve outcomes of patients with cardiac failure (Clancy et al. 2014); natural language processing has been used to extract meaning from captured qualitative data that can impact patient safety and quality and nursing outcomes (Hyun et al. 2009); and healthcare data has been meaningfully used to reduce administrative burden. Nursing-relevant contextual data has been available to practitioners and researchers, and elements of the Nursing Minimum Data Set (MDS), Nursing Management Minimum Data Set (NMMDS) and related clinical terminologies have been updated and then mapped using LOINC (Logical Observation Identifier Names and Codes) and SNOMED-CT (Westra et al. 2016, p. 284). The use of big data from multiple sources, the sociological and clinical influence of big data phenomena, and multiple data science initiatives have changed the landscape of nursing science, practice, and education (Fig. 21.1).

New roles such as data-intensive advanced nurse practitioners (including nurse informaticians; Simpson 2011), nursing scientists working with big data, and nurse data scientists are emerging (Brennan and Bakken 2015). Knowledge, skills, and competencies needed to enact these new roles are now priorities for integration into doctoral nursing education.

Fig. 21.1 Landscape of big data for advance practice nurse, nursing scientist, and data scientist

21.3 Doctoral Education

21.3.1 Context

Current directions of nursing doctoral education in the US are guided by standards issued by the American Association of Colleges of Nursing (AACN), recommendations for integrating emerging areas of nursing science into PhD programs by the Council for the Advancement of Nursing Science (CANS), standards of practice for nursing informatics of American Nurses Association (2014) and AMIA (McCormick et al. 2007), various advanced practice specialty professional groups including the American Association of Nurse Practitioners (AANP), American Association of Nurse Anesthetists (AANA), American College of Nurse-midwives (ACNM), National Organization of Nurse Practitioner Faculties (NONPF), the Institute of Medicine (IOM), and priorities for research at the National Institutes of Health (NIH), especially the National Institute of Nursing Research (NINR). The AACN "Essentials of Doctoral Education for Advanced Nursing Practice" (2006) calls for Doctor of Nursing Practice (DNP) graduates to use information technology and research methods in practice; the "Research-Focused Doctoral Program in Nursing Pathways to Excellence" (2010) states that data, information and knowledge management, processing and analysis are key curricular elements for PhD programs. The CANS Idea Festival Advisory Committee recommendations provided a panoramic view of the current and future environment of nursing research through a presentation of emerging areas of science and the knowledge and skills needed to prepare nursing scientists to advance these areas; big data nursing science was identified as an emerging and priority area (Henly et al. 2015a, b). The IOM (2003) called for all healthcare professionals to use an interprofessional approach to practice and to utilize evidence-based interventions and informatics to guide professional practice, and team science is recognized as critical to the contemporary research paradigm (Begg et al. 2014; Kneipp et al. 2014). Data science, viewed as an "historic opportunity" in the current NIH strategic plan (NIH 2015), is expected to leverage research efforts in all areas and is included in many research and training grants including those at the NINR. Additionally, the seminal *Future of Nursing* report (Institute of Medicine 2010) provided key messages that include nurses improving data collection and usage to drive improvements in health care. Each of these informs the requisite knowledge for big data roles in nursing for the future and shows the need for a harmonized curriculum to prepare advanced practice nurses and nursing scientists to be successful contributors in the era of big data science from their defined roles.

21.3.2 Framework

To create a framework for considering big-data related curricular issues for advanced practice and nursing science education, we considered the essentials for the practice-focused doctorate (AACN 2006; Lis et al. 2014) and the essentials for the research-focused doctorate, along with priority and emerging areas of nursing science (AACN 2010; Wyman and Henly 2015). Big data roles were defined for: (a) data-intensive nurses in practice, who are clinical experts leading teams using evidence from data science; (b) data-intensive nurse scientists, who are nursing scientists leading programs of research informed by data science; and (c) data scientists who are also nurses, and who lead research programs in data science informed by and motivated by issues and problems in nursing science (Brennan and Bakken 2015).

21.3.3 Big Data Knowledge, Skills, and Competencies

21.3.3.1 Common Doctoral Core: DNP and PhD

Education for the practice-focused (DNP) and research-focused (PhD) nursing doctorates is derived from the same substantive core: the health and illness experiences of individuals, families, and communities over time, nursing interventions, and nursing systems and quality of care (Henly et al. 2015c; Zaccagnini and White 2017). Whereas the DNP focuses on application of knowledge, the PhD emphasizes generation of new nursing knowledge. The commonality in substantive focus and variation in emphasis imply similarities and differences in content arising from consideration of big data for DNP and PhD programs in all areas of doctoral education: prerequisite knowledge and skills, admissions criteria, coursework, dissertations and capstone projects, and resources. Familiarity with basic nomenclature related to big data is essential (Table 21.1) for students in both types of programs. The challenge to nurse educators is to integrate big data/data science approaches across curricula (Brennan and Bakken 2015; Westra et al. 2015b).

Table 21.1 Keywords and definitions for big data science basics

Keyword	Definition
Algorithm	A process or set of rules to be followed in calculations or other problem solving operations. A basic (and useful) algorithm in data science is the IF…THEN… algorithm
Classification error	An estimate of the difference between observed values and predicted values when algorithms for classification of data into groups are used
Clinical Decision Support System (CDSS)	Applications that analyze data, both from the individual patient and from current literature, to help healthcare providers make clinical decisions. These complex algorithms are often used at point of care by clinicians
Code	The symbolic arrangement of instructions in a computer program that direct the program as to what data to analyze and the approach to the analysis. Code is also referred to as syntax
Computational biology	Involves mathematical tools for studying dynamic living systems, from molecules to whole organisms, over time. Clinical applications allow prediction of individual patient course and individualization of treatment
Data	Are elements such as measures and symbols which are used to create information when placed in context
Data mining	The practice of examining large databases in order to generate new information. Data mining can range from human extraction of data from medical records to the more commonly assumed approach of machine learning and algorithm-driven data acquisition
Data visualization	An approach to creating knowledge from big data that uses figures, graphs, and other illustrations (e.g., bubble chart or quilt plot)
Data validation	The process of ensuring that data used for analysis are representative, accurate, and consistent. Data validation can occur during multiple phases of data science: at the start of an analysis with review of data source and input, during development of model with frequent model checking in pilot methods development phase of algorithm, and during analysis with additional sensitivity checks for error in classification
Data Warehouse (DW)	Central repositories of integrated data from one or more disparate sources

Table 21.1 (continued)

Keyword	Definition
Data- Information- Knowledge- Wisdom model (DIKW)	An approach to conceptualizing the progressive transformation of raw data into organized information under the direction of the scientist to create knowledge for application to solve problems and to contribute to wisdom for decision making
Discretization	The process of recoding continuous attributes or features of big data with huge numbers of values into a smaller number of categories
Discretization algorithm	Code used to transform continuous attributes into categories for machine learning. Use of class type information (e.g. diagnosis) to produce "IF…THEN…" language that results in partitioning of data
Electronic Health Record (EHR)	Data collected on patient history, test results, care provided, and outcomes from all clinicians involved in their care
International Classification for Nursing Practice (ICNP®)	A formal terminology for nursing practice and a framework into which existing vocabularies and classifications can be cross-mapped to enable comparison of nursing data
Information	In the DIWK model, information is derived from data which is processed to create meaning (knowledge) wisdom (actionable knowledge) (See DIKW)
Knowledge Discovery in Databases (KDD)	Intended outcome of analysis of information in large databases. A priori hypotheses can be confirmed or hypotheses can be refined on a post-hoc basis
Knowledge Discovery Process (KDP)	The interactive and iterative process of discovering new knowledge from data; "making sense" of data, the success of which depends on effective collaboration between data miners and domain experts
Knowledge	Data + meaning, developed using the knowledge discovery process
Logic	Valid reasoning combining thoughts from philosophy and computer science that result in mathematical rules for decision-making that inform the code and algorithms used in data science
Machine learning	Is the use of data to develop predictive models of relationships between variables. Supervised learning can be used when the variables in a relationship are known, and unsupervised learning methods are used when some of the variables are unknown. A way to conceptualize the approach to supervised learning is the approach taken in a Bayesian analysis of using prior data on two variables (an informed prior distribution) to estimate relationships. In contrast, unsupervised learning can be conceptualized as an alternate Bayesian approach when there is no information to guide the estimation of an outcome, and a flat or noninformative prior distribution is used to predict an outcome on which there is no data
Minimum Data Set (MDS)	Part of U. S. initiatives to produce large repositories of data that can be analyzed for quality and financial outcomes in health care. In nursing, the two minimum dataset that are currently being expanded to be included in large data warehouse repositories to provide information of nursing practice are the Nursing Minimum Dataset (NMDS) and the Nursing Management Minimum Dataset (NMMDS)

(continued)

Table 21.1 (continued)

Keyword	Definition
Predictive analytics	A type of data analysis used in data science to provide prediction of an outcome for an individual or group. Big data provides a rich array of elements and variables to develop more comprehensive, and complex, predictive models for probability of disease state, health and illness status, and health trajectories
Pseudocode	A conceptual stepwise description of the logical and sequential process for an algorithm that does not contain any of the required syntax (or code) to run the algorithm. Pseudocode is very useful in a team science approach to the development of analytic algorithms with input from content experts to direct choice of code by the data scientist
Unified Medical Language System (UMLS)	A U.S. resource that integrates terminology and classification across health disciplines to provide interoperability between the datasets. An example is Logical Observation Identifiers Names and Codes (LOINC) which includes ICD codes and laboratory data from patient hospital encounters
Wisdom	Desired endpoint where data provides information that can create knowledge that informs larger decisions of health care resource utilization for present and future science; actionable knowledge (see DIKW)

21.3.3.2 Practice-Focused Doctorate

Prerequisite knowledge, skills, and competencies for DNP programs must focus on strong data management and application to position students for success in later practice as they apply these skills to understand and address complex processes and systems. The AACN (2006) outlined core competency areas for the DNP that require data management knowledge and skills. These competencies include: epidemiology, program evaluation, financial implications, data collection and use. Clinical data management (CDM) incorporates the use of technology, information, and knowledge to provide for better patient outcomes with the application of big data science to practice decisions and population health management upon completion of their studies (Sylvia and Terhaar 2014). The goal of doctor of nursing practice (DNP) programs is to produce nurses who are uniquely prepared to bridge the gap between the discovery of new knowledge and the scholarship of translation, application, and integration of this new knowledge in practice (AACN 2006). Discovery of new knowledge in nursing is inextricably linked with big data (Nguyen et al. 2011). Proficiency in translation, application, and integration of new knowledge into practice will require expertise with big data science of all DNPs (Lilly et al. 2015). Current prerequisites are uneven across DNP programs (Dunbar-Jacob et al. 2013; Mancuso and Udlis 2012), providing opportunity for harmonization of prerequisite preparation for big data science. Competence in prerequisite areas of knowledge and experience appropriate for the DNP course of study and practice needs include basic understanding of numeracy, logic, computer science, and statistical methods (Sylvia and Terhaar 2013). Competency has been defined as: "Habitual and judicious use of communication, knowledge, technical skills, clinical reasoning, emotions, values, and reflection in daily practice for the benefit of the individual and community being served." (Epstein and Hundert 2002). Competency in prerequisite areas can be assessed through prior coursework, testing, or demonstrated experiential learning. Examples of prior coursework that incorporate these competencies are informatics, healthcare technology, and statistical methods. These courses begin the foundational framework for more advanced curriculum in large datasets and clinimetrics—the measurement of clinical occurrences, which are important in CDM. Prerequisite harmonization is an excellent opportunity in a data-intensive nursing educational trajectory to insure that the entrants into a DNP program of study are prepared for the current environment of big data and data science, as DNP faculty and administrators continually struggle with balancing the addition of coursework and experience with priority and emerging areas of curricular focus to an already maximized course credit load for degree programs (Lilly et al. 2015). Assessing if these competencies have been attained prior to the commencement of DNP coursework increases the likelihood of success within the program of study and completion of the program without an overload of coursework.

Admissions criteria to DNP programs vary across programs. DNP programs continue to grow exponentially, with ever increasing numbers of potential applicants (Mundinger and Kennedy 2013). Effective selection of students from the admission pool should incorporate measures of competence in areas of CDM. Some competencies in computer science, information science, numeracy, probability, and logic are measured by the graduate record examination (GRE), but less than 50% of DNP programs include GREs in admissions criteria (Mancuso and Udlis 2012). Alternatively, competency is often developed through informal education such as writing computer code for a computer game or programming electronic devices (Sampson et al. 2015). Admission criteria that assess demonstrated competence with nontraditional education are difficult to standardize, but may also lead to a more diverse and capable cohort of students. Promising applicants who do not meet the prerequisite competencies for data science can still be considered for admission to the program with conditional requirements for admission. These are opportunities for data science competency boot camps, immersions, workshops that will position applicants for program success. As per the NIH website (2016), the NIH and NINR provide boot camps, for both faculty and students, in big data and data science.

Courses of study in DNP programs will need to respond to the big data environment to produce competent advanced practice data intensive nurses with the use and synthesis of data (Mundinger et al. 2009). DNPs will need to know how to lead IT decisions and understand IT implementation, outcomes and limitations. A course in informatics in all DNP programs has been suggested (Jenkins et al. 2007). From a clinical perspective, a data-intensive nurse in practice, such as the DNP, needs to know both how to collect the data to insure validity, as well as how to locate and interpret the data to guide patient treatments. This requires a hands-on approach to data acquisition, management, and analysis through guided learning with worked examples. Sylvia and Terhaar (2014) describe an approach to courses of study that prepare the DNP for practice with CDM competency integration throughout courses of study and DNP projects that meet the AACN (2006) competencies and essentials components. Content modules described focus on data management and analysis application with an identified population project. The modules provide a sequential approach that is common in all data-driven projects, but also provide associated learning objectives for the DNP student. The learning objectives and competencies described by Sylvia and Terhaar (2014) can be applied to the competencies required of DNPs in data intensive roles and integrated into curriculum essentials and prerequisite competencies checklist (Fig. 21.2).

DNP projects require the candidate to address a complex practice, process, or systems problem within a practice setting, using evidence to improve practice, process, or outcomes. With the increasing use of technology and big data to capture events and information about the practice settings, financial reimbursement, utilization of services, patient information, and evaluation for regulatory and quality purposes, a DNP project often will include faculty members from within the data-intensive nurse researcher community or from other related disciplines that focus on improving health outcomes of populations (e.g., sociology, law, engineering). DNP projects can have a great impact on populations with the application of epidemiological data and technological advances in patient care and communication—think texting reminders for medication adherence to homeless HIV clients in an economically disadvantaged area, or social networking site development for patients with orphan diseases in different geographic areas (Waldrop et al. 2014). Examples of successful projects that use big data and data science are available as templates for future DNP projects (McNeill et al. 2013; Rutledge et al. 2014). DNP projects are an opportunity for the advanced practice data-intensive nurse to demonstrate competence in big data science and his/her role in its acquisition, synthesis, evaluation, and application (Waldrop et al. 2014). Waldrop et al. (2014) developed an approach to DNP projects that provides a 5-point evaluation system to determine if projects meet the AACN 2006 essentials. Their paper also provides exemplars of project types and resources that can be accessed by other DNP faculty.

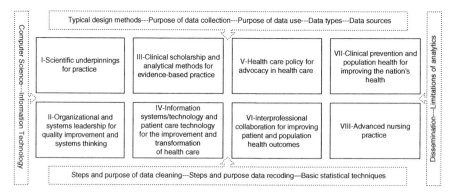

Fig. 21.2 The essentials of doctoral education for advanced nursing practice (2006) and competencies for big data science

21.3.3.3 Research-Focused Doctorate

Prerequisite knowledge needed to undertake the big data-intensive portion of a nursing science doctoral program is in the areas of computer science, mathematics, and statistics. Computer science basics including an appreciation for placing scientific problems into computational frameworks, understanding hardware and software systems, and knowing how computer programs are designed and written using algorithms. Mathematics through college algebra is essential; successful completion of courses in calculus and linear algebra is extremely helpful. Prerequisite statistical knowledge equivalent to an introductory course in probability and descriptive and inferential statistics is needed; some experience applying statistics to real data is optimal. Generally possessing a good "data sense" is needed for study of all quantitative aspects of a nursing science doctoral program.

Admissions criteria for PhD programs in nursing that focus on use of big data should address needed prerequisite knowledge in computer science, statistics, and mathematics. Like students in other graduate programs, nursing PhD students may not have had the opportunity to complete formal coursework in these areas before beginning doctoral study. Scores from tests of quantitative ability may be informative, especially along with documentation of interest and motivation for pursuing quantitative directions in research. A short course at admission covering the mathematical and statistical topics at the level of Fox (2009), augmented with needed computer science content, can position students for success as their big data-intensive training programs in nursing science unfold.

Courses of study in PhD programs in nursing are generally based on nursing inquiry, nursing theory, and research methods. As shown in Fig. 21.3, training for big data nursing science involves refreshing content and instructional approaches in each of these foundational areas.

Nursing inquiry is an evolving process that involves advancement of ideas related to the discipline and practice of nursing and challenges to understandings about health and health care (Thorne 2011); nursing science inquiry enacted as research permits the expansion of knowledge across the domain of nursing (Grady and McIlvane 2016). Technology-enabled research is redefining questions about health and health interventions (Henly et al. 2015a). The evolutionary nature of the nursing inquiry process supports incorporation of new ways of knowing from health and nursing big data as a natural development, and has particular appeal for a practice discipline—because discovery of new knowledge from big data is directed toward discovery of *actionable* knowledge (such as knowledge embedded in clinical decision support systems).

Inquiry is closely linked with theory, so changes in the inquiry process will challenge longstanding beliefs about the nature of theory and theory development (Brennan and Bakken 2015; Henly 2016b). The availability of temporally and spatially dense data from individuals—from sensors and self-reports/ecological momentary assessments of states of health, well being, and health behaviors—offers unprecedented opportunity to move theory development from nomothetics (general-law focused theory) to idiographics (person-centered theory; Henly et al. 2011; also

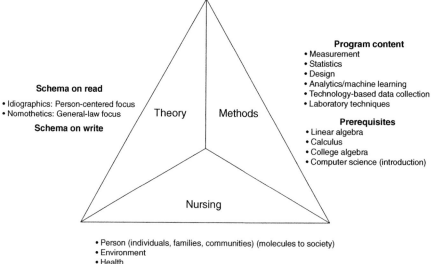

Fig. 21.3 Data science elements for theory, methods, and nursing inquiry in PhD programs

see Molenaar 2004). The person-centered approach to theory is closely linked to foundational ideas in precision medicine initiatives and to extensions of clinical trial designs that allow individualization of conclusions about treatment effectiveness (e.g., multiphase optimization strategies [MOST]; sequential multiple assignment randomized trials [SMART]). The existence of big data along with the hardware and software needed to capture, clean, and process it creates the option to use theory as a lens for interpreting existing data ("schema on read") rather than limiting theory as the boundaries within which research is designed and findings are interpreted ("schema on write") (Brennan and Bakken 2015).

Research methods needed to advance nursing science in the era of big data include traditional measurement, design, and statistical topics along with laboratory techniques for omics work, technology-based data collection in clinical and community settings, and data analytics/machine learning methods (Henly 2016a). Core measurement knowledge includes not only classical test theory and item response theory, from psychometrics, used to develop and assess instrumentation for patient-reported outcomes and other psychosocial and biobehavioral constructs but also biological measurement approaches (for omics and biophysical nursing research); technology-based data collection arises naturally in concert with measurement theory but remains to be included in U.S. PhD programs (Wyman and Henly 2015). Core statistical knowledge includes data descriptive and basics of inference, general linear model, survival analysis, and longitudinal and multilevel modeling (Hayat et al. 2013). Understanding the DIKW perspective, knowing standardized nursing terminologies and the availability, structure, and data elements in nursing relevant databases are first steps in utilizing big data for nursing research. Learning to work effectively with data miners as subject matter experts in the knowledge discovery process (Cios and Nguyen 2016) will be a basic skill for nursing scientists of the future; effective partnerships will require that nursing scientists be schooled in the fundamentals of data mining and data visualization. The span of methods topics relevant to nursing research is broad and no individual student or nursing scientist will master them all; instead, programs should identify core methodological knowledge needed first and then emphasize advanced methods most relevant to the focus of their research training (Henly et al. 2015b).

Dissertations are the independent research projects, completed under the supervision of an advisor, that generate new knowledge and prepare PhD students for a scientific career (AACN 2016; Henly et al. 2015b). Advances in emerging and priority areas of nursing science—omics, health behavior/behavior change, patient-reported outcomes, health economics—rely increasingly on big data and quantitative analysis in both discovery and translation research (Conley et al. 2015; Grady 2015; Henly et al. 2015a). Dissertations based on big data will reflect the many enhancements to inquiry in nursing science, nursing theory, and research methods, creating a new "look and feel" to the PhD research experience. Research questions will push the boundaries of knowledge in novel ways, theory will enter research at different phases of investigation, and new methods of technology-based data collection, study design, and analytics will combine to ensure that dissertation research meets both criteria: generation of new knowledge suitable for advancing nursing in the information age and preparation of students for successful careers in nursing science in the new e-science environment. Versatility, flexibility, and informed enthusiasm in embracing the possibilities in dissertation research engendered by big data will help students and faculty to traverse the challenges of change.

21.3.3.4 Nurse Data Scientist

The nurse data scientist leads programs of research in data science informed by problems in nursing research (Brennan and Bakken 2015). The primary research focus of a nurse data scientist is on methods and analytics as opposed to specific health and illness concerns of individuals, families, communities or populations. Nurse data scientists are educated as nurses and then pursue a research doctorate or post-doctorate in data science field (data analytics, computer science); alternatively, a data scientist might pursue education in nursing. The way to incorporate nurse data scientists into the wider community of nursing science has been paved by past and current generations of nurses with PhD training in quantitative fields (statistics, biostatistics, psychometrics) who have worked in partnership with nursing researchers to advance the science and with advanced practice nurses to optimize patient care. In addition to their research contributions, nurse data scientists can fill areas of instruction need by teaching big data and data science courses and serving on dissertations and doctoral projects exploring or using data science. Consortia of universities with PhD and DNP programs working together can leverage the impact of nurse data scientists.

21.3.3.5   Resources for Implementation

Faculty expertise in big data-based practice and science is a key consideration in the development of doctoral programs of study that incorporate big data. Big data science and findings have emerged recently, so most nurse faculty will not have had systematic training. Collaborations with faculty in engineering, mathematics, informatics, and data science will be needed; in these collaborations, nursing scientists and practitioners have much to offer colleagues in related fields who need their engagement as subject matter experts. For programs pursuing an emphasis on big data in practice or in research, consideration should be given to regular faculty appointments to experts from related fields who are willing and interested in jointly pursuing careers with clinical and science-focused nursing faculty (Henly et al. 2015c).

Information for instructional support is available on school websites, in journal publications and conference proceedings, and at workshops (including those offered by the National Science Foundation and the National Institutes of Health). Key resources for instructional support are listed in Table 21.2.

Table 21.2   Resources for big data science doctoral curriculum development and research

Topic	URL
Clinical practice	
a. The Precision Medicine Initiative The President's 2016 budget includes investments in an emerging field of medicine that takes into account individual differences in people's genes, microbiomes, environments, and lifestyles	https://www.whitehouse.gov/blog/2015/01/30/precision-medicine-initiative-data-driven-treatments-unique-your-own-body
b. NDNQI: The National Database of Nursing Quality Indicators: The National Database of Nursing Quality Indicators (NDNQI®) is the only national nursing database that provides quarterly and annual reporting of structure, process, and outcome indicators to evaluate nursing care at the unit level	http://www.pressganey.com/
c. The Kavli HUMAN Project: study of 10,000 New York City residents in approximately 2500 households over the span of decades by collecting measurements across multiple domains	http://kavlihumanproject.org/
d. PatientsLikeMe: Patient/consumer data	www.patientslikeme.com
e. National Institute of Health Big Data to Knowledge Initiative (BD2K)	http://bd2k.nih.gov/workshops.html
f. Clinical Decision Support Systems (CDDS) are active knowledge systems which use two or more items of patient data to generate case-specific advice	https://healthit.ahrq.gov/ahrq-funded-projects/clinical-decision-support-cds-initiative

Table 21.2 (continued)

Topic	URL
g. IOM: Institute of Medicine*: A component of the US National Academy of Sciences that works outside the framework of government to provide evidence-based research and recommendations for public health and science policy	http://iom.nationalacademies.org/
The "Best of the Best" Applications:	
*Community Apps- Healthy Communities Network	http://iom.nationalacademies.org/activities/publichealth/healthdata/2012-jun-05/awards.aspx#sthash.PuMnjV2k.dpuf
*Consumer Apps—myDrugCosts	http://www.mydrugcosts.com/
*Care Apps—Archimedes IndiGO	https://archimedesmodel.com/indigo
h. A learning healthcare system (LHS): A community of front-line clinicians, patients, and scientists combining comparative effectiveness research with quality improvement to improve population health with outcomes analysis	https://enhancedregistry.org/
i. The Patient-Centered Outcomes Research Institute (PCORI), an independent nonprofit, nongovernmental organization located in Washington, DC	http://www.pcori.org/
j. The COSMIN (Consensus-based Standards for the selection of health Measurement Instruments. Initiative aims to improve the selection of health measurement instruments	http://www.cosmin.nl/
k. Clinimetrics is a methodological discipline with a focus on the quality of measurements in medical research and clinical practice. The quality of measurements includes both	http://www.clinimetrics.nl/
Sharable and comparable nursing data	
a. Gen Bank is the NIH (National Institute of Health) genetic sequence database.	http://www.ncbi.nlm.nih.gov/Genbank/index.html
b. The Omaha System: Is a research-based, comprehensive practice and documentation standardized taxonomy designed to describe client care	http://www.omahasystem.org/
c. ICNP: International Classification for Nursing Practice: The International Classification for Nursing Practice (ICNP®) is defined as a classification of nursing phenomena, nursing actions, and nursing outcomes that describes nursing practice	http://www.icn.ch/
d. National Center for Biotechnology Information	http://www.ncbi.nlm.nih.gov/

(continued)

Table 21.2 (continued)

Topic	URL
e. The Center for Nursing Informatics at the University of Minnesota (Big Data Science)	http://www.nursing.umn.edu/icnp/
f. National Science Foundation: Core techniques and technologies for advancing big data science and engineering	http://www.nsf.gov/
Analytics	
a. WEKA: Free software for algorithms	http://www.cs.waikato.ac.nz/
b. Free Data Mining Software	http://www.predictiveanalyticstoday.com/top-data-analysis-software/
c. Virginia Commonwealth University Data Mining and Biomedical Informatics. Program algorithms from clustering, through discretization	http://www.cioslab.vcu.edu
d. Free Data Visualization Software	http://www.computerworld.com/article/2507728/enterprise-applications/enterprise-applications-22-free-tools-for-data-visualization-and-analysis.ht
e. The NIH Commons: The Commons is a shared virtual space where scientists can work with the digital objects of biomedical research, i.e. it is a system that will allow	https://datascience.nih.gov/commons
Big data bench science	
a. National Science Foundation (NSF)	http://www.nsf.gov/
b. Campus Cyberinfrastructure—Data, Networking, and Innovation Program	http://www.nsf.gov/funding/pgm_summ.jsp?pims_id=504748&org=ACI&from=home
c. Computational and Data Enabled Science and Engineering (CDS&E) program: generalizable computational science including, but not limited to, big data issues	http://www.nsf.gov/funding/pgm_summ.jsp?pims_id=504813
d. Exploiting Parallelism and Scalability (XPS) program: focused on scaling of software	http://www.nsf.gov/funding/pgm_summ.jsp?pims_id=504842

Alternate course delivery methods may be useful approaches to big data-focused instruction in doctoral programs. Boot camps, workshops and intensive on-sites or virtual programs can augment and/or replace traditional delivery of content and demonstration of competence by students and faculty, pointed out for genomics content in PhD programs (Conley et al. 2015). Alternate course delivery methods may enhance access to faculty experts by overcoming geographic locations and competing obligations. Massive online open courses (MOOCS) may be excellent delivery platforms for teaching and learning big data content, methods, and applications. Some alternate learning experiences include participant evaluation but others do not. How competency is documented and whether alternate approaches to learning will be credit-accruing is an issue for the degree-granting institution.

Infrastructure development and maintenance for support of doctoral education is critical. Nurses have been integral to international and national discussions about infrastructure for capacity-building for data science (Brennan and Bakken 2015; Clancy et al. 2014). Decision-makers within schools of nursing offering doctoral programs incorporating big data and data science and their clinical partners require vision and knowledge to advance infrastructure agendas and reach infrastructure goals. The responsibility for developing capacity for data capture, curation, exploration, and information discovery extends well beyond individual units (like schools of nursing or hospitals) to entire organizations (like universities and hospital systems) and beyond to inter-institutional collaborations (e.g., Asakiewicz 2014). Of paramount importance, stakeholders, including those whose personal data are used in research, must work together to develop the infrastructure for ensuring that ethical standards for data sharing and reuse are articulated and met (Antman et al. 2015, pp. 15–6).

21.4 Summary

To paraphrase Brennan and Bakken (2015), big data science needs nurses, and populations need nurses who practice and are expert in data science. Health and the human experience have been transformed, and continue to be transformed, by big data, social networks, and the Internet of Things. The world will not slow down for big data science and healthcare systems to catch up. Nurses are the advocates for the individuals and populations that we serve as practitioners and scientists. Nurses need to lead and participate in the acquisition, structure, use, analysis, protection, and dissemination of human subject data. Advocacy for these populations will be achieved only with informed practitioners and scientists who speak the language of big data. The world has moved ahead voraciously producing and consuming big data—nurses need to pursue their expertise in big data science with the same appetite.

References

American Association of Colleges of Nursing. The essentials of doctoral education for advanced nursing practice. 2006. http://www.aacn.nche.edu/publications/position/DNPEssentials.pdf
American Association of Colleges of Nursing. The essentials of baccalaureate education for professional nursing practice. 2008. http://www.aacn.nche.edu/education-resources/BaccEssentials08.pdf
American Association of Colleges of Nursing. The research-focused doctoral program in nursing: pathways to excellence. 2010. http://www.aacn.nche.edu/education-resources/PhDPosition.pdf
American Association of Colleges of Nursing. The essentials of master's education for advanced practice nursing. 2011. http://www.aacn.nche.edu/education-resources/MastersEssentials11.pdf

American Association of Colleges of Nursing. Outstanding dissertation and DNP project awards. 2016. http://www.aacn.nche.edu/membership/awards/excellence-in-advancing-nursing

American Nurses Association. Nursing informatics: scope and standards of practice. 2nd ed. Maryland: Silver Spring; 2014.

Antman EM, et al. Acquisition, analysis, and sharing of data in 2015 and beyond: a survey of the landscape. A conference report from the American Heart Association Data Summit 2015. J Am Heart Assoc. 2015;4:e002810. doi:10.1161/JAHA.115.002810.

Asakiewicz C. Translational research 2.0: a framework for accelerating collaborative discovery. Pers Med. 2014;11:3. doi:10.2217/PME.14.15.

Azmak O, Bayer H, Caplin A, Chun M, Glimcher P, Koonin S, Patrinos A. Using big data to understand the human condition: the Kavli HUMAN project. Big Data. 2015;3(3):173–88.

Beaty D, Quirk D. The digital revolution. ASHRAE J. 2015;57(5):80–8.

Begg M, Crumley G, Fair A, Martina C, McCormack W, Merchant C, Patino-Sutton C, Umans J. Approaches to preparing young scholars for careers in interdisciplinary team science. J Investig Med. 2014;62(1):14–25.

Bourne PE, Bonazzi V, Dunn M, Green ED, Guyer M, Komatsoulis G, Larkin J, Russell B. The NIH big data to knowledge (BD2K) initiative [editorial]. J Am Med Inform Assoc. 2015;22:1114. doi:10.1093/jamia/ocv136.

Brennan P, Bakken S. Nursing needs big data and big data needs nursing. J Nurs Scholarsh. 2015;47(5):477–84.

Carroll-Scott A, Toy P, Wyn R, Zane JI, Wallace SP. Results from the data & democracy initiative to enhance community-based organization data and research capacity. Am J Public Health. 2012;102(7):1384–91.

Choi J, Boyle D, Dunton N. A standardized measure: NDNQI nursing care hours indicator. West J Nurs Res. 2014;36(1):105–16.

Cios KJ, Nguyen DT. Data mining and data visualization. In: Henly SJ, editor. The Routledge international handbook of advanced quantitative methods in nursing research. Abingdon, Oxon, OX, UK: Routledge/Taylor & Francis; 2016. p. 294–323.

Clancy T, Bowles K, Gelinas L, Androwich I, Delaney C, Matney S, Sensmeier J, Warren J, Welton J, Westra B. A call to action: engage in big data science. Nurs Outlook. 2014;62(1):64–5.

Coffron M, Opelka F. Big promise and big challenges for big heath care data. Bull Am Coll Surg. 2015;100(4):10–6.

Collins L. Dynamic interventions: opportunities and challenges. Paper presented to The National Institutes of Health, Big D.A.T.A. Data and Theory Advancement Workshop. Bethesda, MD; 2013.

Collins F, Varmus H. A new initiative on precision medicine. N Engl J Med. 2015;372(9):793–5.

Conley YP, Heitkemper M, McCarthy D, Anderson CM, Corwin EJ, Daack-Hirsch S, et al. Educating future nursing scientists: recommendations for integrating omics content in PhD programs. Nurs Outlook. 2015;63:417–27. doi:10.1016/j.outlook.2015.06.006.

Dong X, Yambartsev A, Ramsey S, Thomas L, Shulzhenko N, Morgun A. Reverse enGENEering of regulatory networks from big data: a roadmap for biologists. Bioinform Biol Insights. 2015;9:61–74.

Dunbar-Jacob J, Nativio DG, Khalil H. Impact of doctor of nursing practice education in shaping health care systems for the future. J Nurs Educ. 2013;52(8):423–7.

Edelman LS, Jian-Wen G, Fraser A, Beck SL. Linking clinical research data to population databases. Nurs Res. 2013;62(6):438–44. doi:10.1097/NNR.0000000000000002.

Epstein R, Hundert E. Defining and assessing professional competence. JAMA. 2002;287:226–35.

Fox J. A mathematical primer for social statistics. Thousand Oaks, CA: Sage; 2009.

Fulton C, Meek J, Walker P. Faculty and organizational characteristics associated with informatics/health information technology adoption in DNP programs. J Prof Nurs. 2014;30(4):292–9.

Grady PA. National Institute of Nursing Research commentary on the idea restival for nursing science education. Nurs Outlook. 2015;63:432–5. doi:10.1016/j.outlook. 2015.02.006.

Grady PA, McIlvane JM. The domain of nursing science. In: Henly SJ, editor. The Routledge international handbook of advanced quantitative methods in nursing research. Abingdon, Oxon, OX, UK: Routledge/Taylor & Francis; 2016. p. 1–14.

Hayat MJ, Eckardt P, Higgins M, Kim M, Schmiege SJ. Teaching statistics to nursing students: an expert panel consensus. J Nurs Educ. 2013;52:330–4. doi:10.3928/01484834-20130430-01.

Healthdata.gov. Datacatalog. 2016. http://catalog.data.gov/dataset?_organization_limit=0&organization=hhs-gov#topic=health_navigation. Accessed 2 Jan 2016.

Henly SJ, editor. The Routledge international handbook of advanced quantitative methods for nursing research. Abingdon, Oxon, OX, UK: Routledge/Taylor & Francis; 2016a.

Henly SJ. Theorizing in nursing science. In: The Routledge international handbook of advanced quantitative methods in nursing research. Abingdon, Oxon, OX, UK: Routledge/Taylor & Francis; 2016b. p. 14–26.

Henly SJ, Wyman JF, Findorff MJ. Health and illness over time: the trajectory perspective in nursing science. Nurs Res. 2011;60(3 Suppl):S5–S14. doi:10.1097/NNR.0b013e318216dfd3.

Henly SJ, McCarthy DO, Wyman JF, Alt-White AC, Stone PW, McCarthy AM, Moore S. Emerging areas of nursing science and PhD education for the 21st century: response to commentaries. Nurs Outlook. 2015a;63:439–45. doi:10.1016/j.outlook.2015.05.003.

Henly SJ, McCarthy DO, Wyman JF, Heitkemper MM, Redeker NS, Titler MG, et al. Emerging areas of science: recommendations for nursing science education from the Council for the Advancement of Nursing Science Idea Festival. Nurs Outlook. 2015b;63:398–407.

Henly SJ, McCarthy DO, Wyman JF, Stone PW, Redeker NS, McCarthy AM, et al. Integrating emerging areas of nursing science into PhD programs. Nurs Outlook. 2015c;63:408–16.

Higdon R, Haynes W, Stanberry L, Stewart E, Yandl G, Howard C, Broomall W, Kolker N, Kolker E. Unraveling the complexities of life sciences data. Big Data. 2013;1:42–9.

Hong Z, Mei T, Enzhong L, Fujibayashi Y, Lie-Hang S, Yang D. Molecular imaging-guided theranostics and personalized medicine. J Biomed Biotechnol. 2012;2012:1–2.

Jameson JL, Longo DL. Precision medicine—personalized, problematic, and promising. N Engl J Med. 2015;372(23):2229–34. doi:10.1056/NEJMsb1503104.

Jenkins M, Wilson M, Ozbolt J. Informatics in the doctor of nursing practice curriculum. AMIA Annual Symposium Proceedings. AMIA Symposium. 2007;364–68. Available from: MEDLINE with full text, EBSCO*host*. Accessed 29 Jan 2016.

Jordan MI, Mitchell TM. Machine learning: trends, perspectives, and prospects [review]. Science. 2015;349(6245):255–60.

Klein C. Educational innovation. Linking competency-based assessment to successful clinical practice. J Nurs Educ. 2006;45(9):379–83.

Kneipp S, Yarandi H. Complex sampling designs and statistical issues in secondary analysis. West J Nurs Res. 2002;24(5):552–66.

Lei W, Hui-Yi Y, Yong-Qing Z. Personalized medicine of esophageal cancer. J Cancer Res Ther. 2012;8(3):343–7.

Lilly K, Fitzpatrick J, Madigan E. Barriers to integrating information technology content in doctor of nursing practice curricula. J Prof Nurs. 2015;31(3):187–199. EBSCO Education Source. Accessed 29 Jan 2016.

Lis G, Hanson P, Burgermeister D, Banfield B. Transforming graduate nursing education in the context of complex adaptive systems: implications for Master's and DNP curricula. J Prof Nurs. 2014;30(6):456–62.

Magee T, Lee S, Giuliano K, Munro B. Generating new knowledge from existing data: the use of large data sets for nursing research. Nurs Res. 2006;552S:S50–6.

Mancuso J, Udlis K. Doctor of nursing practice programs across the United States: a benchmark of information. Part ii: admission criteria. J Prof Nurs. 2012;28(5):274–83. EBSCO Education Source. Viewed 2 Dec 2015

McCormick K, Delaney C, Brennan P, Effken J, Kendrick K, Murphy J, Skiba D, Warren J, Weaver C, Weiner B, Westra B. Guideposts to the future: an agenda for nursing informatics. J Am Med Inform Assoc. 2007;14(1):19–24.

Molenaar PCM. A manifesto on psychology as idiographic science: bringing the person back into scientific psychology, this time forever. Measurement. 2004;2:201–18. doi:10.1207/s15366359mea0204_1.

Mundinger M, Kennedy E. Why are standards for DNPs who practice comprehensive care so crucial? And what we are doing about it? Clinical Scholars Review. 2013;6(2):82–4. CINAHL Plus with full text, EBSCO*host*. Accessed 2 Dec 2015

Mundinger M, Starck P, Hathaway D, Shaver J, Woods N. The ABCs of the doctor of nursing practice: assessing resources, building a culture of clinical scholarship, curricular models. J Prof Nurs. 2009;25(2):69–74. doi:10.1016/j.profnurs.2008.01.009.

National Institute of Nursing Research. Bringing science to life: NINR strategic plan. 2011. https://www.ninr.nih.gov/sites/default/files/ninr-strategic-plan-2011.pdf

National Institute of Nursing Research. The NINR big data in symptoms research boot camp. 2016. https://www.ninr.nih.gov/training/trainingopportunitiesintramural/bootcamp. Accessed 7 Jan 2016.

National Institutes of Health Research planning and career development. https://researchtraining.nih.gov/programs/training-grants/T32 1. Accessed 1 Jan 2016.

National Institutes of Health strategic plan for fiscal years 2016–2020. http://www.nih.gov/sites/default/files/about-nih/strategic-plan-fy2016-2020-508.pdf. Accessed 28 Dec 2015.

Nguyen D, Zierler B, Nguyen H. A survey of nursing faculty needs for training in use of new technologies for education and practice. J Nurs Educ. 2011;50(4):181–9. doi:10.3928/01484834-20101130-06.

Owens L, Koch R. Understanding quality patient care and the role of the practicing nurse. Nurs Clin North Am. 2015;50(1):33–43.

Precision Medicine Initiative—Report of the 10th Lung Cancer Summit Forum for Chinese Directors of Thoracic Surgery and the 4th Summit Forum for Chinese Lung Cancer Collaborative Group CLCCG. J Thorac Dis. 2015;7(8):E258–61.

Sanders C, Saltzstein S, Schultzel M, Nguyen D, Stafford H, Sadler G. Understanding the limits of large datasets. J Cancer Educ. 2012;27(4):664–9. doi:10.1007/s13187-012-0383-7.

Skiba DJ. The connected age: big data & data visualization. Nurs Educ Perspect. 2014;35(4):267–8.

Sylvia M, Terhaar M. An approach to clinical data management for the doctor of nursing practice curriculum. J Prof Nurs. 2014;30(1):56–62. doi:10.1016/j.profnurs.2013.04.002.

Thorne S. Conceptualizing nursing inquiry [editorial]. Nurs Inq. 2011;18(2):93.

Westra B, Latimer G, Matney S, Park J, Sensmeier J, Simpson R, Swanson M, Warren J, Delaney C. A national action plan for sharable and comparable nursing data to support practice and translational research for transforming health care. J Am Med Inform Assoc. 2015a;22(3):600–7.

Westra B, Pruinelli L, Delaney C. Nursing knowledge: 2015 big data science, computers, informatics, nursing. CIN. 2015b;33(10):427–31.

Westra BL, Monsen KA, Delaney CW. Big data in nursing research. In: Henly SJ, editor. The Routledge international handbook of advanced quantitative methods in nursing research. Abingdon, Oxon, OX, UK: Routledge/Taylor & Francis; 2016. p. 280–393.

Wu Z, Zhang X, Shen L, Xiong Y, Wu X, Huo R, Wei Z, Cai L, Qi G, Xu Q, Cui D, Cui D, Zhao G, He L, Qin S. A systematically combined genotype and functional combination analysis of CYP2E1, CYP2D6, CYP2C9, CYP2C19 in different geographic areas of Mainland China—A basis for personalized therapy. PLoS One. 2013;8(10):8–10. doi:10.1371/journal.pone.0071934.

Zaccagnini ME, White KW. The doctor of nursing practice essentials: a new model for advanced practice nursing. 3rd ed. Burlington, MA: Jones & Bartlett; 2017.

Chapter 22
Global Society & Big Data: Here's the Future We Can Get Ready For

Walter Sermeus

Abstract The world is increasingly global for corporations and individuals, except healthcare, the market for which seems remains more local. The aim of this chapter is to explore if the availability of big data will lead to a more global healthcare system. Four healthcare transformation scenarios, based on Geddes's model of local and global thinking and acting are explored. Within each scenario the role of big data is described.

Keywords Global healthcare system • Big data analytics • Cross-border healthcare • Medical tourism

22.1 Introduction: Are We Moving to a Global Society, Except for Healthcare?

The world is increasingly getting global. Consider that the world economy is dependent on the war in Syria, the cost of oil and the decision of Organization of the Petroleum Exporting Countries (OPEC) to limit their oil production capacity. The same is true for individual citizens. Many people watch the English Premier League on television. The top players of leading soccer teams, such as FC Barcelona or Manchester City, are no longer Spanish or English, but are international. The money used to pay for goods is changing rapidly from local paper money to virtual international credit card money. Wikipedia is used to update our knowledge, Facebook to exchange our ideas and experiences with friends, booking.com to plan and organize our travels and amazon.com to buy books. These activities lead to a global society in which all world citizens have much in common with one another. One does not

W. Sermeus
Department of Public Health & Primary Care, KU Leuven, Leuven, Belgium
e-mail: walter.sermeus@kuleuven.be

© Springer International Publishing AG 2017
C.W. Delaney et al. (eds.), *Big Data-Enabled Nursing*, Health Informatics,
DOI 10.1007/978-3-319-53300-1_22

have to travel to France to drink a French wine; French wines or French restaurants are available all over the world.

These global examples are however in sharp contrast to the situation in healthcare, which maintains a rather local tradition. Hospitals, doctors and nurses mainly serve their local communities. For example, the city of Leuven, Belgium, is the home of many world-class organizations such as AB InBev, KU Leuven University and University Hospitals Leuven. Anheuser-Bush InBev (AB InBev) is the leading global brewer with famous brands such as Stella Artois (where it all begun), Corona, Budweiser and 200 more brands. It is one of the world's top five consumer products companies with a total revenue of 43.2 billion USD in 2013 (AB InBev Annual Report 2015). In 2014, the company produced 400,000,000 hl beer of which 98.8% was sold outside Belgium, while also keeping a leading 56% market share in Belgium. Another Leuven world-class organization, University Hospitals Leuven, is part of KU Leuven University, originated in 1425 as one of the oldest universities in Europe. The university is ranked 35 on the Times University Ranking. More than 18% of its students are international. One of the most famous spin-offs, IMEC, is performing world-leading research in nano-electronics. IMEC, headquartered in Leuven, has offices in the Netherlands, Taiwan, USA, China, India and Japan. The University Hospitals Leuven (UH Leuven) is the leading academic referral hospital in Belgium. The hospital is Joint Commission International (JCI) accredited. The majority of the medical staff is internationally trained and is active in international networks and research. Despite the fact that Brussels, capital of Europe and seat of NATO is nearby, the number of foreign patients is still less than 5%. Although both companies are world leaders in their field, the main striking difference is their market: global for AB InBev, local for University Hospitals Leuven.

The aim of this chapter is to explore if the availability of big data will lead to a more global healthcare system as well. This chapter will describe four phases for healthcare transformation based on the local and global thinking and acting ideas of Patrick Geddes, a Scots town planner and social activist in the beginning of last century. The role of big data will be described within each scenario.

22.1.1 Phase 1: Thinking Local, Acting Local—Healthcare in the Past and Today

Healthcare has a mainly local tradition. Hospitals, doctors and nurses serve their local communities. Fitzgerald et al. (2012) investigated all 183,174 total knee replacements in the US during 2001. Ninety-five percent of the patients underwent surgery at a hospital that was located within 80 km (50 miles) of their residence. A cross-sectional analysis of the 2001 US National Household Travel survey (Probst et al. 2007) showed that the average trip for care in the US entailed 16.4 km (10.2 miles) and 22 min of travel. Although people are traveling much more frequently than before, they want healthcare in their backyard.

Similar data on the volume of patient mobility within the EU is rather limited but are not very different from the US findings. For example, the number of foreign patients in the Belgian healthcare system (De Mars et al. 2011) varies from 0.9% in 2004 to 1.3% in 2008 for inpatients. Similar volumes are confirmed in Germany. The Techniker Krankenkasse (TK) sickness fund in Germany insures approximately 9 million people. Only 1% of all TK members utilize services in other EU Countries (Techniker Krankenkasse 2007). Most patients seem to prefer to be treated near to their homes and families by people they know, trust and understand and in a language they understand (Legido-Quigley et al. 2012).

The fact that healthcare is mainly local has important effects. Since data and information are not exchanged outside the organization or local communities, there is in fact no real need or interest in standards. The lack of comparability and the strong reliance on individual professional competencies and expertise might lead to high variability in medical practice and the quality of care. This phenomenon has been well documented in the literature. An example is the variability in mortality rates for common surgery as reported in the RN4CAST Study (Aiken et al. 2014). Records of 422,730 patients age 50 years or older in 300 hospitals in nine European countries were analyzed. The overall 30-day mortality rate was 1.3% but varied across hospitals from 0% to more than 7%. The variability within countries is much larger than the variability between countries. The rates within a country vary by a factor 15, which means that the outcomes are quite different if you are treated in hospital A compared to hospital B. Some of the reasons for these differences can be explained by differences in patient characteristics such as age, multi-morbidity, and frailty. The RN4CAST-study showed that, if corrected for these patient-related characteristics, unexplained effects related to the qualifications of the nurses and nurse staffing ratios. Other studies refer to the fact that the available evidence is not always known or applied. McGlynn et al. (2003) were tracking the care of a random sample of 6712 adults living in 12 metropolitan areas in the US on 439 indicators of quality of care for 30 acute and chronic conditions. They found that the patients received 54.9% of the care that was recommended based on the available evidence. They found more problems with underuse of care (46.3%) than overuse (11.3%). The report of the National Patient Safety Foundation (2015), "Free from Harm", is clear that 15 years after the Institute of Medicine's (IOM) landmark report, "To Err is Human", healthcare is still facing quality and patient safety problems. One in 10 patients develops an adverse event (AHRQ 2014), 1 out of 2 surgeries had a medication error or adverse event (Nanji et al. 2016). Around the world there are 421 million hospitalizations and 42.7 million adverse events every year (Jha et al. 2013). Berwick and Hackbarth estimated at least 20% of total healthcare expenditures, or around 3% of US GDP, can be considered as inappropriate.

The need for more big data analytics is clear as large studies show major variability in the service and outcome of healthcare. The general hypotheses that healthcare is of equal quality wherever you go doesn't hold. Patients need to know what would be the best and safest place. At the same time, these studies show that data are not routinely or publicly available, and are often not standardized to allow benchmarking and international comparability. RN4CAST can be seen as an example of the different barriers that need to be overcome to analyze and benchmark data across countries. RN4CAST was only able to calculate mortality rate in 9 out of 12 countries. Greece and Poland had no routine hospital discharge data available. Germany had data but the format was not consistent with the mortality data protocol in the other countries. Mapping to different classifications (ICD-9, ICD-10) and DRG schemes was required but not easy. The history of use and extent to which data is validated varied among countries. All these factors are major barriers to link routine patient data across systems and countries.

22.1.2 Phase 2: Thinking Local, Acting Global—Cross-Border Care and Medical Tourism

Although the mainstream of healthcare is local, there are some situations where medical care is being sought abroad. For example, people from Northern European countries tend to spend their retirement in the warmer climates in the south. People are traveling abroad more than before and might need emergency services when abroad. Still more and more people commute weekly between their homes in one country and their jobs in another country. These examples call for available healthcare across national boundaries.

One example to support the care is the framework for cross-border care in the European Union. Although there was an intention that all citizens move freely in the Union, at the same time healthcare was restricted and regarded as a matter for national governments. In 2011, the European Parliament and Council adopted a Directive on Patients' Rights in Cross-Border Healthcare to provide a clear legal framework for cross-border care (European Council 2011). The main issue was reimbursement. Under the new Directive, EU citizens are able to receive reimbursable healthcare in another EU country (Footman et al. 2014). This meant people had the right to seek medical care in another EU-country. This has not created mass flows of patients traveling around Europe. Footman et al. (2014) described a few case studies. The Ardennes cross-border collaboration allows French patients to give birth in a Belgian hospital as it is their closest health facility. The Veneto region of Italy made a significant investment in cross-border health services in response to the high volume of tourists received every year. They offered a wide range of services with a particular focus on chronic conditions such as dialysis services for patients with a chronic kidney disease. The Malta-UK cross-border healthcare collaboration is one of the longest standing in Europe. This agreement gives Maltese patients access to highly specialized care in the UK that is not available locally. In return, UK citizens (working in Malta, UK pensioners) are entitled to free healthcare in Malta. And last an increasing number of patients from a wide range of EU countries seek orthopedic care in Hungary. This service is the combination of well established orthopedic centers offering of thermal spas that attract patients with rheumatic and motion diseases (Kovacs et al. 2014).

Many experts see that the main impact of the European Cross-Border Directive is not in more patients seeking care abroad, but in providing more standardization and transparency of care and the data related to that care. There is a requirement that potential cross-border patients should be able to have an informed choice about what hospital or provider to choose based on the level of quality, service, and costs involved. In most cases, this required transparency is not available for the local patients. The need for this transparency might lead to improvement in the use of guidelines, the development of care pathways, the measurement and publication of quality indicators and changes in funding and reimbursement schemes (Baeten and Jelfs 2012).

These requirements call for a standardized European discharge summary or a standardized electronic health record that can be shared easily within and among the team of health professionals (local as well as international). Knai et al. (2013) studied seven EU countries and found large differences in the way essential elements of care were documented and coded on patient identifiers, hospital identifiers, specialist and/or primary healthcare professional identifiers, admission, clinical data, diagnoses, operations, treatments, procedures, medication, discharge data, follow-up and future management, social and psychological support, contact details of relatives, specific patient preferences.

22.1.3 Phase 3: Thinking Global, Acting Local—Global Healthcare Driven by Networks

There are two major strategies for dealing with the large variability in healthcare. The first strategy is to reduce the variability to zero by optimizing the quality of care to a required minimal level. The second strategy is to accept that there are differences.

As in other industries, such as the airlines, minimal safety levels should be guaranteed. If an airline doesn't comply with these standards, it is banned. So far the EU has banned more than 200 airline carriers to fly within the EU (European Commission, 2016). The result is that, from a passenger perspective, the choice of airline A or B mainly depends on preferences or on given loyalty benefits but not on technical quality and safety. The same is true for healthcare organizations and providers. More and more international accreditation schemes are used to evaluate the quality of care. As an example, the Joint Commission that has accredited US healthcare organizations and programs since 1951, developed international accreditation in1994 (Joint Commission International (JCI)). JCI is currently active on five continents and more than 90 countries for accrediting healthcare organizations. The Canadian Accreditation Canada/Qmentum has a similar history and started in 1953 as the Canadian Commission of Hospital Accreditation. Qmentum International, started in 2010, is also active in five continents. Both accreditation organizations were born out of healthcare organizations such as the American and Canadian

Medical Association. New entities are entering the market with broad experiences in accreditation work. An example is DNV (Det Norske Veritas) Healthcare, a Norwegian accrediting organization active in various sectors such as navy, oil and gas, energy, food, and recently also healthcare.

In addition to accreditation, there is a strong interest in bringing evidence-based healthcare as close to the providers as possible. Buckminster Fuller, in his book "Critical Path" (1982), described the "Knowledge Doubling Curve", explaining that new knowledge, which has doubled every century until 1900, is now estimated to double every 18 months. The pace is getting faster and faster. At this moment, when you finish your education, the knowledge you gained might already be outdated. Traditional pull-systems like guidelines are insufficient. New knowledge should be integrated and embedded in electronic patient records, algorithms and decision support systems. It is interesting to see that the Hearst Health Network, one of the largest media and communication groups in the world, is taking a leading role in healthcare combining forces with strong health care companies as FDB, Map of Medicine, Zynx Health and MCG. FDB (First DataBank) is a UK-based company specializing in integrated drug knowledge to prescribe medication, follow-up drug interactions, improve clinical decision making and patient outcomes. Map of Medicine was created in the UK for clinicians by clinicians with the University College London and the Royal Free Hampstead NHS trust. It offers a web-based visual representation of evidence-based patient journeys covering 28 medical specialties and 390 pathways, driven by medical entrepreneurs, clinical experts and health managers for constantly keeping Map of Medicine up to date. Zynx Health offers a similar US story. A group of clinicians at Cedars-Sinai Medical Center founded Zynx Health in 1996 to provide evidence-based clinical decision support system solutions at the point of care through electronic patient records. At this moment the company serves over 1900 hospitals and outpatient practices, mainly in the US. Zynx Health products and services impact over 50% of hospital discharges in the US. MCG, which produces evidence-based clinical guidelines and software, reviews and rates more than 143,000 references annually. It is widely used in the US, UK and Middle East. The Hearst Health Network reaches 84% of the discharged patients in the US today and is also engaged in the UK, Asia Pacific and Middle East.

Standardized clinical indicators to monitor and benchmark patient outcomes should also be developed. Many indicator systems have been developed worldwide by national governments, professional associations, and international organizations. Three main problems arise: the number of indicators is growing and resulting in administrative overload for organizations and staff; many indicators don't meet the requirements of validity and reliability; and many indicators are built around the availability of data, which results in small differences between systems and impacts the comparability. To strengthen global health, a limited number of standardized, reliable, valid, international, evidence-based indicators need to be developed. A few examples are the OECD health indicator set and ICHOM. The OECD healthcare quality indicator (HCQI) expert group, led by Nick Klazinga, The Netherlands, (Carinci et al. 2015) found consensus on 49 quality indicators for public reporting and benchmarking. These indicators are in the domain of primary care (N = 15), acute care (N = 7), mental health (N = 6), cancer care (N = 3), patient safety (N = 8) and patient experiences (N = 8). The availability of a short standard set of indicators will allow comparisons and benchmarking of health providers and organizations across countries. The International Consortium for Health Outcomes (ICHOM), founded by three Institutions (Harvard Business School represented by Michael Porter, The Boston Consulting Group and Karolinska Institute), aims to transform health care systems worldwide by measuring and reporting patient outcomes in a standardized way for the most relevant medical conditions. Currently 21 standards sets are complete, ten are in progress. ICHOM hopes that health professionals, patients, the public, and policy makers can start communicating on health in a meaningful, data-driven way.

Collaboration and networking are becoming key-strategies. It is more and more accepted that not all healthcare providers are able to work up to the maximum of their competences because too many contextual elements are involved. A major contextual element is volume. If the volume is too low, one might lack the expertise, experience, skills and organization to give the best care. A growing proportion of healthcare is provided by multisite healthcare delivery organizations. Porter and Lee (2013) stated that in 2011 69% of all US hospitals were part of such systems. This gives huge opportunities for defining the scope of services, concentrating volume in fewer locations, choosing the right location for each service line and integrating care for patients across locations. It leads to geographical expansion of leading providers to offer the best services beyond local communities. The models for collaboration are hub-and-spoke and clinical affiliation models. Complex cases are referred to the hub. More simple cases, or long-term follow-up, is done by satellite centers. The staff of doctors and nurses often rotates between hub and satellites so that knowledge and skills are fully developed. These models are already well established for large cancer care centers such as MD Anderson in the US or the Dutch Comprehensive Cancer Organization (IKN) in the Netherlands. In a next phase, these leading multi-site healthcare delivery organizations will start to build international networks and partnerships. An example is Johns Hopkins Medicine International, which has built networks and collaborations in Canada, the Middle East, Europe (Turkey), Asia/Pacific, and Latin America/Caribbean. The main goal is to raise the standard of healthcare in a way that is customized to the local vision but at the same time consistent with Johns Hopkins mission of being the best in medical education and research. A similar aim with the Mayo International Health Program that is supporting Mayo Clinic residents and fellows performing clinical rotations in underserved communities throughout the world. Through this program, the Mayo program serves more than 90,000 patients for free in more than 140 locations in 56 countries around the world.

Individual providers are organized in the international networks and collaborations. Good examples are the International Cancer Genome Consortium (ICGC) and the Human Brain project (www.humanbrainproject.eu) in which research groups all over the world work together. As soon as clinical information is digitalized, it can be shared.

It is clear that the "global thinking, local acting" approach will be driven by big data analytics. Data analytics requires sharing of patient data including, electronic patient records, genome data, insurance data, family history data, environmental data, mhealth data (MIT 2014). The large health networks, such as Kaiser Permanente or Geisinger Health Systems, are already sharing big data and analytics to better advise their patients and to monitor and improve the care given. A good example of the power of analyzing large datasets is the Vioxx-case (rofecoxib). (Vioxx is a NSAID arthritis and pain drug.) Although a clinical trial showed no increased risk of adverse cardiovascular events for the first 18 months of Vioxx-use, a joint analysis of the US FDA and Kaiser Permanente on the Healthconnect database of more than 2 million person-years of follow-up, the drug showed patients to be at risk for heart attacks and sudden cardiac deaths (Graham et al. 2005). After the findings were confirmed in a large meta-analysis, Merck decided to withdraw the drug from the market worldwide in 2004.

Hospitals and networks are sharing data. A few examples of initiatives, and what data analytics can offer, are provided by UK-founded Dr. Foster, the leading provider of healthcare variation analysis and clinical benchmarking solutions, and US-based Premier Health Alliance. Dr. Foster is active in hundreds of leading hospitals in 12 countries, including over half of the English acute hospitals. In 2011 the company started Global Comparators as a major international hospital network. It brings together data from 40 leading hospitals in the US, UK, Europe, Scandinavia and Australia, translated into a common language, enabling comparison of the results within the network. Dr. Foster develops models to determine the risks of mortality and complications based on the large datasets available and then compares the actual mortality and complication rates with the expected model-based rates. Premier is a US healthcare performance improvement alliance of 75% of all US hospitals and 120,000 other healthcare sites, accessing 40% of all US patient data. Premier converts this data into meaningful information for improving quality of care, reducing average length-of-stay, readmission rates, and hospital-acquired conditions, and improving resource utilization and lowering the overall cost of care.

22.1.4 Phase 4: Thinking Global, Acting Global—Discovering the Long Tail in Healthcare

The previous scenarios are making healthcare smarter but not necessarily more global. Information technology, the internet, smart phones, and Google search algorithms have completely changed industries. A good example is the book industry. Publishers were dominating reading behavior. Bestseller lists could be found in any local bookstore. Specialized literature was found in specialized bookstores. One iconic bookshop operating in the 1970s was Foyles at Charing Cross Road 107,

London. Foyles was once listed in the Guinness Book of Records as the world's largest bookshop. The world changed with Amazon, through which one can search online for every book ever printed, read short summaries and customer reviews, and see what else customers bought. Even Foyles had a limit on shelf space showing the titles that sold best. Chris Anderson (2006) calls the distribution the "long tail". A few titles are sold frequently, followed by a long tail of titles that are less frequently sold. In analyzing the distribution, one would expect that the classic 20/80 Pareto law would hold: 20% of the books are good for 80% of revenue. This was reason enough for traditional bookshops to focus on the 20% bestsellers. There was no money to make with the tail. However, when consumers are offered infinite choice, the true shape of demand is revealed. It turns out to be less hit-centric than anyone thought. The sales on Amazon show that 98% of all offered titles are really bought.

Is a similar story possible in healthcare? Hospitals are like bookshops. The limit of bookshelf area is comparable to the limit in beds, waiting times, medical specialties, resources. Hospitals also focus on the bestsellers: high-volume surgical procedures, emergency department admissions, deliveries. But what happens when patients enter with an "unknown" disease? Hospitals and health providers perform a wide range of tests and if they don't find it, they refer the patient to another center where the cycle may start over again. Often the patient record of the first visit is not available or consulted. And the odyssey continues until a diagnosis is made. What would happen if all the world records and databases could be available? For Chris Anderson, this requires three forces: the democratization of production, the democratization of distribution, and developing filters to connect supply and demand.

The democratization of production refers to the people who are making "healthcare". It is the world of medical doctors, nurses, allied health professionals. It is the world of hospitals, healthcare organizations. All health providers have long education tracks and continue doing so. Hospitals have invested in large buildings and technology. Democratization means that these kinds of services might come from patients themselves, non-professionals, technology and digitalization everywhere. This has happened already with iTunes, making the CD-industry superfluous. Other examples are Uber taking over the taxi industry, Airbnb taking over the hotel industry, Booking.com taking over travel agencies, etc. Many thought that the healthcare industry would be resistant to this. However technology, miniaturization, apps, the quantified self, and patient initiatives are becoming important players in healthcare. A few examples illustrate this phenomenon. IMEC (Belgium) and Johns Hopkins University are building a chip-based technology called MiLab, which analyses molecules or cells in bodily fluids, such as DNA, proteins, viruses and blood cells. It can do a complete blood test, a PSA test or check for specific viruses like Ebola or HIV. The expectation is that this will bring a revolution in medical diagnostics. It will replace expensive medical testing in hospitals, which takes days to deliver results while the chip will only take 15 min to analyze the molecules and transmit the results wirelessly to a smartphone. This technology will become available to practitioners. It will take only a while before patients get these technologies in their homes, together with a new range of self-diagnose kits and apps that are coming on the market. The most fascinating example is the "Tricorder" that people may remember from Dr. Spock in Star Trek as a device to diagnose instantly. The Qualcomm Tricorder X Prize was announced in 2011 to develop a portable device that can diagnose patients better than or as well as a panel of board-certified physicians. Prize money is $10 million USD and at this moment 2 finalists are in the final round. The award is expected to be given in the second quarter of 2017. First finalist is the Dynamical Biomarkers Group that developed a Smart Vital-Sense Monitor and Smart Blood-Urine Test Kit to be used by patients. Second finalist is the US Final Frontier Medical Devices group that developed "DxtER" a device to monitor your health and diagnose illnesses in the comfort of your own home. Next, there are the quantified self tools—from Fitbit and Apple watches to wearable sensors and a whole variety of apps. They are collecting a wide range of information on vital signs, physical activity, body movements, sleeping quality, eating habits, calorie intake etc.

The most important new producers in healthcare are patients as shown by initiatives such as PatientsLikeMe, Adrenalnet, Smartpatients. PatientsLikeMe was co-founded in 2004 by three MIT engineers after the brother of two of the founders was diagnosed with amyotrophic lateral sclerosis (ALS) at the age of 29. They conceptualized this data-sharing platform for patients so that they could learn from others' experiences to better manage themselves. They offer members input of real-world data on their conditions, treatment history, side effects, hospitalizations, disease-specific functional scores, and quality of life on an ongoing basis. The result is a longitudinal record that offers insights and identifies patterns. PatientsLikeMe is connecting people all over the world to exchange experiences and learn from each other. Today PatientsLikeMe covers more than 2000 health conditions. The PatientsLikeMe community has 7500 ALS members, making it the largest online population of ALS patients in the world. Adrenalnet is a similar organization for and by patients with adrenal gland disorders and their healthcare providers. The platform is currently only available in the Netherlands (BijnierNet) with the ambition to offer similar services worldwide. The aim is to share experiences among patients, partners, caregivers, parents, and healthcare providers. Smartpatients is an online community for patients and their families and friends to learn about scientific developments on their medical conditions and share questions and concerns with other members. While started mainly for cancer patients, it is now offered to communities for more 70 different conditions.

The democratization of distribution is the way care is delivered. We see that healthcare becomes a hybrid product, in which the part that can be digitized is being distributed freely or at minimal costs via the internet. Apps can be downloaded. People can buy smartwatches to monitor their physical activity and connect their data to their smartphone and, further, to their personal health record. Information is distributed freely on the net. Patient experiences are shared freely. Small devices will be available at low costs, lowering the cost of healthcare. The available information will transform the health provider—patient relationship, from a highly asymmetric relationship to a more peer-to-peer relationship in the future. We see patients enter the consultancy room of a provider after they shared the copy of their personal health records with the most recent data. The personal health record (PHR) is probably the most powerful game-changer tool. At this moment, patients might have several PHRs, given that they are currently determined by the provider and attached to the (professional) electronic patient record. The PHR allows the patient to view the medical record (or at least some elements), to book appointments, to e-mail doctors and nurses, to manage their prescriptions. The disadvantage is that most of these records are not connected. A PHR at hospital A and another at hospital B, combined with more general personal health records (PHR) available, such as Microsoft HealthVault or Dossia, and a growing list of devices and apps and medical images provide a need to be able to share any part of your health record with anyone you want. Next step would be to connect your "My Personal Health Record" in hospital A to my PHR. Patients will gain more control over their health when the PHRs become real personal. It will change traditional healthcare providers to facilitated user networks (as defined by Bohmer 2009), in which professionals and patients are working together. A good example is ParkinsonNet, which is a collaboration of more than 2700 physicians, nurses and allied health professionals in the Netherlands to coordinate the care for Parkinson patients.

Matching supply and demand requires filter tools and health broker functions. Filters are people or software that help find what is wanted in the long tail, driving the demand from bestsellers and routine care to niches and specialist services. All filters have two components. The first component is the matching component. In booking.com you give the region you want to go on holiday, with how many people, if you want to have a hotel or a little apartment, if you want a swimming pool, etc. For health data, it is about patient characteristics, disease, symptoms, problems, specialist advice you are looking for.

The second component is the evaluation component. These are reviews from clients or patients on how happy they were with the product or services. There are many examples of filter services, such as diagnose.me and ZocDoc. Diagnose.me matches second opinions on medical imaging (CT-scan, X-ray, MRI). One selects 1 out of 100 medical specialists (in the field of medical imaging) from more than 20 countries and provides information on symptoms, treatment and medications, and uploads medical images. An independent report is then available in more than 20 languages. ZocDoc is a filter service to search for a doctor nearby. It gives information on qualifications and experience, hospital affiliations, board certification and reviews from previous patients. If a match is found, an appointment can be made. ZocDoc, available as a free app, is used by more than 5 million people per month, covering more than 40% of the US population in more than 2000 cities. ParkinsonNet offers a filter module to guide patients to find the best care throughout the Netherlands. And PatientsLikeMe fulfills all three functions to be successful in the long tail: they add the patient experience as a new production factor to healthcare, offer a distribution platform to manage your disease better by linking apps, devices, and health information, and acts as filter to match patient experiences (who has a similar condition as I have).

22.2   From Local to Global: What Would It Take?

It is clear that there are technical as well as human challenges. Technical challenges include the lack of operability, lack of standardization, IT infrastructure. These are solvable.

There might be an important issue on privacy and confidentiality as the data are based on records of individuals holding often very sensitive information. Most countries require informed consent of the data of patients for purposes other than use in clinical therapeutic relationships. One approach, called privacy by design (PbD), requires a privacy governance structure, coding and uncoding procedures for addressing anonymity involving independent trusted third parties, and new reporting standards.

On the human side we learn from normalization process theory (May et al. 2007) that the impact on health providers and healthcare delivery organizations will be immense. We have seen that new providers such as engineers, informaticians, statisticians, communication and media specialists are entering the healthcare sector. Roles of physicians, nurses and hospitals will change drastically. Nurses' roles will change from a direct care provider to patient knowledge broker. There will be a higher need for advanced nursing roles such as advanced practice nurses and nurse practitioners to guide patients to manage their (mainly chronic) diseases. Hospitals will lack the certainty that patients come to the neighboring hospital as it used to be when they lacked any information to make an informed choice. A study by Culley et al. (2011) on UK fertility travelers shows that patients are willing to travel if they find higher success rates or lower costs abroad. Some patients were willing to travel as they were unsatisfied with the treatments in their own country. Most popular travel destinations for fertility treatments were Spain and the Czech Republic. Patients choices will be based more on qualifications, credentials and reviews. The choice of patients for better quality, lower costs and more service will be the main drivers for changing healthcare.

The required change seems similar to the development of railways in the nineteenth century. The first railway opened in 1830 between Manchester and Liverpool. The railway was primarily built to provide faster transport of coils and steel, but more and more passengers made use of it. The fast traveling within countries and throughout Europe had high impact on the introduction of standard time. Mechanical clocks became widespread in the early nineteenth century, but most cities were keeping to local solar time measured on a sundial. This became increasingly awkward as rail transport and telecommunications improved, as clocks differed between places. Bristol and London are cities 200 km (125 miles) apart. Their local solar time differs by 10 min, making timetables incredibly difficult and complex. The great Western Railway adopted Greenwich Mean Time (GMT) as a standard time in November 1847. The decision seems obvious retrospectively but at that time it was highly controversial as it took 17 years after the opening of the first railway in the UK to adopt a well-known 150-year-old standard. Most public clocks adopted GMT for the whole UK only in 1855 and it took until 1880 that GMT was Britain's legal time. Many old clocks of that time still show the two times: local time and standard "railway" time. The railway story shows many things that are relevant now for the globalization of healthcare. A first conclusion is that change is initiated by technology, facilitated by standards but driven by clients (passengers/patients). The second is that people don't want to give up their standards easily. A third conclusion is that it might take time, even a generation. But change will happen undoubtedly as we cannot imagine a life without trains, standard time, and fast traveling. The same will be true for future healthcare.

References

AB InBev Annual Report. 2015. p. 188. http://www.ab-inbev.com/media/annual-report.html

Agency for Healthcare Research and Quality (AHRQ). Efforts to improve patient safety result in 1.3 million fewer patient harms: interim update on 2013 annual hospital acquired condition rate and estimates of cost savings and deaths averted from 2010 to 2013. Rockville, MD: Agency for Healthcare Research and Quality; 2014. AHRQ Publication No. 15-0011-EF. http://www.psnet.ahrq.gov/resource.aspx?resourceID=28573

Aiken LH, Sloane DM, Bruyneel L, Van den Heede K, Griffiths P, Busse R, Diomidous M, Kinnunen J, Kózka M, Lesaffre E, McHugh MD, Moreno-Casbas MT, Rafferty AM, Schwendimann R, Scott PA, Tishelman C, van Achterberg T, Sermeus W. Nurse staffing and education and hospital mortality in nine European countries: a retrospective observational study. Lancet. 2014;383(9931):1824–30. doi:10.1016/S0140-6736(13)62631-8.

Anderson C. The long tail. Why the future of business is selling less of more. Lebanon, IN: Hachette Books; 2006.

Baeten R, Jelfs E. Simulation on the EU cross-border care directive. Eurohealth. 2012;18(3): 18–20.

Bohmer R. Designing care: aligning the nature and management of healthcare. Boston, MA: Harvard Business Review Press; 2009.

Carinci F, Van Gool K, Mainz J, Veillard J, Pichora EC, Januel JM, Arispe I, Kim SM, Klazinga NS. Towards actionable international comparisons of health system performance: expert revision of the OECD framework and quality indicators. Int J Qual Health Care. 2015;27(2):137–46. doi:10.1093/intqhc/mzv004.

Culley L, Hudson N, Rapport F, Blyth E, Norton W, Pacey AA. Crossing borders for fertility treatment: motivations, destinations and outcomes of UK fertility travellers. Hum Reprod. 2011;26(9):2373–81. doi:10.1093/humrep/der191.

De Mars B, Boulanger K, Schoukens P, Sermeus W, Van de Voorde C, Vrijens F, Vinck I. Geplande zorg voor buitenlandse patiënten: impact op het Belgische gezondheidszorgsysteem. Health Services Research (HSR). Brussels: Federaal Kenniscentrum voor de Gezondheidszorg (KCE). KCE Reports 169A. D/2011/10.273/71; 2011.

European Commission. 2016. http://ec.europa.eu/transport/modes/air/safety/air-ban/index_en.htm

European Council. Directive 2011/24/EU of the European Parliament and of the Council of 9 March 2011 on the application of patients' rights in cross-border healthcare. L 88/45. 2011. http://data.europa.eu/eli/dir/2011/24/oj

FitzGerald JD, Soohoo NF, Losina E, Katz JN. Potential impact on patient residence to hospital travel distance and access to care under a policy of preferential referral to high-volume knee replacement hospitals. Arthritis Care Res (Hoboken). 2012;64(6):890–7. doi:10.1002/acr.21611.

Footman K, Knai C, Baeten R, Glonti K, McKee M. Cross-border health care in Europe. POLICY SUMMARY 14. World Health Organization; 2014.

Fuller BR. Critical path. New York: St. Martin's Griffin; 1982.

Graham DJ, Campen D, Hui R, Spence M, Cheetham C, Levy G, Shoor S, Ray WA. Risk of acute myocardial infarction and sudden cardiac death in patients treated with cyclo-oxygenase 2 selective and non-selective non-steroidal anti-inflammatory drugs: nested case-control study. Lancet. 2005;365(9458):475–81.

Jha AK, Larizgoitia I, Audera-Lopez C, Prasopa-Plaizier N, Waters H, Bates DW. The global burden of unsafe medical care: analytic modelling of observational studies. BMJ Qual Saf. 2013;22(10):809–15. doi:10.1136/bmjqs-2012-001748.

Kovacs E, Szocska G, Knai C. International patients on operation vacation—perspectives of patients travelling to Hungary for orthopedic treatments. Int J Health Policy Manag. 2014;3(6):333–40. doi:10.15171/ijhpm.2014.113.

Knai C, Footman K, Glonti K, Warren E. The role of discharge summaries in improving continuity of care across borders. Eurohealth. 2013;19(4):10–11.

Legido-Quigley H, Glinos IA, Baeten R, et al. Analysing arrangements for cross-border mobility of patients in the European Union: a proposal for a framework. Health Policy. 2012;108:27–36.

May C, Finch T, Mair F, Ballini L, Dowrick C, Eccles M, Gask L, MacFarlane A, Murray E, Rapley T, Rogers A, Treweek S, Wallace P, Anderson G, Burns J, Heaven B. Understanding the implementation of complex interventions in health care: the normalization process model. BMC Health Serv Res. 2007;7:148.

McGlynn EA, Asch SM, Adams J, Keesey J, Hicks J, DeCristofaro A, Kerr EA. The quality of health care delivered to adults in the United States. N Engl J Med. 2003;348(26):2635–45. doi:10.1056/NEJMsa022615.

MIT Technology Review. Data-driven healthcare. 2014. https://www.technologyreview.com/business-report/data-driven-health-care/free/

Nanji KC, Patel A, Shaikh S, Seger DL, Bates DW. Evaluation of perioperative medication errors and adverse drug events. Anesthesiology. 2016;124(1):25–34. doi:10.1097/ALN.0000000000000904.

National Patient Safety Foundation. Free from harm, accelerating patient safety improvements fifteen years after "To Err is Human." 2015. http://www.npsf.org/?page=freefromharm

Porter ME, Lee TH. The strategy that will fix healthcare. Harv Bus Rev. 2013;91(10):50–70.

Probst JC, Laditka SB, Wang JY, Johnson AO. Effects of residence and race on burden of travel for care: cross sectional analysis of the 2001 US National Household Travel Survey. BMC Health Serv Res. 2007;7:40.

Techniker Krankenkasse. TK in Europe: TK analysis of EU cross-border healthcare in 2007. Hamburg: Techniker Krankenkasse; 2007.

Chapter 23
Big-Data Enabled Nursing: Future Possibilities

**Judith J. Warren, Thomas R. Clancy, Connie W. Delaney,
and Charlotte A. Weaver**

Abstract Possible futures for education, partnerships, research, and practice are presented as a result of our immersion in the world of big data and data science. We propose basic and sweeping changes in each of these areas, especially for nursing. Kuhn's episodic model of scientific revolution where suddenly the discovery of a new path changes the way we view and live in the world may explain some of what we see and experience. This new paradigm asks questions of old data, proposes new ways to manage data, and hypothesizes where new data might be found. The paradigm shifts from "puzzle-solving" to changing the rules of the game and the mapping directing new operations and research (Kuhn, 1962, 2012).

Keywords Big data • Data science • Future of healthcare • Precision education • Faculty development • Internet of things • Precision nursing

J.J. Warren, PhD, RN, FAAN, FACMI (✉)
School of Nursing, University of Kansas, Kansas City, KS, USA
e-mail: jjwarren@live.com

T.R. Clancy, PhD, MBA, RN, FAAN • C.W. Delaney, PhD, RN, FAAN, FACMI
School of Nursing, University of Minnesota, Minneapolis, MN, USA

C.A. Weaver, RN, PhD, FAAN, FHIMSS
Retired Healthcare Executive, Board Director, Issaquah, WA, USA

© Springer International Publishing AG 2017   441
C.W. Delaney et al. (eds.), *Big Data-Enabled Nursing*, Health Informatics,
DOI 10.1007/978-3-319-53300-1_23

23.1 Introduction

In closing, the editors have gained insights about the use of big data and data science. We began to envision a future that is enhanced and driven by data, thus requiring new skills of healthcare practitioners, educators, researchers, executives, and entrepreneurs. Based on our readings, deliberations, reviewing chapters and case studies, and immersion in the world of big data, we would like the reader to consider the following possible futures. While analysis of large datasets using statistics has occurred in the past, the enormity of the data that is being accumulated today (Moore 2016) and the projections of the doubling curve of knowledge (Fuller 1981; IBM 2006), plus new computer frameworks, programming languages, analytic approaches, and visualization tools, change the way we view and consume data. It may be likened to Kuhn's episodic model of scientific revolution where suddenly the discovery of a new path changes the way we view and live in the world. Big data and data science may be the "anomalies" leading to new paradigms, bringing us a different view of the world. New paradigms then ask new questions of old data and where new data might be found, move beyond the mere "puzzle-solving" of the previous paradigm, change the rules of the game and the map directing new operations and research (Kuhn 1962, 2012). A further indication of the impact of big data and its analytics in health care is the Gartner Hype Cycle. Big data has been moving though the cycle until 2015. In that year, it was removed from the Hype Cycle because it had become part of every other technology on the cycle—it had become a megatrend! Betsy Burton, Gartner Hype Cycle analyst, declared that "big data…has become prevalent in our lives" (Woodle 2015).

23.2 The Future of Big Data in Education: Implications for Faculty and Students

Judith J. Warren

With big data, data science, and the Internet of Things (IoT), knowledge is growing at an alarming rate with a knowledge-doubling rate predicted to be 12 hours by 2020 (IBM 2006). For the professions, specialized knowledge is a key characteristic (Susskind and Susskind 2015). With machine-learning and other sophisticated approaches to managing large amounts of knowledge, the professional will become more dependent on these sources for knowledge rather than being expected to have the knowledge at their fingertips. However, that knowledge will also be available to the lay person. So the future of the professions is evolving with the question still being, who has control over the profession—the profession or anyone with online access to the specialized knowledge? This future will focus on new ways to share practical experience of the professional. The question, now, becomes how do we educate professionals in this world of online, digitized knowledge to become developers and users of this knowledge?

There are three major trends stimulated and enabled by big data and data science that will affect the future of nursing education: (1) society's demand for data scientists, (2) precision education, and (3) changes in faculty roles.

23.2.1 Demand for Data Scientists

As the value of big data and data science has been acknowledged and embraced, the demand for data scientists has grown. Within health care, this demand has exploded overnight. The need to analyze data and gain insights that enable healthcare delivery organizations to gain competitive edges, improve patient outcomes, and reduce costs is why these organizations have made the recruitment of data scientists a top priority (Gershkoff 2015). Organization-generated data is expected to exceed 240 exabytes daily by 2020, making the need for experts trained in extracting insights from data more important than ever. A McKinsey study identified a need for 490,000 data scientists by 2018. However, our educational systems will produce less than 200,000 data scientists to fill these positions (Manyika et al. 2011). Three strategies have been proposed to relieve this shortage: (1) increase the number of universities offering data science degrees, (2) offer data science degrees for both undergraduate and graduate students, and (3) launch programs that train analysts to become data scientists (Gershkoff 2015).

The University of California (UC) at Berkeley is meeting this challenge by offering their first online graduate degree program—data scientist—to use this format, demonstrating the impact of disruptive technology. The program will focus on working with information in different formats and from different sources—text and physical artifacts, video, data, audio, sensor data collected from computers, and Web clickstream data globally networked through the IoT. UC Berkeley believes they will need this new online education paradigm to support this type of education (Florentine 2013). Their data science degree focuses on teaching students to work with datasets of all sizes; to get them to understand how to ask good questions about data; to teach them how to clean it, extract it, put it together and explore it; as well as how to use statistical and machine-learning tools. Columbia University is beginning their program by offering introductory courses for data scientists (O'Neil and Schutt 2014).

To further complicate the demand for data scientists, the field is still so new that there is no widely accepted definition of what data science is. Neither is there agreement on what should be in the curriculum (see Chap. 3 for a more thorough discussion). Even so, data scientists do agree that they need knowledge and skills in math, statistics, machine learning computer science, programming, experimental design, and some expertise in the domain (for example, nursing) in which they are working to be successful in their jobs. This lack of definition presents three challenges to faculty: (1) designing courses and curricula, (2) recruiting faculty with this expertise or developing faculty so that they gain the expertise, and (3) delivering the education to students who may already be employed in the industry (many organizations are resorting to in-house training to meet their needs for data scientists).

Fortunately, data scientists are very active online discussing what tools to use, methods of analysis, processes for visualizing the data, approaches to wrangling/ cleaning the data, and discussing innovation in the field. They participate in open source development of server frameworks, software, and programming languages. Many post tutorials for learning about using these innovations and some even post how to interview for a data scientist job. There are beginning discussions about competencies needed to be a successful data scientist who works with big data. Faculty can begin to look in these online communities for suggestions and resources for teaching their students about data science (see Case Study 3.1). Once content has been gathered and selected, the decisions about course content and curriculum construction need to be made. Should the content be placed in distinct courses or should the content be threaded throughout the curriculum? Resources need to be obtained for learning experiences—sources of big data, analytic tools, computing power, visualization tools, and programming languages. As these decisions are being made, the selection of who teaches data science needs to be decided.

Since big data and data science are new, recruiting faculty can be a challenge as salaries in industry are far higher than salaries in academia. One approach is to partner with industry and have their data scientists teach in the program and provide field experiences in their organizations. This partnership (see Chap. 4) may also provide a setting for faculty scholarship and research. Another approach is to select from current faculty those who wish to develop expertise in the field. However, if the decision is to integrate big data and data science into an existing curriculum (such as a graduate program in nursing), then all faculty must receive education in this field. Nursing does not want unknowledgeable faculty preparing students to function in highly technical domains. Chapter 19 notes the need for faculty with even basic informatics competencies. Misuse of faculty could jeopardize accreditation of the program regardless of the accreditor because all accreditors have standards requiring faculty expertise to teach their assigned courses.

The last decision to make is the method of course delivery. Will the curriculum be delivered in face-to-face, online, or hybrid formats? Remember big data and data science came into being to manage the massive data deluge of today, to gain insights into processes, and to create data products. As faculty design the delivery methods, they should think about the data they wish to collect about the student, the faculty, and the program. Analyzing their own data will support quality education and the personalization of learning.

23.2.2 Precision Education for Students

Christensen and Eyring identified online education as a major disruptive technology that has challenged and changed higher education since the printing press (2011). With the advent of online education using the IoT comes massive data collected about the student that is not possible in a lecture/paper-based world. Not only are grades collected, but also how the student progresses through an evaluation—learning activities completed prior to the evaluation, search paths, mouse clicks, answers, changes, and time to completion. This data can be analyzed to develop algorithms that predict student success, remediation needs, and educational pathways. Categories of interest for the faculty and educational organization include academic major/minor/courses selected, graduation rates, successful learning activities by learning style, learning and IT resource consumption, counseling/advisement, organization/student financial commitment and many others. For all categories of interest, data will be collected and analyzed—time on task, selection, sequence, completion of task, and engagement with others. The other disruption is a change from evaluating the student by presence in class (exposure to content) and paper-based evaluations to evaluating the student through demonstrated achievement and performance of competencies for the discipline.

Therefore, precision education is based on the three disruptive innovations of online education, competency-based education, and big data analytics. Precision education uses big data from diverse sources—learning management systems, adaptive testing, online discussions, search histories, time on task, order of task engaged, data from social media, wearable sensors, and other yet to be identified sources—to effectively identify a learning profile of an individual student and to develop a customized learning solution. Using big data-generated algorithms, individualized lesson plans can be delivered to a specific student, based on personal preferences and learning styles. Precision education enhances the faculty's ability to detect complex learning problems during initial stages for effective and successful intervention. Faculty can challenge the student to learn through the use of this individualized instructional approach.

Two well-known and successful online programs offer this personalized type of education—Knewton (2016) and the Kahn Academy (2016). Though these two programs mainly address kindergarten to high school education, both are now offering college-level courses, especially in mathematics and programming (two required areas of expertise in data science). Knewton reaches out to faculty to help them with this new teaching strategy by offering the expertise of their team--education specialists, data scientists, and developers. The team works with the educator from initial consultation on content and product design to adaptive quality assurance, including marketing the new course. Kahn Academy focuses on the development and provision of micro lectures, not courses as Knewton does, and an adaptive exercise system that generates problems based on student skill and performance. The mission of the Kahn Academy is to provide free, world-class education for anyone, anywhere, and to offer practice exercises and tools for educators at no cost. Their online platform and adaptive systems are powered by big data and data science.

Higher education is beginning to adapt the tools of online education and the analytics of big data and data science to enhance traditional educational practices (Christensen and Eyring 2011). What will the classroom of the future look like for the student? How will the student's interaction with content, faculty, other students, and socialization activities evolve as more information is provided for reflection and learning? How will learning change for each student when the learning environment is designed precisely for him or her? How will precision learning create student interaction and peer support? What is exciting is that there are plenty of opportunities to transform teaching and learning to better serve each individual student within each school by personalizing and humanizing learning; to undo the factory-model assumptions that dominate our schools. Big data and data science can assist faculty to customize education for each student's distinct needs and create opportunities for more meaningful collaborative work between students and faculty (Horn 2014).

23.2.3 Faculty Role Changes

The Western Governors University (WGU) has adopted the online delivery and competency-based education described by Christensen and Eyring (2011) that is fueled by big data and data science. In the old paradigm the curriculum was developed first, then the course, then the objectives, and finally selection of learning materials. WGU begins with identifying the competencies in a domain, then develops object and performance assessments, then identifies learning materials, then course objectives, then bundling content into courses, and finally the curriculum. They use a team to create the curriculum—domain experts, data scientists, instructional designers, evaluators, IT professionals, and consultants. From the beginning they consider the data and how it is to be collected to insure the analysis for determining algorithms that measure WGU and student success. The next innovation is a disruption of traditional faculty roles into a disaggregated model that provides a team of experts for the student to engage—student mentor, course mentor, and evaluator. The student mentor, via routine telephone contact, provides course and program guidance to assist the student to focus on the goal of graduation (similar to a student advisor). The course mentor is an expert in the content area. He or she assists students to develop new competencies and apply new knowledge and skills. Student mentors and course mentors spend 90% of their time directly interacting with students. The mentors do not develop assessments, nor do they conduct them. Evaluators conduct assessments and provide student feedback on competency attainment. Since they do not interact with students during the learning process, their evaluations are more objective and less subject to bias. This objectivity is further enhanced through the use standardized assessment tools and rubrics. This faculty team, with the help of data scientists, uses big data about their students to insure successful outcomes. The team approach allows each to become expert at advisements, content, and evaluation instead of in the traditional model of each faculty member doing all the tasks (Jones-Schenk 2014). WGU has applied this model and approach to nursing education with positive results including accreditation.

As big data and data science move into education, what other faculty roles will be changed? What will scholarship look like? What will faculty governance look like, especially with mentors spending 90% of their time with students? What data products can a university develop and sell using the data generated for the education of students? How will the traditional role of university as knowledge generator change? How will researchers interact in this new academic environment? How will researchers educate the new generation of researchers—will this disaggregated model be applied?

23.3 Conclusion

The future of education for and with big data is a paradigm shift for our society and the way we think about higher education. For the first time, the innovation of online education can provide massive amounts of data about the student, faculty, and school experience. Some schools are looking at the interaction between students and faculty and developing creative ways to enhance the experience. Partnerships with industry are being explored for more than just financial reasons. Faculty roles are being deconstructed and reformed to provide quality learning for the student. These disruptive innovations lead to new questions about higher education. What is the role of the student? What is the role of the faculty? What is the role of the institution and its physical campus? Yet, we know that universities do well in three areas: discovery and dissemination of new knowledge, remembering and recalling achievement and failures of the past (retaining our history), and mentoring the rising generation of citizens. (Christensen and Eyring 2011). In the spirit of big data and data science, the new charge to higher education is to identify what they do well and then to use that knowledge and the data generated by the educational experience to develop data products that will place higher education on the next level of innovation.

23.4 The Future of Partnerships in Generating Big Data Initiatives, Products, and Services

Thomas R. Clancy
The Institutes of Medicine (IOM) defines a learning healthcare system (LHS) as the alignment of science, informatics, incentives, and culture for continuous improvement and innovation, with best practices seamlessly embedded in the delivery process and new knowledge captured as an integral by-product of the delivery experience (IOM 2016). Key to the development of a LHS is the transformational prospects for large interoperable clinical and administrative datasets to allow real-time discovery on issues ranging from disease etiology to personalized diagnosis and treatment. Already the development of these large-scale datasets is seen through multiple sources that include data from: health insurance claims, electronic health records, genetics, social media, wearable sensors, and other sources. Although these datasets are rich and useful for secondary research purposes, individually they come with many challenges. These include:

- **Non-generalizability of outcomes**: Often the dataset is site specific and biased toward a certain population. For example, Medicare and commercial insurance claims data differ on the populations they serve. Electronic health record data varies by where the health entity is located, the services they provide and populations they serve. The use of streaming data from wearable technology is relatively new and may only represent a small percentage of the dataset. Because these individual data sources are rarely linked, the ability to understand patient comorbidity profiles and the breadth of interaction of patients across the health care system limits their generalizability.

- **Usable data**: Although the formation of large, integrated health systems has enabled the creation of large-scale datasets, data is often extracted from multiple EHR vendors, unstandardized and with frequent gaps in data fields. The same can be said for other data sources such as health insurance claims, streaming and social media. Data scientists estimate that 80% of their time is spent pre-processing big data to make it usable (Wachter 2015).
- **Centralized repository for all source data**: There is no single source for all healthcare data. And each data source has its advantages and disadvantages. For example, claims data are good at capturing medical utilization across care settings, medical co-morbidities, the drugs patients take, their visits, and so forth. However, electronic health record (EHR) data are more effective in capturing clinical detail but often can be confined to just specific sites like hospitals and clinics. Streaming data from wearable sensors often provides serendipitous patterns that can uncover knowledge value hidden in the data. But, because of the high volume, velocity and variety characteristics, the data is challenging to store and process.

As data scientists, healthcare administrators, and policymakers seek ways to discover knowledge value in these disparate data sources, multi-institutional networks are emerging as a potential solution (Wallace et al. 2014). Beginning as research collaborations between academic health centers and integrated health systems, these networks are expanding to include non-traditional partners such as life science companies, health insurance providers, and patient engagement networks. The ultimate vision is to create a LHS that participates as a member in an interacting national and regional research network. Within each network are collaborations of providers or "nodes" that consist of research teams from academia, life science companies, government agencies, and health insurance providers. These "nodes" share and, in some cases, "link" their datasets. The success of these networks depends on the ability of members to align and integrate their research initiatives and develop a financially sustainable model. Already, the formation of multi-institutional research networks has begun. Examples include:

- **PCORnet**: a dataset of 47 million patient lives consisting of EHR and patient reported outcomes data collected from multiple clinical sites. Funding and infrastructure for development comes from the Patient Centered Outcomes Research Institute whose aim is to create a research infrastructure to support patient centered outcomes research (http://www.pcornet.org/).
- **The HealthCare Cost Institute (HCCI)**: a dataset of 40 million lives of claims data developed by a network of large healthcare insurance providers. By contributing data from several insurers the dataset better reflects the general population and is valuable for evaluating the cost of healthcare interventions (www.healthcostinstitute.org).

- **Mini-Sentinel**: a dataset of 193 million lives of claims data provided by a network of health insurance providers. Funding and infrastructure development was provided by the FDA (US Food and Drug Administration) in 2008. Researchers conducting studies for the FDA may access this dataset (www.mini-sentinel.org).
- **OptumLabs**: a dataset of 130 million commercially insured and Medicare Advantage patient lives linked with electronic health records as well as sociodemographic and consumer data. This unique research collaborative, initially developed in a partnership by Optum and the Mayo Clinic in 2013, currently has over 20 partners that range from academic health centers to life science companies to professional organizations. Members collaborate by conducting studies using the OptumLabs Data Warehouse, sharing results at monthly partnership exchanges and serving on various research committees (www.optum.com/optumlabs).

The emergence of multi-institutional networks from multiple data sources is paving the way for what Topel (2015) describes in *The Patient Will See You Now*, as the human graphical information system (GIS). A GIS comprises multiple layers of demographic, physiologic, anatomic, biologic and environmental data about a specific individual. In the fully integrated GIS, these layers of data could include the following:

- **The Social Graph and the Phenome**: Demographics, family history, location, family and social network, education, pictures, videos and other.
- **Sensors and the Physiome**: Electronic health record and biosensor data from wearable technology that could include blood pressure, heart rhythm, respiratory rate, oxygen concentration, sleep patterns, stress levels and a host of other physiological and behavioral data.
- **Imaging and the Anatome**: Radiological images including magnetic resonance, computed tomography, nuclear scanning, ultrasound and other images.
- **Sequencing and the Genome**: The order of DNA nucleotides, or bases, in a genome (i.e., the order of As, Cs, Gs, and Ts that make up an organism's DNA).
- **The Exposome**: Environmental factors such as radiation, air pollution, pollen count, water quality, noise, pesticides in food and so forth.
- **Healthcare Insurance Claims**: Claims data across multiple levels of care (clinic, home, nursing home, hospital and so forth).

The formation of multi-institutional networks from disparate data sources creates the infrastructure to generate a human GIS that represents the medical essence of individual patients. By combining, or "mashing-up" an individual's data layers, healthcare providers create the opportunity to more accurately diagnosis illness, personalize treatment plans and in some cases prevent untoward healthcare events.

In his book, *Future Smart: Managing the Game-Changing Trends That Will Transform your World*, Canton (2015) describes the formation of "innovation ecosystems" as the natural progression of a fully networked society. An innovation ecosystem is a high-performance, collaborative, global web network that is predictive, real-time and mobile. These smart networks of the future would bring together talent, innovation, supply chains, markets, makers, capital and experts to focus on problems of global proportion. For example, domain-specific innovation ecosystems, such as healthcare, would bring together talented researchers, entrepreneurs, funding sources, health systems and others to study a problem, create a solution, and then translate it into practice. The key ingredients for such a future depend upon the ability of multiple, disparate data sources to be linked and shared across emerging research networks of today. However, to achieve that vision, consumers, business and industry, health systems, academia and the government face many challenges. These include issues regarding: privacy and security of health information; storage and processing of big data; standardization and harmonization of data; interoperability and health information exchange across systems; intellectual property rights and commercialization; and cost issues, just to name a few. Yet through the collaboration and partnership of these many organizations, the opportunities to improve the health and well-being of individuals is enormous.

23.5   Big Data Through the Research Lens

Connie White Delaney

The pace of science in the last 200 years has indeed increased: wind, water, animal, and steam power; railroad, telegraph, telephone, electricity; automobiles, airplane, radio, television; computers, fiber optics. The needs of society influence our ultimate promise and goal of translating scientific and technological discoveries to improve human lives, our environment, health, and sustainability. A clear example of the inter–relationship among the pace of science, promises, discoveries and societal needs resides in the structures of health care. While some discoveries reduce human risks and costs, others have reverse effects on the supply/demand curves; hide actual costs from the users, increase government regulation, and challenge individual discernments in life and death situations (The National Academies of Sciences, Engineering, and Medicine 1993). While science and technology cannot alone solve societal and health problems, they are necessary. Thus we ask—what are the forces affecting big data and related discoveries in nursing and health care? what do we anticipate for the future because of big data? and what is nursing's call to action on big data and data science now?

23.5.1   Forces Affecting Big Data and Related Discoveries in Nursing and Health Care

Multiple national health system and research transformation initiatives in the U.S. have major enabling effects on the promises of big data and data science to advance health. At the root of these imperatives is the overwhelming evidence of the unsustainable percent of GDP expenditures on health, high infant mortality, large number of deaths from medical errors, and the very low ranking on high-functioning health system in comparison to the top developed nations.

The national commitment to create a Learning Health System (LHS) (IOM 2007) where there is a seamless cyclical flow across care, study, and training is intended to substantively decrease the time (estimated to be 17 years) from a discovery to use and impact on clinical care. The LHS would foster trust and value across all stakeholders, be adaptable, capable of health improvement, and sustainable. The LHS agenda is closely articulated with the research enterprise in the U.S. The U.S. research enterprise is largely driven by the National Institutes of Health (NIH) which is charged with ensuring its research and training programs are thoughtfully selected, effectively pursued, and responsive to NIH's research mission, national health concerns, and the need to prepare the next generation of scientists as noted in *Enhancing the Vitality of the National Institutes of Health: Organizational Change to Meet New Challenges* (National Academies 2003).

In 2002, the Director of NIH announced bold initiatives designed to transform medical research capabilities and speed the movement of research from the laboratory bench to the patient's bedside, following a series of meetings and reports. NIH launched the NIH Roadmap (2004). The purpose of the Roadmap was to identify major opportunities and gaps in biomedical and behavioral research that no one institute at NIH could undertake alone. The Roadmap's structure, comprising 28 initiatives, focuses on three main areas: (1) new pathways to discovery, (2) research teams of the future, and (3) re-engineering the clinical research enterprise.

Drawing from experience of the NIH Roadmap for Medical Research, extensive community input, and the IOM's multiple reports, the Clinical Translational Science Award (CTSAs) program was created to build a definable academic home for clinical and translational research. CTSA institutions work to transform the local, regional, and national environment to increase the efficiency and speed of clinical and translational research across the country (CTSA 2006). The CTSA program was designed to devise innovative and far-reaching approaches to build academic homes for clinical and translational science which would:

- Develop better designs for clinical trials to ensure that patients with rare as well as common diseases benefit from new medical therapies;
- Produce enriched environments to educate and develop the next generation of researchers trained in the complexities of translating research discoveries into clinical trials and ultimately into practice;
- Design new and improved clinical research informatics tools;
- Expand outreach efforts to minority and medically underserved communities;
- Assemble interdisciplinary teams that cover the complete spectrum of research—biology, clinical medicine, dentistry, nursing, biomedical engineering, genomics, and population sciences; and
- Forge new partnerships with private and public health care organizations.

As Director Zerhouni noted, "The development of this consortium represents the first systematic change in our approach to clinical research in 50 years. Working together, these sites will serve as discovery engines that will improve medical care by applying new scientific advances to real world practice. We expect to see new approaches reach underserved populations, local community organizations, and health care providers to ensure that medical advances are reaching the people who need them" (NIH 2006).

By 2013, the CTSA Consortium had expanded to 62 medical research institutions located throughout the nation, linking them together to energize the discipline of clinical and translational science. Further, the impact of the CTSA consortium was reaching far beyond individual organizations to include innovation, integration, inclusion, and dissemination across all organizations involved in health care throughout the country (NIH 2006). Interdisciplinary research and science teams, also key in the transformation of NIH and support for the LHS evolution, are priorities of the CTSAs. Clearly, integrating the analytical strengths of two or more often disparate scientific disciplines to solve health problems is a priority.

PCORnet is another major initiative empowering research and a LHS. PCORnet is a national patient-centered clinical research network developed by the nonprofit Patient-Centered Outcomes Research Institute (PCORI). PCORnet seeks to improve the nation's capacity to conduct clinical research by creating a large, highly representative, national patient-centered network that supports more efficient clinical trials and observational studies. PCORnet brings together the expertise, populations, resources, and data of its participating organizations to create a national infrastructure that enables more efficient, patient-centered research. Specifically, for research, PCORnet is intended to strengthen the quality and nature of research by providing researchers access to patient-centered data, reduce the time needed to conduct large-scale research projects, and make researchers' proposals more competitive in the current funding environment. PCORnet is comprised of two networks: the 13 Clinical Data Research Networks (CDRNs) and the 20 Patient Powered Research Networks (PPRNs). The PCORnet, like the CTSAs, is clearly addressing reciprocal IRBs, data sharing agreements, master research partnership contracts, and collaborative research groups.

Academic research and corporate/industry partnerships are also critical to the LHS and research driving improved health. The dynamic system of interconnected institutions, persons, and policies that are necessary to propel technological and economic development is commonly referred to as the U.S. innovation ecosystem. One regularly informing national entity is the President's Council of Advisors on Science and Technology (PCAST) (PCAST 2008). Multiple reports have been issued over the years, which addressed trends in Federal funding of research and development (R&D), Federal-State partnerships for R&D, and mechanisms for enhanced technology transfer. Although many successful research partnerships exist among a range of participants from the public and private sectors, there are several new trends that PCAST identified that fall specifically within the context of university-private sector research partnerships.

The first of these funding trends is the growing imbalance between the academic research capacity and the Federal research budget. The second development of note is the reduction in basic research performed by the industrial sector, including the disappearance of research labs. Private foundations are expanding their capacity to fund research, another trend that is expected to be important in the future. And last, the accelerating speed of technological development requires new methods of knowledge exchange between universities and industry to capture the societal and economic benefits of these innovations.

To recognize fully the importance of university-private sector partnerships and their role in the rapidly globalizing innovation ecosystem, PCAST has identified several models of university-private sector research partnerships and models of private foundations that fund such partnerships. Barriers to these partnerships encompass the following areas: basic research and innovation; economic and regulatory policies impacting U.S. innovation and research partnerships; network models of open innovation; connection points between partners in the innovation ecosystem; and measuring and assessing innovation. As mentioned above, one successful example is the OptumLabs open, collaborative research and innovation center, founded in 2013 by a partnership between Optum and the Mayo Clinic.

The OptumLabs research collaborative is made up of over 20 partners, including the University of Minnesota School of Nursing, whose research teams conduct secondary data analysis using the OptumLabs Data Warehouse (OLDW). The team science benefits of academic/corporate partnerships such as the OptumLabs research collaborative are readily apparent. Academic/corporate partnerships can complement the strengths of different institutions and tap the opportunities hidden in big data.

23.5.2 Anticipating the Future with Big Data

The transformation toward the LHS and the fast-paced movement of the CTSAs and PCORI initiatives support the evolution of science conducted through collaboration, within interdisciplinary teams, and engaging in partnerships across societal sectors, health systems and academia, and expansion of the inter-dependencies among all components. These partnerships are creating the infrastructure to foster big data collection/access and scientific inquiry by creating teams comprised of researchers, statisticians, computer scientists, and data visual designers to conduct research in this innovative way. Moreover, precision medicine, an emerging approach for disease treatment and prevention, takes into account individual variability in genes, environment, and lifestyle for each person (NIH 2006). This future adds the rich data generated from wearable sensor devices and social media as it pertains to the health and well-being of the individual.

In summary, the Learning Health System, precision medicine and person-centric care, connected communities, CTSA and PCORI research initiatives, and global connectivity of the Internet of Things will continue to reform research- and knowledge-generation toward a team science not bound by organization or local resource.

23.5.3 Nursing's Call to Action for Big Data and Data Science

The transformation to maximizing the use of big data to solve many of the most challenging questions has implications for the future of nursing in all missions: education, research, practice, and policy. The education of all nurses, from entry level to nurse scientists, will change. Creating entry-level nurses who are adept at using data to inform real-time, knowledge-based practice that encompasses population analytics will be the norm. What will be the implications for curriculum development, for all degree programs, to insure knowledge of big data and data science? The demand for increased data resources and analytics to drive organizational decision-making and policy is clear.

Is nursing data encompassing patient observations, patient problems, nursing interventions, nursing sensitive outcomes, and context of care available from new sources as we consider data collection from a big data perspective? Are nurse scientists prepared to be data scientists, to understand the assumptions concerning data in a big data environment as opposed to statistical assumptions of data, and to engage data science methodology? Are nurse scientists prepared to function in data science teams? These are the challenges and opportunities for nursing to embrace as we move to the completion of the second decade in the twenty-first century.

23.6 Healthcare in 2020: Looking at Big Data Through the Clinical Executive's Lens

Charlotte A. Weaver

Busy clinical healthcare executives question what the noise about "big data" means and why they should take the time and make the effort to learn about it. After all, they have been keen users of data and analytics, believers in the principle that "if you can't measure it, you can't manage it". Organizations have taken steps to be able to store, aggregate and report on the vast volumes of data generated by their electronic health records (EHR) over the past decade or more. And they have already invested heavily in capturing and generating the mandatory quality metrics required by regulatory agencies and third party payers, such as CMS (Centers for Medicaid and Medicare Services) and insurance companies. So what's new here and where's the business imperative to convince nurse executives that there is a paradigm shift happening?

In healthcare over the past five to 10 years, two technology-based, game-changing trends have happened. First, web technologies, smartphones, mobile devices and cheap data storage and processing entered the market and offered new options for consumers as well as new health applications that generated massive volumes of data. Second, government mandates required the use of EHRs in hospitals and ambulatory practices. By the end of 2015, an estimated 85% of the acute/ambulatory care market was using EHRs (Health IT Gov. 2016), and healthcare organizations found that the magnitude of EHR data being generated started having a doubling time of months rather than years (IBM 2016). Adding to EHR data volumes are health data from mobile devices, smartphones, mobile sensors like Fitbits, medical devices in the home, patient portals and an ever-growing body of web-based applications built to support managed care, patient engagement, clinical decision support, and consumer engagement (Fortini 2015; O'Connor et al. 2016; Austin 2016; Phillips 2016). To meet the new care delivery models and business imperatives, healthcare organizations needed to be able to include these disparate data in their analytics, and thus, the imperative for new database tools and different processing tools. Additionally, a plethora of web applications, the Internet of Things, have emerged in response to organizations finding that they needed more robust functionality to support their population health management efforts—registries and care coordination with extensions into the community—than is available in the architectures of the traditional electronic health record systems. Taken together, this second decade of the twenty-first century has ushered in rapid health system transformation at the same time that a major revolution in data management is happening.

Lest the clinical executive think that healthcare is unique in these ever-changing technology challenges, it may help to see how much big data and the new product solutions, services and technologies enabling these powerful tools are with us today in every aspect of business, from grocery stores to Wall Street (Alter and Russell 2016; Cardwell 2015; Costa 2016). For example, when ordering anything from Amazon, customers are recipients of an on-line store that captures their preference data in real-time to push suggestions on what else they might like to buy. This business approach was so much more effective than the volume and inventory analytics used by the brick and mortar bookstores that Amazon effectively put them out of business. Similarly, Win-Co grocery stores display to shoppers how its prices compare to other stores in the vicinity on any given item. Also, Win-Co gathers price data from its competitors in real-time so that they can change their prices to always be the lowest price. In the entertainment world, Netflix uses the same customer-preference functions as Amazon to suggest other TV programs or movies that one might want to view. Netflix and other streaming services companies like them have so effectively competed with the cable companies on price and services that millions have fled the cable market since 2014 (Ingram 2015). Amazon, Win-Co and Netflix are just a few examples of data-enabled organizations that successfully use big data processing and analytic tools to invent new business models that win consumers' business and loyalty. However, to move from these anecdotal examples to research-based evidence, there is evidence that organizations that are the best users of big data are the most successful in their sector. In an MIT-led study of 330 public companies that evaluated the relationship between use of data for decision-making and performance, the researchers found a strong correlation between companies that described themselves as "data-driven" and their financial/operational results. Those companies scoring in the top third of their industry as using data-driven, decision-making were found to have also outperformed their competitors on productivity and profitability, regardless of industry type (McAfee and Brynjolfsson 2012).

For healthcare, these data trends show a paradigm shift happening in data volumes and complexity. The evidence from other industries suggest that, for organizations to navigate structural and reimbursement changes, the most successful will be those that use big data analytics to inform its decisions, strategies and care delivery programs. To be optimally prepared, nurse executives will want to be fluent in the use of big data analytics and to have their business and care delivery strategies informed from analytics and data from the widest possible data sources.

23.6.1   Healthcare's Journey into Big Data

For healthcare, a harbinger of the game-changing power of these new technologies quietly entered the public domain in late 2012 (Kaelber et al. 2012). A group of informatists and researchers from The MetroHealth System and Case Western Reserve School of Medicine, working with an innovative, big data analytics start-up out of the Cleveland Clinic called Explorys, demonstrated the ability to pull data from multiple EHR systems and produce an analysis in a time that shocked the informatics world. The study used nearly a million patients from a 13-year period to identify and profile those most at risk for developing blood clots in the extremities and lungs. What would have taken years to perform using traditional data management tools took this five-author, part-time team just 125 hours over an 11-week period to complete. None too surprising, since its publication this work has been referenced more for its use of new data analytics technologies and the minimal resource and time required to complete the study than for its clinical findings. The important takeaway for the industry was that a new set of tools is now available. These new big data wrangling/cleaning solutions allow data analysts to build and manage complex databases for rapid, low-resource time required, while remaining user-friendly for business users and scientists (Haight and Park 2016). The visualization and analytics tools in these solutions place data directly into hands of end-users who will need to perform the analyses. Tools allow for structured data, unstructured data, local and standardized terminologies, streaming data and single-point data to be analyzed, thus moving away from traditional statistical methods and into graph and cluster visual analyses showing the interconnectivity of relationships and trends (Haight and Park 2016).

In the intervening 4 years since the publication of the Cleveland study over 30 of the largest integrated healthcare systems have purchased the same big data system and its analytic services (Explorys 2016). Just as important to the healthcare industry is that IBM purchased this system in April 2015 to launch its new analytic product line under "Watson Health" (IBM 2015). As an indication of how fast the big data/analytics vendor healthcare market is moving, IBM added four more acquisitions to its Watson Health offering in the year following its 2015 launch (IBM 2016). Its latest acquisition was Truven Health Analytics in April 2016, adding 200 million lives of data to IBM's 100 million lives. In addition to IBM Watson Health, other major companies in this big data analytics market include Optum Health, Health Catalyst, Mede Analytics and SCIO Health Analytics. The 2016 report on the top 20 companies by Healthcare Tech Outlook calls out how rapidly the market is changing year over year, with rapid growth as well as consolidation (see http://www.healthcaretechoutlook.com/vendors/20special2). These healthcare big data companies offer analytic services using their database tools to do data mining, research, benchmarking and segmentation of populations. Segmentation means groupings derived from analytics that identify specific risk populations.

These big data companies' analytic services use their databases to do data mining, research, benchmarking, and segmentation of populations into pre-condition, at-risk, high risk, highest resource consumption, profiles and risk factors for any type of query requested by the client. These analytics are core to population health management initiatives and to profiling high-risk groups with high utilization or readmission rates. And subsequently when programs are launched to address risk, analytics are needed for evaluating the on-going effectiveness, financial metrics and quality indicators of the program.

In 2016, healthcare has an estimated 20% of its organizations using big data solutions and services (Explorys 2016), however, indications are that the large, integrated delivery networks (IDN) organizations are actively engaging at a much quicker pace than small systems. Over 50% of IDNs are estimated to have purchased solutions and services from specialist entities in the big data space other than from their EHR vendor (Bresnick 2015; Explorys 2016) and this is projected to change rapidly with broader adoption occurring across all segments by 2020.

23.6.2 Looking at Care Delivery in 2020

In 2016, we have a number of organizations engaged in population health management. Third-party payers like Anthem, United Health, Medicare, Medicaid, self-insured labor union, etc., Accountable Care Organization partnerships, and integrated healthcare systems are moving rapidly to use the best solutions to help them identify, engage, track and report outcomes on given segments of the individuals under their coverage and care. In ambulatory settings, primary care providers are using analytic tools to identify groupings by diagnosis or conditions. These disease groupings are further segmented into subgroups by evidence-based risk indicators and intervention protocols are assigned to each segmented group. Often this identification and grouping can be done within the organization's EHR analytic tools. But increasingly, as patient's engage in their own health and use other patient health record tools and mobile devices, organizations need new cloud and web-platform solutions that can serve as data storage and data processing solutions for data coming in from web applications, mobile sensing devices and multiple EHRs.

As care delivery models are planned to be patient-centered that extend into the community and home, it is important for the nurse executive to be aware that these new big data, web/cloud-based technologies and services are available. Registry and dashboard products are entering the marketplace that are more robust in their ability to support frontline clinicians' workflow in population management and community-based care coordination. These types of solutions use the EHR database to access information for the following purposes: to develop algorithms to push clinical decision support and workflow to frontline clinicians or managers; display comprehensive dashboards using information from multiple sources; support team communications across disciples and settings; and push best practice options.

23.6.3   Population Health Managed Care—An Example from Bon Secours Medical Group (BSMG)

In this example, a large ambulatory medical practice of an IDN health system committed to an ACO (Accountable Care Organization) partnership, and redefined its ambulatory care delivery model to shift to Patient Centered Medical Home and population health management. They have a committed leadership that lead significant process and role changes, and have adopted the use of leading-edge IT solutions with data analytics and evidence-based developed protocols (IHT² 2014). Chief Clinical Officer Robert Fortini, RN, MSN, leads these transformative initiatives. Starting with a population health management approach in their ambulatory care practice, they targeted post-discharge care of their high-risk, readmission patients (Fortini 2015). In an effort to do better post-discharge follow-up, handoffs and care coordination for high-risk patient groups, the BSMG ambulatory practice of 600 physicians invested in role and process changes, as well as new IT infrastructure. Nurse navigators, coordinating between hospital and post-acute care services and primary care/specialist teams, used a new registry system that could do the data analytics to identify the risk populations as well as to do automatic calling and send patient reminders. This registry functionality enabled time and resource savings such that it paid for additional resources needed for the program, and the post-discharge follow-up and wellness visits made the clinic eligible for additional Medicare reimbursement (Fortini 2015). Fortini stresses that it was "good old-fashioned nursing-based case management" using Phytel's registry and care management tools that delivered the transformative changes (IHT² 2014).

Embedded in a Patient Centered Medical Home model with nurse-based case management, Fortini describes using the registry and care management tools of Phytel (acquired by IBM Watson Health in 2016, http://www3.phytel.com/our-platform/phytel-outreach) in conjunction with their Epic EHR system to support their shift to a population care management model (IHT² 2014). In addition to developing 35 protocols for chronic and preventive care, Fortini describes an initial focus in the population health management on their high-risk readmission patients that yielded significant quality improvements and additional reimbursement.

23.6.4   Looking at Near-Term Future Examples

The following examples are projections for what we might expect to see healthcare systems routinely using data analytic tools and methods to do in operations and care delivery analyses:

23.6.4.1 Operations

- Analyze workflow of Emergency Department or Operating Theaters to understand patient flow, discern equipment/room usage patterns, identify trends and bottlenecks, and use in continuous quality improvement methods to improve staff and resource allocation linked to quality improvement measures and patient satisfaction scores.
- Identify hospitalized patients at high-risk for readmission early with links to post discharge program for medical home/home health care that involves handoff and data exchange at discharge and 30, 60, and 90 days post discharge tracking for key quality indicators, including readmission.
- Look at patient population by groupings that include age, gender, ethnicity, location, payment type to detect access to care, utilization patterns, biases or outliers.

23.6.4.2 Care Delivery

- Personalize treatment for the cancer (think NIH Cancer Moonshot project) patient by having a merged view of all EHR data including genomic and socio-determinants of health information. Care is supported by live feeds of clinical trials and research findings and drug approvals that combine insights to suggest new treatment opportunities for specific patient groups.
- React to remote patient monitoring devices (FitBit, meds, home monitors for BP, WTs vital signs) as alerts to potential problems within the full context of the patient's historical record in support of the community-based care team/Patient Centered Medical Home.
- Proactively identify individuals at risk for developing high-cost/quality-of-life conditions, and actively engage them in health promotion programs within population health management constructs.

23.6.5 Looking Forward

Trends that we are currently seeing in the vendor service offerings and uses by healthcare organizations heavily target managing risk. Risk management is a key capability going forward. Directors of managed care programs talk about early identification, patient engagement within best practices, and care protocols that allow for patient preferences. They plan to use new web-based tools and mobile devices for monitoring and close communication with patient and family that generate data needed for their tracking and outcomes. Ideally in these programs, the earlier the identification of risk can happen, the sooner interventions can be deployed to offset risk of poor outcomes and costly chronic disease. Some solutions even aim to apply this predictive segmentation into risk scores at an individual, family or community level (http://www.ibm.com/smarterplanet/us/en/ibmwatson/health/)

that employers, health systems and managed care insurers will use increasingly to go upstream to "proactively" target individuals for interventions before disease develops.

23.6.6   Personalization of Care

In health care there is a move expected away from a one-size-fits-most protocol approach to one of personalized treatment based off the individual's genetic, socio-health determinants, and personal monitoring data. And this means that clinicians will need to have a single view into a patient's full medical record using the most effective care delivery models; and, organizations will have to have solutions in place that can unite disparate data sources and provide the needed displays and functionality. The goal of "personalization" is to allow for the single most effective treatment to be selected for a given patient, avoiding mistreatments, complications, poor outcomes, and high costs.

To position themselves to be able to do personalized, patient-centered care with optimal outcomes and cost efficiencies benchmarked against CMS expectations/levels, healthcare systems will need to take on capabilities, solutions and expertise to conduct data analytics using the most robust tools and services available. This means that by 2020, health systems should have service contracts in place for the data processing solutions provider who will do the work of bringing together operational data with an assortment of health information systems. In the context of population health management, health systems will need to be able to use social media, demographic and census data for population-wide studies and disease management programs. For individualized care, population health, risk management, and for operational and business improvement, healthcare systems must get to the point where they can combine practice-based data with external third-party data—personal health record systems, remote mobile monitoring, claims data, retail pharmacy, multiple EMRs, and unstructured notes. How much a health system outsources to technology service providers versus brings in-house is for each entity to evaluate.

As important as these technology tools are for the health system transformation that we are facing in the next 5 years, they are not sufficient. Strong leadership and a committed, functionally healthy leadership team are essential for an organization to successfully navigate these profound changes. As seen in Fortini's work at Bon Secours and Kaiser-Permanente's story in Chap. 16, the IT analytic tools are just tools used by good leaders. In both instances, IT is used in conjunction with cultural and structural changes that redefine roles and how disciplines work together, as well as create new roles, new tasks and workflow. Making cultural change happen is hard work that requires leadership and sustained commitment to make it stick. So while this book and chapter happen to lopsidedly emphasize the new and emerging use of IT analytic tools, in no way do we mean to minimize the most essential ingredient to organizational success—courageous and strong leadership.

23.7 Final Thoughts About the Future with Big Data

As with any new innovation, the world moves quickly and, if technology is involved, the movement is at light speed. Nursing's challenges are many. The education of faculty, administrators, and nursing data scientists is key to enabling nursing to continue to be at the table when big data is being discussed and analytics are driving decisions. The determination of new data sources is imperative—images, text, sound, motion, geolocation, environment (weather), wearable sensors, Internet of Things devices and apps, genomics, social media, medical devices, and many more—to capture more information and patient-nurse interactions. The invention of new data products is essential to gather information about and for nurses. Creating new partnerships is critical to building new data science teams that will leverage the nursing perspective in health care delivery. Engaging in big data research is needed to generate new nursing knowledge. The phenomena of big data and data science moves nursing from analyses of what happened (traditional research and interactive dashboards) to predictions of what will happen (machine learning) and recommendations of what to do (data products). The world is getting ready to shift before us!

Big data and data science move nursing care into the world of precision nursing! Precision nursing uses big data from diverse sources to outline a detailed picture of the patient and recommend a customized healthcare solution. Using big data, customized treatment plans can be tailored to specific individuals and their preferences. Precision nursing enhances the nurse's ability to detect complex nursing problems during initial stages which is imperative for effective and successful treatment and to offer treatment for lifestyle-related diseases by intrinsically analyzing data pertaining to lifestyle patterns of patients. Big data analytics facilitate the capture of the patient's story and offer patient care recommendations unique to the patient. The challenge is to prepare nurses in this new world to give patient care that makes a difference.

References

Alter A, Russell K. Moneyball for book publishers: a detailed look at how we read. 2016. http://www.nytimes.com/2016/03/15/business/media/moneyball-for-book-publishers-for-a-detailed-look-at-how-we-read.html. Accessed 20 May 2016.

Austin G. OneCare, LLC. 2016. http://www.onecare.me. Accessed 15 Mar 2016.

Bresnick J. How to use healthcare Big Data analytics for Accountable Care. Hlt IT Anal. 2015, September 21. Available at: http://healthitanalytics.com/news/how-to-use-healthcare-big-data-analytics-for-accountable-care. Accessed April 5, 2017

Canton J. Future smart: managing the game-changing trends that will transform your world. Boston: De Capo Press; 2015.

Cardwell D. A light bulb goes on, over the mall. 2015. http://www.nytimes.com/2015/07/20/technology/a-light-bulb-goes-on-over-the-mall.html. Accessed 25 May 2016.

Christensen CM, Eyring HJ. The innovative university: changing the DNA of higher education from the inside out. San Francisco: Jossey-Bass; 2011.

Costa B. Golfers join the rest of world, use data. 2016. http://www.wsj.com/articles/golfers-join-rest-of-world-use-data-1463434676. Accessed 25 May 2016.

CTSA. 2006. https://www.nih.gov/news-events/news-releases/nih-launches-national-consortium-transform-clinical-research. Accessed 24 May 2016.

Explorys. 2016. https://www.explorys.com/the-platform.html. Accessed 20 May 2016.

Florentine S. Who's training the next generation of data scientists? 2013. http://www.cio.com/article/2382080/careers-staffing/who-s-training-the-next-generation-of-data-scientists-.html. Accessed 15 May 2016.

Fortini RJ. Helping value-based care delivery pay for itself. Hlt Fin Mgt. 2015;69(1):42–5.

Fuller RB. Critical path. New York: St. Martin's Press; 1981.

Gershkoff A. How to stem the global shortage of data scientists. 2015. http://techcrunch.com/2015/12/31/how-to-stem-the-global-shortage-of-data-scientists. Accessed 15 May 2016.

Haight J, Park H. Bridging the data preparation gap: Healthcare. Blue Hill Research Report No#A0205. 2016. http://www.datawatch.com/wp-content/uploads/2016/01/RT-A0205a-DatawatchHealth-HP.pdf. Accessed 1 May 2016.

Health IT Gov. 2016. http://dashboard.healthit.gov/quickstats/quickstats.php. Accessed 15 May 2016.

Healthcare Cost Institute. http://www.healthcostinstitute.org. Accessed 3 Feb 2016.

Healthcare Tech Outlook. 20 most promising healthcare analytics health providers 2015. http://www.healthcaretechoutlook.com/vendors/20special2. Accessed 20 May 2016.

Horn M. Disruptive innovation and education. 2014. http://www.forbes.com/sites/michael-horn/2014/07/02/disruptive-innovation-and-education. Accessed 15 Mar 2016.

IBM. The toxic terabyte: how data-dumping threatens business efficiency. 2006. http://www-935.ibm.com/services/no/cio/leverage/levinfo_wp_gts_thetoxic.pdf. Accessed 15 May 2016.

IBM. IBM acquires Explorys to accelerate cognitive insights for health and wellness. 2015. https://www-03.ibm.com/press/us/en/pressrelease/46585.wss. Accessed 20 May 2016.

IBM. IBM Watson Health closes acquisition of Truven Health Analytics. 2016. http://www.prnewswire.com/news-releases/ibm-watson-health-closes-acquisition-of-truven-health-analytics-300248222.html. Accessed 24 May 2016.

IHT[2]. Institute for Health Technology Transformation: An interview with Robert J Fortini, RN, MSN, VP and Chief Clinical Officer, Bon Secours Medical Group. 2014. http://ihealthtran.com/wordpress/2014/02/interview-with-robert-fortini-r-n-m-s-n-vp-and-chief-clinical-officer-bon-secours-medical-group. Accessed 24 May 2016.

Ingram M. Pay TV industry loses subscribers to cord cutting. Fortune. 2015. http://fortune.com/2015/11/10/tv-industry-loses-subscribers-to-cord-cutting. Accessed 20 May 2016.

Institute of Medicine. Learning health system series: continuous improvement and innovation in health and health care. 2016. http://iom.nationalacademies.org/~/media/Files/Activity%20Files/Quality/VSRT/Core%20Documents/LearningHealthSystem.pdf. Accessed 3 Feb 2016.

Institute of Medicine (IOM). The learning healthcare system: workshop summary roundtable on evidence-based medicine. In: Olsen LA, Aisner D, McGinnig J, editors. Washington, DC: National Academies Press; 2007.

Jones-Schenk J. Nursing education at Western Governors University: a modern, disruptive approach. J Prof Nrsg. 2014;30(2):168–74.

Kaelber DC, Foster W, Gilder J, Love TE, Jain AK. Patient characteristics associated with venous thromboembolic events: a cohort study using pooled electronic health record data. J Am Med Inform Assoc. 2012;19(6):965–72. doi:10.1136/amiajnl-2011-000782.

Kahn Academy. 2016. https://www.khanacademy.org. Accessed 15 May 2016.

Knewton. 2016. https://www.knewton.com. Accessed 15 May 2016.

Kuhn TS. The structure of scientific revolutions. 1st ed. Chicago: University of Chicago Press; 1962.

Kuhn TS. The structure of scientific revolutions: 50th anniversary edition. Chicago: University of Chicago Press; 2012.

Manyika J, Chui M, Brown B, Bughin J, Dobbs R, Roxburgh C, Byers AH. Big data: the next frontier for innovation, competition, and productivity. 2011. http://www.mckinsey.com/

business-functions/business-technology/our-insights/big-data-the-next-frontier-for-innovation. Accessed 15 May 2016.

McAfee A, Brynjolfsson E. Big data: the management revolution. 2012. https://hbr.org/2012/10/big-data-the-management-revolution/ar. Accessed 15 Apr 2016.

Mini–Sentinel. http://www.mini-sentinel.org. Accessed 3 Feb 2016.

Moore's Law. 2016. http://www.mooreslaw.org. Accessed 18 Apr 2016.

National Academies. Enhancing the vitality of the National Institutes of Health: organizational change to meet new challenges. 2003. http://www.nap.edu/catalog/10779/enhancing-the-vitality-of-the-national-institutes-of-health-organizational. Accessed 22 May 2016.

National Academies of Sciences, Engineering, and Medicine. Science, technology, and the federal government: national goals for a new era. Washington, DC: National Academy of Science; 1993.

NIH. NIH roadmap in 2004. http://www.niehs.nih.gov/funding/grants/announcements/roadmap. Accessed 22 May 2016.

NIH. Precision medicine. 2006. https://www.nih.gov/precision-medicine-initiative-cohort-program. Accessed 22 May 2016.

O'Connor PJ, Sperl-Hillen JM, Fazio CJ, Averbeck M, Rank BH, Margolis KL. Review article: outpatient diabetes clinical decision support: current status and future directions. Diabet Med. 2016;33:734–41. doi:10.1111/dme.13090.

O'Neil C, Schutt R. Doing data science: straight talk from the frontline. Sebastopol, CA: O'Reilly Media; 2014.

OptumLabs. http://www.optum.com/optumlabs. Accessed 3 Feb 2016.

PCAST. University-private sector research partnerships in the innovation ecosystem—report of the president's council of advisors on science and technology. 2008. https://www.whitehouse.gov/administration/eop/ostp/pcast/docsreports/archives. Accessed 4 Jun 2016.

Pcornet. http://www.pcornet.org/. Accessed 3 Feb 2016.

Phillips M. Konnarock Healthcare. 2016. http://www.konnarockhealthcare.com/; http://msepartners.com/healthcareforum/speaker/mark-phillips-konnarock-healthcare. Accessed 27 May 2016.

Susskind R, Susskind D. The futures of the professions: how technology will transform the work of human experts. Oxford: Oxford Press; 2015.

Topel E. The patient will see you now: the future of medicine is in your hands. New York: Basic Books; 2015.

Wachter R. The digital doctor: hope, hype, and harm at the dawn of medicine's computer age. New York: McGraw Hill; 2015.

Wallace PJ, Shah ND, Dennen T, Bleicher PA, Crown WH. Optum labs: building a novel node in the learning health care system. Health Aff. 2014;33(7):1187–94.

Woodle A. Why Gartner dropped big data off the Hype Cycle. 2015. http://www.datanami.com/2015/08/26/why-gartner-dropped-big-data-off-the-hype-curve. Accessed 15 May 2016.

Glossary

Aggregation is a process of finding, gathering and merging data.

Algorithm is a mathematical formula that gives the computer a set of rules to follow to perform data analysis. Think of an algorithm as a set of directions or a recipe for combining data to get a solution.

Anomaly detection is the search for data in a data set that does not match a projected pattern. Anomalies are also known as outliers. They may provide critical and actionable information.

Anonymization is the process of making data anonymous; no ability to attribute the data to a specific individual; removing all data that could identify a person.

Application program interface (API) is a set of routines, protocols, and tools for building software applications. It specifies how the software components should interact and how to build the graphical user interface (GUI) so that it interacts with the software.

Big data is a term for data sets that are so large or complex that traditional data processing applications are inadequate. The data is characterized by volume, velocity, and variety. These very large data sets may be analyzed to reveal patterns and relationships, particularly about human behavior and interactions.

Causality is the relationship between cause and effect. This is often the goal of research.

Classification analysis is a process for obtaining information about data; also called metadata.

Cloud computing/storage is a distributed computing system over a network of remote servers hosted in the Internet rather than on a local device; used for storing data off site; saving a file to the cloud ensures access with any computer that has an Internet connection.

Clustering analysis is a statistical process for identifying objects that are similar to one another and to cluster them to reveal both similarities and differences.

Commodification is the transformation of data, ideas, services, and products into objects of trade. These data, ideas, services, and products become commodities in the marketplace.

Comparative analysis is a specified process of comparisons and calculations to detect patterns within very large data sets.

© Springer International Publishing AG 2017
C.W. Delaney et al. (eds.), *Big Data-Enabled Nursing*, Health Informatics,
DOI 10.1007/978-3-319-53300-1

Complex structured data is composed of two or more complex, complicated, and interrelated parts that cannot be easily interpreted by structured query languages (SQL) and tools.

Computer generated data is simply data that is generated by a computer as it does its calculations, e.g. log files, time stamps, algorithm checking.

Correlation analysis is a statistical process to determine the relationship between variables; the relationships may be positive or negative.

Dashboard is graphical representation(s) of one or more analyses performed by an algorithm; only the results are shown, not the data or the calculations.

Data refers to a description of something that allows it to be recorded, analyzed, and reorganized; the observation and measurement of a phenomena created data.

Data analyst is someone who cleans/wrangles, analyzes, models, and processes data.

Data ethical guidelines guide an organization in making data management transparent. This is part of insuring privacy and security for the data.

Data feed is a live streaming of data. This is used by Twitter, news feeds, and RSS (really simple syndications) feeds.

Data governance is the management of the availability, usability, integrity, and security of the data owned by an organization. A program of data governance includes a governing body, policies, procedures, and plans to execute the procedures.

Data lake(s) refer to a massive, easily accessible data repository designed to retain all data attributes and built on relatively inexpensive computer hardware for storing big data.

Data modeling is the analysis of data objects using data modeling techniques (such as Unified Modeling language [UML]) to create insights concerning the data.

Data science. While there is no widely accepted definition of data science, several experts have made an effort. Loukides (2012) says that using data isn't, by itself, data science. Data science is using data to create a data application that acquires value from the data itself and creates more data or a data product. Dumbill says that big data and data science create "the challenges of massive data flows, and the erosion of hierarchy and boundaries, will lead us to the statistical approaches, systems thinking and machine learning we need to cope with the future we're inventing" (p.17, 2012). O'Neil and Schutt (2014) add the following skills for data science: computer science, math, statistics, machine learning, domain expertise, communication and presentation skills, and data visualization. A further distinction about data science is that the product of engaging in data science is creating a **data product** that feeds data back into the system for another iteration of analysis, a practical endeavor not traditional research.

Data scientist is a person who is able to search for data and develop algorithms to process the data. This may involve programming and statistics.

Data set is a collection of data.

Data visualization is the representation of data in a visual format that is a complex graph that includes many variables while remaining understandable and readable.

Data wrangling/cleaning/munging are synonyms for the process of reviewing and revising data to delete duplicates, correct errors, deal with missing data, provide consistency, and standardize formats.

Datafication turns a phenomenon into a quantified format so it can be tabulated and analyzed. The earliest foundation of datafication is the measuring and recording which facilitated the creation of data.

Database is a digital collection of data stored using specified techniques depending on the type of database. Databases can be hierarchical, relational, object, graphical, or a hybrid.

Digitization makes analog information readable to computers; makes it easier to store and process. Digitization is the process of converting analog information into the zeros and ones of binary code so computers can handle it.

Distributed file system is a system that stores, analyzes, and processes data from many sites.

Electronic Health Record (EHR) is a longitudinal record, stored in a database, of a patient's health information from their encounters in all care settings. It includes demographics, health history, problem list, medications, progress notes, check lists, immunizations, laboratory and diagnostic tests, images, vital signs, consultations, and therapies received by the patient.

Epigenetics is the study of inheritable changes (either mitotically or meiotically) that alter gene expression and phenotypes, but are independent from the underlying DNA sequence.

Exploratory data analysis was proposed by John Tukey in 1977. The procedure describes the data and finds its main characteristics. It also finds patterns in the data without standard procedures or methods.

Exposome describes the complementary environmental component of the gene-environment interaction indicative of complex traits and diseases.

Extract, transform, and load (ETL) is a database process that identifies and moves a set of data from one database to another. It is also used for the same purpose in data warehouses.

Fault-tolerance design is used to design computer systems that will continue to work if part of the system fails.

Funding opportunity announcements (FOAs) are announcements posted by the federal government, foundations, or other funding bodies. These FOAs solicit program or research proposals for specified target areas of research or services.

Genetic risk scores (GRS) are developed using algorithms about genetic risk (based on big data) to predict the risk for a specific individual. Genetic risk is the probability that a trait will occur in a family. The probability is based on the genetic pattern of transmission.

Grid computing is connecting different computer systems from different locations. The connection is often done via the Cloud.

Hadoop is an open-source Java-based programming framework that supports the processing of large data sets in a distributed computing environment. It is part of the Apache project sponsored by the Apache Software Foundation.

HBase is a NoSQL database designed to work with Hadoop when the volume of data exceeds the capacity of a relational database.

Hadoop distributed file system (HDFS) is a file system designed to work with Hadoop. HDFS is a file system that stores data on multiple computers or servers. The design of HDFS facilitates a high throughput and scalable processing of data.

Health disparity is a particular type of health difference that is closely linked with social, economic, and/or environmental disadvantage.

Health equity is the attainment of the highest level of health for all people. Achieving health equity requires valuing everyone equally with focused and ongoing societal efforts to address avoidable inequities, historical and contemporary injustices, and the elimination of health and health care disparities.

In-memory database is a database management system that stores data in the main memory of the computer instead of on a disk. This characteristic facilitates very fast processing, storing and loading of data.

Internet of Things (IoT) are devices with sensors that connect to the Internet. The devices generate data and can be analyzed for relationships.

Interoperability is the ability of health information systems to share data and information within and across organizational boundaries to promote effective health care.

Location data is the data generated by Geo-Positioning Satellites (GPS). The data is recorded in longitude and latitude and describes a geographical location.

Log file is a file that is generated by a computer that documents all events taking place in the computer while it is operational.

Machine data is data created by machines by sensors or algorithms.

Machine learning is a subset of artificial intelligence. Through algorithms machines learn from what they are doing and become more efficient over time. Machine learning is a key component of data science and is used in big data analysis.

MapReduce, invented by Google, is software for processing very large amounts of data. The MapReduce algorithm is used to divide a large query into multiple smaller queries. Then it sends those queries (the Map) to different processing nodes and then combines (the Reduce) those results back into one query.

Massively parallel processing (MPP) uses numerous processes, located in many separate computers, to perform computational tasks at the same time.

Mathematical model is an abstract model that uses mathematical language to describe the behavior of a system.

Metadata is data about data; giving information about the characteristics of the data.

News feed(s) refer to continuous transmission of data (consisting of news updates) to websites through a syndicated news service provider. Subscribers receive the news feed(s) or web feed(s) as summaries or links to the original news source.

NoSQL databases are used when the volume of data exceeds the capacity of a relational database.

Nurse data scientists are educated as nurses and then pursue a research doctorate or post-doctorate in a data science field (data analytics, computer science). The primary research focus of a nurse data scientist is on methods and analytics as opposed to specific health and illness concerns of individuals, families, communities or populations.

Nursing informatics is defined by the American Nurses Association as the specialty that combines nursing science, computer science, and information science to manage and communicate data, information, knowledge, and wisdom in nursing practice.

Omics is the application of powerful high through-put molecular techniques to generate a comprehensive understanding of DNA, RNA, proteins, intermediary metabolites, micronutrients and so forth involved in biological pathways resulting in phenotypes.

Ontology, from a computer science perspective, is created to represent knowledge as a set of concepts and their relationships with one another within a domain. Ontologies limit complexity and organize information, thus they can be used to solve problems.

Outlier is a piece(s) of data that deviate significantly from the other data in the data set. It is important to detect these during data wrangling/cleaning and exploratory analysis since it might indicate something useful happening.

Pattern recognition is identifying patterns within the data using algorithms. It is used to make predictions about new data coming from the same source.

Population health refers to the management and improvement of health outcomes for a group of individuals, including the distribution of such outcomes within the group. Managed Care under Population Health organizes populations and panels under the care of delivery systems, practices and physicians with accountability for the health of all enrollees and for the resources and costs of providing this care.

Portability is the ability of different types of hardware that allow software to operate on a variety of platforms employing different operating systems.

Precision medicine/personalized healthcare is a medical model that proposes to customize healthcare by incorporating medical decisions, practices, and products that are based on individual variability in genes, environment, and lifestyle.

Precision nursing uses big data from diverse sources; genetic records, medical and insurance records, data from social media, and wearable sensors are effectively harnessed to outline a detailed picture of the patient and offer a customized healthcare solution. Using big data customized treatment plans to specific individuals based on their preferences are provided. Precision nursing enhances the nurse's ability to detect complex nursing problems during initial stages which is imperative for effective and successful treatment and to offer treatment for lifestyle-related diseases by intrinsically analyzing data pertaining to lifestyle patterns of patients.

Public health refers to the function of state and local governments to provide services for preventing epidemics, containing environmental hazards, and encouraging healthy behaviors. The Future of the Public's Health in the 21st Century calls for significant movement in "building a new generation of intersectoral partnerships that draw on the perspectives and resources of diverse communities and actively engage them in health action."

Python is a general purpose programming language created in the late 1980s, and named after Monty Python. It is considered to be the optimal language used in data science by people with a computer science background.

Quantified self is a movement to use applications to track an individual's every move (activity) during the day to better understand one's behavior and health.

Query is asking a question of the data to gain information to answer a question. It is done through a query language, e.g. SQL, Hive, or Pig.

R is a programming language for statistical computing and graphics. It is supported by the R Foundation for Statistical Computing. The R language is widely used among statisticians and data miners for developing statistical software and data analysis.

Re-identification is a process for re-identifying an individual from an anonymized data set.

Radio Frequency Identification (RFID) is a sensor that uses a wireless non-contact radio-frequency electromagnetic field to transfer data.

Real-time data is data that is created, processed, stored, analyzed, and visualized in milliseconds.

Schema-on-read is a data analysis strategy in new data-handling tools like Hadoop and other more involved database technologies. In schema-on-read, data is applied to a plan or schema as it is extracted out of a stored location, rather than as it is entered.

Schema-on-write has been the standard in relational databases. Before any data is entered, the structure of that data is strictly defined, and that metadata stored and tracked. Irrelevant data is discarded and data types, lengths and positions are all defined and enforced with constraints.

Secondary use of health data applies to patient data used, not for the delivery of care, but for other purposes. These purposes may be for research, quality and safety measurement, public health, billing/payment, provider credentialing, marketing, and other entrepreneurial applications.

Semi-structured data is a form of structured data that does not have a formal structure like structured data (using a set of standards or terminology to specify meaning), but does have tags, metadata, or other markers to enforce a hierarchy of records.

Sentiment analysis uses algorithms to determine how people feel about certain topics.

Signal analysis is the extraction of information from complex signals in the presence of noise, generally by conversion of the signals into digital form followed by analysis using various algorithms. This is important when testing sensor data that uses time or other varying physical quantities.

Social determinants of health are conditions that shape a person's health: where they are born, grow, live, work and age, including the health system, and distribution of resources at global, national and local levels.

Structured data is data that is identifiable as it is organized in rows and columns. The data resides in fixed fields within a record/file (as in a relational database).

Transactional data is dynamic data that changes over time.

Transparency is a process that data owners provide consumers that generate the data to inform the consumer how the data is being used.

Unstructured data is data that is usually text in nature, though numbers and dates may be included. There is no known location for the data as there is for structured data.

Value is generated from data by the decision made and the products produced from that data.

Variability occurs when the meaning of the data can change rapidly. For example, in the same tweet a word can have more than one meaning.

Variety indicates the many different formats that data has in the big data world. The data is not ordered, due to its source or collection strategy, and it is not ready for processing. Even the data sources are highly diverse: text data from social networks, images, or raw data from a sensor. Big data is known as messy data with error and inconsistency.

Velocity is the speed at which data is created, stored, analyzed and visualized. Data flows into systems and is processed in batch, periodic, near real time, or real time.

Veracity is the correctness or integrity of the data. This should be established before analysis is performed.

Volume is the amount of data. Data volume is quantified by a unit of storage that holds a single character, or one byte.

Wisdom and clairvoyance, in big data, is the ability to predict and correct before a user knows something is wrong. Traditionally, wisdom is the ability to think and act using knowledge, experience, understanding, common sense, and insight.

YARN (Yet Another Resource Negotiator) is a management system that keeps track of CPU, RAM, and disk space and insures that processing runs smoothly.

Index

© Springer International Publishing AG 2017
C.W. Delaney et al. (eds.), *Big Data-Enabled Nursing*, Health Informatics,
DOI 10.1007/978-3-319-53300-1